McGRAW-HILL YEARBOOK OF
Science &
Technology

2007

McGRAW-HILL YEARBOOK OF
Science &
Technology

2007

Comprehensive coverage of recent events and research as compiled by the staff of the McGraw-Hill Encyclopedia of Science & Technology

McGraw-Hill
New York   Chicago   San Francisco   Lisbon   London   Madrid   Mexico City   Milan
New Delhi   San Juan   Seoul   Singapore   Sydney   Toronto

The McGraw·Hill Companies

Library of Congress Cataloging in Publication data

McGraw-Hill yearbook of science and technology.
1962–   . New York, McGraw-Hill.

v. illus. 26 cm.
Vols. for 1962–   compiled by the staff of the
McGraw-Hill encyclopedia of science and technology.
1. Science—Yearbooks.   2. Technology—
Yearbooks.   1. McGraw-Hill encyclopedia of
science and technology.
Q1.M13   505.8   62-12028

ISBN 0-07-148647-X
ISSN 0076-2016

This book was printed on acid-free paper.

*It was set in Garamond Book and Neue Helvetica Black Condensed by
TechBooks, Falls Church, Virginia. The art was prepared by TechBooks.
The book was printed and bound by Quebecor World/Versailles.*

Contents

Editorial Staff

Mark D. Licker, Publisher

Elizabeth Geller, Managing Editor

Jonathan Weil, Senior Staff Editor

David Blumel, Editor

Stefan Malmoli, Editor

Jessa Netting, Editor

Charles Wagner, Manager, Digital Content

Renee Taylor, Editorial Assistant

Editing, Design, and Production Staff

Roger Kasunic, Vice President—Editing, Design, and Production

Joe Faulk, Editing Manager

Frank Kotowski, Jr., Senior Editing Supervisor

Ron Lane, Art Director

Thomas G. Kowalczyk, Production Manager

and Professor, Laboratory Medicine and Pathobiology, University of Toronto, Mount Sinai Hospital, Toronto, Ontario, Canada. MEDICINE AND PATHOLOGY.

Prof. Justin Revenaugh. *Department of Geology and Geophysics, University of Minnesota, Minneapolis.* GEOPHYSICS.

Dr. Roger M. Rowell. *USDA–Forest Service, Forest Products Laboratory, Madison, Wisconsin.* FORESTRY.

Prof. Ali M. Sadegh. *Center for Advanced Engineering and Design, The City College School of Engineering, New York.* MECHANICAL ENGINEERING.

Dr. Andrew P. Sage. *Founding Dean Emeritus and First American Bank Professor, University Professor, School of Information Technology and Engineering, George Mason University, Fairfax, Virginia.* CONTROL AND INFORMATION SYSTEMS.

Dr. Alfred S. Schlachter. *Advanced Light Source, Lawrence Berkeley National Laboratory, Berkeley, California.* ATOMIC AND MOLECULAR PHYSICS.

Prof. Ivan K. Schuller. *Department of Physics, University of California–San Diego, La Jolla, California.* CONDENSED-MATTER PHYSICS.

Dr. David M. Sherman. *Department of Earth Sciences, University of Bristol, United Kingdom.* MINERALOGY.

Prof. Arthur A. Spector. *Department of Biochemistry, University of Iowa, Iowa City.* BIOCHEMISTRY.

Prof. Anthony P. Stanton. *Carnegie Mellon University, Pittsburgh, Pennsylvania.* GRAPHIC ARTS AND PHOTOGRAPHY.

Prof. Trent Stephens. *Department of Biological Sciences, Idaho State University, Pocatello.* DEVELOPMENTAL BIOLOGY.

Prof. John F. Timoney. *Department of Veterinary Science, University of Kentucky, Lexington.* VETERINARY MEDICINE.

Dr. Daniel A. Vallero. *Adjunct Professor of Engineering Ethics, Pratt School of Engineering, Duke University, Durham, North Carolina.* ENVIRONMENTAL ENGINEERING.

Dr. Sally E. Walker. *Associate Professor of Geology and Marine Science, University of Georgia, Athens.* INVERTEBRATE PALEONTOLOGY.

Prof. Pao K. Wang. *Department of Atmospheric and Oceanic Sciences, University of Wisconsin-Madison.* METEOROLOGY AND CLIMATOLOGY.

Dr. Nicole Y. Weekes. *Associate Professor of Psychology, Pomona College, Claremont, California.* NEUROPSYCHOLOGY.

Prof. Mary Anne White. *Killam Research Professor in Materials Science, Department of Chemistry, Dalhousie University, Halifax, Nova Scotia, Canada.* MATERIALS SCIENCE AND METALLURGIC ENGINEERING.

Prof. Thomas A. Wikle. *Department of Geography, and Associate Dean, College of Arts and Sciences, Oklahoma State University, Stillwater.* PHYSICAL GEOGRAPHY.

Article Titles and Authors

The 2007 *McGraw-Hill Yearbook of Science & Technology* provides a broad overview of important recent developments in science, technology, and engineering as selected by our distinguished board of consulting editors. At the same time, it satisfies the nonspecialist reader's need to stay informed about important trends in research and development that will advance our knowledge in fields ranging from astrophysics to zoology and lead to important new practical applications. Readers of the *McGraw-Hill Encyclopedia of Science & Technology* also will find the *Yearbook* to be a valuable companion publication, complementing the content of that work.

In the 2007 edition, we continue to chronicle the rapid advances in cell and molecular biology with articles on topics such as DNA repair, oxygen and the evolution of complex life, and P-bodies. Reviews in topical areas of biomedicine, such as bioinformatics tools, clinical forensic nursing, monitoring bioterrorism and biowarfare agents, multiple sclerosis, and reducing human error in medicine, are presented. In chemistry, we report on biological-synthetic hybrid polymers, gold-catalyzed reactions, and multicomponent coupling in organic synthesis. Advances in computing and communication are documented in articles on intelligent search engines, optical character recognition, and spyware. Noteworthy developments in engineering and technology are reported in reviews of air-inflated fabric structures, designing for and mitigating earthquakes, high-throughput materials chemistry, metabolic engineering, precast and prestressed concrete, printable semiconductors for flexible electronics, "smart skin," and sport biomechanics. In the physical sciences and astronomy, we report on astrobiology, dark energy, dark matter, and Pluto. And reviews on arctic sea-ice monitoring, atmospheric electricity, ecosystem valuation, precipitation scavenging, remote sensing of fish populations, and scientific drilling in hotspot volcanoes are among the articles in the earth and environmental sciences.

Each contribution to the *Yearbook* is a concise yet authoritative article authored by one or more authorities in the field. We are pleased that noted researchers have been supporting the *Yearbook* since its first edition in 1962 by taking time to share their knowledge with our readers. The topics are selected by our consulting editors, in conjunction with our editorial staff, based on present significance and potential applications. McGraw-Hill strives to make each article as readily understandable as possible for the nonspecialist reader through careful editing and the extensive use of graphics, most of which are prepared specially for the *Yearbook*.

Librarians, students, teachers, the scientific community, journalists and writers, and the general reader continue to find in the *McGraw-Hill Yearbook of Science & Technology* the information they need in order to follow the rapid pace of advances in science and technology and to understand the developments in these fields that will shape the world of the twenty-first century.

Mark D. Licker

PUBLISHER

McGRAW-HILL YEARBOOK OF
Science &
Technology

2007

African mammals

Africa offers an enormous range of natural habitats. Partly because of its abundance of forests, deserts, rivers, lakes, mountains, and savannas, Africa is the birthplace of some of the most remarkable mammals that the Earth has ever known. The familiar "safari" animals include gazelles, rhinoceroses, elephants, lions, cheetahs, giraffes, and hippos. Because they are too small, nocturnal, aquatic, or rare, aardvarks, elephant shrews, golden moles, tenrecs, and sea cows usually go unnoticed. Not all of these mammals are unique to Africa; most have been in parts of Asia, at least during recorded history.

Evolutionary descent. Even though many of these "African" mammals occur elsewhere, several share a unique bond of evolutionary descent that has only recently been recognized in the scientific community. In August 1998, Michael Stanhope, Mark Springer, and several other molecular biologists published a paper recognizing the very intimate evolutionary relationship shared by elephants, sea cows, hyraxes, aardvarks, elephant shrews, golden moles, and tenrecs, and named this group the Afrotheria (**Fig. 1**). Unlike a giraffe or a lion, which have many other close evolutionary cousins widely distributed elsewhere (for example, cows and bears), afrotherians share a single common ancestor to the exclusion of all other living mammals. This means, for example, that the burrowing golden moles found in the suburban gardens of Western Cape Province (*Chrysochloris asiatica*) have a closer evolutionary relationship with elephants than they do with European moles (*Talpa europaea*), in spite of the fact that the two resemble each other in external appearance much more closely than either does an elephant. The "core" of the Afrotheria has actually been recognized for many years by zoologists. In 1945 George Simpson gave the name Paenungulata to elephants, sea cows, and hyraxes. However, the inclusion of animals such as golden moles and tenrecs in the Afrotheria is a very new development in evolutionary science.

Recognizing evolutionary relationships. Many interesting details of mammalian evolutionary history have been well known for over a century, such as the fact that very different looking animals, such as humans and the aye-aye of Madagascar (*Daubentonia madagascariensis*, a small prosimian animal), belong together in the order Primates, or that the Tasmanian wolf is more closely related to a koala than to a dog. By searching the anatomical makeup of various mammals (for example, bony composition of the skull or reproductive anatomy), biologists have been able to recognize the extent to which various animals share a common evolutionary history. However, different sources of data may present contradictory information regarding hypotheses of common descent.

Such has been the case with Afrotheria. Two groups, the tenrecs and golden moles, show a few conspicuous anatomical similarities to northern insectivoran mammals such as hedgehogs (Erinaceidae), moles (Talpidae), and shrews (Soricidae). In particular, tenrecs and golden moles share with the three insectivorans a narrow or even absent connection between each half of the pelvis (pubic symphysis). These animals also possess a simplified small intestine, and a peculiar extension of the maxilla (a bone bearing upper teeth in the mammalian skull) into the bones of the anterior orbit (**Fig. 2**). For most

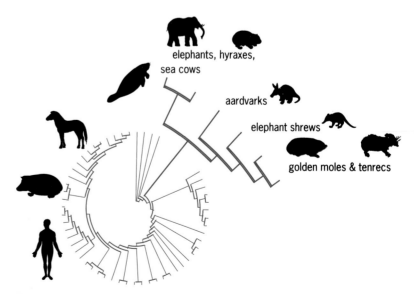

Fig. 1. Abridged mammalian tree of life showing the approximate positions of the seven major afrotherian groups relative to hedgehogs, horses, and humans.

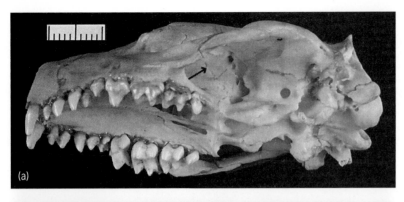

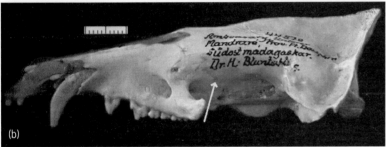

Fig. 2. Skulls of (*a*) hedgehog and (*b*) tenrec. Arrows point to the extension of the maxilla into the orbit in both animals.

of the twentieth century, these similarities led most biologists to the conclusion that golden moles and tenrecs shared an evolutionary relationship with the various northern insectivoran groups.

Similarly, the posterior grinding teeth of hyraxes and horses look very similar. The teeth possess unique "pi" shape, in which the enamel on the outer edge of the tooth comprises a ridge joined by two transversely oriented ridges. Living hyraxes are referred to in the *New International Version of the English Bible* as the coney, and are designated in Leviticus chapter 11 as nonkosher because they "divideth not the hoof." This is an accurate observation insofar as the animal's weight is carried by the middle of the three central digits of the hand and foot. As in horses, the animal's weight is carried on a central digit of the fore- and hindlimbs, in contrast to the cloven hoof of cows, sheep, and deer, which divides the animal's weight fairly evenly across two digits. Based on such similarities in the teeth and skeleton, some biologists believed until very recently that hyraxes and horses (including tapirs and rhinoceroses in the order Perissodactyla) were very closely related.

Since the early 1990s, when the development of polymerase chain reaction (PCR) made it relatively easy to acquire data on the genetic structure of particular animals, biologists have been able to use these data in reconstructing evolutionary relationships on a large scale. The building blocks of any given mammalian gene consist of four nucleotides: adenine, thymine, guanine, and cytosine (A, T, G, and C). By 1997, enough was known about the genetic structure of endemic African mammals to allow a fairly rigorous comparison across mammals; since then, the database of mammalian genetic

diversity has grown exponentially. To date, every study which was designed to evaluate how endemic African mammals relate among themselves and to other mammalian groups (for example, hedgehogs or horses), and which incorporates molecular data, yields a fairly strong signal in favor of the Afrotheria and against competing hypotheses such as "Insectivora" or hyrax+horse. This is due not only to the very similar pattern of nucleotides among afrotherian genes, but also to the shared pattern of indels, that is, those sites in a gene that show either inserted genetic material or a stretch of nucleotides present in other mammals that is lacking among afrotherians (**Fig. 3**).

In many cases, patterns of genetic diversity have supported theories of relationship based on anatomical data. For example, DNA sequences support the idea that aye-ayes and humans are both primates. As mentioned previously, the "core" of Afrotheria (elephants, sea cows, and hyraxes) was familiar to many biologists throughout the twentieth century based on their anatomical similarities, although this was a bit more controversial due to other similarities between hyraxes and horses. The advent of large-scale DNA sequencing not only has confirmed many previously existing hypotheses of mammalian interrelationships (for example, Primates) and settled long-standing arguments (such as dispelling the hyrax-horse clade), but also has presented a few surprises, such as the close relationship among tenrecs, golden moles, and other endemic mammals in the Afrotheria.

Are afrotherians African? The reader might ask if Caribbean manatees (a kind of sea cow) and Indian elephants can also be part of Afrotheria. Yes, they can; Afrotheria is an evolutionary or phylogenetic term, not a geographic one. Hence, no matter where you live, if it turns out that you share a common ancestor with a golden mole or elephant to the exclusion of humans and other mammals, your taxonomic affiliation will instantly change from the Primates (within which your former Linnean name, *Homo sapiens*, belongs) to Afrotheria. Like other animal groups which have originated in a particular place (such as our own genus *Homo* in Africa) and gone on to colonize a variety of other continents and habitats, certain afrotherians have an extensive, non-African distribution (for example, sea cows in the Caribbean and Amazon), particularly when their fossil record is taken into account (for example, mammoths and mastodons throughout northern Eurasia and North America). Incidentally, all golden moles such as *Chrysochloris asiatica* are known only from continental Africa. The species name *asiatica* results from the mislabeled specimen that Linnaeus had at his disposal when he named this species in 1758; apparently he believed it came from Siberia.

A couple of enigmatic fossil mammals, such as hyopsodontids and phenacodontids, have non-African representatives that are older than any undisputed afrotherian. In some published phylogenetic trees, one or more of these non-African fossils appear

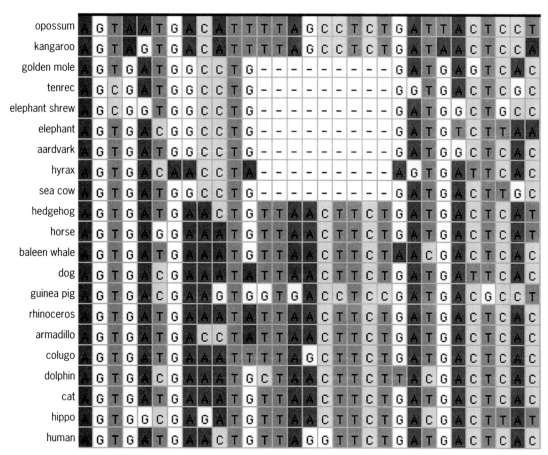

Fig. 3. Small segment of a DNA sequence from the nuclear gene BRCA1 (breast and ovarian cancer susceptibility gene 1) showing the nine-nucleotide deletion (represented by dashes) common to afrotherian mammals.

within the living afrotherian radiation, leading some scientists to wonder whether or not the first afrotherians were truly African. Regardless of where they came from, living members of the "African clade" are predominantly from Africa, and Afrotheria is the taxonomically legitimate term by which we can recognize the common heritage shared by this interesting assemblage.

For background information *see* AFRICA; ANIMAL EVOLUTION; DENTITION; MAMMALIA; PHYLOGENY; THERIA in the McGraw-Hill Encyclopedia of Science & Technology.   Robert Asher

Bibliography. R. J. Asher, M. J. Novacek, and J. H. Geisler, Relationships of endemic African mammals and their fossil relatives based on morphological and molecular evidence, *J. Mammal. Evol.*, 10:131–194, 2003; T. J. Robinson and E. R. Seiffert, Afrotherian origins and interrelationships: New views and future prospects, *Curr. Top. Develop. Biol.*, 63:37–60, 2004; G. G. Simpson, The principles of classification and a classification of mammals, *Bull. Amer. Mus. Nat. Hist.*, 85:1–350, 1945; M. S. Springer et al., Molecules consolidate the placental mammal tree, *Trends Ecol. Evol.*, 19:430–438, 2004; M. J. Stanhope et al., Molecular evidence for multiple origins of the Insectivora and for a new order of endemic African insectivore mammals, *Proc. Nat. Acad. Sci. USA*, 95:9967–9972, 1998.

Air-inflated fabric structures

Air-inflated fabric structures are categorized as pretensioned structures. They are capable of many advantages not available with traditional structures, including lighter-weight design, rapid and self-erecting deployment, enhanced mobility, large deployed-to-packaged volume ratios, fail-safe collapse, and optional rigidification.

Research and development of air-inflated structures can be traced to space, military, commercial, and marine applications. Examples include air ships, weather balloons, shelters, pneumatic artificial muscles, inflatable boats, bridging, and energy absorbers such as automotive air bags and landing cushions for space vehicles. Recent advances in high-performance fibers and improved textile manufacturing methods have renewed interest in air-inflated fabric structures, which are increasingly being designed as reliable alternatives to conventional structures. Several examples demonstrating substantial load-carrying capacities and self-deployment are shown in **Figs. 1–4**.

Description. Air-inflated fabric structures are constructed of lightweight fabric skins, internal elastomeric bladders, inflation valves, and optional pressure relief valves. The bladder contains the air and transfers the pressure to the fabric. Once sufficiently

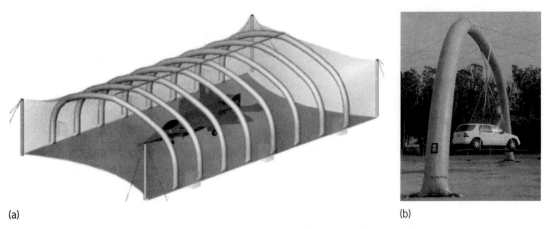

(a)   (b)

Fig. 1. Inflated fabric structures. (*a*) Arches used in aircraft shelters. (*b*) Demonstration of arch load-carrying capability. (*Courtesy of Vertigo Inc.*)

(a)   (b)

Fig. 2. Impact attenuation system for *Mars Pathfinder*. (*a*) Inflated unit. (*b*) Deflated units after impact. (*Courtesy of ILC Dover Inc.*)

pressurized, the fabric becomes pretensioned, providing the structure with a plurality of stiffnesses, including axial, bending, shear, and torsional.

Fabrics in air-inflated structures are typically formed from textile architectures. The various textile architectures have specific design, manufacturing, and performance advantages, and their responses to applied forces vary (**Fig. 5**). For example, the weave develops extensional stiffness when tensile force F_y is applied, but lacks rotational stiffness when shear force F_x is applied. Braided fabrics, in contrast, generate rotational stiffness in the presence of F_x but lack extensional stiffness in the presence of F_y. Here, θ is the braid, or bias angle. The triaxial braid and strap-reinforced braid architectures provide extensional stiffness when loaded with F_y and rotational stiffness when loaded with F_x.

During inflation, the bladder expands until resisted by the fabric. A biaxial pretensioning stress develops in the fabric, enabling the structure to achieve static equilibrium. The pretension stress allows the structure to generate its intended shape, stiffness to resist deformations, and stability against collapse from external forces. Stiffness of the structure is primarily a function of inflation pressure.

Upon application of external forces, a redistribution of stresses balances the forces and maintains equilibrium. Stability is ensured when no regions of the fabric experience a net loss in tensile stress. Otherwise, wrinkling will occur, which decreases the structure's load-carrying capability. Continued loading of a wrinkled structure will ultimately lead to collapse. Of the unique advantages afforded by inflated fabric structures, two relate to stability. First, a collapse does not necessarily damage the fabric. When an overload condition is removed, the structure may restore itself to its original configuration. This is known as fail-safe collapse. Second, since

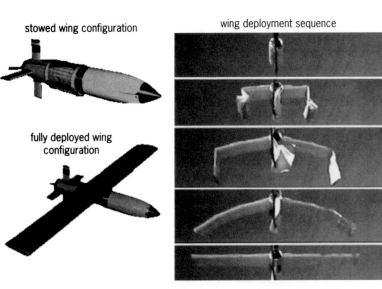

stowed wing configuration

fully deployed wing configuration

wing deployment sequence

Fig. 3. Inflatable wings on a gun-launched observation vehicle. (*Courtesy of Vertigo Inc.*)

(a)

(b)

Fig. 4. Self-deploying woven fabric air beam system. (a) Shelter deployed for use. (b) Inflated frame. (*Courtesy of Federal Fabrics-Fibers Inc.*)

wrinkling can be visually detected, it can serve as a warning prior to collapse.

Yarn fibers. Many of today's fabric structures use yarns constructed of high-performance continuous fibers such as Vectran® (liquid crystal polymer), PEN® [poly(ethylene naphthalate)], and DSP® (dimensionally stable polyester). These fibers provide high strength, low elongation, high flex-fold fatigue, low creep, and enhanced environmental resistance to ultraviolet radiation, heat, humidity, moisture, abrasion, and chemicals. Other fibers used include Kevlar®, Dacron®, nylon, Spectra® (ultrahigh-molecular-weight polyethylene), and polyester.

Continuous manufacturing and seamless fabrics. Prior to the continuous circular weaving and braiding processes, air-inflated fabric structures were constructed using adhesively bonded, piece-cut manufacturing methods. These methods were limited to relatively low pressures because of fabric failures and air leakage through the seams. Continuous circular weaving and braiding processes can eliminate or minimize the number of seams, resulting in improved reliability, significantly higher pressure limits, and greater load-carrying capacities (**Fig. 6**).

Improved damage tolerance. Assorted methods are used to enhance the reliability of air-inflated fabric structures against damage. Resistance to punctures, impacts, tears, and abrasion can be improved by using high-density weaves, rip-stop construction, and coatings. High-density weaves are less susceptible to penetration and provide greater coverage protection for bladders. Rip-stop fabrics have periodic

inclusions of high-tenacity yarns woven in a cellular arrangement (**Fig. 7**). These cells surround fractures of the basic yarns and prevent them from propagating.

Coatings protect the fabric against environmental exposure to ultraviolet radiation, moisture, fire, and chemicals. Coating materials such as urethane, PVC [poly(vinyl chloride)], neoprene, and EPDM (ethylene propylene diene monomer) rubber are commonly used. Additives such as Hypalon® (chlorosulfonated polyethylene) further enhance a coating's resistance to ultraviolet light and abrasion. Coatings generally increase the extensional and rotational stiffnesses of the fabric but remain sufficiently flexible to allow stowage of the structure.

Rigidification. Air-inflated fabric structures can be rigidified using coatings such as thermoplastics, thermosets, and shape memory polymers. Prior to inflation, these coatings are applied to the fabric and remain initially uncured. After the structure is inflated and properly erected, a phase change is triggered in the coating by a controlled chemical reaction (curing process) activated by exposure to elevated temperature, ultraviolet light, pressure, or diffusion. Once the phase change is fully developed, the coating binds the yarns together, stiffens the fabric in tension, compression, and shear, and behaves similar to a matrix material found in traditional fiber-reinforced composites. The fabric structure is now rigidified and no longer requires inflation pressure to maintain its shape and stiffness. Depending upon the coating used, the transition process may be permanent or reversible. Reversible rigidification is especially suited

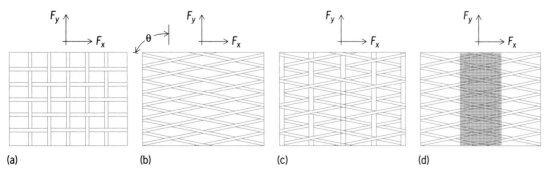
(a)   (b)   (c)   (d)

Fig. 5. Textile architectures used in air-inflated fabric structures. (a) Weave. (b) Braid. (c) Triaxial braid. (d) Strap-reinforced braid.

(a)

(b)

Fig. 6. Continuous manufacturing methods. (*a*) Continuous circular weaving (*courtesy of Federal Fabrics-Fibers Inc.*). (*b*) Braiding (*courtesy of Vertigo Inc.*)

for applications requiring multiple long-term deployments. The rigidified structure may have different failure modes than its pressurized counterpart, including shell buckling and compression, rather than wrinkling. Rigidification is of particular interest for space structures because of restrictions on payload volumes. Shape memory composites are a current focus for rigidizing deployable space frames.

Drop-stitched fabrics. Drop-stitch technology, originally developed by the aerospace industry, extends the shapes that air-inflated fabric structures can achieve to include flat and curved panels, with mod-

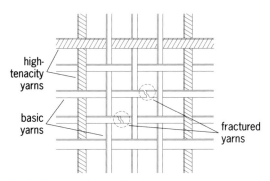

Fig. 7. Rip-stop fabric architecture.

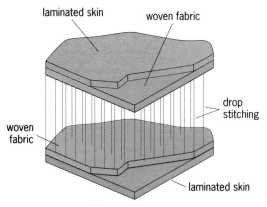

Fig. 8. Section view of a drop-stitched fabric.

erate to large aspect ratios and variable thickness. Drop-stitched fabric construction consists of external skins laminated to a pair of intermediate woven fabric layers separated by a length of perpendicularly aligned fibers (**Fig. 8**). During the weaving process of the intermediate layers, fibers are dropped between the layers. Upon inflation, the intermediate layers separate, forming a panel of thickness controlled by the drop-stitched fiber lengths. Flatness of the inflated panel can be achieved with a sufficient distribution of drop stitching. The external skins can be membranes or coated fabrics that serve as impermeable barriers to prevent air leakage, eliminating the need for a separate bladder. Air-inflated structures incorporating drop-stitched fabrics include floors for inflatable boats, energy-absorbing walls, temporary barriers, lightweight foundation forms, and most recently an inflatable body armor system.

Air beams. Air beams are fundamental examples of inflated structures and are constructed of fabric skins with internal bladders. They support a variety of loads similar to conventional beams. Seamless air beams have been constructed using continuous manufacturing methods achieving diameters up to 42 in. (1.07 m). Air beams have circular cross sections, and their lengths can be straight, tapered, or curved such as an arch. The ends are closed using various termination methods such as bonding, stitching, or clamping, depending upon pressure and loading requirements.

In woven air beams, weft yarns, running along the hoop (cylindrical) direction, spiral through the weave nearly 90° to warp yarns (**Fig. 9**). The warp yarns, aligned on the longitudinal axis, resist longitudinal forces and bending moments, and the weft yarns provide stability against collapse by maintaining the circular cross section. Once the structure is pressurized, the weft-yarn tension per unit length of air beam equals Pr, where P is the pressure and r is the radius. The warp-yarn tension per unit circumference equals $Pr/2$. Hence, the ratio of hoop stress per unit length to the longitudinal stress per unit circumference is 2:1.

Consider a woven air beam subjected to bending. The pretension and bending forces superimpose so that compressive bending forces subtract from the pretension forces, while tensile bending forces add to the pretension forces. The instant any point along the air beam develops a zero net longitudinal tensile stress, the onset of wrinkling has occurred. The corresponding bending moment is the wrinkling moment M_w. Once wrinkling develops, the moment-curvature relationship behaves nonlinearly. With further loading, the cross section loses bending stiffness and eventually collapses. The spread of wrinkling around the circumference is similar to the plastic flow in metal beams subjected to bending.

The wrinkling moment, derived from a simple balance of the longitudinal stresses due to inflation and bending, is shown in Eq. (1) and is valid only for wo-

$$M_w = \frac{P\pi r^3}{2} \tag{1}$$

ven air beams. Note that M_w is independent of the fabric properties, where P is the inflation pressure and r is the radius. **Figure 10** shows the laboratory testing results of a 6.0-in.-diameter (15.24-cm) woven fabric air beam using a 4-point bend arrangement and force versus midspan deflection, and demonstrates the influence of pressure on bending behavior.

Effects of air compressibility. The load-deflection response may depend upon stiffening sources other than the initial inflation pressure. If appreciable changes in pressure or volume occur, as for energy absorbers, work is performed on the air through compressibility. Air compressibility can be modeled from thermodynamic principles according to the ideal gas law of Eq. (2), where P is the absolute pressure, V the volume, m the mass of air, R the gas

$$PV = mRT \tag{2}$$

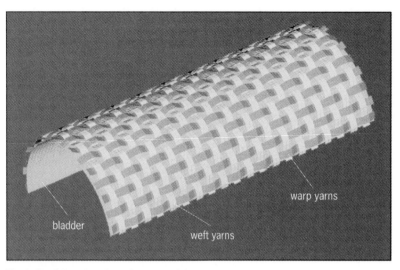

Fig. 9. Partial section view of a woven air beam.

constant for air, and T the absolute temperature (K). For a quasi-static, isothermal process, the work done on the air is shown in Eq. (3). The total energy, E_{total},

$$W_{air} = \int_{v_1}^{v_2} P\,dV = \int_{v_1}^{v_2} \frac{mRT}{V} = mRT \ln\frac{V_2}{V_1} \tag{3}$$

is the work done by external forces W_{ext}, which is related to the total strain energy of the fabric U_f, and W_{air} is as shown in the energy balance of Eq. (4). If

$$E_{total} = W_{ext} = U_f + W_{air} \tag{4}$$

volume changes are large, U_f may be negligible, W_{air} will dominate the energy balance, and deflections can be readily computed.

For background information *see* BEAM; BENDING MOMENT; COMPOSITE MATERIAL; CREEP (MATERIALS); MANUFACTURED FIBER; SHEAR; STRESS AND STRAIN; STRUCTURAL DESIGN; STRUCTURAL STABILITY; TEXTILE; TORSION in the McGraw-Hill Encyclopedia of Science & Technology.　　　　Paul V. Cavallaro

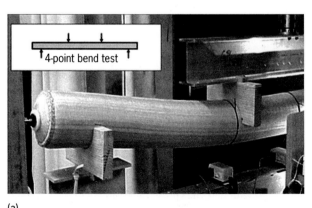

(a)

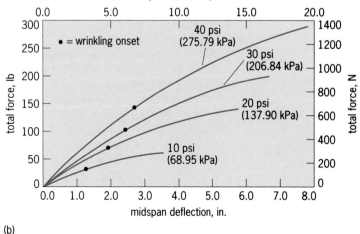

(b)

Fig. 10. A 4-point bend test of a 6.0-in.-diameter (15.24-cm) woven fabric air beam. (*a*) Partial text view. (*b*) Graph of midspan deflection.

Bibliography. P. S. Bulson, Design principles of pneumatic structures, *Struc. Engineer*, vol. 51, no. 6, June 1973; P. Cavallaro et al., Effects of coupled biaxial tension and shear stresses on decrimping behavior in pressurized woven fabrics, *2004 ASME International Mechanical Engineering Congress and Exposition*, Anaheim, CA, IMECE2004-59848, Nov. 13, 2004; P. Cavallaro, A. Sadegh, and M. Johnson, Mechanics of plain-woven fabrics for inflated structures, *J. Composite Struc.*, 61(4):375–393, 2003; W. B. Fichter, *A Theory for Inflated Thin-wall Cylindrical Beams*, NASA Tech. Note D-3466, National Aeronautics and Space Administration, Washington, DC, 1966; W. D. Freeston, M. M. Platt, and M. M. Schoppee, Mechanics of elastic performance of textile materials, Part XVIII: Stress-strain response of fabrics under two-dimensional loading, *Textile Res. J.*, 37:948–975, 1967; W. Fung, *Coated and Laminated Textiles*, Woodhead, 2002; J. W. S. Hearle, *High Performance Fibres*, Woodhead, 2001; J. W. S. Hearle et al., *Structural Mechanics of Fibers, Yarns and Fabrics*, Wiley, 1969; E. C. Steeves, *The Structural Behavior of Pressure Stabilized Arches*, Tech. Rep. 78/018 (AD-A063263), United States Army, Soldier Biological Chemical Command, Natick, MA, 1978.

Antibiotic resistance in soil

The clinical introduction of antibiotics has dramatically affected our approach to treating infectious diseases. Surprisingly, soil-dwelling bacteria have played a central role in this undertaking. Approximately two-thirds of all known antibiotics are synthesized as natural products of the bacterial class Actinomycetes, many of which belong to the *Streptomyces* genus. Evolving in an environment of antibiotic production, these remarkably resilient bacteria must develop diverse ways to survive the toxic antimicrobial compounds synthesized around them. As a result, they have evolved robust resistance to diverse classes of antibiotics, mechanisms that are often homologous to those identified in clinical pathogens. Thus their resistance mechanisms may help us glimpse into the future of clinical resistance to antibiotics, serving as an early-warning system of resistance that might emerge clinically.

Soil. Soil is quite complex: more biological processes occur in the first few inches below the surface than one could begin to imagine. This profoundly complex and often fragile ecosystem has affected our society in astounding ways.

The ever-expanding field of phylogenetics is revealing the immense diversity of subterranean microbial life. It is estimated that a gram of soil can be inhabited by up to 10^9 organisms and approximately 60,000 bacterial species. The complex web of interactions serves a multitude of purposes—decomposition, nutrient recycling and shuffling, toxin sequestration, disease suppression—many of which have been applied by researchers for other invaluable purposes in society.

Antibiotic biosynthesis. Competing for nutrients and survival in a diverse environment of microorganisms, soil bacteria have evolved a seemingly limitless number of tactics to impair the growth of neighboring strains. They naturally synthesize antibacterial agents, encoded by biosynthetic gene clusters, which include some of the most well-known antibiotics, such as tetracyclines, penicillins, glycopeptides (such as vancomycin), and macrolides (such as erythromycin).

For this reason, the soil has been extensively explored for decades in search of organisms that produce antibiotics with clinical potential, an exploration that has changed the way we treat bacterial infections. To date, approximately 80% of clinically implemented antibiotics are derived from soil bacteria, over half of which are synthesized by the genus *Streptomyces* (gram-positive filamentous bacteria). These immensely resilient sporulating bacteria synthesize antibacterial compounds as part of a diverse array of secondary metabolites that also include antifungal, insecticidal, and anticancer agents.

During the introduction of new antibiotics to medicine, little attention was paid to the possible correlation between clinical resistance and resistance in the soil. It was believed that clinical and environmental resistance generally evolved independently, and that there was no implication of utilizing environmentally produced antibiotics in medicine and agriculture. However, evidence from recent decades has suggested otherwise.

Link with clinically implemented natural product antibiotics. Despite the seemingly limitless potential and invaluable benefit of utilizing natural products of soil bacteria as clinical antibiotics, there was an underrecognized downfall: millions of years of antimicrobial exposure have provided soil bacteria with an equal amount of time to evolve sophisticated mechanisms of resistance. Since soil-dwelling bacteria not only produce antibiotics but also are exposed to a myriad of antibiotics produced by surrounding strains, they must develop multiple tactics to survive. Furthermore, it has been demonstrated that a number of these mechanisms, some of which are discussed below, bear stunning resemblance to mechanisms identified in clinical pathogens.

For decades, the natural product antibiotic vancomycin was a commonly prescribed antibiotic for drug-resistant staphylococcal infections, often considered as a drug of last resort for the sickest patients. Its mechanism of action is distinctive in that it acts on an essential molecular component of the bacterial cell wall—the D-Ala-D-Ala peptidoglycan terminus—rather than on a protein or nucleic acid sequence. Thus it appeared that resistance by mutation of the amino acid or nucleic acid sequence, a common resistance mechanism, was not possible. What eventually emerged clinically was a more sophisticated mechanism, a trio of enzymes encoded by a gene cluster that reconstructed the peptidoglycan termini to a form with an impaired ability to bind the drug. Years later, it was discovered that this resistance mechanism in the clinic was identical to that found in actinomycetes that produce members of this antibiotic class, as well as some nonproducing soil bacteria.

Clinically, aminoglycosides play an important role in the treatment of severe sepsis due to enterobacterial infection, as well as infections caused by selected gram-positive aerobic bacilli. They are naturally occurring or semisynthetic polycationic antibiotics that bind to the 16S ribosomal RNA of the 30S ribosome in the aminoacyl–transfer RNA site (A-site), causing misreading and consequently inhibition of translation. The most common mechanism of clinical resistance to aminoglycosides is mediated by antibiotic-modifying enzymes such as kinases that confer a high level of resistance. However, aminoglycoside kinases, enzymes that modify the antibiotic by the transfer of a phosphate group from ATP (adenosine triphosphate), have also been identified in soil-dwelling antibiotic-producing actinomycetes with sequence homology to the enzymes found in clinical pathogens. For these antibiotic producers, as well as for the glycopeptide producers, resistance likely evolved as a means of self-protection.

Early warning system. These findings have a number of implications for both the scientific and medical communities. First, study of resistance mechanisms harbored by soil bacteria—the soil antibiotic resistome—as a subset of all environmental resistance can be suggestive of mechanisms that may emerge in the future as clinical problems. Second, establishing a systematic screen for resistance frequencies exhibited by soil bacteria can help guide the development of next-generation therapies. However, only recently have the density and diversity of this resistance been quantified.

In a study published in the journal *Science* in 2006, researchers conducted a survey of environmental resistance to antibiotics, focusing on streptomycetes and close relatives, a possible reservoir of resistance mechanisms. Researchers constructed a library of spore-forming bacterial strains isolated from diversely located soil samples, and screened 480 strains for resistance to 21 clinically relevant antibiotics, including natural products, their semisynthetic derivatives, and completely synthetic antibiotics. The levels of resistance uncovered were significantly higher than anticipated. At high drug concentrations, not only were isolates resistant to an average of 7 or 8 antibiotics, but every strain was found to be multidrug-resistant. Resistance was observed to all major classes of antibiotic, regardless of whether the compounds were naturally produced or completely synthetic. This screen has the potential to benefit the development of new antibiotics, complementing traditional in-vitro studies of clinical pathogens. Should a new antibiotic show promise clinically, it can be screened against the archive of soil isolates to assess the drug's long-term potential. High levels of resistance could be suggestive of widespread environmental resistance that with time could emerge clinically. This information could be invaluable, as pharmaceutical companies can use this information to make chemical modifications to the drug in an attempt to minimize resistance levels.

The study also delved deeper into resistance in the soil by investigating particular mechanisms of resistance. The predominant mechanism of resistance to vancomycin was determined to be identical to that found in clinics, illuminating the value of studying resistance in the soil to rationally anticipate future clinical resistance. In addition, the research focused on identifying one of the most common tactics implemented by bacteria to circumvent the cytotoxic effects of antibiotics—to enzymatically break down or modify the antibiotic. In altering a region of the drug important for action, the effective binding of the antibiotic to its bacterial target can be hindered. From this, isolates were identified that were capable of enzymatically modifying two recently Food and Drug Administration–approved antibiotics to inactive forms, mechanisms which have yet to emerge clinically for these drugs.

In anticipating future mechanisms of clinical resistance, sufficient warning can be given to hospital microbiology laboratories, infectious disease clinicians, and the drug discovery sector to allow for the development of diagnostic techniques and alternative therapies. Thus, the soil not only is a vast reservoir of antimicrobial agents but also harbors clues to their eventual demise. A thorough comprehension of antibiotic synthesis and evasion in the environment will enable us to better understand these agents which have become so vital to human health.

For background information *see* ACTINOMYCETES; ANTIBIOTIC; BACTERIA; DRUG RESISTANCE; MEDICAL BACTERIOLOGY; SOIL; SOIL ECOLOGY; SOIL MICROBIOLOGY in the McGraw-Hill Encyclopedia of Science & Technology.   Vanessa D'Costa; Gerard Wright

Bibliography. J. Davies, Inactivation of antibiotics and the dissemination of resistance genes, *Science*, 264:375–382, 1994; V. M. D'Costa et al., Sampling the antibiotic resistome, *Science*, 311:374–377, 2006; C. G. Marshall et al., Glycopeptide antibiotic resistance genes in glycopeptide-producing organisms, *Antimicrob. Agents Chemother.*, 42:2215–2220, 1998; A. Tomasz, Weapons of microbial drug resistance abound in soil flora, *Science*, 311:342–343, 2006; C. T. Walsh, *Antibiotics Actions, Origins, Resistance*, ASM Press, Washington, DC, 2003.

Antifungal agents

The area of antifungal agents, especially systemic agents used to treat life-threatening infections, has undergone rapid and significant change since 2000. The commercial availability of three echinocandins plus the addition of newer-generation triazole compounds has radically changed the landscape and management of invasive fungal infections. This review will focus primarily on the newer antifungal agents for systemic fungal infections that have become available since 2000. It will also address the role of the amphotericin B products in the current armamentarium, as many guidelines and authorities in the field still consider these agents to be first-line therapy.

Azoles. The systemic azole class (comprising organic compounds having a five-membered heterocyclic ring with two double bonds) is best represented by four drugs: fluconazole, itraconazole, voriconazole, and posaconazole. The first three are currently commercially available in the United States, and the fourth is near to being so. The azoles have very similar mechanisms of action in molds and yeasts. This class of drugs inhibits 14-α-lanosterol demethylase, thereby inhibiting the formation of ergosterol, which is a critical component of the fungal cell membrane. However, binding affinities for the target enzymes differ dramatically across the class and result in the different spectrums that are seen for these compounds.

Fluconazole has the most narrow spectrum of the class and is currently the most widely used systemic antifungal in the world. It is now generic and inexpensive, making its use even more attractive and cost-effective. Fluconazole is an excellent compound against many *Candida* spp. with the notable exception of *C. krusei* and approximately 25% of strains of *C. glabrata. Candida glabrata* is potentially problematic for all azoles, however, due to the ability of its efflux systems to expel all of the clinically useful systemic azoles, including posaconazole. Itraconazole has activity against *Candida* spp. but unlike fluconazole is also active against many molds, including *Aspergillus* spp., though it has been largely replaced in its anti-aspergillus roles by voriconazole. The availability of generic fluconazole as well as its more extensive data and better tolerability has also relegated itraconazole to a nonplayer in the arena of *Candida* spp. infections. Thus itraconazole is now mostly used for other less commonly encountered fungi.

Voriconazole, which has been commercially available in the United States since 2001, has become the gold standard for therapy of invasive aspergillosis (a fungus infection of humans and animals caused by several species of *Aspergillus*) secondary to a randomized, blinded, controlled study in which the comparator was conventional amphotericin B (the active component of amphotericin, which is an antifungal antibiotic). Regardless of how the data were viewed, voriconazole outperformed amphotericin B across the board for proven/probable invasive aspergillosis. Whether high-dose lipid amphotericin B would have performed better has been the subject of debate; however, the results of a recent study cast significant doubt on the hypothesis that high-dose lipid amphotericin B would have been any more effective. Voriconazole is also very potent against many *Candida* spp. and is currently the oral option of choice for *C. krusei* infections due to its ability to bind to the target in this organism much more avidly than fluconazole.

Limitations of voriconazole are its frequent cross resistance in *C. glabrata* isolates that are resistant to fluconazole, its considerable drug interaction profile, and its ability to cause liver function test abnormalities and visual disturbances in a significant percentage of patients. The strengths of the compound are the invasive aspergillus data, its intravenous and oral formulations, and its high bioavailability in the oral formulation. Another advantage is the ability to perform drug level tests (which detect the presence and the amount of specific drugs in the blood in order to ensure that drug levels are within an effective range and not toxic) on this compound, which can be highly advantageous due to the high degree of interpatient variability in its pharmacokinetics and the increased risk of liver function test abnormalities in patients with elevated voriconazole serum concentrations.

Posaconazole represents a valuable addition to the antifungal armamentarium and is currently in late-stage development in the United States and is commercially available in Europe. Studies have recently been reported that show posaconazole to be a very reasonable choice for antifungal prophylaxis in patients at high risk of invasive fungal infections [that is, patients with graft versus host disease (GVHD; a complication of bone marrow transplants where T cells in the donor bone marrow graft attack the host's tissues) or allogeneic stem cell transplants]. One of these studies found a mortality benefit in patients receiving posaconazole compared to patients in the fluconazole arm. Though this finding is not surprising given the fact that most of the failures in the fluconazole arm were due to aspergillus, against which fluconazole has little to no activity, it is still a first for studies in this area. A major limitation of posaconazole is the lack of an intravenous formulation which is lagging behind the oral in development. In many respects, posaconazole is very similar to voriconazole in spectrum with the notable exception that it appears to provide reasonable activity against some zygomycetes. These organisms have become a significant problem in some centers in which extensive voriconazole use has occurred, though causality is somewhat difficult to prove. There are reasonable clinical data with posaconazole for the use of zygomycetes in the setting of salvage therapy, though data for primary therapy are lacking. A study of posaconazole versus an amphotericin B–based product for primary therapy of zygomycetes infection would be a very welcome addition to the literature, but is unlikely to occur prior to the availability of the intravenous formulation. In its current oral formulation, posaconazole appears to be well tolerated, with a drug interaction profile that may be slightly improved over voriconazole. However, a potential issue with posaconazole, in addition to the lack of an intravenous formulation, is the serum levels and absorption of the oral suspension. First the oral requires food, preferably high-fat food, to optimize absorption. Second, the posaconazole area under the curve and serum concentrations appear to be optimized when the drug is given in increasingly frequent smaller doses (that is, 200 mg every 6 h appears better than 400 mg every 12 h). Finally, the drug has saturable absorption past 800 mg per day such that going higher results in no further increase in serum concentrations. In several studies the maximum serum concentration is less than 1.0 μg/mL, which is the minimal inhibitory concentration (MIC) of many pathogens that posaconazole could be used to treat against or to provide prophylaxis. While the clinical success of posaconazole hints that serum concentrations are not the entire story and the relevance of mold MICs is hotly debated, the inability to escalate dose past 800 mg per day orally and achieve higher serum concentrations creates a potentially worrisome situation.

Echinocandins. The first echinocandin to be marketed was caspofungin. Ironically this drug was at first indicated only for refractory/intolerant aspergillus infections, though it is now abundantly clear that this class of agents will be a first-line therapy for life-threatening invasive candida infections for the foreseeable future. This class of agents is particularly valuable because of their exceptional tolerability profile and their lack of drug-drug interaction profile when compared to the azole antifungals. Echinocandins exert their antifungal activity by inhibiting glucan synthase in susceptible fungi, blocking the formation of $1,3$-β-D-glucan, which is vital to the formation of the fungal cell wall. The minimal side effect profile is secondary to no analogous mammalian target for the drug, creating what has been referred to as penicillin for fungi. The arrival of micafungin and now anidulafungin also creates an interesting pharmaco-economic situation since distinguishing between these agents with the currently available data is difficult and the least expensive agent will likely be chosen by many health care systems. The clinical data for these drugs are rapidly evolving, and our knowledge of them is far from complete. Therefore the remainder of this review will focus on the available clinical data for each compound, paying special attention to data in candidemia (bloodstream infection with *Candida*, a yeastlike fungus).

Echinocandins in candidemia. The first randomized double-blind study to be published in this area was for caspofungin in comparison with conventional amphotericin B. This study revealed that caspofungin was safe and effective for candidemia and that conventional amphotericin B, while effective, was also toxic. The differences in outcome in the study largely favored caspofungin, although the vast majority of the difference was attributable to the toxicity and not microbiologic efficacy of conventional amphotericin B. Another important point from this study was the recognition of a potential hole in the candida spectrum of the echinocandin class. Patients with *C. parapsilosis*, which has higher MICs against all echinocandins, who received caspofungin therapy appeared to have prolonged fungemia (presence of fungi in the blood). Further randomized, blinded studies with echinocandins for candidemia were presented at the Interscience Conference on Antimicrobial Agents and Chemotherapy (ICAAC) meeting in December 2005. These studies evaluated micafungin at 100 mg per day compared to liposomal amphotericin B, and anidulafungin given as a 200 mg loading dose followed by 100 mg per day compared to fluconazole. Micafungin was found to be noninferior to liposomal amphotericin B, although sur-

prisingly anidulafungin appeared to be superior to fluconazole. The results of the anidulafungin versus fluconazole study appear to support the opinion that the rapid candidacidal activity of echinocandins offers a clinical benefit. The benefit of a candidacidal drug had been hinted at in fluconazole versus amphotericin B studies but had never reached statistical significance. The anidulafungin versus fluconazole study needs to be presented in full peer-reviewed form, but the results presented at ICAAC suggest that we may soon see a changing of the guard for life-threatening *Candida* spp. infections.

Another area where the echinocandins are receiving a great deal of attention is for combination therapy of invasive *Aspergillus* spp. infections. These extremely dangerous infections have a very high mortality rate even with voriconazole monotherapy, which is considered the current gold standard. Preliminary data have emerged in animal models and in retrospective case series that suggest that the addition of an echinocandin to a newer-generation azole may improve anti-aspergillus activity compared to either drug alone. Although these data are encouraging, prospective studies are desperately needed to clarify whether combination therapy truly results in improved outcomes for patients. Currently a protocol for voriconazole, plus one of the licensed echinocandins versus voriconazole alone for invasive aspergillosis, is under development. Results of this study will be eagerly awaited.

Amphotericin B, which interacts with the ergosterol in fungal cell membranes resulting in permeability changes and rapid cell death, has historically been considered the gold standard of systemic antifungal therapy. However, conventional amphotericin B (C-AmB) use has greatly decreased in the past few years due to its increasingly recognized toxicities and the availability of the lipid formulations of amphotericin B (L-AmB) as well as the newer systemic antifungals mentioned above. While C-AmB may retain some use, most authorities now recommend the use of L-AmB products for the vast majority of indications where amphotericin B is still considered a first-line compound. Though acquisition costs for L-AmB formulations are considerably higher than C-AmB, several studies have shown that the cost savings is offset by increased costs due to toxicity. The use of amphotericin B compounds will likely continue to decrease in the coming years as data with the newer less toxic agents become available.

In conclusion, systemic antifungal therapy, especially for *Candida* spp. and *Aspergillus* spp. infections, has undergone rapid change since 2000. Newer agents have largely taken away the emphasis on amphotericin B products for life-threatening infections due to the aforementioned organisms, and as data continue to emerge this trend will likely continue. *Candida glabrata* remains problematic for the azole group of antifungals, and *C. parapsilosis* may represent the Achilles' heel of the echinocandins. Death rates remain unacceptably high for invasive aspergillus infections, but the emergence of voriconazole has dramatically changed the treatment paradigm for this organism. Emergence of other molds in an era of high rates of voriconazole use will have to be closely watched.

For background information *see* ANTIBIOTIC; ANTIMICROBIAL AGENTS; FUNGAL BIOTECHNOLOGY; FUNGI; FUNGISTAT AND FUNGISIDE; MEDICAL MYCOLOGY; PLANT PATHOLOGY; YEAST INFECTION in the McGraw-Hill Encyclopedia of Science & Technology.

James S. Lewis

Bibliography. O. A. Cornely et al., Posaconazole (POS) vs standard azoles as antifungal prophylaxis in neutropenic patients with acute myelogenous leukemia (AML) or myelodysplastic syndrome (MDS): Impact on mortality, *45th Interscience Conference on Antimicrobial Agents and Chemotherapy*, Washington, DC, Abstract M-722b, Dec. 16–19, 2005; O. A. Cornely et al., Liposomal amphotericin B (L-AmB) as initial therapy for invasive filamentous fungal infections (IFFI): A randomized, prospective trial of a high loading regimen vs. standard dosing (AmBiLoad Trial), *Blood*, 106(11):3222, 2005; F. Ezzet et al., Oral bioavailability of posaconazole in fasted healthy subjects, *Clin. Pharmacokinetics*, 44:211–220, 2005; R. N. Greenberg et al., Posaconazole as salvage therapy for zygomycosis, *Antimicrob. Agents Chemother.*, 50:126–133, 2006; D. P. Kontoyiannis et al., Zygomycosis in a tertiary-care cancer center in the era of *Aspergillus*-active antifungal therapy: A case-control observational study of 27 recent cases, *J. Infect. Dis.*, 191:1350–1360, 2005; K. A. Marr et al., Combination antifungal therapy for invasive aspergillosis, *Clin. Infect. Dis.*, 39:797–802, 2004; M. A. Pfaller et al., In vitro activities of voriconazole, posaconazole, and fluconazole against 4,169 clinical isolates of *Candida* spp. and *Cryptococcus neoformans* collected during 2001 and 2002 in the ARTEMIS global antifungal surveillance program, *Diag. Microbiol. Infect. Dis.*, 48:201–205, 2004; J. Smith et al., Voriconazole therapeutic drug monitoring, *Antimicrob. Agents Chemother.*, 50:1570–1572, 2006; A. J. Ullmann et al., Posaconazole (POS) vs fluconazole (FLU) for prophylaxis of invasive fungal infections (IFIs) in allogeneic hematopoietic stem cell transplant (HSCT) recipients with graft-versus-host disease (GVHD): Results of a multicenter trial, *45th Interscience Conference on Antimicrobial Agents and Chemotherapy*, Washington, DC, Abstract M-716, Dec. 16–19, 2005; A. J. Ullmann et al., Prospective study of amphotericin B formulations in immunocompromised patients in 4 European countries, *Clin. Infect. Dis.*, 43:e29–e38, 2006.

Arctic sea-ice monitoring

Arctic sea ice has undergone drastic change in recent years, emphasizing the critical role of satellite remote sensing to monitor Arctic change with a frequent coverage over large regions. Optical and multispectral sensors require clear sky and sufficient solar lighting. Microwave radiometers measure ice concentration, the fraction of ice-covered area over the total area, but cannot accurately detect different sea-ice classes. The satellite scatterometer, defined as a stable and accurate radar, is emerging as a versatile sensor for Earth observations and providing new capabilities for sea-ice classification and melt detection regardless of cloud cover and darkness.

Satellite scatterometer. The National Aeronautics and Space Administration (NASA) launched the Sea-Winds scatterometer aboard the *QuikSCAT* satellite on June 19, 1999. The scatterometer, called QSCAT, uses a rotating dish antenna with two polarized beams sweeping across a swath as wide as 1800 km (1120 mi), covering Arctic regions twice per day (**Fig. 1**). QSCAT transmits 13.4-GHz pulses from a Sun-synchronous orbit at an altitude of 803 km (500 mi). The sensor measures backscatter, representing the fraction of radar power scattered back to the radar by geophysical media on ocean, land, and ice. Originally designed for wind measurement over ice-free ocean, QSCAT can monitor sea ice using innovative algorithms.

The capability of QSCAT to map sea ice has been demonstrated with a robust algorithm over a wide range of wind speed on the ocean surface. The QSCAT algorithm detects the presence of sea ice based on differences from the open water in polarization signature, azimuth symmetry, and backscatter stability and consistency. Within the ice cover, different ice classes have distinctive backscatter signatures. There are two major classes of Arctic sea ice. (1) Perennial or multiyear sea ice is defined as sea ice that survives at least one summer and can be as thick as 3 m (10 ft) or more. (2) Seasonal or first-year sea ice grows in the current freezing season and is significantly thinner (0.3–2 m; 1–6.6 ft) than perennial ice. QSCAT ice classification is based on statistical analysis that shows two distinctive peaks in the backscatter data for seasonal and perennial sea ice in freezing seasons. Backscatter of the two major ice classes has an overlapping range, which is used to identify the mixed sea-ice class, consisting of different forms of first-year and multiyear ice. First-year ice, compressed by strong winds, becomes thicker, rougher, and less saline, rendering it similar to multiyear ice; thus, this ice type is also included in the mixed ice class. Furthermore, QSCAT backscatter is sensitive to melt water on sea ice, and the diurnal backscatter difference is used to monitor the melt process from melt onset to freeze-up.

Minimal extent of summer sea ice. QSCAT data has tracked the Arctic sea ice over 7 years so far, fortuitously spanning a period of extreme change. Summer melt decreases the sea-ice extent to the smallest area around the fall equinox. The extent of sea ice on the fall equinox is presented for each year from 2000 to 2005 in **Fig. 2**. In 2002, the area of sea-ice extent was reduced to a minimum, compared to QSCAT results in previous years. In the same year, QSCAT concordantly delineated the most extensive melt area over the Greenland ice sheet in summer. In 2003 and 2004, the equinox sea-ice area was also reduced to a minimal extent close to that in 2002. In 2005, the sea-ice extent observed by QSCAT further decreased to a record minimum, compared to results from all satellite datasets since the 1970s. Note the disappearance of most sea ice east of Greenland in 2001–2004 compared to more normal conditions in 2000 and in 2005, when the sea ice reached much farther south in the Greenland Sea (Fig. 2). In 2002–2005, QSCAT revealed a significant reduction of sea ice in the Beaufort Sea and the East Siberian Sea, compared to 2000–2001. Thus, over the recent years QSCAT observations show a consistent trend in the reduction of summer sea-ice extent in the Arctic Ocean, with a record minimum of 5.8×10^6 km^2 (2.2×10^6 mi^2) in 2005.

Reduction of perennial sea ice. In addition to summer ice observations, monitoring the distribution of perennial, seasonal, and mixed sea ice in winter and spring provides important insights into physical processes to address changes in Arctic sea-ice mass balance. Observed by QSCAT between the winters of 2005 and 2006, perennial sea-ice extent suffered an abrupt loss, larger than the size of Texas. This perennial ice loss was caused by wind-forcing of sea ice to the south out of the Fram Strait between Greenland and Spitsbergen. From winter to spring 2006, perennial sea ice not only continued to be reduced but also to be depleted from the East Arctic Ocean (0–180°E). The distribution of different ice classes is mapped from QSCAT data over the Arctic Ocean on the spring equinox for each year from 2001 to 2006 in **Fig. 3**. The ice classification maps show a trend of reduced perennial ice in the East Siberian Sea from 2001 to 2006. A drastic change occurred in 2006, when perennial sea ice was confined mostly to the West Arctic Ocean (0–180°W), leaving the East Arctic Ocean dominated by seasonal sea ice (Fig. 3). By spring 2006, perennial ice had retreated well away from the north of the Laptev Sea toward the North Pole. On the spring equinox, perennial ice extent in the East Arctic had abruptly decreased from 1.3×10^6 km^2 (500,000 mi^2) in 2005 down to 630,000 km^2 (240,000 mi^2) in 2006, while the perennial ice in the West was stable with 3.2×10^6 km^2 (1.2×10^6 mi^2) in 2005 compared to 3.5×10^6 km^2 (1.4×10^6 mi^2) in 2006.

Sea-ice melting process. The Arctic sea-ice melting process is complex and involves solar radiation, ice surface albedo, cloud cover, atmospheric temperature, and ocean interactions. QSCAT tracks not only active melt zones but also refrozen areas on a daily basis. For example, on May 25, 2005, extensive areas showed active and recent melting over the Beaufort, Chukchi, East Siberian, Kara, and Greenland seas, and in the Hudson and Baffin bays (**Fig. 4**). Results over the complete melt cycle in 2005 indicated that the melt season started about day 100 (April 10, 2005) and ended about day 258 (September 15, 2005). However, the dominant melting in terms of the largest total area of active and recent melt zones did not occur at the temporal center of the melt season, but took place around day 162 (June 11, 2005). This asymmetry in the melt season is caused by two factors. First, ice albedo switches from a high to low value right after an

Fig. 1. Artist's rendering of the NASA SeaWinds scatterometer on the *QuikSCAT* satellite. The dish antenna spins around the vertical axis at a rate of 18 revolutions per minute. (*Courtesy of NASA/JPL*)

area is melted, initiating a sharp increase in solar radiation absorption. Second, the insolation becomes stronger and approaches the peak in the first half of the melt season, around the summer solstice in June. This result suggests that the timing of melt onset and incident solar radiation are more important factors in the ice melting process than the melt season length.

Implications of diminished sea ice. The recent drastic reduction of sea ice may profoundly affect the Arctic environment and ecosystem, transportation and commerce, and resource development. As perennial ice declines from the East Arctic Ocean, a warm summer may significantly melt seasonal ice and force the ice edge away from the extensive continental shelf in the East Arctic Ocean, enhancing ocean mixing and melting. Since the albedo of seasonal ice is larger than that of perennial ice, a larger seasonal ice extent can increase the total insolation. Consequently, summer melt would be accelerated, fall freeze-up impeded, and the ice cover further diminished. Assuming perennial ice is kept confined in the West Arctic Ocean, a strong melt season would open a vast ice-free region over the East Arctic Ocean, allowing a new sea route. This potential seasonal

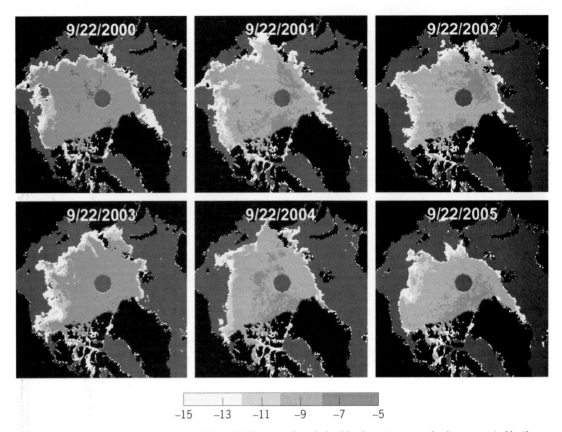

Fig. 2. Sea-ice extent on the fall equinox in 2000–2005. Horizontally polarized backscatter on sea ice is represented by the scale corresponding to values from −15 to −10 dB.

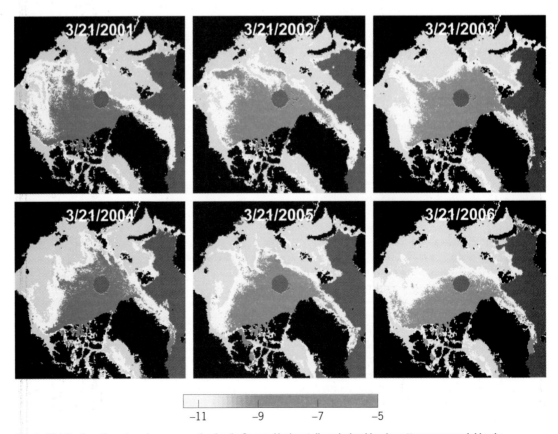

Fig. 3. Distribution of sea-ice classes over the Arctic Ocean. Horizontally polarized backscatter on perennial ice is represented by the scale corresponding to values from −11.5 to −5.0 dB.

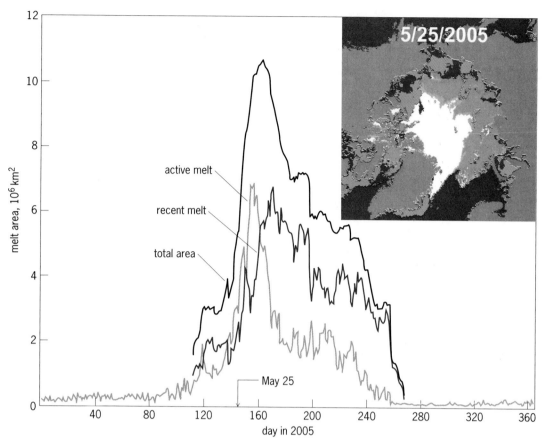

Fig. 4. Melt area on Arctic sea ice. The inset map presents melt conditions over sea ice on May 25, 2005. Light blue is for currently active melt, dark blue for reduced melt or refrozen area that had recent active melt in the previous 10 days. The curves show active melt area, recent melt area, and total area of active and recent melt zones. Results for recent melt start when its area is first equal that of active melta, and end 10 days after the last significant melt in the ice pack.

sea route, called the Polar Passage, would be shorter over deeper bathymetry and further away from the Northwest Passage along the Canadian coast and the Northern Sea Route along the Siberian coast. Such a sea route may significantly change global commercial maritime and transportation across the Arctic Ocean.

For background information *see* ALBEDO; ALGORITHM; ARCTIC AND SUBARCTIC ISLANDS; ARCTIC OCEAN; INSOLATION; METEOROLOGICAL SATELLITES; RADAR; RADIOMETRY; REMOTE SENSING; SEA ICE in the McGraw-Hill Encyclopedia of Science & Technology.   S. V. Nghiem; G. Neumann

Bibliography. T. Armstrong, B. Roberts, and C. Swithinbank, *Illustrated Glossary of Snow and Ice*, 2d ed., Scott Polar Press, Yorkshire, 1973; S. V. Nghiem et al., Depletion of perennial sea ice in the East Arctic Ocean, *Geophys. Res. Lett.*, 33:L17501, doi:10.1029/2006GL027198, 2006; S. V. Nghiem, M. L. Van Woert, and G. Neumann, Rapid formation of a sea ice barrier east of Svalbard, *J. Geophys. Res.*, vol. 110, doi:10.1029/2004JC002654, 2005; M. Sturm, D. K. Perovich, and M. C. Serreze, Meltdown in the North, *Sci. Amer.*, 288:60–67, 2003; W.-Y. Tsai et al., Polarimetric scatterometry: A promising technique for improving ocean surface wind measurements, *IEEE Trans. Geosci. Remote Sens.*, 38:1903–1921, 2000.

Astrobiology

Astrobiology is the scientific discipline that studies the origin, evolution, distribution, and future of life in the universe. It is an interdisciplinary science integrating contributions from biology, geology, astronomy, paleontology, and planetary science, among others. The first use of the term appears to have been in 1941 when it was defined as the subject of life in the universe other than on Earth. The U.S. National Aeronautics and Space Administration (NASA) adopted the term in 1996 and expanded the meaning to include the origin and history of life on Earth, and the future of life in the universe. In 1998, the NASA Astrobiology Institute was established, solidifying this new discipline within the science community.

Astrobiology is perhaps best understood by starting with one of its most tangible areas of study, the multiplicity of living organisms that surround us here on Earth. It is natural to ask how such diversity evolved, and when during the Earth's early history did life itself begin? To address these questions one must understand the composition of prebiotic chemicals which were present during this period, as well as the physical and environmental conditions of the early Earth in which these chemicals interacted. Investigating these areas will naturally lead one to pose broader questions, "How and when during life's origin and earliest evolution were these critical prebiotic materials formed or delivered to the Earth?" To understand the origin and evolution of life on the early Earth, one must also look to other examples of planet formation both within our solar system and as part of other solar systems throughout our galaxy. What are the factors that make a planet such as the Earth habitable? Are habitable planets common or quite rare in the universe? If life does exist on such habitable planets, how can it be detected? Such is the thread that ties all of the cosmos together, both inanimate and living, into the new discipline that is astrobiology.

Life on early Earth. The first record of life on Earth has been observed in the form of cyanobacteria-like microfossils believed to have been laid down approximately 3.5 billion years ago during the Archaean (3.8 to 2.5 billion years ago). This stage of life on Earth appears to have already evolved to the level of complex interacting microbial communities, and seems to have done so in a relatively brief period of time. There is compelling evidence indicating that until about 3.8 billion years ago the Earth was subject to an intense meteorite bombardment of debris left over from the formation of the solar system, and the intensity of these impacts would have likely sterilized the planet—destroying any nascent life. The geologic record suggests, therefore, that only about 300 million years was required for the first life to appear and then to evolve to a high level of complexity. This initial life was microbial, which continued as the only form of life for the next billion years of the Earth's history. Much of the current research in astrobiology is involved with understanding the mechanisms of adaptation and evolution which shaped this microbial life on Earth and tracing (mostly) microbial life back to it earliest forms. Comparing the genetic information carried within each organism, in the form of the nucleic acids RNA (ribonucleic acid) and DNA (deoxyribonucleic acid), astrobiologists have been able to resolve the "tree of life" to great detail, and have begun to understand the nature and metabolism of the earliest microbes. This biological information, together with geological evidence of the physical Earth during this early period of its history, sets the conditions for understanding the origin of life on Earth.

Chemical origin of life. While a rigorous definition of life has proved elusive, there is general agreement among astrobiologists that three conditions must be present for life to exist: liquid water, organic compounds as a source of nutrients, and a biologically useful source of energy. There are a variety of potential sources of energy that astrobiologists are examining, including radiation, photochemical products, minerals, and reduced gases. Life seems quite opportunistic in this regard; it has been found in nearly every possible habitat on Earth, including the polar regions, beneath the subsurface, in desert environments, and even associated with the extreme conditions of hydrothermal vent systems. Liquid water is the one factor that is present in all of these environments, and to the best of our knowledge it appears to be a required factor, one which

may limit the distribution of life elsewhere in the universe.

Given that these conditions were met on early Earth, how then did life begin? To begin to address this, researchers have acknowledged that all life on Earth shares a common set of biochemical machinery to orchestrate the processes of life.

These processes involve the use of DNA to store genetic information, a universal genetic code enabling a DNA sequence to specify a protein, protein enzymes to act as chemical catalysts which drive the cells complex chemical processes, and additional protein synthesis machinery based on RNA. Identifying the evolutionary origin of DNA and its associated proteins has proven very difficult, since each requires the other for its own synthesis. Researchers have been faced with the dilemma of which came first? One possible scenario which has been proposed that could bypass this circularity is the concept of an entirely "RNA world" that might have been the basis for the first forms of life on Earth. RNA molecules can replicate and carry information (like DNA), but they can also act as catalysts (ribozymes) much as proteins do. These abilities suggest that RNA could have originally provided both the genetic code and the catalytic function, and these roles could subsequently have been taken over by DNA and proteins. This stage of life's early history, however, continues to be controversial as well as an area of intense study.

Although the concept of an RNA world has reached some level of acceptance, it is unlikely that RNA itself, with the associated complexity of its four nucleotide bases and a ribose phosphate backbone, represents the earliest prebiotic molecule. Research on nucleic acids with hexoses (six-carbon monosaccharide sugars), for example, suggests that a wide variety of related informational macromolecules are possible. Nonsugar alternatives have also been proposed, such as peptide nucleic acid (PNA).

In any case, it is clear that the prebiotic chemistry which led to life at some point must have been segregated from the changing surrounding environment to achieve the stability necessary for continued evolution. Among present-day organisms this segregation is achieved by means of a membrane composed of phospholipids and proteins. A fundamental area of research related to prebiotic chemistry, therefore, concerns the origin of the first cell membrane, which made possible a regulated vesicle for the chemistry and metabolism of life.

Source of life's building blocks. To fully understand the biochemical mechanisms leading to the origin of life it is necessary for astrobiologists to search back further in time, and identify what building blocks, in the form of the earliest organic compounds, were present on the early Earth. In characterizing this early material, researchers must address the related question, "How were these critical compounds delivered to our planet?" A significant fraction of this precursor material is believed to have been acquired as a late-accreting layer from impacts of C-type asteroids and comets during the period of heavy bombardment of the inner solar system 4.5 to 3.8 billion years ago. In addition to simple volatile molecules such as water and short-chain hydrocarbons, these C-type asteroids and comets are also rich in complex organic chemicals.

An important part of the puzzle, the idea of a cometary source for prebiotic organics on Earth, in some respects simply moves the question out into the cosmos, leading astrobiologists to ask, "What is the ultimate source and distribution of organic matter in the universe?" One place they are looking at are the dense interstellar clouds, the birth sites of stars and star systems which include planets. Interstellar ices are composed primarily of water, but they have also been shown to contain some ammonia, carbon monoxide, carbon dioxide, and the simplest alcohol, methanol. It has been known for some time that when similar icy solids are exposed to conditions of ultraviolet irradiation, more complex chemicals can be produced than those originally present in the ice, and astrobiologists speculate that some of these chemicals might have played an important role in the chemistry of the early Earth and in the prebiotic chemistry of planets in general.

Life in extreme environments. Life on Earth has been found in nearly every environment in which water is present, including extremes of temperature, oxygen, pH, salinity, radiation, light intensity, and pressure. The study of life in these extreme environments is one of the most active areas of research in astrobiology, focusing on the environmental limits at which life can be found, the types of microorganisms which can be found under these conditions, how they are able to adapt, and where they occur in the "tree of life." Viable organisms have been discovered frozen within permafrost over geological time scales as well as associated with hot springs and hydrothermal fluids venting from Earth's subsurface crust onto the deep seafloor at temperatures near 120°C (248°F). Other organisms, such as the primitive unicellular red alga *Cyanidium caldarium* or the acidiphilic archaebacterium *Sulpholobus acidocaldarius*, can tolerate acidic conditions down to a pH of 1–2. Halophiles such as *Halobacterium* and *Halococcus*, which can tolerate high salinity, have been recorded at salt concentrations as high as 1.5–4 M (moles per liter of solution). This aspect of astrobiological research has greatly expanded our view of the versatility of life. The results of this work have broadened the scope of possible environments within our solar system which could harbor life as well as formed the basis for searches for particular sites, for example on Mars or Europa, which should be the focus of future space missions.

Biosignatures. The chemistry of life is inherently different from abiotic chemistry, and all organisms leave evidence of this in the environments with which they interact. Physical properties of an environment indicative of previous life are termed biosignatures, and their characterization and use for detecting possible life elsewhere in the universe is a major component of astrobiological research. Biosignatures are generally of two types, geologic or astronomical. Those associated with geological samples include the distribution of organic compounds, the chirality of the molecules (whether they are left or right handed), the isotopic composition of biological elements, biomineralization, and even the presence of microfossils. Astronomical biosignatures are also being investigated now that it is possible to characterize the chemical composition of planets and other extraterrestrial bodies both within and outside our solar system. Such biosignatures are often associated with atmospheric composition but can also reflect the chemical make-up of the surface as well. Spectral biosignatures of the Earth, for example, would indicate the presence of liquid water, chlorophyll on the surface, and abundant oxygen in the atmosphere. More generally, the simultaneous presence of both strongly reduced gases and oxidized gases which are not in chemical equilibrium (such as oxygen and methane in the Earth's atmosphere) is considered a reliable biosignature for many types of planetary atmospheres.

Habitability of other planets. One of the fundamental questions in astrobiology asks, "Where among what is likely to be a vast number of planets and moons are conditions suitable for life as we know it on Earth?" In other words, "What parts of the universe are habitable?" Recalling the three ingredients upon which life depends—liquid water, organic molecules (or carbon), and a suitable energy source—leads to the conclusion that life is a planetary phenomenon. Initially, the habitable zone of planets around stars was thought to encompass planets whose orbits are close enough to their sun for solar energy to drive the chemistry of life, but not so close as to boil off water or break down the organic molecules on which life depends. Within our own solar system this region would encompass perhaps Venus and Mars in addition to the Earth. More recently, however, astrobiologists have come to realize that the habitable zone may be much larger than originally conceived, since other localized sources of heat and energy could be present at great distances from the Sun. The strong gravitational pull caused by large planets, for example, may produce enough energy to sufficiently heat the cores of moons orbiting planets at great distances from their star. This situation is believed to be occurring between Jupiter and its moon Europa, where evidence suggests that a large, salty ocean is present beneath the surface.

The likelihood of planets or other bodies being present in the habitable zone of solar systems other than our own is dependent on, among other factors, the frequency with which planets form around stars in general. After centuries of speculation, scientists have confirmed just since 1995 that there are indeed planets orbiting other stars, and more than 200 extrasolar planets had already been discovered by late 2006. While almost all of these extrasolar planets seem to be gas giants, like Jupiter, it is likely that Earth-like worlds also orbit other stars, and instrumentation with sufficient precision to detect a world as small as Earth should be available within a decade.

For background information *see* ARCHEBACTERIA; BIOSPHERE; HALOPHILISM (MICROBIOLOGY); INTERSTELLAR MATTER; MOLECULAR CLOUD; NUCLEIC

ACID; PEPTIDE; PLANET; PREBIOTIC ORGANIC SYN-
THESIS; RIBONUCLEIC ACID (RNA); RIBOZYME in the
McGraw-Hill Encyclopedia of Science & Technology.

Edward M. Goolish

Bibliography. L. J. LaFleur, Astrobiology, *Astronomical Society of the Pacific Leaflet Series*, Leaflet
no. 143, pp. 333–340, 1941.

Atmospheric electricity and effects on clouds

Atmospheric electricity shows modulation by solar
activity and by changes in global convective activity,
in the form of changes in the current density, which
flows downward from the ionosphere to the land
and ocean surface. The ionosphere is charged to a
potential V_i, usually in the range 200–300 kV, by positive charge flowing upward from the tops of strongly
electrified convective clouds, which are likely to be
associated with thunderstorms. The charge flows
outward over the global ionosphere and then down
as the current density J_z through the thickness of
the atmosphere, with magnitude 1–6 pA m^{-2}. The
illustration is a schematic circuit diagram of a section through the global circuit. There are of the order
of a thousand thunderstorms operating simultaneously around the globe, supplying about 1 ampere
each to the ionosphere, with the global return path
resistance of order 250 ohms.

The distribution of the current density over the
globe is not uniform in time or space. The value of
J_z is given by Ohm's law as $J_z = V_i/R$, where V_i
depends on the global total of convective activity
producing highly electrified clouds. The convective
activity depends, in turn, on diurnal heating cycles
and day-to-day and seasonal changes in surface heating and air masses at low latitudes.

The resistance R is that of a column of air of unit
cross-sectional area, extending between the ionosphere and the surface, and R is the sum of the tropospheric column resistance T and the stratospheric
component S. The primary source of conductivity in
the atmosphere is ionization produced by galactic
cosmic rays (GCR). With normal stratospheric conductivity, most of the column resistance is contained
in T; but at higher latitudes, S can be comparable
with T when there is a high concentration of ultrafine
aerosol particles in the stratosphere. This is only for
a few years following large explosive volcanic eruptions, and when sources of stratospheric ionization
other than galactic cosmic rays are absent.

Solar activity effects. The changes in J_z due to solar
activity result from several inputs. The flux of galactic cosmic rays varies because of variable magnetic
fields in the solar wind, which attenuate the incom-
ing lower-energy particles in the region from the
Earth out beyond the farthest planets, where the
solar wind encounters the interstellar medium. On
the 11-year solar cycle and during short-term Forbush
decreases, the value of T at high magnetic latitudes
varies by about 10%, with corresponding J_z changes.
Solar activity also affects J_z at high magnetic latitudes

due to the precipitation of solar megaelectronvolt
protons following solar flares. This increases J_z, but
the effects last for less than a day and occur only a
few times a year.

At middle and high latitudes, large increases in
stratospheric ionization rates are caused by precipitating megaelectronvolt electrons from the radiation
belts, dependent on the solar wind velocity. The precipitation shows a solar cycle modulation, and short-
term reductions for periods of a few days, when the
precipitation rate falls by up to an order of magnitude. These events occur twice every 27-day solar
rotation period when the Earth passes through the
extension of the coronal streamer belt with lower
solar wind velocity. These reductions cause a decrease in S and, during the years when there is a high
concentration of stratospheric ultrafine aerosols,
also a decrease in J_z.

The solar wind also affects V_i in the magnetic
polar cap regions by the inward propagation of solar
wind electric fields (see illustration). The interplanetary magnetic field B_y (east-west component) value
changes with a 27-day period, causing correlated
changes in J_z across the polar caps. The B_z (north-
south) component changes cause diurnal variations
in J_z near the edges of the polar caps.

Of all the solar wind effects on J_z, the most important for climate change appears to be the large
reduction in R at high latitudes due to the increases
in the flux of lower energy GCR at times of extended
solar minima, lasting tens of decades. The measurements of ^{10}Be in polar ice cores indicate an increase
of order 100% in the GCR flux and in J_z during the
Maunder minimum, from about A. D. 1650 to 1715.

**Correlations of weather and climate parameters with
J_z.** As noted, both solar activity and changes in low-
latitude convective activity affect V_i and/or R, and
so modulate J_z. There have been many reports of
correlations of weather and climate parameters with
solar activity, including the association of a series of
very cold winters in northern Europe with the Maunder minimum. It is significant that a number of the

Schematic diagram of a section through the global
atmospheric electric circuit, in the dawn-dusk magnetic
meridian. The geometry is essentially plane-parallel. Shown
are discrete column resistances, representing the
atmospheric continuum at equatorial (E), low (L), high (H),
and polar (P) latitudes. Each total column resistance R is
the sum of the tropospheric (T) and stratospheric (S)
column resistances. The main generator is tropical
thunderclouds, which charge the ionosphere to a potential
of order of 250 kV, with a superimposed dawn-dusk
potential difference in the polar ionospheres, and potential
difference between polar ionospheres, produced by the
solar wind-magnetosphere-ionosphere interactions. The
polar and high-latitude regions have much larger changes
in column resistance, R, due to energetic particle influx,
than the low and equatorial latitude regions. The variable
high-latitude solar wind generators and the low-latitude
thunderstorm generators, together with the energetic
particle and volcanic aerosol effects on R, all modulate the
ionosphere-earth current density J_z at various locations.
GCR = galactic cosmic rays. B_y = east-west component of
the interplanetary magnetic field. B_z = north-south
component of the interplanetary magnetic field. nT =
nanotesla. (*After B. A. Tinsley and L. Zhou, 2006, by
permission of American Geophysical Union*)

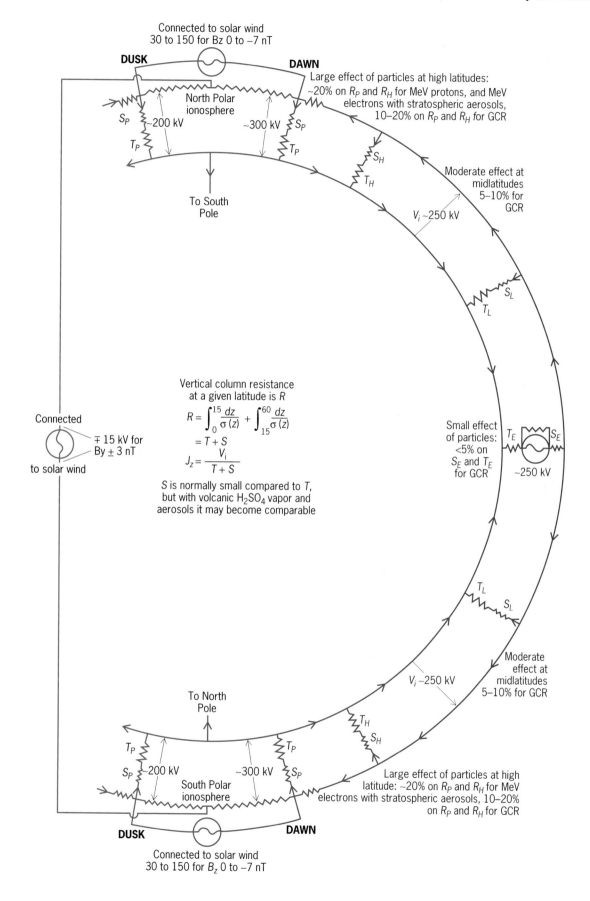

Connected to solar wind
30 to 150 for Bz 0 to −7 nT

DUSK   **DAWN**

North Polar
ionosphere

S_P ~200 kV   ~300 kV S_P

T_P T_P

To South
Pole

Large effect of particles at high latitudes:
~20% on R_P and R_H for MeV protons, and MeV
electrons with stratospheric aerosols,
10–20% on R_P and R_H for GCR

S_H

T_H

Moderate effect at
midlatitudes
5–10% for
GCR

V_i ~250 kV

S_L

T_L

Connected

∓ 15 kV for
By ± 3 nT

to solar wind

Vertical column resistance
at a given latitude is R

$$R = \int_0^{15} \frac{dz}{\sigma(z)} + \int_{15}^{60} \frac{dz}{\sigma(z)}$$
$$= T + S$$
$$J_z = \frac{V_i}{T + S}$$

S is normally small compared to T,
but with volcanic H_2SO_4 vapor and
aerosols it may become comparable

Small effect
of particles:
<5% on
S_E and T_E
for GCR

T_E S_E

~250 kV

T_L S_L

Moderate
effect at
midlatitudes
5–10% for GCR

V_i ~250 kV

To North
Pole

T_P T_P

S_P ~200 kV   ~300 kV S_P

South Polar
ionosphere

T_H
S_H

Large effect of particles at high
latitude: ~20% on R_P and R_H for MeV
electrons with stratospheric aerosols, 10–20%
on R_P and R_H for GCR

DUSK   **DAWN**

Connected to solar wind
30 to 150 for B_z 0 to −7 nT

correlations are for cloud cover and/or precipitation with J_z, because electrically induced changes in cloud microphysical processes are capable of explaining all the weather and climate correlations with solar activity. Correlations on the day-to-day and decadal time scales have been summarized by G. B. Burns and coworkers and B. A. Tinsley and F. Yu, with the most recent work showing correlation of surface pressure on the Antarctic ice cap with changes in daily average J_z. These J_z changes are evidently caused by day-to-day changes in V_i that are caused by changes in the activity of highly electrified, largely low-latitude convective cloud systems.

On multidecadal to millennial time scales, high sun-climate correlations are exhibited in terms of proxies for climate change such as $\Delta^{18}O$ in polar ice cores and cave stalactites, rafted glacial debris in ocean cores, and organic deposits in lake cores, in comparison with proxies for galactic cosmic ray changes such as ^{10}Be and ^{14}C in the same geological repositories. The galactic cosmic ray changes imply J_z changes. Thus, there is evidence ranging from day-to-day changes through millennial time scales that J_z affects weather and climate.

Cloud microphysics responding to J_z. For nonconvective stratus-type clouds of large horizontal extent, the gradients in conductivity at the cloud boundaries, together with J_z flowing through the cloud, set up gradients of electric field at the cloud boundaries. These electric-field gradients entail positive space charge at cloud tops in accordance with Gauss's law, and negative space charge at cloud bases. It is the enhancement by the space charge of scavenging of cloud condensation nuclei and ice-forming nuclei on droplets that appears to be the best candidate for the observed global circuit effects on clouds and climate.

For cloud tops in the temperature range from about 0 to $-15°C$ (32 to 5°F), electrically enhanced scavenging of ice-forming nuclei by the supercooled droplets results in droplet freezing (contact ice nucleation) and then rapid growth of ice crystals from vapor that is supersaturated with respect to ice but subsaturated with respect to liquid water. So ice crystals grow large enough to fall as precipitation, while the surrounding droplets evaporate. This effect can increase precipitation in storm clouds, while decreasing cloud cover in thin cloud layers.

For both cold and warm clouds, the scavenging of cloud condensation nuclei can be increased or decreased, depending on the size of the nuclei. The electrical force between charged droplets and aerosol particles consists of a long-range Coulomb-type force that is repulsive for particles and droplets with like charges, and a short-range force due to induced dipole moment that is always attractive. Trajectory calculations show electrically induced decreases in scavenging rates (below rates due to Brownian diffusion) for the smaller particles, as well as increases for the larger particles. The changes in scavenging lead to greater numbers of small cloud condensation nuclei and fewer large nuclei for the next cycle of cloud formation. The result is the indirect aerosol effect, and leads to less precipitation and more cloud cover.

The above processes provide qualitative explanations for correlations of weather and climate with both externally and internally forced atmospheric electrical changes. Much work is required to evaluate such processes quantitatively.

For background information *see* AERONOMY; AEROSOL; AIR MASS; ATMOSPHERE; ATMOSPHERIC ELECTRICITY; CLOUD; CLOUD PHYSICS; COSMIC RAYS; CURRENT DENSITY; IONOSPHERE; SOLAR WIND; SPACE CHARGE; STRATOSPHERE; SUN; THUNDERSTORM; TROPOSPHERE in the McGraw-Hill Encyclopedia of Science & Technology.　　　　Brian A. Tinsley

Bibliography. G. B. Bond et al., Persistent solar influence on North Atlantic climate during the Holocene, *Science*, 294:2130–2136, 2001; G. B. Burns et al., Interplanetary magnetic field and atmospheric electric circuit influences on ground level pressure at Vostok, *J. Geophys. Res.*, in press, 2006; M. Stuiver, P. M. Grootes, and T Braziunas, The GISP2 $\Delta^{18}O$ climate record of the past 16,000 years and the role of the sun, ocean and volcanoes, *Quatenary Res.*, 44(3): 341–354, 1995; B. A. Tinsley and F. Yu, Atmospheric ionization and clouds as links between solar activity and climate, in *Solar Variability and Its Effects on Climate*, ed. by J. Pap and P. Fox, pp. 321–339, Geophys. Monogr. 141, AGU Press, Washington, DC, 2004; B. A. Tinsley and L. Zhou, Initial results of a global circuit model with stratospheric and tropospheric aerosols, *J. Geophys. Res.*, 111:D16205, 2006; B. A. Tinsley, L. Zhou, and A. Plemmons, Changes in scavenging of particles by droplets due to weak electrification in clouds, *Atm. Res.*, 79:266–295, 2006.

Avian influenza (bird flu)

Avian influenza or "bird flu" first made headlines in 1997, when the H5N1 virus caused an outbreak among poultry in Hong Kong. Within 3 days, Hong Kong's entire poultry population of 1.5 million birds was culled, and the outbreak was temporarily contained. The virus reappeared in Asia in December 2003. By 2004, several Southeast Asian countries were reporting outbreaks in poultry, and human cases were reported from Thailand and Vietnam. Despite massive culling operations, the disease continued to spread and by July 2005 was reported in Siberia, Kazakhstan, Mongolia, and Croatia. The number of human cases also continued to rise. In addition, cases of fatal avian influenza were reported in civets, zoo leopards and tigers, and domestic cats—species that were not previously thought to be susceptible. In 2006, disease spread to several countries in Western Europe and appeared in Africa for the first time. The human case count has now exceeded 200 (as of September 2006), with mortality of approximately 50%. The virus continues to mutate and evolve, and two distinct strains or clades have emerged—the original index strain that initiated the outbreak (Clade 1) and is circulating in Cambodia, Laos, Malaysia, Thailand, and Vietnam; and another strain (Clade 2) called the Indonesia strain that is circulating in Indonesia, China, Japan, and South Korea.

A brief summary of influenza viruses and H5N1 background follows:

1. H5N1 virus is an Influenza A virus.

2. Influenza A viruses are responsible for the annual seasonal outbreaks of influenza in temperate climates or the "flu season."

3. Influenza A can cause disease in a number of species—most commonly birds, humans, and pigs.

4. Influenza A viruses are further classified into subtypes on the basis of the two surface proteins—hemagglutinin (HA) and neuraminidase (NA). Hemagglutinin allows the virus to "stick" to a cell and initiate infection, while neuraminidase enables newly formed viruses to exit the host cell.

5. The term H5N1 refers to the fact that the virus has the hemagglutinin type 5 and the neuraminidase type 1 on its surface.

6. Influenza viruses lack a proofreading mechanism that corrects for errors as they replicate. Hence each generation of virus is slightly different from the previous one. This process, called antigenic drift, is one of the ways by which the virus evades the body's natural immune defenses and the reason why a new flu vaccine is needed every year.

7. Influenza viruses can also exchange genetic material, resulting in a completely new virus. This abrupt change is called antigenic shift and results in a completely new virus to which humans may have no preexisting immunity.

Human H5N1 disease. Most cases of documented H5N1 infection have occurred in previously healthy children and young adults. The disease follows an unusually aggressive clinical course, with rapid deterioration and high fatality. Symptoms typically begin 2–5 days after exposure. Most patients have high fever [>38°C (100°F)], shortness of breath, and cough. On chest x-rays, patients are seen to have a prominent pneumonia early in the course of disease. Some patients may also have bleeding from the nose or gums or watery diarrhea. The disease progresses rapidly to multiorgan failure, and approximately half of those infected die. Death typically occurs in the second week of illness. It should be kept in mind that the clinical spectrum of H5N1 in humans is based on descriptions of hospitalized patients. It is not known if or how often there are milder illnesses perhaps not reported to medical/public health authorities.

A significant feature has been the high death rate in healthy persons. Infection with the H5N1 virus induces a very strong inflammatory response. Infected cells release inflammatory substances called cytokines as part of their self-defense mechanism, causing significant damage to tissues as they battle the virus. Thus, healthy persons who have the ability to produce greater levels of cytokines may actually have poorer outcomes due to this "collateral damage" than older, weaker persons with less robust immune mechanisms.

Transmission. Infected birds excrete very large amounts of virus in droppings. It is estimated that 1 g (0.035 oz) of bird droppings has enough virus to infect 1 million birds. The virus is readily transmitted from farm to farm by the movement of live birds, people, and contaminated vehicles and equipment. The virus can survive as long as 35 days in the environment.

Migratory waterfowl are now carrying the virus over long distances and introducing infection to poultry flocks in areas along migratory routes with shared water sources.

Humans acquire the disease mainly by direct contact with infected poultry or with surfaces or objects contaminated with bird droppings. Exposure is most likely during slaughter and preparation of poultry for cooking. Poultry consumption in an affected country is not a risk factor, provided the food is thoroughly cooked, as heat effectively kills the virus.

Sustained human-to-human transmission has not been recorded. Limited human-to-human transmission has been reported, but only after prolonged and very close contact. Unlike other influenza viruses which replicate in the nose and upper airways of infected persons, it is thought that the avian influenza virus mainly replicates in the lower airways, which limits transmissibility.

Diagnosis. The gold standard in diagnosis is culture of the virus from infected patients. However, this is difficult, takes several days, and must be performed under strict laboratory safety conditions in order to prevent infection of laboratory workers. Rapid antigen tests that detect a portion of the virus provide results quickly but are notoriously insensitive, missing many cases. Serologic tests measure antibodies in blood that develop after several days in response to infection, so these tests are helpful only in retrospectively confirming the diagnosis. Polymerase chain reaction (PCR) tests are the most promising. These tests use molecular biology techniques to amplify a portion of the genetic elements of the virus and can provide very accurate results in a matter of hours.

Treatment. Two antiviral drugs that are active against seasonal influenza have shown activity against the H5N1 virus in the laboratory: oseltamivir (Tamiflu®) and zanamivir (Relenza®). The H5N1 virus is resistant to two older influenza drugs, amantadine and rimantadine. For the drugs to be effective in seasonal influenza, they need to be given within 48 hours of infection, and this is assumed to be true for avian influenza also. The World Health Organization also suggests that oseltamivir for avian influenza be used at twice the dose used for seasonal influenza and for twice as long.

H5N1 and the pandemic threat. A pandemic is a global outbreak of an infectious disease that causes significant illness and death. The H5N1 virus has demonstrated considerable pandemic potential. The three prerequisites for a pandemic are: (1) A novel virus subtype must emerge to which the general population will have little or no immunity. (2) The virus must be able to cause serious illness in humans. (3) The new virus must be efficiently transmitted from one person to another.

In the case of H5N1, the first two requisites have been met; the only feature the virus lacks is efficient human-to-human transmission. With every new human infection, the chances increase that the virus will acquire this ability either by reassortment or by mutation.

Reassortment is the exchange of genetic material between two different viruses. This could happen when a person or animal is infected simultaneously with both the human and the avian virus. Reassortment could result in a fully transmissible pandemic virus, announced by a sudden surge of cases with explosive spread.

Adaptive mutation is a more gradual process. Through small cumulative changes in its genetic makeup, the virus slowly increases its ability to pass from person to person. Adaptive mutation would be associated with small clusters of human cases with some evidence of human-to-human transmission. This would allow time for defensive action, if detected sufficiently early.

Parallels have been drawn between the greatest influenza pandemic of the twentieth century—the "Spanish flu" of 1918—and the current H5N1 outbreak. The similarities between the H5N1 virus and the H1N1 virus that caused the 1918 pandemic are summarized: (1) Both are influenza A viruses entirely of avian origin. (2) Both cause severe respiratory illness. (3) Both have high death rates. (4) The death rate is highest in young healthy persons compared to other age groups. (5) As in 1918, the human population lacks immunity to the H5N1 virus.

The 1918 pandemic swept across the globe in a matter of months and caused an estimated 40 million deaths. If the H5N1 virus were to acquire the ability to pass easily between humans and retained its current lethality, it could be even more devastating. However, as viruses acquire increased transmissibility they generally lose some virulence or the ability to cause serious illness, a possibility that might hold true for H5N1. However, given the current rate of global travel, disease would probably spread much more rapidly across the world than it did in 1918.

We are better prepared today to deal with a pandemic than in 1918. First, we recognize the potential that infections can cause significant loss of human life. The global outbreak of SARS (severe acute respiratory syndrome) in 2003 brought home the fact that no region of the world is immune and that international cooperation and public health measures can be effective in limiting the spread of disease. The media and the Internet make sharing of medical information easier, and the public is better informed. Antiviral drugs and superior medical technology will result in better supportive care. However, in a pandemic, the medical infrastructure will likely be quickly overwhelmed by the huge numbers of ill people, so the overall outcome may not differ greatly from that of 1918.

Vaccine prospects. There has been considerable interest in developing vaccines against avian influenza. The most promising vaccine to date has been a killed vaccine (that is, a vaccine made by taking the real, disease-causing virus and inactivating it) derived from a Clade 1 virus. This vaccine was safe and had few side effects in healthy volunteers. However, large doses of antigen (the active part of the vaccine) were required in two injections given 4 weeks apart, and the vaccine induced protective antibodies in only half of the volunteers.

There are several limitations to this vaccine in the event of an H5N1 pandemic: (1) The large dose of antigen (approximately 12 times the dose used in the seasonal influenza vaccine) will require production of large quantities of the vaccine. The worldwide manufacturing capacity for the current seasonal influenza vaccine is 900 million doses per year. Given the higher dosage required, only 75 million individuals or 1.25% of the world's population could receive vaccine, and only half of these people would be protected. (2) The vaccine protects against only the Clade 1 virus and is ineffective against the Clade 2 virus. Both clades currently have the potential to cause a pandemic.

Studies are currently under way to determine safety and efficacy of this vaccine in other age groups. Antigen sparing methods are also being studied—administration by the intradermal instead of intramuscular route, and addition of adjuvants to the vaccine to make it more likely to produce protective antibodies.

Research areas. Research is urgently needed to better define the possible genetic or immunological factors that might enhance the likelihood of human infection with H5N1.

More effective drugs against the virus and increased manufacturing capacity are needed. Better diagnostic tests are vital. If we were able to make a rapid diagnosis, antiviral drugs could be administered earlier. Most importantly, persons infected with avian influenza could be isolated in hospitals; people in contact with them could be given preventive antivirals, thus limiting further spread.

Vaccine research needs to continue. A vaccine that uses a part of the virus that does not change with time would be ideal, as this universal vaccine would eliminate the need to constantly make new vaccines that keep up with viral mutations.

International cooperation to respond quickly with diagnostic tests, antivirals, vaccines, protective equipment, and quarantine measures will be crucial. While this may not prevent a pandemic altogether, it may buy the world enough time to ramp up production of vaccines and antivirals and may save countless lives.

For background information *see* ANIMAL VIRUS; EPIDEMIC; EPIDEMIOLOGY; INFECTIOUS DISEASE; INFLUENZA; VACCINATION; VIRULENCE; ZOONOSES in the McGraw-Hill Encyclopedia of Science & Technology.   Priya Sampathkumar

Bibliography. J. H. Beigel et al., Avian influenza A (H5N1) infection in humans, *New Eng. J. Med.*, 353(13):1374–1385, 2005; M. Munch et al., Detection and subtyping (H5 and H7) of avian type A influenza virus by reverse transcription-PCR and PCR-ELISA, *Arch. Virol.*, 146(1):87–97, 2001; P. Sampathkumar and D. G. Maki, Avian H5N1 influenza—Are we inching closer to a global pandemic?, *Mayo Clinic Proc.*, 80(12):1552–1555, 2005; K. Shinya et al., Avian flu: Influenza virus receptors in the human airway, *Nature*, 440(7083): 435–436, 2006; J. J. Treanor et al., Safety and immunogenicity of an inactivated subvirion influenza A (H5N1) vaccine [see comment], *New Eng. J. Med.*, 354(13):1343-1351, 2006; K. Ungchusak et al., Probable person-to-person transmission of avian influenza A (H5N1) [see comment], *New Eng. J. Med.*, 352(4):333-340, 2005.

Baylisascariasis

Baylisascaris procyonis is the raccoon (*Procyon lotor*) parasitic roundworm belonging to the phylum Nematoda. This large parasite is responsible for the disease baylisascariasis in animals and humans. The disease is manifested by severe or fatal visceral larval migrans, ocular larval migrans, and neurological disease (neurological larval migrans) in both animals and humans. Larval migrans is defined as the prolonged migration and persistence of helminth larvae in the organs and tissues of animals or humans. This infection is considered by the Centers for Disease Control and Prevention as an emerging zoonotic (referring to a disease of animals that is transmissible to humans) infection in the United States. It was first reported in the United States in 1931 from a raccoon in the New York Zoological Park. The infection appears to be common in raccoons from California, the Midwest, Northeast, Middle Atlantic region, and the mountainous areas of the southeast. The first human case was reported in 1984 in a 10-month-old child in Pennsylvania. To date (September 2006), 13 either severe or fatal cases have been reported and well documented in humans. The prevalence of subclinical cases is unknown.

Populations of raccoons harboring *B. procyonis* in and around major urban areas hold particular potential for zoonotic spread to humans. One reason for this is that scavenging raccoons are bold animals that adapt readily to human habitation and therefore tend to defecate in proximity to homes, potentially putting millions of infectious eggs daily in the immediate environment (dirt) of children and others playing or working in day care centers, yards, parks, and playgrounds. This is important because much of human exposure to this roundworm is through the fecal-oral route and depends on the number of eggs in the environment. Eggs can remain infective for years under ideal conditions, so once an area is contaminated it is nearly impossible to decontaminate it.

Life cycle. The infection in the raccoon is usually subclinical; adult nematodes (9–22 cm or 3.5-8.7 in. long) are confined to the small intestine (see **illustration**) with a mean burden of 50 worms. An infected raccoon can shed millions of *B. procyonis* eggs daily, leading to widespread and heavy environmental contamination. Young raccoons are infected in the first few months of life by ingestion of infected eggs. After ingestion, larvae hatch from the eggs in the small intestine and mature into egg-laying adult worms. The same eggs can be ingested by many different hosts (over 50 species of birds and mammals, especially rodents, have been identified as intermediate hosts). In these intermediate hosts, the eggs

hatch and larvae penetrate the small intestine wall and migrate to various tissues, where they encyst. The life cycle is completed when raccoons eat these hosts. The encysted worms excyst in the small intestine and develop into egg-laying adult worms.

Humans become accidentally infected when they ingest infective eggs from the environment; typically this occurs in young children playing in the dirt. After ingestion, the eggs hatch in the small intestine, and the larvae penetrate the gut wall and migrate to a wide variety of tissues (brain, eyes, heart, lungs) and cause visceral larval migrans, ocular larval migrans, and neurological larval migrans.

Pathology. Human infections can be asymptomatic; however, because the larvae are large, continue to grow, do not readily die, and wander widely in the human host, infections often result in severe disease manifestation. The larvae of *B. procyonis* have a tendency to migrate to the spinal cord, eye, and brain of humans, resulting in permanent neurological damage, blindness, or even death. The clinical presentation of this infection in humans is determined primarily by the number of eggs ingested, which in turn determines the number of larvae entering various tissues. Because there is no widely available definitive diagnostic test for humans infected

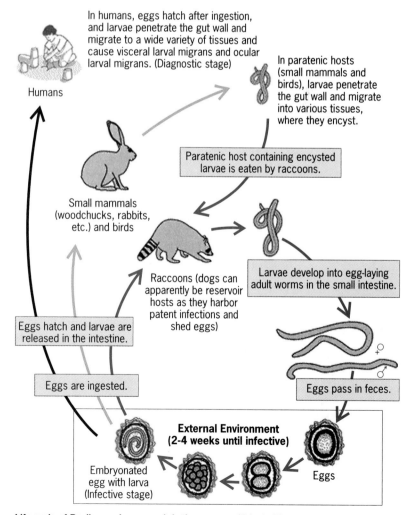

In humans, eggs hatch after ingestion, and larvae penetrate the gut wall and migrate to a wide variety of tissues and cause visceral larval migrans and ocular larval migrans. (Diagnostic stage)

Humans

In paratenic hosts (small mammals and birds), larvae penetrate the gut wall and migrate into various tissues, where they encyst.

Paratenic host containing encysted larvae is eaten by raccoons.

Small mammals (woodchucks, rabbits, etc.) and birds

Larvae develop into egg-laying adult worms in the small intestine.

Raccoons (dogs can apparently be reservoir hosts as they harbor patent infections and shed eggs)

Eggs hatch and larvae are released in the intestine.

Eggs pass in feces.

Eggs are ingested.

External Environment (2-4 weeks until infective)

Embryonated egg with larva (Infective stage)

Eggs

Life cycle of *Baylisascaris procyonis* in the raccoon. (*Adapted from http:/dpd.cdc.gov/DPDx/HTML/Baylisascariasis.htm*)

with this parasite, many cases are probably not diagnosed immediately or correctly.

Diagnosis. There is no definitive diagnostic test for baylisascariasis. The combination of encephalopathy, peripheral eosinophilia, and diffuse white matter disease on neuroimaging, with or without eye disease, should strongly suggest the diagnosis of baylisascariasis. Results from a complete blood count and cerebral spinal fluid examination would be consistent with a parasitic infection. A history of exposure to raccoons or their feces should also be sought. Ocular examinations revealing a migrating larva, larval tracts, or lesions consistent with a nematode larva are often the most significant clue to infection with *Baylisascaris*. Currently, the mainstay of diagnosis is serology. Demonstration of anti–*B. procyonis* antibodies in serum and cerebral spinal fluid, particularly in the setting of epidemiological history (raccoon association), may rule out other confounding diagnoses.

Treatment. No drugs have been demonstrated to be totally effective for the treatment of baylisascariasis. Drugs such as albendazole or mebendazole have been recommended for specific cases. If these drugs should be started as soon as possible (up to 3 days after possible infection), they might prevent clinical disease, and are definitely recommended for children with known exposure (ingestion of raccoon stool or contaminated soil). Steroid therapy may be helpful, especially in the eye and central nervous system infections. Ocular baylisascariasis has been successfully treated using laser photocoagulation therapy to destroy the intraretinal larvae.

Prevention and control. In the absence of an effective treatment and early diagnosis, prevention of a *B. procyonis* infection remains the best practice. For example, education of the public regarding the potential dangers of contact with raccoons or their feces is the most important first step. The risk of infection is greatest with children (infants and toddlers) who have a propensity for pica (an appetite or a craving for soil) and come into contact with raccoon latrines or an environment contaminated by infected raccoon feces. Raccoons should not be encouraged to visit homes or yards for food, and the keeping of pet raccoons should be strongly discouraged. Raccoon latrines in and around homes should be cleaned up and decontaminated. The use of direct flames from a propane gun or torch is the preferred method. Chemical disinfection with bleach is not effective. Finally, if there is no food or shelter to support raccoons, most of these wild animals will simply go away.

Outlook. Baylisascariasis is a potentially fatal, neurologically devastating infection primarily in infants and toddlers and many wild birds and mammals, especially rodents. Although well-documented cases are rare, the increase of raccoon populations in many parts of the United States indicates an increasing number of them will be infected with *B. procyonis* and the likelihood of human exposure and infection will continue to increase. This is especially true for those raccoons living in proximity to human residen-

tial populations where there are infants and toddlers. While much has been learned about the epidemiology and pathogenesis of baylisascariasis, more research is needed. Large population-based serology studies should be done to help define the full clinical spectrum of infection in humans and identify the greatest at-risk individuals most likely to be helped from targeted public health interventions. Until an effective treatment is found, prevention remains the most important tool. Education of the public in general, especially parents with infants and toddlers, to the potential dangers of this nematode infection is the most important first step.

For background information *see* ASCARIDIDA; NEMATA (NEMATODA); PARASITOLOGY; RACCOON; ZOONOSES in the McGraw-Hill Encyclopedia of Science & Technology.   John P. Harley

Bibliography. P. Beaver, The nature of visceral larval migrans, *J. Parasitol.*, 55:3–12, 1969; M. L. Eberhard et al., *Baylisascaris procyonis* in the metropolitan Atlanta area, *Emerg. Infect. Dis.*, 9:1636–1637, 2003; P. J. Gavin, K. Kazacos, and S. T. Shulman, Baylisascariasis, *Clin. Microbiol. Rev.*, 18(4):703–718, 2005; P. J. Gavin et al., Neural larval migrans caused by the raccoon roundworm *Baylisascaris procyonis*, *Pediat. Infect. Dis. J.*, 21:971–975, 2002; K. R. Kazacos, Protecting children from helminthic zoonoses, *Contemp. Pediat.*, 17(suppl.):1–24, 2000.

Bioinformatics tools

Bioinformatics tools aim to exploit the potential of computers to model and understand complex biological systems and phenomena. Initially, the approach used by most bioinformatics tools was more analytical and engineering-like, separating DNA from all other factors and isolating the sequence for a better understanding of the DNA structure. Today, the genomics approach combines various elements, while taking into consideration their relationships as well as evolutionary aspects in space and time. A more global approach is being developed that accounts for biomolecules in their environment, including intra- and intermolecular interactions at the cell, tissue, organ, and organism level.

DNA sequencing first appeared in 1977, with the enzymatic method of Frederick Sanger becoming the standard. In the laboratory, sequencing was done on a gene-by-gene basis. In 1985, the idea of identifying the three billion base pairs of the human genome was born, with the objective of decoding our genetic inheritance and applying the information to scientific and medical outcomes. Considerable international effort and funding were dedicated to the research on instruments. The first sequencing instrument (sequencer) was commercialized in 1987.

The development of the polymerase chain reaction (PCR) technique in 1984 resulted in a tool essential to both research and industry. Other advances in molecular biology techniques have led to automation and data production with increasingly greater reliability and lower costs.

The application of computers to genomics represents a revolution at several levels. Fundamentally, it offers processes to manage information systems, the capacity for modeling and treatment of massive genomic data, and the possibility of automating repetitive operations (as in sequencing).

Two major obstacles stand in the way of formalizing and modeling biological systems in computer science: the diversity and complexity of living organisms. Web tools have become available for assessing genetic and protein data in an extremely convenient manner.

Designing a user interface for bioinformatics tools is not trivial, since it must be effective in performing a task, conveying information, or organizing a body of data, information, and knowledge. The complexity and diversity surrounding usability of bioinformatics systems is twofold. First, there is a wide array of different tools. For example, the NCBI (National Center for Biotechnology Information) site is an information portal with access to specialized tools for sequence alignment and molecular visualization. Second, users accessing and interacting with these tools have different goals. The diversity in user goals ranges from simple information gathering, such as with article searching in PubMed, to using a specific molecular biology tool such as BLAST (Basic Local Alignment Search Tool). This lack of usability is especially true for visualization tools.

Visualization tools. Bioinformatics visualization tools are used for the representation, analysis, and mining of molecular structures and genomic databases.

Visualization of molecular structures. The idea that visual representations aid our understanding of an artifact, such as a genomic or molecular structure, underlies the research undertaken since the early 1990s to visualize abstract data in order to provide insights into the data that would otherwise be impossible to gain. The transition from manual techniques to virtual reality applications for visualizing complex data or artifacts is a consequence of the increasing need to analyze huge amount of scientific data.

Virtual reality is a form of simulation in which computer graphics is used to create a synthetic world with which the user is able to interact. While incorporating many of the advantages of two- and three-dimensional visualization, virtual reality offers potential solutions to many problems. For example, two-dimensional (2D) visualizations are unable to capture multiple relationships simultaneously, and suffer from a limited number of graphical dimensions available for encoding information (such as position, size, and color). Although three-dimensional (3D) visualizations have made progress in solving these problems, such visualizations easily disorient a user and do not handle large-scale scientific problems. The esthetic qualities of virtual reality make working with it more pleasant. In addition, virtual reality technology may offer opportunities to create highly customizable visualizations representing combinations of different visual metaphors that give simultaneous access to different aspects of visualized data for exploratory research. Virtual reality applications allow perceiving information at different hierarchical levels and have a huge potential for scientific visualization, especially in bioinformatics.

In bioinformatics, virtual reality can help deepen users' perception of DNA or protein structures. 3D visualization of molecules constitutes a great part of stereochemistry. 3D visualization offers a global point of view on the protein or DNA molecule under study and makes possible the capture of its shape. Such visualization is very important in depicting a phenomenon or simulating a biological mechanism.

Visualization can use data from two different methods. The first consists of observing the biological element (for example, DNA or protein) and establishing the complete 3D cartography that will be used to visualize this element. The second method is based on a model of a 3D structure of DNA from which one can calculate the spatial coordinates of each base pair of any sequence.

Whatever the observation method used, the result obtained is stored in a standardized file format [Protein Data Bank (PDB)]. For the visualization of these files, the two most popular applications are RasMol and MDL® Chime.

ADN-Viewer software offers visualization of the complex spatial trajectory of huge naked DNA, and is augmented by genetic annotated information provided by the GenoMEDIA platform (**Fig. 1**). Three-dimensional DNA visualization allows biologists to go from a genetic approach (studying targeted genes) to a genomic one (studying the whole genome). In this kind of 3D visualization, biologists interact more directly and intuitively with the objects of interest, which facilitates the exploration of large genomes.

Other software exists to visualize several proteins, to superimpose them to deduce some information from structural alignments, and to compare their active sites.

Exploration of factual and textual genomic data. Many databases are available for biologists to access and process genomic information, including sequencing, alignment, and transcription. Examples include GenBank, NCBI, and SwissProt. These data are heterogeneous, huge in quantity, geographically distributed, recorded within public or private databanks, and constitute an important factual data source (GenBank, SwissProt, and Decrypthon). However, genome knowledge is not limited to DNA or annotated protein sequences. There is a significant quantity of information relating to these genes recorded in an unstructured format within many publications. These data are stored as structured or semistructured databanks, and constitute an immense factual and textual data source that is not easily understandable by humans. One major challenge is to make data and tools widely shared and easily accessible. This requires research on data and tool integration as well as data exploration.

Genome3DExplorer is a new immersive modeling tool for exploring textual and numerical genomic data (**Fig. 2**). Genome3DExplorer offers biologists a user-friendly visualization of data within a virtual

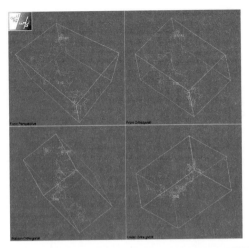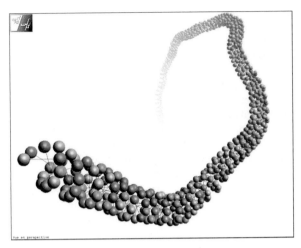

Fig. 1. ADN-Viewer software for the visualization of DNA 3D structure.

reality environment, using a well-adapted graphical paradigm (dynamical graph), which allows biologists to explore huge sets of genomic data and their relationships.

SequenceWord represents numerical genomic databases in a virtual reality context. In order to use the human orientation capabilities, its authors present a solution where genes are positioned in a landscape, allowing observers to explore the data. The genes can be gathered spatially according to several clustering criteria by using the BLAST algorithm to calculate an alignment score. In the virtual world, the distance between objects represents this score.

Biobibliometrics has been proposed for 2D visualization of textual and numerical biological information, using a mutual information coefficient to measure bibliographical proximity between two gene names or aliases. This gene bibliographical information is visualized by a graph. The integration of the textual and numerical data is treated in an interesting way, and it would be beneficial to look further into the exploration of these textual databases.

Shift from features-oriented to biologist-centric tools. Most current tools are feature-oriented as opposed to human-centric, creating a conceptual gap between biologists' expectations and experiences versus how the tools work and how the features are accessed. In the human-centric approach, the challenge is to design tools for both novice and expert users, while applying a separate set of usability principles. This could range from focusing on learnability for novice users using a wizard approach to efficiency for expert users using a performance electronic support system.

A roadmap for designing human-centric visualization tools and bioinformatics systems includes a better understanding of biologists' behaviors and experiences, their tasks, and the context of their work.

There is considerable challenge in designing a bioinformatics Web site. The NCBI site, for example, is a well-established site with a large community of users and a vast amount of information. NCBI was established in 1988 in Maryland, as a division of the National Library of Medicine. The role of the NCBI is to advance scientific knowledge of the underlying molecular and genetic processes surrounding health and disease.

Ethnographic interviews with NCBI site users, mainly bioinformatics researchers from three different academic and government-run research labs from North America and Europe, resulted in the following discoveries. The NCBI site is by far the most popular bioinformatics information provider. It is rich in information and provides access to nucleotide, protein, and literature databases. It contains computerized information processing methods and tools that are used on a daily basis by some biomedical researchers. Users pointed out a number of problems with the site, including difficulty in finding desired information, poor site organization, information overload, and inconveniences due to long response delays.

Modeling biologists' experiences using empirical research methods. The redesigned NCBI site used the

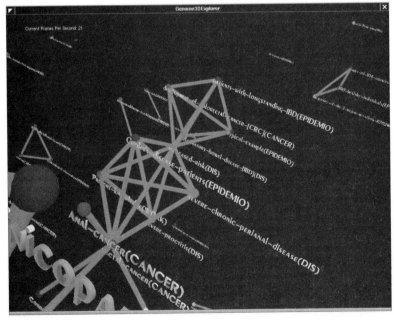

Fig. 2. Genome3DExplorer for the visualization of factual and textual genomic data.

concept of persona as a way to place the user at the center of the design process. Persona is a medium to capture the educational background, work habits, and tasks performed by bioinformatics researchers. For example, two personae were constructed to describe the users of the NCBI site, as well as their basic characteristics. The process of building these personae helps identify potential users for evaluating the usability of the existing site. The observations made in the usability evaluation were used as feedback for two purposes: (1) To further enhance the personae and determine more precise interaction behavior. These refined personae, along with context information, gave a clearer picture of the context of use. (2) To determine usability issues and existing problems with the site. Both these points helped to choose appropriate patterns for user interface design.

In order to increase their effectiveness, personae are based on empirical evidence and studies. To create the personae, domain analysis and ethnographic interview results were used to postulate the user characteristics of the NCBI site, as well as the kind of experiences these users may have. A biomedical expert provided advice on domain-specific information. Psychometric assessment, through objective questionnaires administered to users, is another technique used to build personae and to model the interaction between the biologists and the tools. In this case, questionnaires were used to gather user perceptions in a systematic way. They are analogous to structured interviews in that questions were presented in the same way to all respondents.

The second part of the questionnaire contained specific questions to quantify user experiences with

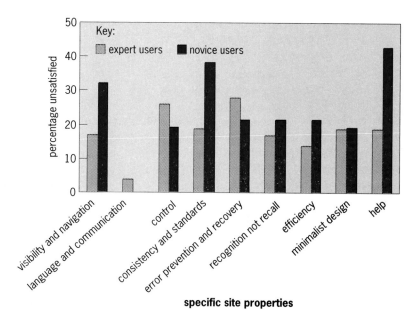

specific site properties

Fig. 3. Questionnaire results of novice and expert users.

certain properties of the site. When analyzed, the two user groups had differing results with relation to visibility and navigation, consistency and standards, and help (**Fig. 3**). The only property of the site that seemed to be satisfy both user groups was language and communication.

Transforming personae into human-oriented conceptual designs using patterns. Based on the above tests, the personae were refined to include new information about user behavior and experiences. In the case study, these enhanced personae included only

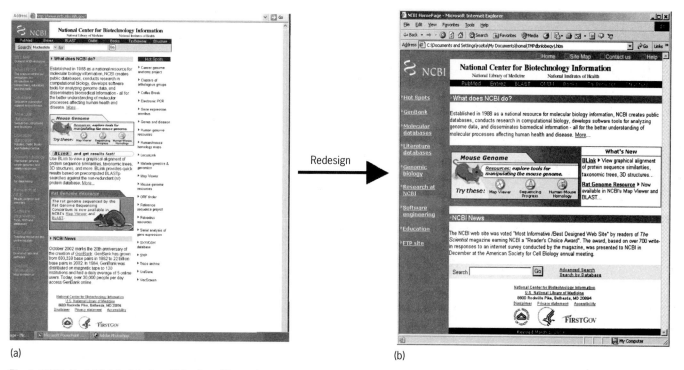

(a)   (b)

Fig. 4. NCBI site. (*a*) Original design. (*b*) Design with user inputs.

two types of users: novice and expert users. Each type was characterized by personal characteristics, computer and bioinformatics skills, interaction behavior in using the NCBI site, and different tasks performed with the site. From the information gathered, pattern-based redesign was applied to close the gap between actual user experiences and the features offered by such a complex Web site. Patterns were associated with each type of desired task and user behavior, and then combined to redesign the user interface in **Fig. 4**.

Patterns in human-computer interaction (HCI) were introduced as a tool for capturing and disseminating the best user experiences in terms of interaction and to facilitate the design of more usable systems via the reuse of well-established design knowledge and principles. Pattern-oriented design (POD) is another approach for incorporating human experiences to achieve design solutions in which patterns aim to capture and communicate the best practices of user interface design with a focus on the user's experience and the context of use.

Figure 5 illustrates a skeleton on which a site design was improved. All the patterns are well known in the HCI community. Such a skeleton could be used as the building blocks to improve, for example, the original NCBI design. Much better, it could be used by bioinformatics tool builders to create a new open-source portal that would serve as a front end to the various parts of NCBI site and similar Web sites.

Outlook. It is crucial to design bioinformatics visualization tools so that they are usable and effective in achieving their purpose of conveying information or organizing a body of data, information, and knowledge. Users accessing and interacting with these

tools have different goals and experiences, which are not necessarily taken into account by bioinformatics designers. User-centered design methodologies can lead to more usable and accessible tools, while closing the gap between user experiences and the tools. An effective design method includes a better understanding of biologists' behaviors and experiences, their tasks, and the context of their work. Such information is captured using personae, which are based on empirical studies on all aspects of user interactions with tools. Personae are then refined and used to identify patterns and to develop patterns-oriented design.

[Acknowledgement: This work was supported partially by different grants. We thank NSERC (National Sciences and Engineering Research Council of Canada), FCAR and the Faculty of Engineering via the Concordia Research Chair programs, as well as GENOPOLE© at Evry (France) via the ATIGE funding program.]

For background information *see* COMPUTER GRAPHICS; DEOXYRIBONUCLEIC ACID (DNA); GENOMICS; HUMAN-COMPUTER INTERACTION; HUMAN GENOME PROJECT; MODEL THEORY; MOLECULAR BIOLOGY; VIRTUAL REALITY in the McGraw-Hill Encyclopedia of Science & Technology.

Ahmed Seffah; Rachid Gherbi

Bibliography. A. Cooper, *The Inmates Are Running the Asylum: Why High Tech Products Drive Us Crazy and How To Restore the Sanity*, SAMS Publishing, 1999; N. Férey et al., Exploration by visualization of numerical and textual genomic data, *J. Biol. Phys. Chem.*, 4:102–110, 2004; Y. Q. Guan et al., Application of virtual reality in volumetric cellular visualization, *Proceedings of the 2004 ACM SIGGRAPH International Conference on Virtual Reality Continuum and Its Applications in Industry*, Singapore, pp 65–71, 2004; N. Guex and M.-C. Peitsch, Swiss-PdbViewer: A fast and easy-to-use PDB viewer for Macintosh and PC, *Protein Data Bank Quart. Newsl.*, 77:7, 1996; International Organization for Stancards, *ISO-13407 Standard: Human-Centered Design Processes for Interactive Systems*, 1999; H. Javahery and A. Seffah, A model for usability pattern-oriented design, *Proceedings of the 1st International Workshop on Task Models and Diagrams for User Interface Design*, Bucharest, July 18–19, 2002; D. J. Mayhew, *Usability Engineering Lifecycle: A Practitioner's Handbook for User Interface Design*, Morgan Kauffman, 1999; I. Rojdestvenski, Virtual reality and knowledge spaces: Examples and applications in molecular biology, *Proceedings of the 2004 ACM SIGGRAPH International Conference on Virtual Reality Continuum and Its Applications in Industry*, Singapore, pp 49–56, 2004; I. Rojdestvenski et al., World: A genetics database in virtual reality, *International Conference on Information Visualization*, 2000; B. Stolk et al., Mining the human genome using virtual reality, *ACM International Conference Proceedings of the 4th Eurographics Workshop on Parallel Graphics and Visualization*, Blaubeuren, Germany, 29:17–21, 2002; M. V. Welie and H. Traetteberg, Interaction patterns

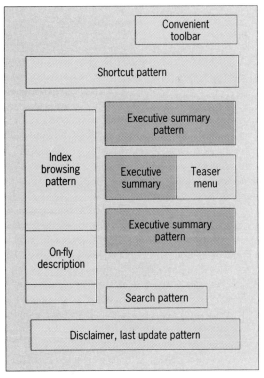

Fig. 5. Skeleton for the pattern oriented design.

in user interfaces, *Proceedings of the 7th Pattern Languages of Programs Conference*, Allerton Park, IL, August 13–16, 2000.

Biological-synthetic hybrid polymers

One traditional approach to categorizing polymers is by their origin. Polymers can be roughly divided into synthetic polymers and naturally occurring polymers, such as polysaccharides, polynucleotides (DNA and RNA), and proteins. There are important distinctions between these two classes of macromolecules. Synthetic polymers can be made with a wide variety of monomer building blocks, which can be arranged in a large number of different topologies. Synthetic polymers are very adaptable, and their properties can be fine-tuned for specific applications. However, a synthetic polymer always consists of a mixture of polymer chains with a distribution in chain length. One drawback of synthetic polymers is that there is no absolute control over the molecular weight of the polymer. Even more importantly, there is no control over the specific order of the monomers within the polymer chain. Control of molecular weight and the specific order of the monomers within the polymer chain are prominently present in biomacromolecules such as proteins. Because the sequence of amino acids in a protein is perfectly controlled, information can be stored in the linear polypeptide chain, which leads to a well-defined folding process of the biopolymer and the introduction of structure and functionality in the resulting protein. However, biomacromolecules are limited with respect to the number of monomers that can be incorporated, and they can easily loose function, for example, due to unfolding or degradation.

Many polymer scientists have investigated the possibilities of creating biological-synthetic polymer hybrids in order to combine the versatility and adaptability of synthetic polymers with the structural and functional control of biopolymers. Because of recent developments in controlled polymerization and peptide and protein chemistry, as well as the emergence of improved techniques for coupling synthetic and natural polymers, the activities in hybrid polymer synthesis have increased greatly.

Protein-polymer conjugates. One of the earliest developments in polymer-protein hybrids is based on the coupling of a poly(ethylene glycol) [PEG] chain to a protein—a process known as PEGylation. Proteins that are used as drugs and are introduced into the human body have a short circulation time because they are easily removed or degraded by the immune system. It was found that attaching a PEG chain enhances a drug's lifetime, and therefore its biological activity, tremendously. This is explained by the fact that the synthetic polymer shields the protein from interactions with the immune system. One of the first commercial PEGylated proteins is a conjugate of PEG with bovine adenosine deaminase, which received FDA approval in 1990 and is used to treat severe combined immunodeficiency disease. The modification of this enzyme with multiple PEG chains extends the circulating half-life by a factor of 6.4. Another example is PEGylated interferon, which is used to treat hepatitis. It is expected that the worldwide PEGylated biopharmaceutical market will be over $40 billon by 2010.

One of the shortcomings of present-day PEGylated products is that coupling of PEG to the protein occurs in a random fashion to the available functional groups on the protein surface, such as the amino groups of lysines. This can lead to undesired modifications which diminish the protein's biological activity and can make it difficult to produce drug batches with reproducible properties. Improved synthetic tools have now created the opportunity to make a new generation of PEGylated drugs with a high

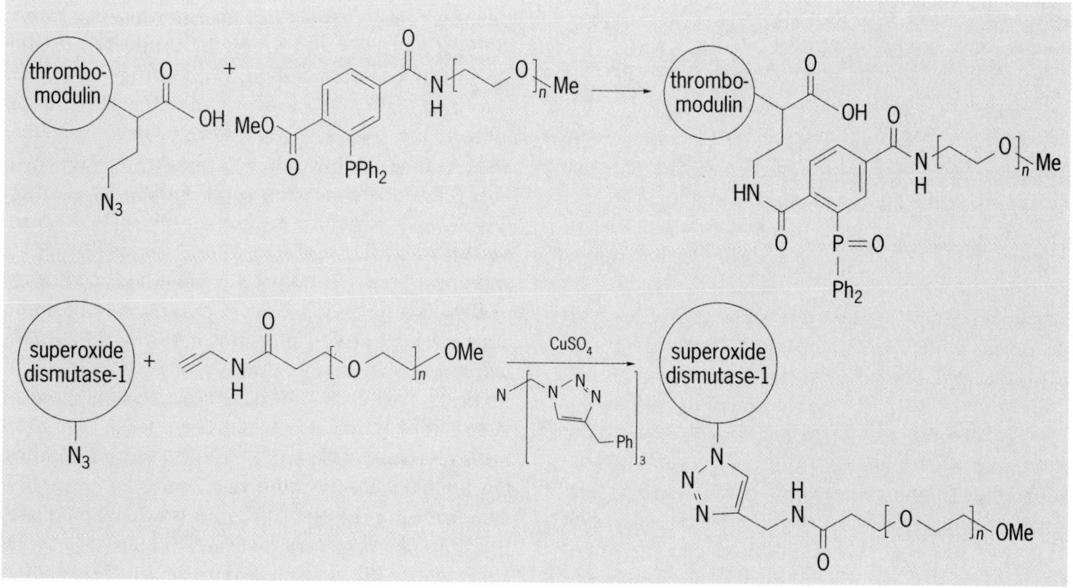

Fig. 1. Two examples of PEGylation of proteins via an azide-functional nonnatural amino acid.

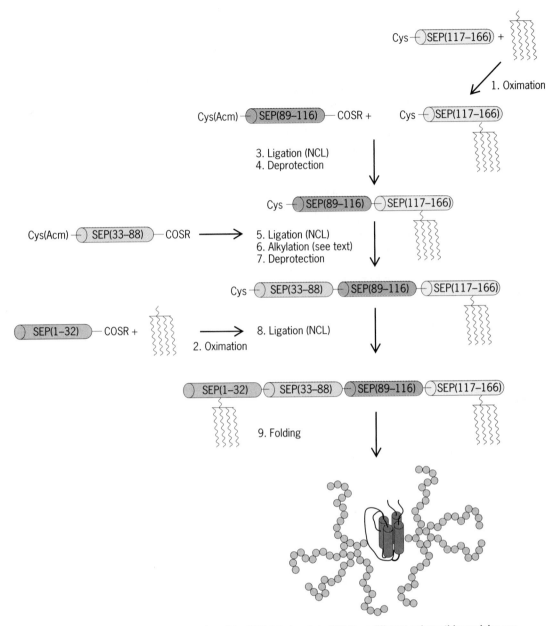

Fig. 2. All-synthetic method for the preparation of the PEGylated protein SEP. Four different polypeptide modules are synthesized and coupled via ligation (steps 3, 5, and 8). The PEG chains are introduced via a process called oximation (steps 1 and 2). *(Reprinted with permission from G. G. Kochendoerfer et al., Science, 299(5608):884–887, 2003. Copyright 2003 AAAS)*

level of control over what protein sites the polymer chains are connected. New coupling chemistry has been developed that targets amino acids that are rare at the surface of proteins (for example, tyrosines), so that only a few connection sites are available. Using molecular biology techniques, proteins can be constructed that contain only one reactive amino acid at a well-defined position. This can be a natural amino acid such as a thiol-containing cysteine, or even amino acids that are normally not encountered in proteins such as azide- and alkyne-modified amino acids (**Fig. 1**). These nonnatural moieties enable site-specific coupling of PEG to them without interference with any of the other functionalities present in the protein, yielding much better defined polymer-protein conjugates.

In another approach, protein-PEG conjugates were built up entirely via synthetic chemistry. This was accomplished for synthetic erythropoiesis protein (SEP), which is a conjugate consisting of a 166-α-amino acid polypeptide and two branched PEG-like polymers (**Fig. 2**). The polypeptide was built up by solid-phase peptide chemistry in four modules, two of them site-selectively labeled with PEG. Using a process called native chemical ligation, the four modules were coupled to yield the active protein.

Conjugation of polymers to proteins is nowadays not limited to PEG. With the advances in controlled radical polymerization, a wide variety of well-defined polymers can be made which are functionalized with a reactive chain end for conjugation to a protein. An interesting example is poly(*N*-isopropyl acrylamide), or poly(NIPAAM). This polymer displays lower critical solution temperature (LCST) behavior, which means that it aggregates above a certain temperature and is soluble below the LCST. The process is thermally reversible. Using this thermoresponsive principle, poly(NIPAAM)-enzyme conjugates could be recovered from a reaction mixture by simple heating above the LCST. In another application, the temperature-induced collapse of the polymer chain could be used to switch off enzyme activity when the chain was positioned in the proximity of the active site.

Polymeric drug delivery systems. Another important class of biological-synthetic hybrid polymers consists of conjugates in which the bioactive component is connected to the side chain of the polymer (**Fig. 3**). The polymer therefore functions as a carrier. These hybrid systems have been extensively investigated for targeted drug delivery, especially for cancer treatment. Most cancer drugs are so toxic that the maximum adjustable dose is quite limited so that anticancer therapy is often accompanied by severe side effects. There is a strong demand to improve the selectivity of the drugs by targeting them to the tumor tissue, which can be accomplished by connecting the drugs to a polymer carrier. Tumor tissue structure is markedly different from healthy tissue; in particular its cardiovascular system is more open and leaky. Polymers can therefore more easily penetrate and accumulate in tumor tissue. When drugs are connected to a polymer carrier, this will lead to a higher effective concentration at the target site. A further improvement can be made by connecting drugs via a peptide spacer to the polymer carrier. By choosing a peptide which is preferentially cleaved by

enzymes that are active in tumor tissue, drugs will be released only in the proximity of the tumor, which enhances the specificity of the polymeric drug. A third improvement is the additional attachment of a homing device to the polymer carrier. This can be an antibody which recognizes specific tissue types. Although polymer-drug carrier systems have experienced a long developmental pathway, the first examples are now in the final stages of clinical trials and approach commercialization.

Polymer scaffolds for multiple bioactive moieties. In biological processes, a response is often triggered by the interaction of multiple peptides. When such peptides are coupled to the side chains of a polymer, a scaffold is created with a high concentration of these bioactive moieties, which is an efficient method to mimic natural binding and recognition processes. This has been demonstrated by connecting multiple copies of two peptide cell-binding domains to a polymer backbone. The hybrid polymer structure showed an enhanced ability to cell binding when compared to single peptide moieties. A polymer to which multiple antigenic peptides were connected displayed a stronger immune response than the loose antigens. These polymers have potential applications in biomedicine, such as cell culturing and tissue engineering. New controlled polymerization methods will improve the versatility of this approach even further.

Structural peptide-polymer hybrids. Besides function, biomolecules can introduce structure in hybrid polymers. Relatively short peptide fragments already display important folding motifs, such as the β-sheet or α-helix structure. When these peptides are connected to synthetic polymers, either as side chains or as part of the main chain, the resulting polymers exhibit the properties that accompany the structural peptide. In case of β-sheets, it has been demonstrated that their natural ability to stack in ribbonlike structures was maintained upon attachment of a PEG chain. Hybrid polymers containing α-helices form stable hydrogels and polymer aggregates. A special case of the α-helical conformation, the coiled-coil structure, has been used to bind DNA for gene delivery. Structural peptide-polymer hybrids can now be made in a well-defined way via a number of synthetic techniques, and recently the scientific challenge has shifted from mere synthesis to application.

Outlook. Biological-synthetic hybrid polymers have experienced growing interest since new synthetic methods have improved the production of well-defined conjugates. It is expected that besides the already successfully applied PEGylated proteins and polymer drug carriers, in the near future other materials will be made with a unique set of properties which cannot be obtained by either natural or purely synthetic polymer materials.

For background information see AMINO ACIDS; ANTIBODY; ANTIGEN; BIOPOLYMER; CANCER (MEDICINE); CHEMOTHERAPY; DRUG DELIVERY SYSTEMS; MOLECULAR BIOLOGY; PEPTIDE; POLY(ETHYLENE GLYCOL); POLYMER; PROTEIN in the McGraw-Hill Encyclopedia of Science & Technology.   Jan van Hest

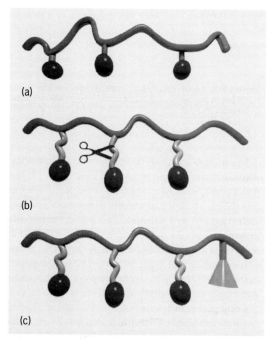

Fig. 3. Three levels of polymer drug delivery. (a) A polymer is used as carrier of drugs, (b) The drugs are connected to the polymer via cleavable peptide linkers. (c) A homing device is added to target specific tissue types.

Bibliography. A. Deiters et al., Site-specific PEGylation of proteins containing unnatural amino acids, *Bioorg. Med. Chem. Lett.*, 14(23):5743–5745, 2004; R. Duncan, The dawning era of polymer therapeutics, *Nat. Rev. Drug Disc.*, 2(5):347–360, 2003; G. G. Kochendoerfer et al., Design and chemical synthesis of a homogeneous polymer-modified erythropoiesis protein, *Science*, 299(5608):884–887, 2003; H. D. Maynard, S. Y. Okada, and R. H. Grubbs, Inhibition of cell adhesion to fibronectin by oligopeptide-substituted polynorbornenes, *J. Amer. Chem. Soc.*, 123(7):1275–1279, 2001; J. M. Smeenk et al., Controlled assembly of macromolecular beta-sheet fibrils, *Angew. Chem. Int. Ed.*, 44(13):1968–1971, 2005; P. S. Stayton et al., Control of protein-ligand recognition using a stimuli-responsive polymer, *Nature*, 378(6556):472–474, 1995.

Box jellyfish

Like other jellyfish, the box jellyfish have neither a head nor a central brain; but unlike other jellyfish, they have well-developed eyes and display visually guided behavior. Recent findings on their optics indicate that they are at an early phase of eye evolution, where excellent visual performance does not have the same meaning as it does in other animals.

Nasty stingers. The box jellyfish, or Cubozoa, have a box-shaped body, with stinging tentacles attached at the four lower corners of the animal (**Fig. 1**). Across the world there are some 20 described species, and at least the same number probably remain to be described. Most are warm-water animals from tropical and subtropical seas. Although some species of box jellyfish are harmless, others are among the most venomous animals known. At the tropical beaches along the coast of northern Australia, public swimming is banned in the warmer half of the year because of the presence of a large (maximum width 30 cm or 12 in.) species of box jellyfish, *Chironex fleckeri*, which fishes for prey close to the shoreline. Humans that get extensive stings of this species suffer cardiac arrest and die within minutes after exposure. The 68 recorded deaths of *Chironex* stings in Australia probably are only a small percentage of the actual number, as *Chironex fleckeri* or equally lethal species exist throughout southeastern Asia. In addition, there is a smaller species (1 cm or 0.4 in.) of box jellyfish, *Carukia barnesi*, that gives a hardly noticeable sting that induces severe systemic effects starting some 20 minutes later. In Australia this is known as the Irukandji syndrome, and the victims may suffer for days or weeks from severe general body pain, deep anxiety, and extremely elevated blood pressure. Yet another species of box jellyfish (*Carybdea alata*) occurs in large numbers once every 4 weeks on Hawaiian beaches, keeping swimmers out of the water for a few days on each lunar cycle.

Behavior and vision. The behavior of box jellyfish is strikingly fishlike. They display rapid directional swimming and skillfully avoid colliding with obstacles. All known species are strictly coastal, and many inhabit the shallow area next to the shoreline on sandy beaches, rocky shores, estuaries, or mangrove swamps. Box jellyfish are distinguished by a high metabolic rate, and they need to find and remain in places where food is abundant. Crustaceans and fish aggregating close to the shore are the typical diet for most species. Other types of jellyfish are open ocean drifters, and when currents carry them close to the shore they run a high risk of getting washed up or damaged in the surf. Box jellyfish are specialized to navigate and survive in the dangerous zone close to the shore.

This lifestyle is the probably why they have well-developed eyes, whereas other jellyfish at best have ocelli (simple invertebrate eyes composed of photoreceptor and pigment cells). The eyes of box jellyfish are impressive in design and number. On each of the four sides of a box jellyfish, there is a sensory club, termed a rhopalium, carrying one large-lens and one small-lens eye, and two pairs of pit eyes (**Fig. 2**). This makes a total of 24 eyes of four types in each individual. However, the eyes are not huge: sensory clubs range in length from 0.3 to 1.2 mm (0.012 to 0.05 in.) depending on species, and the large-lens eyes vary between 0.15 and 0.8 mm (0.006 and 0.03 in.). Despite the small size, the lens eyes are fully equipped with a retina, vitreous, lens, cornea, and a mobile iris. The sensory clubs are attached and suspended by a thin flexible stalk, and a crystalline weight allows the clubs to retain a vertical orientation irrespective of the orientation of the animal. There are indications that the sensory clubs also have an olfactory patch opposite the eyes, and there are mechanoreceptors in the stalk. The sensory clubs are thus multisensory structures serving vision,

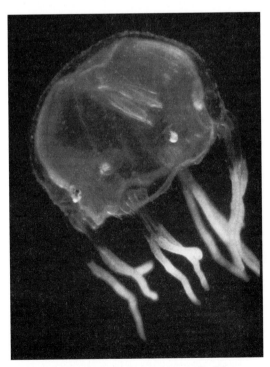

Fig. 1. Box jellyfish *Tripedalia cystophora* with all four sensory clubs visible. (© *Dan-E. Nilsson*)

smell, and balance. The interior of the sensory club contains a small brain, which communicates with its neighboring sensory clubs via a ring nerve. Branches from the ring nerve form a diffuse nerve net innervating the muscles.

These are the main components of the sensory and nervous system that endows box jellyfish with an almost fishlike behavior. There is no central brain, but all information processing and muscle coordination must result from the joint action of the identical brains in the four sensory clubs, and the ring nerve that joins them. This is an unusual and amazingly simple system for an animal capable of navigation, precise positioning, and even courtship behavior in some species. Even though the different species of box jellyfish inhabit very different coastal locations, the eyes and nervous systems hardly vary at all. All have the same set of 24 eyes of four types arranged in a stereotyped way on the sensory club. Evidently, inconspicuous variations on the common theme account for the differing behavior and choice of environment in different species.

Visual physiology. Recent investigations of the lens eyes of the best model species of box jellyfish, *Tripedalia cystophora*, have revealed that they rely on graded index optics, where the center of the lens has a much higher refractive index than the periphery. It produces high-power lenses that are free of spherical aberration and are much superior to homogeneous lenses. This type of advanced biological lens has long been known to exist in the eyes of fish and invertebrates such as octopus and squid, but it was a surprise to find the principle in an animal as primitive as a jellyfish. It was even more surprising to find that the good lenses are not used for acute vision in box jellyfish. The retina is too close to the lens to pick up a sharp image, and the consequence is blurred vision. At first glance this may seem like a poorly designed eye, but careful mapping of receptive fields of individual photoreceptor cells in the retina revealed that the blurring generates receptive fields similar to those of large field motion detector neurons in the brain of other animals. Findings that box jellyfish eyes are colorblind add to the similarities with large field motion detector neurons. The obvious explanation is that the lens eyes of box jellyfish are devoted to large field motion detection and are tuned to extract information about self-motion and the presence of large nearby objects.

More eye than brain. All evidence now suggests that the optical apparatus of box jellyfish eyes is specialized for single visual tasks, and that this allows them to remove redundant visual information before it reaches the retina. Eyes of most other animals serve a multitude of visual tasks and need to pick up as much visual information as possible even though each task uses only limited subsets of the information. Having a general-purpose eye requires the neural capacity to separate and extract the information used for each visual task. The jellyfish way of having one eye type for each visual task lifts a burden from the nervous system, and it suddenly becomes understandable how these animals can generate complex behavior with their simple nervous system. Evidence is accumulating that the largest of the two lens eyes serves large field motion detection used in monitoring self-motion and warning for collisions. It is not yet clear which visual tasks are served by the remaining three types of eye.

Evolutionary relationships. Box jellyfish belong to the phylum Cnidaria, which is a sister group to all bilaterally symmetric animals (Bilateria). Complex visual systems are common among Bilateria, but among cnidarians they exist only in box jellyfish. As to whether the common ancestor to Cnidaria and Bilateria had eyes and vision, the probable answer is both yes and no. The biochemistry of photoreceptor cells seems to date back to a common ancestor, although we have yet to learn whether jellyfish photoreceptors are in all respects equivalent to those of Bilateria. In contrast, advanced eyes and visual processing have clearly formed independently in the two groups. The visual system of box jellyfish is of unique interest not only because of its largely independent evolution but also because it offers our only insight into how visually guided behavior can be realized in a radially symmetric creature.

For background information *see* COELENTERATA; CUBOMEDUSAE; EYE (INVERTEBRATE); NERVOUS SYSTEM (INVERTEBRATE); PHOTORECEPTION; VISION in the McGraw-Hill Encyclopedia of Science & Technology.   Dan-E. Nilsson

Bibliography. P. A. V. Anderson, Cnidarian neurobiology: What does the future hold?, *Hydrobiologia*, 530/531:107–116, 2004; M. F. Land and D.-E. Nilsson, *Animal Eyes*, Oxford University Press, 2002; D.-E. Nilsson, Eye evolution: A question of genetic promiscuity, *Curr. Opin. Neurobiol.*, 14:407–414, 2004.

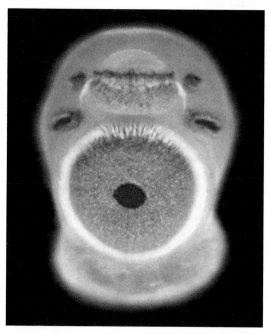

Fig. 2. Sensory club, 1 mm or 0.04 in. tall, of *Chiropsalmus* sp., with the typical set of six eyes. (© *Dan-E. Nilsson*)

Brominated flame retardants in the environment

Brominated flame retardants (BFRs) are materials added to or treatments applied to products or materials that significantly increase their resistance to catching fire. BFRs are used in a variety of consumer products, and several of those are produced in large quantities. Some BFRs have been detected in environmental matrices and human tissue, which can be attributed to the anthropogenic uses of these compounds.

Fire is a major source of damage to properties and loss of life. In the United States, every year over 3,000,000 fires are reported, resulting in 29,000 injuries, 4500 deaths, and direct losses of over $8 billion. Recent advances in technology have resulted in increased use of synthetic polymers, electronic equipment, and other ignitable materials in commercial and residential buildings. This has drastically contributed to fire hazard. To reduce the chances of ignition and burning of these materials, the use of flame retardants has increased.

The idea of flame retardant materials dates back to about 450 B.C., when the Egyptians used alum to reduce the flammability of wood. About 200 B.C., the Romans used a mixture of alum and vinegar to reduce the combustibility of wood. Today, there are more than 175 chemicals classified as flame retardants. Flame retardants are divided into four major groups: inorganic, halogenated organic, organophosphorus, and nitrogen-based, which account for 50%, 25%, 20%, and <5% of the annual production, respectively.

Mechanisms. To meet fire safety regulations, flame retardants are applied to combustible materials such as wood, paper, plastics, and textiles. To understand the action of flame retardants, it is essential to understand the combustion process. Combustion is a gas-phase reaction involving a fuel source and oxygen. There are four steps in the combustion process: preheating, volatilization/decomposition, combustion, and propagation. Depending on the mode of their action, flame retardants can act on any of these four steps to interrupt the combustion cycle. For examples, free radicals—highly oxidizing agents that are essential elements for the flame to propagate—are produced during the combustion process. Halogens are very effective in trapping free radicals, thereby removing the capability of the flame to propagate. All the halogens are effective at trapping free radicals; however their trapping efficiency is I > Br > Cl > F. Organohalogen compounds are a good vehicle for storing and delivering halogens in polymers, but not all organohalogen compounds are suitable as flame retardants. Fluorinated compounds are very stable and decompose at much higher temperature than that at which the polymers burn, while iodinated compounds are not very stable and decompose at slightly elevated temperatures. Consequently, only organochlorine and organobromine compounds can be used as flame retardants. With higher trapping efficiency and lower decomposing

temperature, organobromine flame retardants have become more popular than their organochlorine counterparts. Since the bromine content is the main ingredient of a BFR, there is no particular restriction on the structure of the compound's backbone. As the result, more than 75 different aliphatic, aromatic, and cycloaliphatic compounds are used as brominated flame retardants.

Types. BFRs are divided into three subgroups depending on the mode of incorporation into polymers: monomers, reactive, and additive. A brominated monomer, such as brominated styrene or brominated butadiene, is used in the production of brominated polymers. The brominated polymers are blended with nonhalogenated polymers or introduced into the feed mixture prior to polymerization, resulting in a polymer that contains both brominated and nonbrominated monomers. Reactive flame retardants, such as tetrabromobisphenol A (TBBPA), are chemically bonded to the plastics. Additive flame retardants, which include polybrominated diphenyl ethers (PBDEs) and hexabromocylododecane (HBCD), are simply blended with the polymers and are more likely to leach out of them.

Polybrominated biphenyls (PBBs). Polybrominated biphenyls were introduced as flame retardants in the early 1970s. The commercial production of PBBs in the form of hexabromobiphenyl flame retardants in the United States continued until 1976. Approximately 6071 metric tons of PBBs were produced during this period. In 1973, hexabromobiphenyl flame retardants were unintentionally mixed into cattle feed at a production site and distributed in rural Michigan. The widespread contamination of Michigan farm products that resulted from this accident led to their ban in the United States in 1974. Nevertheless, the production of octabromobiphenyl and decabromobiphenyl formulations continued until 1979. In Europe a mixture of highly brominated PBBs was in production in Germany until 1985, and in France decabromobiphenyl flame was in production until 2000.

Polybrominated diphenyl ethers (PBDEs). Polybrominated diphenyl ethers are the most widely studied BFR, and have been detected in all environmental compartments (that is, biota, air, water, and land). PBDEs are produced by bromination of diphenyl ether in the presence of a Friedel-Craft catalyst in an organic solvent such as dibromomethane. Diphenyl ether molecules contain 10 hydrogen atoms, any of which can be exchanged with bromine, resulting in 209 possible congeners. PBDEs are produced at three different degrees of bromination—penta-PBDE, octa-PBDE, and deca-PBDE formulations—and classified according to their average bromine content.

In 1979, the presence of deca-BDE in soil and sludge was detected in the areas surrounding plants where PBDEs were manufactured in the United States. Two years later, Ö. Andersson and G. Blomkvist reported the presence of PBDEs in samples collected along Visken River in Sweden. B. Jansson and coworkers first reported that PBDEs are global contaminants by demonstrating their presence in fish-eating birds and marine mammals in samples collected from the Baltic Sea, North Sea, and Arctic Ocean. PBDE congeners were also found in marine fish, shellfish, and sediment, as well as in air particulate from Japan and Taiwan. PBDEs were also reported in cod liver and herring from the North Sea and in eels from freshwater systems in the Netherlands. C. J. Stafford reported the presence of PBDEs in eggs and tissues of fish-eating birds from six states in the United States and from Ontario, Canada.

PBDEs are lipophilic compounds that have been shown to bioaccumulate through the food web. PBDEs have also been detected in human adipose tissue. In 1998, K. Norén and D. Meironyté showed that the concentration of PBDEs in breast milk doubled every 5 years over the past 25 years. Recent studies have indicated that the levels of PBDEs are on the rise in North American environment. However, studies from Sweden and Japan have reported a decline in the concentrations of PBDEs in biota.

On December 5, 2002, the European Parliament issued Council Directive 76/769/EEC, restricting the marking and use of penta- and octa-PBDE formulations. In the United States, California was the first state to ban the use of penta- and octa-PBDEs. Currently seven other states have similar legislation. In Japan, as the result of a voluntary industry ban, the use of penta-PBDE formulation was phased out in 1990, and treatments using other PBDE formulation have been reduced.

In 2003, the BFR manufacturers announced a voluntary ban on the production of penta- and octa-PBDEs formulations by the end of 2004.

Hexabromocyclododecane (HBCD). Hexabromocyclododecane is a white crystalline powder, containing 74.7% bromine by weight. The main application of HBCD is for polystyrene foam that is used in building construction. The global market demand for HBCD increased from 15,900 metric tons in 1999 to 21,951 metric tons in 2003. HBCD has been detected in various environmental compartments, including arctic air and biota. Recent data indicate that the levels of HBCDs are on the rise in the environment.

Tetrabromobisphenol A (TBBPA). Tetrabromobisphenol A is a reactive flame retardant with a global market of more than 145,000 metric tons in 2003, which makes it the highest-volume BFR. Approximately 90% of TBBPA is used as a reactive intermediate in the production of epoxy and polycarbonate resins. The main application of epoxy resins is in the manufacturing of printed circuit boards, which contain approximately 20% bromine by weight. The remaining 10% of TBBPA is transformed into derivatives, such as tetrabromobisphenol A-bis(2-hydroxyethyl ether), which are used as flame retardants for paper and textile adhesives and coatings. Despite the primary use of TBBPA as a reactive flame retardant (covalently bonded to the polymer), TBBPA has been observed in the environment. TBBPA has been detected in sewage sludge from three sewage treatment plants in Sweden. The concentrations of TBBPA ranged between 3.6 and 45 nanograms per gram of sludge;

similar results have been observed in Canada. The key concern with TBBPA is its similarity in chemical structure to thyroxine. I. A. T. M. Meerts and coworkers showed that TBBPA has stronger affinity for binding with the thyroid hormone transport protein transthyretin than the natural ligand thyroxine.

Others. Researchers have recently reported the detection of other BFRs in environmental matrices. Occurrence of decabromodiphenyl ethane (DBDPEt) in sewage sludge was reported by A. Kierkegaard and coworkers and subsequently in sewage sludge from North America by A. Konstinov and coworkers. Occurrence of tribromophenoxy ethane (TBE) in air and sediment from the Great Lakes was reported by E. Hoh and coworkers.

Recently the number of reports on occurrence of BFRs in the environment has increased exponentially. An increase in the number of BFRs detected in the environment is expected to increase with the number of studies on BFRs.

For background information *see* BROMINE; COMBUSTION; ENVIRONMENTAL TOXICOLOGY; FIRE TECHNOLOGY; FLAME; FLAMEPROOFING; FREE RADICAL; HALOGENATED HYDROCARBON; POLYMER; THYROXINE; TOXICOLOGY; TROPHIC ECOLOGY in the McGraw-Hill Encyclopedia of Science & Technology.

Mehran Alaee

Bibliography. A. Covaci et al., Hexabromocyclododecanes (HBCDs) in the environment and humans: A review, *Environ. Sci. Technol.*, 40(12):3679–3688, 2006; C. A. de Wit, M. Alaee, and D. C. G. Muir, Levels and trends of brominated flame retardants in the Arctic, *Chemosphere*, 64(2):209–233, 2006; E. Hoh, L. Zhu, and R. A. Hites, Novel flame retardants, 1,2-bis(2,4,6-tribromophenoxy)-ethane and 2,3,4,5,6-pentabromoethylbenzene, in United States' environmental samples, *Environ. Sci. Technol.*, 39:2472–2477, 2005; A. Kierkegaard, J. Björklund, and U. Fridén, Identification of the flame retardant decabromodiphenyl ethane in the environment, *Environ. Sci. Technol.*, 38:3247–3253, 2004; A. Konstantinov et al., Characterization of mass-labeled [^{13}C$_{14}$]-decabromodiphenylethane and its use as a surrogate standard in the analysis of sewage sludge samples, *Chemosphere*, 64(2):245–249, 2006; R. J. Law et al., Levels and trends of brominated flame retardants in the European environment, *Chemosphere*, 64(2):187–208, 2006; M. Lebeuf et al., Polybrominated diphenyl ethers (PBDEs) in blubber of beluga whales (*Delphinapterus leucas*) from the St. Lawrence estuary, Canada, *Environ. Sci. Technol.*, 38:2971–2977, 2004; I. A. T. M. Meerts et al., Potential competitive interactions of some brominated flame retardants and related compounds with human transthyretin *in vitro*, *Toxicol. Sci.*, 56:95–104, 2000; R. J. Norstrom et al., Geographical distribution (2000) and temporal trends (1981 to 2000) of brominated diphenyl ethers in Great Lakes herring gull eggs, *Environ Sci. Technol.*, 36:4783–4789, 2002; O. Watanabe and S. Sakai, Environmental release and behavior of brominated flame retardants, *Environ. Int.*, 29:665–682, 2003.

Carbonate sedimentology

Carbonate sedimentary rocks, irrespective of their age, are economically important because they act as reservoirs for oil and gas, and host ores of metals such as lead and zinc, and are used for building stone, aggregates, and cement production. Limestone, formed mainly of calcium carbonate [CaCO$_3$], and dolostone, formed mainly of magnesium-calcium carbonate [CaMg(CO$_3$)$_2$], are the two main rock types. Despite their chemical simplicity, these rocks have commonly undergone complex chemical transformations during diagenesis. Although marine carbonates are typically equated with shallow, tropical marine settings (**Fig. 1**), vast areas of carbonate sediments are also forming in cool water temperature seas, such as those found off the south coast of Australia. Nonmarine carbonates form in freshwater lakes, streams, caves, and springs (**Fig. 2**).

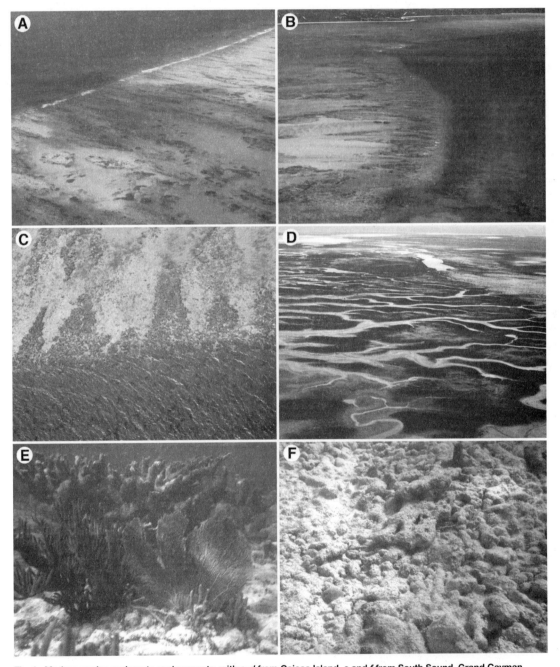

Fig. 1. Modern marine carbonate environments, with *a–d* from Caicos Island, *e* and *f* from South Sound, Grand Cayman. (*a*) Shallow lagoon (on right) separated from deep ocean by coral reef (marked by white breaking waves). Note scattered patch reefs in area behind reef crest. (*b*) Different view of area in *a*, showing shallow lagoon (left) separated from deep ocean by coral reef. (*c*) Almost vertical view down on reef crest and back-reef zone from area in *b*. (*d*) Intertidal and supratidal flats covered with well-developed algal mats and cut by meandering tidal channels. (*e*) Patch reefs developed in back-reef area, showing large *Acropora palmata* colonies (background) and Gorgonian sea fan (foreground). Water ~2 m (6 ft) deep. (*f*) Mixture of broken coral fragments and coarse sand in area behind reef crest. Most coral fragments have been encrusted by coralline algae. Water ~2 m deep.

Fig. 2. Upper part of spring deposits, formed of calcite at Pamukkale, Turkey, showing rimstone dams and large pools.

In nature, $CaCO_3$ is precipitated as calcite (trigonal crystal system), aragonite (othorhomic crystal system), and vaterite (hexagonal crystal system). Vaterite is rare. At the low temperatures and pressures operative in sedimentary systems, calcite is stable whereas aragonite is metastable. Nevertheless, many organisms precipitate aragonitic skeletons, including corals, bivalves, gastropods, and calcareous algae. With time, aragonite will convert to calcite.

The origin of dolomite $[CaMg(CO_3)_2]$ remains controversial because it typically forms through the replacement of calcite in the subsurface realm where direct observation is impossible. Dolomite has yet to be precipitated in the laboratory at low temperatures and low pressures.

Limestone components. Marine limestones are formed of allochems (chemical or biochemical objects typically formed close to their place of deposition), matrix (sediment mechanically deposited between allochems), or cement (crystals precipitated from waters passing between the allochems). These components form the basis for both R. L. Folk's and R. J. Dunham's limestone classifications. Allochems include coated grains, pellets/peloids, bioclasts, and lithoclasts. Coated grains, up to 0.25 m (1.6 ft) in diameter with a nucleus (sand grain, shell, bone) encased by a cortex of concentrically nested laminae include ooids (<2 mm diameter), pisoids (>2 mm diameter), oncoids (formed by filamentous cyanobacteria), and rhodolites (formed by coralline algae). Pellets are ovate, <1 mm long, and formed of carbonate mud. They are produced by animals (such as gastropods and worms) that ingest sediment on the sea floor, extract organic matter as it passes through their intestinal tract, and then expel the unwanted sediment as fecal pellets. Peloids refer to similar grains that formed through micritization of other allochems, such as ooids and bioclasts. Bioclasts, derived from the skeletons of animals or plants, are compositionally heterogeneous because their calcite-aragonite content depends on their parent biotas. Lithoclasts, derived from semilithified or lithified sediments, can be common in some carbonate sediments.

Nonmarine limestones commonly contain allochems akin to those found in marine carbonates. Many cave and spring deposits contain ooids, pisoids, and oncoids. Tufa and travertine are common in many nonmarine environments. Tufa, characterized by high porosity, forms as calcite is precipitated around plants and plant roots. Travertines, which generally have a lower porosity than tufa, are typically formed of calcite with bizarre crystal morphologies (Fig. 2). Dendrite crystals seem to form from spring waters that have undergone rapid carbon dioxide (CO_2) degassing. The precise factors that control the vast array of crystal morphologies found in travertines are still being debated.

Evolution of carbonate depositional successions. The balance between submergence during sea-level highstands and exposure during sea-level lowstands fundamentally controls the temporal development of a carbonate rock succession. During highstands, the "carbonate factory," which is at maximum efficiency in the photic zone, becomes established as plants and animals colonize the sea floor (Fig. 1e).

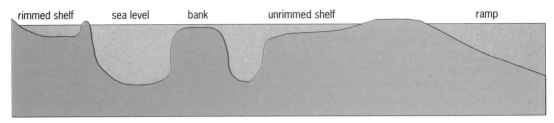

Fig. 3. Diagrammatic settings where carbonate sediments will form and accumulate, provided there is no siliciclastic input from rivers.

The biota start to influence sedimentation by constructing reefs (Fig. 1*a–c*), contributing their CaCO₃ skeletons to the sediment, binding and baffling and trapping sediment, and burrowing and boring into substrates. Sediments form and accumulate in the subtidal (Fig. 1*a–c*), intertidal, and supratidal realms (Fig. 1*d*). If there is no significant input of siliciclastic sediments from adjacent land areas, carbonate sediment accumulate in lagoons, on open shelves, on shelves rimmed by reefs or islands, on isolated banks, and on ramps that have a sea floor sloping gradually into deeper water (**Fig. 3**). Sediments, including bioclastic sands, peloidal sands, and calcareous mud accumulate in accord with local conditions. Oolitic sands commonly characterize high-energy zones, which are typically found along the seaward margin and in tidal channels. In most tropical settings, sediments accumulate under low-energy conditions, and there is little transport of sediment from one area to another. However, aperiodic storms and hurricanes radically alter that scenario by quickly moving large volumes of sediment

and killing and severely damaging the biota.

During lowstand conditions, marine carbonate sediments are exposed to the atmosphere and rapidly become lithified as their constituents react to the newly imposed meteoric conditions. Metastable aragonite, which is especially prone to change in such settings, is dissolved or converted to calcite. The former process leads to an increase in porosity, commonly through the preferential dissolution of aragonitic skeletons of corals, bivalves, and gastropods. The sediment-rock transition and/or the modification of existing lithified rock is controlled largely by climatic conditions. Wet climates typically generate karst terrains (for example, caves), with dissolution and precipitation being common. Dry climates will usually lead to caliche (calcrete) formation. Soils commonly develop and accumulate on limestone surfaces that are exposed to the atmosphere. The recognition of a paleo-land surface (unconformity) in the rock succession is extremely important because it denotes a significant stage in the evolution of an area (**Fig. 4**). Such surfaces

Fig. 4. Field photograph showing the Cayman Unconformity that forms the boundary between the Cayman Formation (Miocene) and Pedro Castle Formation (Pliocene) in Pedro Castle Quarry on the south coast of Grand Cayman. There the unconformity is ∼9 m (30 ft) above sea level; approximately 2 km (1.25 mi) to the north, this unconformity is ∼26 m (85 ft) below sea level. This unconformity, which is a sequence boundary, formed during the sea-level lowstand at the end of the Miocene. The paleocaves formed in conjunction with the Cayman Unconformity.

are commonly used to correlate geographically distant areas and to divide a succession into sequences that provide the basis for sequence stratigraphy.

Limestone diagenesis. Limestones are prone to diagenesis (alteration of preexisting sediments/rocks under low-temperature and low-pressure conditions), especially given the susceptibility of calcite and aragonite to chemical alteration. Most diagenesis is poorly understood because it takes place in the subsurface where it cannot be directly monitored or observed. Diagenesis, which is typically triggered by temporal changes in the environmental setting of limestone, involves dissolution, recrystallization (change in texture of rock, but no change in mineralogy), replacement (change of mineralogy), and cementation (precipitation of minerals). These processes are ultimately controlled by microenvironmental conditions in the rock, which may be operative on the scale of individual pores with neighboring pores being subjected to different processes. Dissolution usually leads to an increase in porosity. Many limestones are characterized by fossil-moldic porosity that can be attributed to the preferential dissolution of aragonitic bioclasts. The resultant pores reflect the size and shape of the original bioclast; for example, those formed by dissolution of foraminifera skeletons are typically less than 0.5 cm (0.2 in.) long, whereas those formed by the dissolution of colonial corals may be 2–3 m (6–10 ft) in diameter.

Cements may be precipitated from pore fluids that are supersaturated with respect to either calcite or aragonite. Such cementation is important because it binds the allochems together and commonly reduces the porosity that had been created during earlier diagenetic phases. Zoned cements, with zones defined by slight differences in their trace element (such as Fe and Mn) content provide a record of progressive changes in fluid compositions.

Dolomitization. Dolomitization has been attributed to many different processes that are operative under a wide range of conditions. Common to all models is the need to (1) supply large quantities of Mg, (2) constantly deliver Mg to the site of dolomite formation, and (3) have a microenvironmental niche suitable for dolomite precipitation. These processes must be operative on a large scale, given that limestones hundreds of meters thick and covering vast areas are commonly subjected to pervasive dolomitization. Old dolostones are commonly difficult to interpret because they have undergone a complex diagenetic history over hundreds of millions of years that may be impossible to fully decipher. Therefore, one approach to the dolomitization problem has been to study Tertiary dolostones found on isolated oceanic islands in the Pacific Ocean and Caribbean Sea. Attention has focused on these successions because they immediately allow some of the potential causes (for example, deep burial) of dolomitization to be eliminated. Pure dolomite, $CaMg(CO_3)_2$, is characterized by a 50-50 Ca:Mg ratio with the calcium and magnesium ions being orderly arranged into distinct planes that are sandwiched between CO_3 planes. Much of the dolomite found in older sedimentary sequences closely approximates ideal dolomite. Dolomite found in younger successions is commonly disordered and contains 49–62 mol % $CaCO_3$. The reason(s) for the variance in the amount of $CaCO_3$ and the exact location of the excess Ca cations in the dolomite latticework are unknown. Determining the amount of Ca in a dolomite is not an easy matter and must rely on bulk x-ray diffraction analysis, backscattered electron imaging on an electron microprobe, or elemental analysis on the electron microprobe. Recent research has shown that dolostones in many Tertiary successions are formed of oscillatory zoned crystals, which have alternating zones of low-calcium dolomite (<55 mol % $CaCO_3$) and high-calcium dolomite (>55 mol % $CaCO_3$).

Outlook. Many aspects of carbonate rocks remain enigmatic despite extensive research and the need to better understand them because of economic necessities. Problems related to sequence stratigraphy, diagenesis, and dolomitization remain to be resolved.

For background information *see* ARAGONITE; CALCITE; CARBONATE MINERALS; DIAGENESIS; DOLOMITE; DOLOMITE ROCK; GEOLOGIC TIME SCALE; KARST TOPOGRAPHY; LIMESTONE; MARINE SEDIMENTS; SEDIMENTARY ROCKS; SEQUENCE STRATIGRAPHY; STRATIGRAPHY; TERTIARY; TRAVERTINE; TUFA; UNCONFORMITY; X-RAY DIFFRACTION in the McGraw-Hill Encyclopedia of Science & Technology.

Brian Jones

Bibliography. R. G. C. Bathurst, *Carbonate Sediments and Their Diagenesis: Developments in Sedimentology*, Elsevier, Amsterdam, 1957; R. G. Loucks and J. F. Sarg, *Carbonate Sequence Stratigraphy*, American Association of Petroleum Geologists, 1993; B. Purser, M. Tucker, and D. Zenger, *Dolomites—A Volume in Honour of Dolomieu*, International Association of Sedimentologists, Spec. Publ. 21, 1994; P. A. Scholle and D. S. Ulmer-Scholle, *A Color Guide to the Petrography of Carbonate Rocks: Grains, Textures, Porosity, Diagenesis*, American Association of Petroleum Geologists, 2003; P. A. Scholle, D. G. Bebout, and C. H. Moore (eds.), *Carbonate Depositional Environments*, American Association of Petroleum Geologists, 1983; M. E. Tucker and V. P. Wright, *Carbonate Sedimentology*, Blackwell Scientific, Oxford, England, 1990; R. G. Walker and N. P. James, *Facies Models: Response to Sea Level Change*, Geological Association of Canada, 1992.

Chinese historical documents and climate change

As late as the nineteenth century, it was commonly believed that climate represented a kind of averaged weather over time at a geographical location and hardly changed, at least for the period comparable to the length of human history. It has been demonstrated that climate changes at millennial, centennial, and even decadal scales, as many studies in the twentieth century have revealed. Such studies were largely done by examining environmental data such as tree rings, pollen assemblages, lake sediment, and ice cores. These data are objective and usually continuous, but they are often difficult to interpret. For example, a narrow tree ring could mean either a dry spell or cold spring, or both. Consequently, conclusions obtained this way are often associated with considerable uncertainties or ambiguities. Furthermore, environmental data usually have low time resolution and hence are not suitable for high-resolution analysis.

Historical climate records. Many historical documents contain records of direct human observations of climatic conditions (cold, warm, wet, and dry) that can be used directly for studying climatic changes. In addition, human activities are influenced by climate, and by carefully scrutinizing the descriptions in historical documents detailing these activities, it is possible to decipher what the climate must have been. These documents, if available, can serve as a very useful source of data for climate studies.

The main advantage of human-recorded climate descriptions, as opposed to those inferred from environmental data, is that the climatic conditions are presumably direct observations, which are easier to interpret, at least qualitatively. For example, it is hard to imagine that an extremely cold winter would be recorded as warm. Of course, missing or discontinuous records may be a serious problem as documents could become lost or destroyed due to dislocation, war, and other causes (such as earthquake and floods). To be useful for climatic studies, it is necessary to have long and continuous records.

China is one of the few countries that has kept a huge amount of historical documents, thanks to its long and continuous history and antiquarian culture. There were governmental offices that were responsible for watching unusual environmental phenomena and recording them in daily logs, and many of these phenomena related to weather and climate. Furthermore, China's main economical activity has been agriculture, which is greatly influenced by climate. As a result, a large amount of agriculture-related records of floods, droughts, famine, harvest, price of grain, damages to fruit trees, and so on have been kept in archives that can be used as a data source for climate studies. Then there are phenological records, such as the freezing and thawing dates of lakes and rivers, blossoming dates of certain flowers, and arrival dates of certain species of migrating birds, which are indicators of climatic conditions. In addition, there are some unusual weather phenomena (for example, thunderstorms in winter), which were considered as bad omens out of superstition and were recorded systematically in official histories that can also serve as sources of climate data. Indeed, all these records have been used successfully for reconstructing past climate conditions in China.

Given that there are so many different kinds of data, one would expect to see many discrepancies among conclusions derived from different data series. Instead, there is a surprising degree of consistency among them, at least in the broad sense. One typical example is shown in the **illustration**. Here two different series are plotted: a winter temperature series and a winter thunderstorm frequency series. It is obvious that the two series have good correlation, despite the fact that they are of completely different nature.

Temperature series. In the following, we will focus on discussing the temperature series only. This series, deduced mostly from phenological phenomena as described in historical documents, shows immediately that there had been obvious warm and cold periods in China in the historical time. Several notable broad periods are discussed, for which the ranges are approximate.

A.D. 200–600. This was a cold period (as compared to the current situation) roughly corresponding to the Three-Kingdoms, Wei, Jin and North-South Dynasties in Chinese history. This was a time of relative turmoil when many kingdoms coexisted. Wars within each kingdom and between different

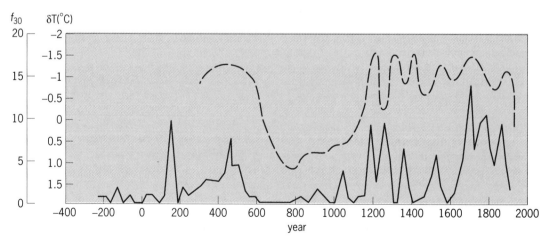

Winter temperature (dashed curve) and winter thunderstorm frequency (solid curve) of China in historical time. The winter temperature series was derived from phenological records (K. C. Chu, 1972), and the winter thunderstorm series was compiled from the Twenty-Four Official Histories of China (P. K. Wang, 1980). The winter temperature series is plotted as the deviation from the modern (1900 to 1960s) mean temperature (the peaks represent cold deviation). Winter thunderstorm frequency represents the number of winter thunderstorm reports in a 30-year interval.

kingdoms were frequent. Previous to this period during the Han dynasty, bamboo, a semitropical plant, was still spread widely in northern China. During this colder period, bamboo became less populated, and it is virtually extinct in nature in northern China today. Similarly, the once widespread tangerine orchards became nearly extinct because of the colder climate. One report at the beginning of the third century says that the tangerine trees merely flowered but failed to bear fruit in Henan Province. Cold spells in winter occurred more frequently. In 225, a military exercise at Guangling (33.5°N, 119°E), a town near Huai River, was halted due to a sudden freeze of the river. During the later part of this period, called the South-North Dynasties, China was divided into north and south empires, separated by the Yangtze River. An imperial "icehouse" was established in Nanjing (32N°, 119°E) by the southern empires to store ice obtained during the winter for the imperial family's use. This was the first time in history that the imperial icehouse was built at such a southern latitude. This feat would be impossible today because the winter would not be cold enough to obtain thick chunks of ice there.

600–1000. A relatively warm climate characterized this period and, coincidentally or perhaps consequentially, China enjoyed a relatively stable political climate as well. The Tang Dynasty, often regarded as the golden age of Chinese culture, began at 618 and ended in 907. In 650, 669, and 678, it was reported that there was no snow or ice in winter in the national capital Changan (today's Xian, 34.2°N, 109°E). Orange and tangerine trees were again widely planted in Changan, and "the fruits tasted as good as those tributes from southern China." Another indication of the warm climate was the wide existence of plum trees at the time, as indicated in many poems.

Near the end of this period, however, the warm climate turned into a colder period. The once stable Tang Empire fragmented into several kingdoms and entered the Five Dynasties and Ten Countries period. This again was a fairly chaotic time, and wars between kingdoms ensued. As a rule in Chinese history, it appears that the cold period usually coincided with more frequent wars between kingdoms. (This coincidence is not explained here.)

After 1000 and until at least the beginning of the twentieth century, it appears that the climate of China fluctuated more rapidly than in previous periods. Cold and warm periods seemed to alternate in intervals of roughly 100–200 years, but in general the mean winter temperature appears to be lower than that in the 600–1000 period and moderately lower than that in the twentieth century. At present it is unclear whether this rapid fluctuation is real or due to artifacts in the data.

1000–1200. This generally cold period culminated around 1150. An initially affluent Song Empire (960–1279) was gradually weakened by the rising nomad rivals in the north—the Liao, Jin, and eventually the Yuan (Mongolian) empires. The Song Empire shrank to become the Southern Song Empire in 1127, occupying only lands to the south of the Yangtze River. Kublai Khan defeated Jin and then the Southern Song Empire in 1279. The Lake Tai [Tai Hu, area 2250 km^2 (870 mi^2), located at 31.2°N] was reported to freeze for the first time in history in 1111, and the cold spell wiped out all citrus trees in this area. Snowfall records indicated that the average last snow dates in 1131–1260 was the ninth day of the fourth month (roughly corresponding to May in the Gregorian calendar), nearly a full month later than that in 1100–1110, indicating a colder trend. There were also many reports of damage to other fruit plants (notably, *Nephelium litchi*, commonly known as lichee) in parts of China even farther south because of the cold climate in the twelfth century.

1200–1300. This was a relatively warm period. An indication was the reinstating of government offices overseeing the bamboo production in Shanxi and Henan provinces under the Mongolian rule. These offices were disbanded previously in the twelfth century by Song rulers because the cold climate greatly reduced production.

1300–1400. In this cold period, according to the Mongolian poet Nai-Xian (1309–1352), icing in the Yellow River occurred earlier by as much as a full month as compared to the present. The poet also lamented that swallows "were merely short visitors" due to the colder climate.

1400–1900. The climate became somewhat warmer near 1400, although still colder than the present. In general, this was a colder period but seemed to approach the present warm condition near 1900. Ironically, the records in this period become too numerous to easily decipher the climatic conditions without ambiguities. Many records are probably noise rather than signals, which will take much research effort to sort out. But certain severe climate-caused disasters stand out as clear signals. One such example was the adverse climatic conditions (severe winters and droughts) near the end of Ming Dynasty (1368–1644), which were often attributed to the collapse of the Ming Empire and succession by the Qing (Manchurian) Dynasty (1644–1911).

For background information *see* ASIA; BAMBOO; CLIMATE HISTORY; CLIMATE MODIFICATION; CLIMATIC PREDICTION; GLOBAL CLIMATE CHANGE; LYCHEE; PLANT GEOGRAPHY in the McGraw-Hill Encyclopedia of Science & Technology.   Pao K. Wang

Bibliography. K. C. Chu, A preliminary study on the climatic fluctuations during the last 5000 years in China, *Scientia Sinica*, 16:226–256, 1973; P. K. Wang, On the possible relationship between winter thunder and climatic changes in China over the past 2,200 years. *Clim. Change*, 3:37–46, 1980; P. K. Wang and D. Zhang, An introduction of some historical governmental weather records in the 18th and 19th centuries of China, *Bull. Amer. Meteorol. Soc.*, 69:753–758, 1988; P. K. Wang and D. Zhang, Recent studies of the reconstruction of East Asian monsoon climate in the past using historical literature of China, *J. Meteorol. Soc. Jap.*, 70:423–446, 1992; P. K. Wang and D. Zhang, Reconstruction of the 18th century precipitation of Nanjing, Suzhou, and Hangzhou using the Clear and Rain Records, in *Climate Since 1500 AD*, ed. by R. S. Bradley and P. D. Jones, pp. 184–209, Routledge, London, 1991; M. Winkler and P. K. Wang, The late Pleistocene and Holocene climate of China: A review of biogeologic evidence and a comparison with GCM climate simulations, in *Global Climates since Last Glacial Maximum*, ed. by H. E. Wright et al., pp. 221–264, 1994.

Cholesteryl ester transfer protein (CETP)

Cholesteryl ester transfer protein is a 74-kilodalton hydrophobic glycoprotein that plays a central role in human high-density lipoprotein (HDL) metabolism and reverse cholesterol transport. It mediates the exchange of neutral lipid (both heteroexchange, which is the net transfer of cholesteryl ester between lipoproteins in exchange for triglyceride, and homoexchange, which is the bidirectional transfer of the same lipid) between apolipoprotein (apo) A-I- and apoB-containing lipoproteins (**Fig. 1**).

Reverse cholesterol transport. HDL has a functional role in the protection against atherosclerosis, and its plasma concentration is inversely correlated to the risk of cardiovascular disease. One of the protective actions of HDL involves its ability to act as an acceptor of cholesterol from peripheral cells, including arterial macrophage foam cells (which constitute atherosclerotic type II lesions, including the fatty streak lesions—the first grossly visible lesions of atherosclerosis). The adenosine 5′-triphosphate (ATP)–binding cassette transporter A1 (ABCA1) mediates phospholipid and cholesterol efflux to lipid-free or lipid-poor apoA-I, whereas other proteins, including scavenger receptor class BI (SR-BI), ABCG1, and ABCG4, enhance cholesterol efflux to larger lipidated HDL particles. Cholesterol extracted by efflux from peripheral tissues is esterified within HDL by lecithin:cholesterol acyltransferase (LCAT). CETP has an established role in mediating neutral lipid

transport between lipoproteins. Thus CETP mediates a net transfer of cholesteryl ester from HDL to apoB-containing lipoproteins. Under conditions of efficient hepatic apoB lipoprotein clearance, CETP may promote cholesterol transport from HDL to the liver for subsequent biliary secretion via this pathway (**Fig. 2**). However, high plasma concentrations of CETP are associated with low levels of HDL-cholesterol, which are commonly associated with increased atherosclerosis risk.

HDL–cholesteryl ester is also directly returned to the liver by a process known as selective uptake. This involves the reversible incorporation of HDL-derived cholesteryl ester into a plasma membrane pool, followed by transfer of the lipid to an inaccessible pool by mechanisms not involving coated pit-mediated endocytosis (coated pits are cell surface depressions coated with clathrin, a protein, that aid in internalization of extracellular materials into the cytoplasm) [**Fig. 3**]. SR-BI has been shown to be the primary receptor responsible for hepatic HDL-cholesterol clearance in the mouse, a species that intrinsically lacks CETP. The human homologue of SR-BI, CLA-1, was cloned several years ago, but has not yet been established as the primary receptor for HDL-cholesteryl ester selective uptake in humans.

Regulation of CETP gene expression. CETP messenger ribonucleic acid is predominantly expressed in the liver, spleen, and adipose tissue in humans, and is upregulated by cholesterol by a mechanism that involves both a cholesterol response element between nucleotides -361 and -138, which binds transcription factors YY1 (Yin Yang 1) and SREBP-1 (sterol response element binding protein-1), and a DR-4 element (a direct repeat element spaced by 4 base pairs), which binds LXR/RXR (liver X receptor/retinoid X receptor, which are nuclear hormone receptors). Notably, plasma concentrations of CETP are increased in various hyperlipidemic states and in response to cholesterol intake. Dietary or pharmacological treatments which lower plasma lipids, such as statin therapy, also reduce circulating CETP levels.

Genetic variation at CETP gene locus. Several mutations have been identified in the CETP gene, which is located on chromosome 16q21. Of particular interest are variants which result in either a premature stop codon or splicing defects leading to altered protein structure. The most common of these is the G+1A/In splicing defect, which occurs in about 2% of the Japanese population. Homozygotes have no detectable CETP protein or activity and have a marked increase in plasma concentrations of HDL-cholesteryl ester. Although plasma concentrations of HDL are elevated in this disorder, complete CETP deficiency does not clearly confer protection against coronary heart disease. The D442G polymorphism is also common in Japan and results in partial loss of plasma CETP activity. A number of polymorphisms in the coding sequence or 5′-flanking region of CETP have been shown to have reciprocal effects on plasma CETP and HDL-cholesterol concentrations. The best-studied is the *Taq*1B polymorphism,

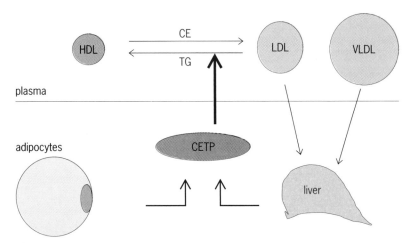

Fig. 1. CETP as a neutral lipid transfer protein. CETP is synthesized and secreted by hepatocytes and adipocytes as well as other cell types. CETP mediates the heteroexchange of CE (cholesteryl ester) and TG (triglyceride) between apoA-I- and apoB-containing lipoproteins resulting in net transfer of CE to VLDL (very low density lipoprotein) and LDL, and reciprocal TG enrichment of HDL.

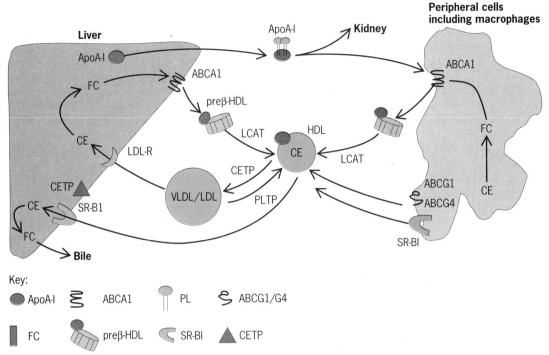

Fig. 2. Reverse cholesterol transport. ABCA1 mediates phospholipid and cholesterol efflux to lipid-poor apoA-I. Cholesterol efflux to lipidated HDL occurs by diffusional efflux and by SR-BI, ABCG1, and ABCG4. Free cholesterol is esterified within HDL by LCAT. CETP mediates net transfer of cholesteryl esters from HDL to VLDL and LDL. Under conditions of efficient hepatic apoB lipoprotein clearance, CETP may promote cholesterol transport from HDL to the liver or subsequent biliary secretion via this pathway. However, by decreasing plasma HDL-C (HDL-cholesterol) concentrations, high CETP activity may increase atherosclerosis risk.

which is in strong linkage disequilibrium (occurrence in a population of certain combinations of linked alleles in greater proportion than expected from the allele frequencies at the loci) with functional polymorphisms in the 5′-flanking sequence, including −629C/A, −971G/A, and −1337C/T, which interact to determine CETP gene expression. Homozygotes for the B2B2 polymorphism have lower CETP mass and activity and higher HDL-cholesterol concentrations as compared to B1B1 homozygotes. Most studies have suggested that B2B2 homozygotes are at reduced risk for cardiovascular disease. Another common polymorphism results in an amino acid change at amino acid 405 (I405V). The 405V allele is present in 25% of the population and is asso-

ciated with lower plasma CETP activity and higher HDL-cholesterol concentrations. However, no consistent relationship of the I405V variant to coronary artery disease risk has been established.

Animal studies. Mice do not express CETP, and introduction of the human transgene results in a marked reduction in HDL-cholesterol concentrations. The effect on atherosclerosis susceptibility is highly contingent on the level of CETP expression and on the genetic and metabolic context. Thus, both apoE and low density lipoprotein (LDL) receptor knockout mice develop increased atherosclerosis when crossed with CETP transgenic mice. In contrast, introduction of the CETP transgene results in decreased atherosclerosis in mice deficient

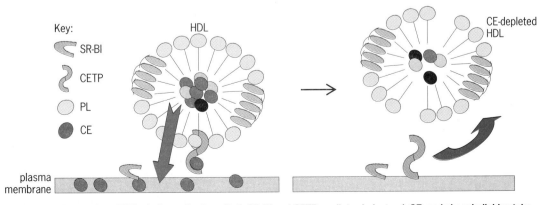

Fig. 3. Selective uptake of HDL-cholesteryl esters. Both SR-BI and CETP mediate cholesterol, CE, and phospholipid uptake via a pathway that does not involve significant degradation of the HDL particle, a process known as selective uptake.

in either apoCIII or LCAT. Plasma CETP concentrations in cholesterol-fed rabbits are 10-fold that of humans, and inhibition of CETP expression or function using antisense oligonucleotides, a CETP vaccine, or a CETP inhibitor reduces atherosclerosis in this model.

CETP inhibitors in treatment of low-HDL states. Pharmacological drug treatments which increase HDL-cholesterol include inhibition of CETP activity. One of these compounds, torcetrapib, has been shown to increase the association of CETP with its lipoprotein substrates, creating a nonfunctional complex. In a multidose study in healthy individuals, torcetrapib resulted in a dose-dependent inhibition of CETP activity and marked increase in HDL-cholesterol (+16% to +91%) and a decrease in LDL-cholesterol (−21% to −42%). Ongoing clinical studies are determining the effects of torcetrapib on progression of atherosclerosis and on clinical cardiovascular events in coronary artery disease patients receiving concomitant atorvastatin (an oral drug that lowers the level of cholesterol in the blood) therapy.

CETP in direct clearance of HDL-derived cholesteryl ester (selective uptake). In addition to its role in neutral lipid transport in the plasma compartment, CETP directly mediates selective uptake of HDL-derived cholesteryl ester by hepatocytes, which are cells of fundamental importance in HDL metabolism. This effect occurs by a mechanism that does not involve SR-BI and does not require transfer of cholesteryl ester to apoB-containing lipoproteins and subsequent uptake via a member of the LDL receptor gene family. Hepatic expression of CETP in vivo in mice results in a marked decrease in cholesterol in particles in the HDL density range consistent with a physiological role for hepatocyte CETP in selective uptake analogous to that of the well-established HDL receptor, SR-BI. These findings have significant implications for the role of CETP in the final steps of reverse cholesterol transport.

CETP-mediated selective uptake is only partially inhibited by torcetrapib, suggesting the existence of two distinct pathways. First, CETP may mediate selective uptake by shuttling cholesteryl ester directly from HDL to the plasma membrane. This process may require the participation of another protein on the cell surface but clearly does not require SR-BI, members of the LDL receptor family, or an intact heparin sulfate proteoglycan matrix. CETP may deliver lipids to particular membrane structures such as microvilli or protrusions that have a high curvature and therefore may, at a molecular level, be comparable to a lipoprotein. Second, cell-associated CETP appears to mediate selective uptake by a mechanism that is not inhibited by CETP and thus does not involve CETP-mediated shuttling of cholesteryl ester from HDL to the plasma membrane, but may represent direct interaction of HDL with CETP on the cell surface. It is plausible that cell-associated CETP may mediate the transient fusion of the HDL amphipathic (having a polar region separate from the nonpolar region) coat with the membrane outer leaflet, allowing cholesteryl ester movement into the membrane, without HDL particle uptake. CETP contains a C-terminal peptide that has a tilted orientation relative to the lipid–water interface, and this peptide has fusogenic properties similar to those of viral fusion peptides in a lipid-mixing assay. Indeed, CETP has been shown to mediate lipoprotein fusion under certain circumstances. Further studies will be required to dissect the molecular mechanisms of CETP-mediated selective uptake and the relative importance of these two pathways under different circumstances.

Summary. Although active debate on the role of CETP in modifying coronary heart disease risk continues, very high plasma concentrations of CETP may have unfavorable effects on plasma lipids and possibly on atherogenesis, whereas complete CETP deficiency may not be of universal benefit, consistent with a direct role for CETP in mediating the hepatic clearance of HDL-cholesteryl ester.

For background information see ARTERIOSCLEROSIS; CHOLESTEROL; HEART DISORDERS; LIPID METABOLISM; LIPOPROTEIN; METABOLIC DISORDERS in the McGraw-Hill Encyclopedia of Science & Technology.   Ruth McPherson

Bibliography. P. J. Barter and J. J. P. Kastelein, Targeting cholesteryl ester transfer protein for the prevention and management of cardiovascular disease, *J. Amer. Coll. Cardiol.*, 47(3):492–499, 2006; S. M. Boekholdt and J. F. Thompson, Natural genetic variation as a tool in understanding the role of CETP in lipid levels and disease, *J. Lipid Res.*, 44(6):1080–1093, 2003; A. Gauthier et al., Cholesteryl ester transfer protein directly mediates selective uptake of high density lipoprotein cholesteryl esters by the liver, *Arterios. Thromb. Vasc. Biol.*, 25(10):2177–2184, 2005; P. C. N. Rensen and L. M. Havekes, Cholesteryl ester transfer protein inhibition: Effect on reverse cholesterol transport?, *Arterios. Thromb. Vasc. Biol.*, 26(4):681–684, 2006; G. Vassiliou and R. McPherson, Role of cholesteryl ester transfer protein in selective uptake of high density lipoprotein cholesteryl esters by adipocytes, *J. Lipid Res.*, 45(9):1683–1693, 2004.

Chronic wasting disease

The transmissible spongiform encephalopathies (TSEs; also known as prion diseases) are a group of fatal neurodegenerative diseases affecting both human beings and animals. Within the human spectrum are Creutzfeldt-Jakob disease and its variant form, kuru, fatal familial insomnia, and Gerstmann-Sträussler-Scheinker syndrome. Animal diseases include scrapie in sheep and goats, bovine spongiform encephalopathy (mad cow disease), and chronic wasting disease (CWD) of cervids (members of the ruminant family Cervidae, characterized by antlers, including deer and elk). A number of shared features provide the unifying links that delineate this group of disorders: similarities in neuropathology with vacuolation (spongiform change) and neuronal loss in the central nervous system; conversion of the ubiquitously expressed, normal prion protein (PrPc) into abnormally folded, protease-resistant forms (PrPres; a generic term for any misfolded, disease-associated form of the prion protein) with accumulation in the brain and sometimes other organs; and transmissibility of the disease. The infectious nature of the various prion diseases stands in stark contrast to all the other age-related neurodegenerative diseases, such as Alzheimer's disease, that also feature brain cell loss in association with misfolding and aggregation of endogenously produced central nervous system neuronal proteins.

Epidemiology, surveillance, and containment measures. Chronic wasting disease was confirmed as a transmissible spongiform encephalopathy in the late 1970s, although the disease had been recognized as a syndrome in captive mule deer in the northeastern Colorado region for many years prior to this. The origin of CWD remains uncertain, but contentious theories include transmission of sheep scrapie to cervids. In contrast to other animal transmissible spongiform encephalopathies, CWD is unique because it is found in both captive or farmed cervids as well as free-ranging animals. To date, CWD has been found to naturally affect four species of Cervidae: mule deer (*Odocoileus hemionus*), white-tailed deer (*O. virginianus*), Rocky Mountain elk (*Cervuselaphus nelsoni*), and moose. Natural horizontal transmission of CWD within these three species, and across the three species, can occur.

It is currently estimated that there are more than 22 million captive and free-range deer and elk in the United States. The northeastern region of Colorado and the adjoining southeastern part of Wyoming were the first areas of the United States in which CWD was recognized. While these areas continue to constitute a continuing high prevalence or endemic focus, the geographical distribution of the disease has broadened considerably to include a number of midwestern states, extending from New Mexico to Montana, Minnesota, and Wisconsin, with reports in early 2005 of CWD in a small number of captive and free-ranging white-tailed deer in New York State. The basis of the spread of CWD to free-ranging cervids across such a wide geographical expanse is unresolved, but is postulated to represent transmission from affected farmed herds, perhaps in association with unauthorized movement of diseased animals. CWD was originally recognized in farmed cervids in Canada in 1996, and overall the disease remains essentially restricted to North America. Only rarely has CWD occurred outside North America, such as in cervids exported from Canada to South Korea.

Active surveillance programs for CWD in captive and free-ranging animals exist within a number of countries. Comprehensive surveillance surveys have reported highly variable prevalence of CWD, at least partly influenced by factors such as the specific geographical region sampled; symptomatic and asymptomatic CWD varies from <1% to 100% in captive herds, and from <1% to 30% in free-ranging cervids, with an overall disease prevalence in the United States estimated to be up to 5%.

Accurate quantification of incubation periods in naturally occurring CWD, especially in free-ranging cervids, is very difficult to estimate because of the lack of certainty regarding the precise contamination or transmission event; however, it is believed to average between 2 and 4 years. Direct animal-to-animal transmission of CWD occurs, albeit through poorly defined mechanisms. Probably important to the high efficiency of natural horizontal transmission compared to other animal transmissible spongiform encephalopathies (such as bovine spongiform encephalopathy) is the tissue distribution of the abnormal, misfolded, infectious form of the cervid prion protein (PrP^{CWD}; denoting cervid form of PrPres). PrP^{CWD} has a relatively widespread distribution, present throughout lymphoid tissues from early in the incubation period, and as the disease progresses also found in the central nervous system, and in mule deer in some additional peripheral locations. The gastrointestinal tract harbors considerable lymphoid tissue, and PrP^{CWD} can be observed in these locations by 6 weeks following oral exposure in experimental models. The deposition of PrP^{CWD} within peripheral lymphoid tissues is similar to what is observed in scrapie, and this relatively ready accessibility (such as the palatine tonsil) has been utilized for confirmation of presymptomatically infected and symptomatic animals. Despite very recent experiments failing to demonstrate transmission through repeated oral exposure to urine and feces from CWD-affected animals, it is conjectured that the presence of PrP^{CWD} within the gastrointestinal tract lymphoid tissue may lead to fecal shedding and consequent environmental contamination with the infectious agent (principally composed of PrP^{CWD}), allowing subsequent ingestion by grazing or foraging susceptible animals. Other mechanisms of environmental transmission could be through decomposing carcasses of CWD-affected animals, and from secretions such as saliva, recently proven capable of transmitting disease.

In addition to the active surveillance programs, a range of guidelines and regulatory measures to contain and eradicate CWD have evolved at both the local and federal level in North America. Such measures include targeted depopulation of cervids in high-prevalence regions, quarantining and potentially culling entire herds once a member animal is confirmed to have the disease, and restricting the geographical movement of farmed cervids and advising against the transport of the head and spinal column of cervids (body parts known to harbor the highest levels of infectivity in affected animals) from higher-prevalence areas. In some Canadian provinces, farmed cervids cannot be released for commercial use until tested and proven negative for CWD. Although these measures are of merit for captive cervids, they are clearly of more limited applicability and utility for controlling CWD in free-ranging cervids, especially if environmental contamination by PrP^{CWD} represents an important mechanism of natural horizontal transmission. Given the likely inherent long-term stability of infectious

PrP^{CWD} even in an environmental location, programs to eradicate CWD from free-ranging cervids face considerably greater challenges.

Clinical features. CWD typically causes progressive weight loss to a state of emaciation, usually accompanied by behavioral and locomotion abnormalities, with illness duration generally less than 4 months, although survival for 12 months and sudden death early in the disease associated with intercurrent environmental stress may occur. Symptomatic deer and elk are usually older than 2 years (only rarely are much younger animals such as yearlings affected), with no sex predominance. Prior to overt weight loss, other early clinical features are nonspecific and relatively subtle, especially the behavioral changes, which may require good knowledge of the animal's normal demeanor. Postural and motor abnormalities (more common in elk than deer), which often become more obvious as the illness progresses, include lowered head, flaccid facial muscles, head tremor, and ataxia (lack of muscular coordination). Esophageal dilatation and regurgitation may develop and predispose to aspiration pneumonia, particularly in the later stages of the disease. In the advanced stages of the disease, animals may also manifest polyuria (excessive urine output) and polydipsia (excessive drinking), altered awareness, hyperexcitability, repetitive walking, and altered stance. A rough, dry hair coat, which has been incompletely shed, are possible features, but hair loss because of pruritus (itching) and consequent scratching or rubbing, characteristic of scrapie in sheep, is not a feature in CWD.

At present there is no proven effective therapy for any of the animal or human forms of prion diseases, including CWD.

Pathology and tissue distribution of PrP^{CWD}. The histopathological abnormalities found in CWD are similar to those found in scrapie and bovine spongiform encephalopathy, and are confined to the central nervous system. Spongiform change is a salient feature in the central nervous system and appears to develop concomitantly with the onset of overt clinical disease. It is observed in neuronal cell bodies and their processes, and is symmetrically widespread but most prominent in the diencephalon, olfactory cortex, and medulla oblongata (especially the dorsal vagal nucleus). The pattern of brain involvement in CWD is quite stereotyped, regardless of whether naturally or experimentally acquired, or observed in free-ranging or captive cervids, and largely independent of duration of clinical illness. The pattern of neuropathological changes is sufficiently characteristic that for confirmation of CWD in symptomatic cervids a single site in the brainstem is all that is required for sampling and examination, and as such offers improved convenience for large-scale surveillance programs. Central nervous system neuronal loss is less conspicuous than the spongiform change, and inflammation within the brain and spinal cord is not observed in uncomplicated CWD. Amyloid plaques (best shown with special chemical stains) may also be observed in the central nervous system,

most commonly in white-tailed deer; similar to what is seen in the variant form of Creutzfeldt-Jakob disease, they can be surrounded by vacuoles giving rise to distinctive florid plaques. The amyloid plaques are principally composed of PrP^{CWD} and are particularly well displayed by using tissue staining methods employing antibodies that have high affinity for this abnormal form of the cervid prion protein.

As mentioned, PrP^{CWD} is also typically found throughout lymphoid tissues such as spleen, lymph nodes, and tonsils, including during the incubation period, and may also be found in peripheral nerves, pancreas, and adrenal medulla; however, in contrast to the central nervous system, the deposition of the misfolded form of the prion protein is not accompanied by any other abnormalities. Only rarely in deer and uncommonly in elk (approximately 10%) is PrP^{CWD} found in the brain but absent in peripheral lymphoid organs. The PrP^{CWD} in lymphoid organs is found in the germinal centers in close association with specific immune cells, including follicular dendritic cells. To date, PrP^{CWD} has not been detected in the skeletal muscles of cervids affected with CWD. However, given the importance of this issue to human health and the presence of PrPres in muscles in a range of experimental and other naturally occurring forms of transmissible spongiform encephalopathy, investigations are continuing.

Cross-species transmission and concerns for human health. Acknowledging the species-transgressing precedent of bovine spongiform encephalopathy to human beings as the variant form of Creutzfeldt-Jakob disease, there are concerns that CWD may also pose a similar zoonotic threat, particularly if the prevalence continues to rise. At present there is no epidemiological evidence to suggest that CWD can cause disease in human beings under any circumstances. Various scientific studies to assess the threat to human health have been undertaken, including transmission experiments. For example, using transgenic mice engineered to overexpress the elk PrP gene, brain homogenates made from cervids with CWD successfully transmitted disease to these mice, while transgenic mice overexpressing the human PrP gene did not. Other experiments have used cell-free techniques to assess the efficiency with which recombinant forms of prion proteins equivalent to those found in human beings, cattle, sheep, and cervids can be converted by PrP^{CWD} into the abnormal misfolded forms found in disease. Only the cervid form of recombinant PrPc could be converted efficiently by PrP^{CWD}. In line with our current understanding that scrapie does not transmit to humans, these results offer some reassurance that the species barrier between cervids and human beings may be so large that CWD does not constitute a direct threat to human health. Until more research is completed, total reassurance cannot be given, and maintenance of the current advice for humans to avoid consumption of cervids incubating and/or manifesting CWD appears prudent.

Previously of considerable concern to livestock industries, and possibly posing an indirect threat to human health, was the possibility of transmission of CWD to other commercial or farmed animals. CWD can be experimentally transmitted to cattle and sheep through direct intracerebral inoculation of brain tissue from sick cervids. In contrast, a range of experiments performed to date have demonstrated that CWD appears very unlikely to be able to naturally transmit to cattle and sheep, particularly through cohabitation or sharing grazing areas, or to be transmitted through the oral route.

Future research. A number of important priorities exist for ongoing and future research. These include continuing assessments of the potential threat of CWD to human health; a better understanding of natural CWD transmission mechanisms; whether some normal variations (polymorphisms) in the genes coding for cervid forms of PrPc offer greater resistance to CWD; development of effective treatments; and more convenient but highly specific confirmatory tests (such as detection of PrP^{CWD} through a blood test) to facilitate ongoing mass screening and surveillance programs.

For background information *see* AGRICULTURAL SCIENCE (ANIMAL); CLINICAL PATHOLOGY; DEER; NERVOUS SYSTEM DISORDERS; PRION DISEASE; SCRAPIE; ZOONOSES in the McGraw-Hill Encyclopedia of Science & Technology.   Steven Collins

Bibliography. S. Collins, V. Lawson, and C. L. Masters, Transmissible spongiform encephalopathies, *Lancet*, 363:51–61, 2004; S. B. Prusiner, Neurodegenerative diseases and prions, *N. Engl. J. Med.*, 344:1516–1526, 2001; C. J. Sigurdson and M. W. Miller, Other animal prion diseases, *Brit. Med. Bull.*, 66:199–212, 2003; E. S. Williams, Chronic wasting disease, *Vet. Pathol.*, 42:530–549, 2005.

Clinical forensic nursing

Forensic nursing is the application of forensic science combined with the bio-psychological education and experience of the registered nurse to the medical-legal investigation of injury or death of victims of violence, criminal activity, and traumatic accidents. The forensic nurse functions as a staff nurse, nurse scientist, nurse investigator, or independent consulting nurse specialist to public or private operatives or individuals. There are several subspecialties within forensic nursing, including sexual assault nurse examiners, nurse death investigators, forensic psychiatric nurses, legal nurse consultants, and clinical forensic nurses.

Forensic nursing was formally recognized as a nursing specialty by the American Academy of Forensic Sciences in 1991 and by the American Nurses Association in 1995. Among its several subspecialties, clinical forensic nursing has the potential to have the greatest impact on health care delivery, encompassing a wide range of medical-legal issues within hospitals and domiciliary or custodial treatment facilities.

Evolution. Historically, forensic pathologists have been solely concerned with death investigations.

In forensic science, the newest field is called living forensics, which relates to the identification and collection of evidence from living patients. Clinical forensic practice focuses on the civil and criminal investigation of injuries or deaths associated with trauma, violent behavior, or other scenarios with legal implications. Clinical forensic nurses care for both living victims and perpetrators of crime, including those already in custody. The practice arenas include hospitals, correctional facilities, and domiciliary treatment centers such as psychiatric facilities or nursing homes. Clinical forensic nurses focus on known medical-legal cases and work within the facilities' systems to identify environmental factors and human behavior that suggest potential harm to patients, personnel, and others, as well as to effect preventive strategies.

Many hospitals are designating personnel to manage forensic cases and ensure that a systematic approach is used to obtain, safeguard, and transmit evidentiary materials required by the coroner, medical examiner, or law enforcement officers. This individual is ideally a clinical forensic nurse (CFN). The CFN possesses knowledge in the forensic sciences and of the judicial system, including forensic wound identification, evidentiary examinations of sexual assault and other victims of human abuse, investigative techniques, and fact and expert-witness testimony.

The CFN serves as a role model by increasing staff awareness to potential medico-legal implications of patient care. They work closely with those investigating patient complaints, suspicious patient events, unexplained deaths, and adverse trends. Nurses are typically the first and last caregiver to see a patient such as in triage in the emergency department, when responding to a clinical emergency, and when a death occurs. Nurses observe nonverbal communication and hear interactions among patients and their visitors. They are in a unique position to assess support systems, potential threats, and dysfunctional relationships that might affect the patient.

Adverse events may cause serious harm to patients or personnel, but most are not criminal in nature. The precise and timely identification and management of facts, data, and medical evidence are critical, whether criminal or not.

Today, complex healthcare protocols and escalating lawsuits make it extremely important that all clinicians recognize the medico-legal implications or potential forensic issues in patient care. Until all clinicians are trained in forensic issues, the CFN must lead by applying forensic expertise to assessments, documentation, and treatment plans to ensure that the required reporting and referrals are accomplished. The CFN must also lead in evidence collecting and safeguarding to ensure that these activities do not compromise or interfere with other aspects of patient care.

Science versus advocacy. Patient care is not ordinarily within the scope of practice for forensic scientists, but the CFN who participates in direct patient care has a unique opportunity to apply the knowledge and skills that could influence the outcome of medical-legal scenarios. When a nurse's primary role is a caregiver (patient advocate), the first duty is to ensure that physical and emotional needs are met. Another nurse must assume the forensic roles, such as case review, analysis, and investigation, or serve as a liaison with law enforcement or judicial authorities.

Forensic nursing's earliest roots in the United States were in caring for victims of sexual assault, human abuse, and domestic violence. As a result, many nurses assumed roles of victim advocacy. The sensitivity, caring, protecting, and nurturing behaviors so inherent within nursing actually was counterproductive, since a court of law seeks truth based on scientific evidence and not emotion or empathy. Leaders within forensic nursing soon realized that if forensic nurses were to be taken seriously, they needed to align themselves with science. It is virtually impossible for nurses to profess objectivity and impartiality when they are viewed as an arm of the prosecution.

When a nurse's job is primarily of a forensic nature, a nursing background enhances this function. For example, nurses make excellent forensic examiners of sexual assault victims because the basic skills of a nurse and forensic examiner are similar. Nurses have experience in handling and touching patients and the interviewing skills to elicit sensitive information from distraught people. Nurses understand the concepts of clean and sterile and what contamination means. This translates into forensic nurse examiners being ideal candidates for collecting medical evidence and preserving it per protocol for law enforcement and the crime labs. They understand the principles required for obtaining evidentiary specimens while preserving the scene, and documenting their findings and observations in a way that will withstand legal scrutiny in a court of law.

The CFN must never compromise scientific objectivity with the subjectivity and bias required in the role of the patient advocate. Some forensic nurses struggle with the dilemma of how to be caring and compassionate caregivers while balancing the nuances of being a scientist. In many instances, it is impossible to do both. But to be an effective forensic nurse and to be taken seriously as a scientist, one must be willing to set aside most aspects of advocacy.

Scenarios. Numerous forensic scenarios occur in a hospital setting, such as caregivers or family members who may decide to hasten death to end suffering or to exercise power and control over a vulnerable person. The methods used vary, but often include the administration of a medication that depresses respirations enough to cause death. Infant abduction, drug diversion, medical record tampering, therapeutic misadventures, patient assault, and neglect by caregivers are other scenarios (see **table**). A clinician who has the forensic training to recognize such forensic scenarios is an incredible resource for the hospital.

Collateral responsibilities. There are two primary collateral functions that a CFN should be able to fulfill in any hospital setting, thereby providing another line of protection against opportunities for misconduct. The first is in responding to every patient death whether it is expected or not. In this case, the nurse would be paged to the bedside of the deceased, where a standing order for a standardized set of specimens would be obtained to provide baseline information for what occurred at the time of death. This includes any medication levels specific to the patient's treatment regime or those that might be suspected of abuse. The CFN would also obtain information on the staff for the last 24 hours, visitors, changes in doctors' orders, the patient's last meal, any specific patient complaints, and any information pertinent to the case. This information would be reviewed and archived. If a death were suspicious or unexpected, the critical evidence would be preserved. Otherwise, the information would be available for later quality management review. If a trend were spotted at a later time, critical information would be preserved for investigators.

The second collateral function of a CFN response is to respond to all clinical emergencies. The CFN would arrive with the response team, but instead of participating directly in medical activities the CFN would observe and document general scene information and extraneous conditions or events that might not ordinarily be recorded. Many nuances that might later become important can be captured by an impartial observer who is alert to multiple variables in the environment.

Outlook. The clinical forensic nurse will continue to pioneer in regard to how this role fits into the functioning of the hospital and legal system. Boundaries or obstacles must be determined before rules and regulations are standardized. In the meantime, some specific questions need to be addressed if the role of the CFN is to be fully realized:

What is the legal position of a nurse working in a hospital in terms of evidentiary specimens?

If a nurse suspects something suspicious or recognizes an inconsistency in the course of a particular patient's care, can that nurse collect specimens from the patient without a doctor's order even if the patient approves it?

Does law enforcement override a doctor's orders?

How much information should be entered into the medical record?

Who must be notified and how or where are these notifications documented?

How soon does the hospital attorney need to be involved?

At what point does the nurse consult an attorney or law enforcement on her own?

What guidelines do nurses have if they are discouraged or forbidden to gather evidence that they believe is in the patient's best interest?

Is photography appropriate in the health care settings, even when it is in the patient's best interest as issues of confidentiality and patient consent are blurred with the identification and management of medical evidence?

With the increase of serial murders committed by health care providers as well as other situations involving patients where foul play has occurred before

Forensic scenarios in clinical applications

Scenario	Examples
Alleged sexual assault	Nursing home care unit patient
	Any private/single-patient room
	Psychiatric unit assaults (psychotic female on a primarily male unit)
	ER admittance: date rape/substance-abuse involved
Unexpected hospital deaths	Assisted suicides
	Homicides
	Psychopathic caregiver
	Real accident due to medication error
	Negligence on long-term care unit
Medical record tampering	Patient continues to complain of pain despite increased medications charted
	EKG strips are removed from monitor or chart
	Alteration/removal of death certificates
	Removal of pages in paper chart/delayed entries
	Misappropriate computer records (deleted certain sections; entered data of another patient)
	Destruction of automatic documentation or of storage memory
Drug tampering	Drug diversion for staffs own use
	Increase in medication's to hasten death
	A hazardous substance is mixed with an intended medication (IV or PO)
Nonaccidental injuries to patients/caregivers	Poison (injectables, inhaled substances, food, drink, etc.)
	Restraints not applied properly, resulting in asphyxiation
	Patient improperly restrained during behavioral emergency
	Negative or positive airflow is improperly secured in isolation area or OR, respectively
	Needle sticks (nonaccidental)
	Instruments or sponges left inside surgical patient post-wound-closure
	Cancer medication diverted from IV drug solution supply
	Elderly/confused patient wanders off unit and is "lost" within hospital
Screening for human abuse, neglect, and violence	Primary care: routine visits when the nurse sees the patient before the physician
	Emergency department: child presents with suspected nonaccidental injuries
	Nursing home admission: upon intake, assessment reveals gross neglect, bruises
Caregiver-associated deaths	Death in the OR/procedure room
	Restraint or accidental death (psychiatric unit/nursing home)
	Leather restraints improperly applied: suicidal patient is able to asphyxiate him/herself
Other	Computer fraud
	An individual is court-ordered to receive a 30-day psychiatric evaluation in a locked facility. A staff member may see the vulnerability of this patient and attempt to gain something from this situation by threatening to enter notes in the medical record about this patient that would not lead to a fair or just outcome.
	Investigation of employee-related extortion (for example, Workman's Compensation fraud)
	Food-borne illness outbreaks (food poisoning)

SOURCE: M. K. Sullivan, Forensic nursing in the hospital setting, in V. A. Lynch and J. Border (eds.), *Forensic Nursing*, Elsevier/Mosby, 2005.

or after the patient enters the hospital, the collection and documentation of forensic evidence is a vital factor. The forensic nursing profession and health care delivery systems must consider these questions carefully.

For background information *see* CRIMINALISTICS; FORENSIC BIOLOGY; FORENSIC EVIDENCE; FORENSIC MEDICINE; MEDICINE; NURSING in the McGraw-Hill Encyclopedia of Science & Technology.

Mary K. Sullivan

Bibliography. J. Barber, Forensic nursing's past, present and future, *Forensic Nurse Mag.*, 2(3):14, 2003; J. B. Duval, C. A. Dougherty, and M. K. Sullivan, Forensic nursing, in S. S. James and J. J. Nordby (eds.), *Forensic Science: An Introduction to Scientific and Investigative Techniques*, 2d ed., CRC Press, 2005; International Association of Forensic Nurses, *Scope and Standards of Forensic Nursing Practice*, American Nurses Publishing, Washington, DC, 1997; V. A. Lynch, The specialty of forensic nursing, in V. A. Lynch and J. Barber, J. (eds.), *Forensic Nursing*, Elsevier/Mosby, 2005; M. K. Sullivan, Clinical forensic nursing: A higher standard of care, *Forensic Nurse Mag.*, 1(1):8, 2002; M. K. Sullivan, Forensic nursing in the hospital setting, in V. A. Lynch and J. Barber (eds.), *Forensic Nursing*, Elsevier/Mosby, 2005.

Conveyor design and engineering (mining)

A conveyor is a moving belt that transports objects from one place to another. Commonly seen on a small scale at grocery store checkouts or airport baggage claims, belt conveyors have thousands of uses in most industries worldwide. In the mining industry, 95% of all the materials mined are transported from one place to another on conveyors.

Conveyor design and engineering for mining applications can be traced back to Thomas Edison's developments at the Ogden iron ore mine near Odgenburg, New Jersey, around 1900. Edison understood the benefits of automation and the need to move

Fig. 1. Long-distance conveyor at a phosphate mine in Vernal, Utah.

large volumes of iron ore reliably, and turned to conveyors. In the last 100 years, all mining has turned to conveyors, which have developed into huge complex machines carrying up to 27,000 metric tons per hour (30,000 tons) over many miles. Today, belt conveyors used in mining have the distinction of being the longest machines ever designed and built in the world (**Fig. 1**).

Long machines. Economic efficiency is a term for conditions that create the greatest possible profit with the smallest possible costs. The drive for higher economic efficiency is driving conveyors to be longer and transport higher volumes. By 1980, conveyor lengths up to 8–10 km (5–6 mi) were fairly common. But long lengths brought a new challenge as the desired routes were seldom straight. The answer was the development of horizontally curved conveyors, and lengths began to grow again (**Fig. 2**).

Horizontally curved conveyors. In 1989, a 21-km-long (13-mi) conveyor system with two horizontal curves was installed in Western Australia. In 1996, a 16-km (10-mi) curved conveyor was installed in Zimbabwe. In 1999, a 24-km (15-mi) overland conveying system replaced rail haulage in a Colorado molybdenum mine. The longest flight required 11,000 horsepower.

Conveyor lengths now were requiring huge components, which began to challenge the limits of manufacturing from motors to belts. The answer was the development of remote power distribution.

Distributed power. Although positioning of drives at both ends of long conveyors has been around a long time, the idea has expanded to position drive power wherever it is most needed (**Fig. 3**). By splitting the total required power into smaller increments and distributing the drives at critical locations along the conveyor length, components can be reduced as much as 60–70%, which can significantly reduce the capital cost as well as many other operating and maintenance concerns. Intermediate drive technology is well accepted today in underground mining where space constraints have dictated design decisions for many years, but now the technology is being used to allow longer conveyors in surface mines.

In 2005, the longest single-flight conveyor in the world was commissioned in the United States at 19.1 km (11.9 mi) in length with a 1500-hp main drive, a 1500-hp intermediate drive, and a 750-hp tail drive; the route included 11 horizontal curves. In 2006, a 20-km (12-mi) conveyor with four horizontal curves and one intermediate drive was commissioned in Australia. New applications of 50 and 100 km (30 and 60 mi) are being considered.

Complexity. Designing and building a machine this long with power distributed over many miles is a complex process. Today, components might be manufactured in several different countries. And with machines of this size, the finished system cannot be tested until all the components arrive and are assembled on site. In addition, the greater complexity of all individual components increases the amount of knowledge, information, and communication required to optimize a design. At each step in the design development, the detailed options multiply and interact, and they all have to be evaluated iteratively against the product performance, quality, and cost requirements. For engineers to cope with these conflicting demands, a new approach to the design process was needed that was capable of delivering not just higher engineering productivity but a design environment that actually supports and promotes innovation.

Systems engineering. A system is a set of interrelated components working together toward some common purpose. The properties and behavior of

Fig. 2. Horizontally curved conveyor at a coal-fired power plant in Gillette, Wyoming.

each component of the system has an effect on the performance of the whole system, and the performance of each component of the system depends on the properties and behavior of the system as a whole. There are many individual discipline specialists in the fields of mechanical, electrical, and civil engineering, as well as many components specialists in rubber, bearings, motors, and controls who are knowledgeable in their respective fields. However, seldom are these specialists knowledgeable about the system in which their expertise is used. The field of systems engineering was established, and is growing out of the need for a system management function in the design of complex machines. The influence of the systems engineer is critical during the early stages of the design process, when the emphasis is on optimizing the system and not the individual components.

Since a system test is not possible, the systems engineer must usually rely on mathematical models to ensure the machine will perform as expected. The quality of the mathematical tools used is directly related to long-term performance and reliability.

Virtual prototyping. Virtual prototyping (or digital mockup) has been used for years in the automotive and aerospace industries, where it is recognized as the only way to manage the complexity inherent in their products and engineering networks. Today, this approach is being adopted in many aspects of belt conveyor design.

An example uses the finite element method to simulate the transient conditions of starting and stopping, taking the elasticity of the belt into consideration. Starting and stopping the huge inertia of a 19-km (12-mi) loaded machine with 38 km (24 mi) of elastic band connecting all the pieces is a significant challenge. When considering the many loading conditions possible, sometime more than 50 simulations are necessary to understand all the operating possibilities.

Another example where virtual prototyping has proven most beneficial is at the loading point, which is generally accepted as the riskiest area of design.

Mass flow at transfer points. A transfer point is a point on a conveyor where material is loaded or unloaded. Many of the most difficult problems associated with belt conveyors center around transfer points. The transfer chute is often cited as the highest-maintenance area of the conveyor, and many significant production risks are centered there. A special type of virtual prototyping is used to help in this area.

Discrete element method (DEM). This method is a family of numerical modeling techniques and equations specifically designed to solve problems in engineering and applied science such as bulk material flow. The discrete element method explicitly models the dynamic motion and mechanical interactions of a system of bodies and provides a detailed description of the positions, velocities, and forces acting on each body and/or particle at discrete points in time during the analysis. Simply put, the DEM allows engineers to analyze transfer-chute design quantitatively and visually. Particles falling through a transfer chute can

Fig. 3. Construction site of the Seven Oaks Dam in San Bernardino, California, showing a conveyor that has an intermediate drive to evenly distribute the power.

be visualized, and the chute design can be analyzed (**Fig. 4**). Particle positions, forces, and velocities are calculated using DEM, providing designers with the quantitative data needed to evaluate designs. Perhaps the greatest benefit derived from the use of these tools is the feeling that an experienced engineer can develop by visualizing performance prior to

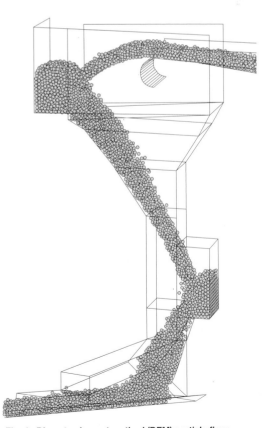

Fig. 4. Discrete element method (DEM) particle flow through virtual transfer chute.

transfer house

Fig. 5. Transfer house connecting the PC2 and PC3 overland conveyor systems in Henderson, Colorado.

building. From this, the designer can arrange the components to eliminate unwanted behavior.

Example of modern conveyor design. One of the more interesting conveyor systems installed in the recent years was at the Climax Molybdenum Mine at Henderson, Colorado. This project consisted of a 24-km (15-mi) overland conveying system, replacing a 30-year-old rail haulage system.

The longest conveyor in the system (PC2) is 16.28 km (10.12 mi) in length with 475 m (1560 ft) of lift (**Fig. 5**). The most important system fact on this flight is that 50% of the operating power is required to turn an empty belt; therefore, power efficiency is critical. Very close attention was focused on the idlers, belt cover rubber, and alignment. This close attention contributed a significant saving in capital cost of the equipment. After commissioning, the actual measured results over six operating shifts showed a 30% lower power demand than expected, resulting in an additional $100,000 savings per year in electricity costs alone. The third flight (PC3) is 6.4 km (4 mi) long and included 11 vertical and 9 horizontal curves as it wound over and around the Rocky Mountains (Fig. 5).

The Henderson conveyor system exemplifies all aspects of modern conveyor design. As innovation drives technology forward, conveyors will continue to get longer, faster, and more efficient, increasing the productivity and efficiency of the mining industry as a whole.

For background information *see* BULK-HANDLING MACHINES; COAL MINING; CONVEYOR; MINING; OPEN-PIT MINING; SYSTEMS ENGINEERING; UNDERGROUND MINING; VIRTUAL MANUFACTURING in the McGraw-Hill Encyclopedia of Science & Technology.

Mark A. Alspaugh; Paul Ormsbee

Bibliography. M. Alspaugh, The evolution of intermediate driven belt conveyor technology, *Bulk Solids Handling*, 23(2), 2003; Conveyor Equipment Manufacturers Association, *Belt Conveyors for Bulk Materials*, 6th ed., 2005; G. Dewicki, Bulk material handling and processing: Numerical techniques and simulation of granular material, *Bulk Solids Handling*, 23(2), 2003; W. Kung, The Henderson coarse ore conveying system: A review of commissioning, start-up, and operation, *Bulk Material Handling by Belt Conveyor V*, Society for Mining, Metallurgy and Exploration, 2004; I. G. Mulani, *Engineering Science and Application Design for Belt Conveyors*, 2002; A. Reicks and E. A. Viren (eds.), *Bulk Material Handling by Conveyor Belt IV*, Society for Mining, Metallurgy, and Exploration, 2004; R. H. Wohlbier, *The Best of Powder Handling & Processing: Bulk Solids Handling Belt Conveyor Technology, I/2000*, 2000.

Cougar (mountain lion)

The cougar once had the largest range of any mammal in the Western Hemisphere. It is also known as puma, mountain lion, catamount, deer tiger, Mexican lion, panther, painter, chim blea, Leon, and leopardo in various parts of its range in Canada, the United States, Mexico, and Central and South America. In addition, there are over 30 Native American names for this animal in the United States, such as the Cherokee *tlv-da-tsi*. Although extirpated from much of its range in the United States and parts of South America as a consequence of habitat destruction and human depredation, this large cat is still common locally from the Rocky Mountains westward, along the Mexican border in Texas, in South Dakota, in southern Florida, and possibly from central Maine into Nova Scotia and New Brunswick. The small Florida population numbered just 30–50 animals in 1990 and was too small to survive in the wild. Three major threats faced the Florida panther: inbreeding, elevated levels of mercury, and habitat loss. Calculations indicated that the panther was experiencing a population decline of 6 to 10% a year, with a current rate of loss of genetic diversity of 3 to 7% per generation. This was predicted to increase as the population grew smaller. Total extirpation of the Florida panther was predicted in 25 to 40 years if there was no intervention. Amid considerable controversy, 10 male cougars from Texas were introduced into the Florida population in an attempt to increase the genetic diversity. This experiment has proven successful and has allowed the population to increase to approximately 75 animals in 2005. Additional efforts are under way to lower mercury levels and increase suitable habitat.

Characteristics. Some 32 separate subspecies of cougar have been described based on geographic and morphometric criteria. However, an extensive

Cougar (*Puma concolor*) in Zion National Park, Utah. (*Photo by Gerald and Buff Corsi; © 2002 California Academy of Sciences*)

DNA investigation (M. Culver et al., 2000) has shown that there are only six phylogeographic groupings or subspecies, and that the entire North American population is genetically homogeneous in overall variation relative to Central and South American populations.

Cougars have an elongate body, a small head, a short face, and a long neck and tail (see **illustration**). The pelage is short and ranges in color from tawny to slaty gray. They have powerful limbs with the hindlegs being larger than the front legs. The ears are small, short, and rounded. Males have a head and body length of 1050–1959 mm (40–76 in.), a tail length of 660–774 mm (26–30 in.), and a weight of 67–103 kg (146–225 lb). Females are smaller in overall size and weight.

Threats and conservation. As the Mid-Atlantic states were settled, the cougar was systematically killed off due to its predatory habits. The last verified cougar in Virginia was killed in 1882, while the last verified cougar in West Virginia was recorded in 1887. Cougars in Maryland, Pennsylvania, North Carolina, and Tennessee survived into the 1920s. Up until the 1960s, few cougar sightings were reported in the eastern United States. Since that time, however, the presence of these big cats has been reported from numerous areas.

The question always arises as to the origin of these cougars. Some believe that they are part of the original population. Others feel that they may be members of the Florida population that made their way northward through western Florida and Alabama and into the mountains. Others feel that the animals being seen are the result of escape or intentional release from captivity. Since recent DNA studies have shown that all cougars within the United States have a similar genetic makeup, it may never be possible to definitively answer this question. The sighting of an individual cougar in Pennsylvania, Virginia, or Tennessee does not prove that a breeding population exists. What is needed is evidence of reproduction.

Cougars are among the most adaptable of the big cats. They can live any place where adequate cover exists for them to stalk game. Contrary to popular myth, most of their kills are made by stalking, not ambush. They live and prosper in swampy jungles, tropical forests, open grasslands, dry bush, semideserts, and alpine areas. Unlike the cougar, many large cats such as the tiger, cheetah, and lion may well be gone from the wild within the next 25–50 years. The cougar is unique among the big cats in that its future, primarily in the West, is directly linked with the success of managing large and growing populations, rather than shrinking populations and ranges. Human attitudes toward the cougar have undergone a radical transition in the last 40–50 years, and this has greatly benefited the big cat. In North America and Canada, mountain lions were considered unprotected predators and could be killed on sight until the mid-1960s. They did not attain game status in the 12 Western states until 1965. Game animal status was granted by Alberta in 1969 and by British Columbia in 1970. Since then, a steady rise in the strictness of game regulations has occurred. Bounty laws in all of the states except Arizona had been repealed by 1972. East of the Mississippi River, the cougar now has full legal protection based on the Endangered Species Act of 1973, which supersedes all state statutes and regulations.

Since the early 1970s, because of legal protection, this species has enjoyed an unprecedented increase in both numbers and geographic range. For example, in 2004 a young male cougar that had been collared as part of a study in South Dakota was hit by a train near the Kansas-Oklahoma border, having traveled an estimated 950 mi (1530 km). Another young male from South Dakota traveled through North Dakota and Minnesota before radio contact was lost in Manitoba. In a straight line, he traveled at least 350 mi (564 km). Young male cougars—known as dispersers—move along wooded river corridors and can travel 50–100 mi (80 to 160 km) in a night. As the numbers of cougars increase in Colorado, South Dakota, and other states and exceed the carrying capacity of the land, they will continue to keep moving east. This is the same pattern of dispersal followed by the coyote. There are, however, several distinct differences: the coyote benefited from the loss of forest, the spread of agriculture, and the elimination of competitors such as the wolf. The cougar will have a more difficult time repeating the coyote's feat due to its lower reproductive rate, the tendency for females not to disperse, its dependency on large prey, and the fear that most people will exhibit toward this species, especially in areas where it has been absent for more than a century.

Predators are an integral part of the food web. They are the natural method of controlling various prey populations. One can begin to see the effects of natural predator control with the return of the wolf to the Yellowstone ecosystem. The elimination of wolves many years ago allowed the elk population to increase dramatically, which in turn overbrowsed much of the vegetation. Forested areas have disappeared and stream banks that were formerly covered with aspen trees and other vegetation have become bare. Now that approximately 150 wolves are present and the vegetation is returning, beaver and other forms of wildlife are once again able to live in the Lamar Valley in Yellowstone National Park.

Some say that the cougar is going to come back in the East on its own as it is doing in the Midwest. There is certainly an abundance of deer, the primary food of cougars. There are more white-tailed deer in the United States now than before European settlement (estimates now range from 20 to 33 million), with huge and increasingly unhunted populations in rural and suburban areas east of the Mississippi. Modern suburban America—without intending to do so— has transformed itself into superb wildlife habitat. Since the 1920s, extensive forested areas have gradually recovered from earlier logging operations to provide necessary habitat. In addition, the establishment of national forests and national parks has provided a measure of protection to many of these areas. No one knows the size of a territory that a cougar would require in the Mid-Atlantic region. An estimate would be 25–50 mi^2 (65–130 km^2), which is considerably less than the 300–400 mi^2 (777–1036 km^2) that have been reported for some Western studies. The size of the territory is dependent on the food supply, and food should not be a problem in the eastern United States. Corridors connecting blocks of suitable habitat would be advantageous. Land trusts and other groups could help secure easements for large blocks of private lands that could become parts of protected corridors.

Outlook. The cougar is an intelligent, highly adaptable predator that, along with wolves and bears, adds an excitement to the wildlands that simply would not be there in their absence. Humans live and interact with black bear populations. Persons who live or hike in black bear habitats recognize that there are inherent risks. The challenge for the future will be to recognize that the cougar is a highly efficient, dynamic, potentially dangerous predator, and then to manage this species for its own well-being to the best extent possible without allowing condemnation of specific individual cats (in negative incidents) to endanger the current public acceptance of the animal. As long as these cats are treated with the respect they deserve, they can once again become a valuable member of the eastern fauna.

For background information *see* CARNIVORA; CAT; ENDANGERED SPECIES; EXTINCTION (BIOLOGY); MAMMALIA; POPULATION DISPERSAL in the McGraw-Hill Encyclopedia of Science & Technology.

Donald W. Linzey

Bibliography. M. Culver et al., Genomic ancestry of the American puma (*Puma concolor*), *J. Hered.*, 91:186–197, 2000; G. A. Feldhamer, B. C. Thompson, and J. A. Chapman, *Wild Mammals of North America: Biology, Management, and Conservation*, 2d ed., Johns Hopkins University Press, 2003; R. M. Nowak, *Walker's Mammals of the World*, 6th ed., Johns Hopkins University Press, 1999; D. E. Wilson and S. Ruff (eds.), *The Smithsonian Book of North American Mammals*, Smithsonian Institution Press, 1999.

Cretaceous bird radiation

The recent report of the eleventh specimen or tenth skeleton of the Late Jurassic *Archaeopteryx* not only provided new evidence linking birds and theropod dinosaurs, but also reaffirmed its unique position in avian evolution as the oldest and most primitive known bird. In addition, in the last two decades many more birds have been unearthed from the Cretaceous than in all time before. (The Cretaceous was the last period of the Mesozoic, preceded by the Jurassic and followed by the Tertiary Period; it extended from 145 million years to 65 million years before the present.) In particular, hundreds of completely preserved skeletons in association with beautiful feathers, referable to over 20 new avian orders and families, have been reported from Lower Cretaceous lake deposits in Liaoning (northeast China), Spain, and other regions. These discoveries constitute the best evidence for the earliest bird radiation in history. Thanks to these new fossils, our knowledge of early avian diversity, phylogeny, flight, development, and diet, size, and habitat variations has been significantly improved.

New Cretaceous discoveries contain much additional information about the ancestry of birds by revealing more skeletal and integumental resemblance between these primitive birds, *Archaeopteryx*, and their dinosaurian ancestors. They not only provide important clues to the debate on the flight origin of birds—that is, "trees-down" versus "ground-up" hypothesis—but also shed new light on the discussion of the issues of when modern bird groups (Neornithes) first appeared. Furthermore, our understanding of the paleophysiology of these birds has also increased and may help to answer the questions of when and how endothermy (utilization of metabolic heat for thermoregulation) first appeared in birds.

Archaeopteryx remains the only known bird genus from the Jurassic, while by the Cretaceous there were already many different lineages. For instance, Ornithurae, which include the well-known *Hesperornis* and *Ichthyornis* from Late Cretaceous marine deposits of North America, comprise advanced birds in the Cretaceous, and include the ancestors of all living birds. Since 1981, the group Enantiornithes was introduced, based on some fragmentary discoveries from the Late Cretaceous of Argentina. With the recent explosive discoveries of birds from the Early Cretaceous, Enantiornithes now comprises a lot more taxa from the Early Cretaceous than from the Late Cretaceous. Enantiornithes

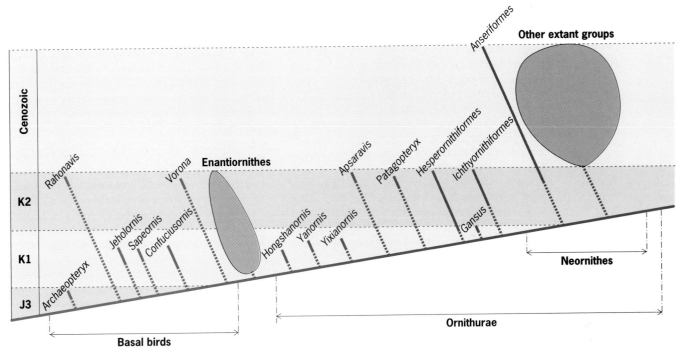

Cladogram showing major bird lineages in the Cretaceous radiation.

undoubtedly represents the predominating bird group in the Cretaceous, becoming extinct by the end of the Cretaceous.

Many newly discovered Early Cretaceous birds cannot be referred to either Enantiornithes or Ornithurae. Phylogenetically, they occupy a transitional position between *Archaeopteryx* and Enantiornithes. They represent the most primitive birds in the Cretaceous and are tentatively categorized as "basal birds" to distinguish them from enantiornithine birds and ornithurine birds. A brief introduction of each of these three categories is provided here, with a discussion of their implications for their role in the radiation of birds throughout the Cretaceous.

Basal birds. Several Cretaceous birds have been discovered from the Cretaceous that are only slightly more advanced than *Archaeopteryx*. Most notable among them are two long-tailed birds: *Jeholornis* from the Early Cretaceous of China, and *Rahonavis* from the Late Cretaceous of Madagascar. Phylogenetically, they also represent the most primitive birds known except *Archaeopteryx* (see **illustration**). *Rahonavis* has a large sickle claw that is typical of some theropod dinosaurs. *Jeholornis* retains an even longer skeletal tail than *Archaeopteryx*. It also has fan-shaped tail feathers that are attached only to the distal caudal vertebrae much as in the feathered dinosaurs *Caudipteryx* and *Microraptor*, and unlike in *Archaeopteryx*. It has a sternum comprising a pair of sternal plates, and a phalangeal formula of 2-3-4 as in *Archaeopteryx* and theropod dinosaurs. Nevertheless, *Jeholornis* has a more derived flight capability than *Archaeopteryx* despite the presence of a more primitive tail.

Jeholornis also represents the earliest known bird

with a specialized seed-eating dietary adaptation, as it has over 50 seeds preserved in its stomach. *Sapeornis* from the Early Cretaceous of China has preserved gizzard stones in its stomach region, suggesting it was a herbivorous bird. This is also the largest bird known from the Early Cretaceous. It has a wing much longer than the leg, suggesting that it might have had a soaring flight style.

Confuciusornis is another important bird from the Early Cretaceous of China. This bird was probably omnivorous, and there is also evidence for its preying on fish. It shares with *Archaeopteryx* and *Jeholornis* a primitive manual phalangeal formula of 2-3-4. Their large wing claws could help in the tree climbing of these basal birds. Unlike most Mesozoic birds, *Confuciusornis* has no teeth but possesses a horny beak, representing one of the earliest known birds with a beak.

In addition to *Rahonavis*, *Vorona* represents another basal bird from the Late Cretaceous of Madagascar which retains many primitive features in the leg.

All these basal birds are approximately the size of a chicken or a turkey. However, they share similar foot structure that is adapted for arboreal life. The foot claws are large and very curved. It is noteworthy that in both *Archaeopteryx* and *Jeholornis* the first pedal digit (hallux) is not completely reversed as compared to the other three digits. This must be explained as a feature transitional between theropod dinosaurs and more advanced birds. However, the presence of an incompletely reversed hallux would probably not be a problem for adaptation of tree climbing, through which the fully reversed hallux indicative of improved perching capability finally evolved in more advanced birds.

Enantiornithes. Enantiornithes is a diverse Mesozoic bird group represented by over 20 species, with a global distribution in the Cretaceous. Although first reported from the Late Cretaceous, most of its known species are now from Early Cretaceous lake deposits, particularly in China and Spain. Early Cretaceous enantiornithines are small compared with contemporaneous ornithurines and basal birds. Many of them are about the size of a sparrow, such as *Sinornis* from China and *Eoalulavis* from Spain. Late Cretaceous enantiornithines, for example, *Enantiornis*, became comparatively larger.

Enantiornithines are better fliers than basal birds such as *Jeholornis* and *Confuciusornis*. Many of the Early Cretaceous enantiornithines are represented by complete skeletons and feathers. That they are better fliers than the basal birds is evidenced by a more derived pectoral girdle, a keeled sternum, and the presence of an alula or bastard wing (a tuft of small stiff feathers on the first digit of a bird wing), which also suggests the possession of a sophisticated flight skill.

Nearly all enantiornithines are arboreal as shown by their large and curved foot claws. They probably lived in or near the forest and mainly fed on insects. Almost all enantiornithines had teeth, with only a few exceptions such as *Gobipteryx* from the Late Cretaceous of Mongolia. *Longipteryx* and *Longirostravis* are two specialized enantiornithines from the Early Cretaceous of China. *Longipteryx* has elongated and toothed jaws and elongated wings compared with the leg. It was probably a piscivorous (fish-eating) bird living near the forest. *Longirostravis* has an extended jaw with a pointed rostral end with teeth restricted to the rostralmost end of the jaws. It probably had a probing feeding behavior unknown in any other Cretaceous birds.

Obviously by the Early Cretaceous, enantiornithines already diversified significantly in morphology, flight, diet, and niches. By the Late Cretaceous, enantiornithines became larger and more specialized, and were extinct by the end of the Cretaceous together with dinosaurs.

Ornithurae. *Hesperornis* and *Ichthyornis* represent the best-known members of the Ornithurae. The former was a specialized diving bird, and the latter presumably a fish-eating bird with strong flying ability. Many recent new fossil discoveries from both the Early Cretaceous and Late Cretaceous have significantly increased our understanding of this most advanced bird group in the Cretaceous.

It is now known that the earliest ornithurines had appeared in the Early Cretaceous. *Hongshanornis* and *Liaoningornis* represent the two earliest ornithurines from about 125 million years ago (Ma), which is only slightly younger than the earliest enantiornithine (*Protopteryx*) dated as 131 Ma from Hebei, China. Both of them are small and about the size of contemporaneous enantiornithines.

Yanornis and *Yixianornis*, also from the Early Cretaceous of China, are slightly younger (120 Ma) and more advanced. Both waded near the water. Both were toothed birds and probably fed mainly on fish. In fact, one *Yanornis* specimen preserved fish remains in the stomach. *Ambiortus* from the Early Cretaceous of Mongolia was probably from the same age as *Yanornis*. *Gansus* is an even more advanced ornithurine from the Early Cretaceous of Gansu, China, which further confirms that aquatic adaptation played a key role in the early evolution of ornithurines.

All ornithurines shared many advanced flight related features such as elongate sternum with a deep keel, as well as coracoid (one of the paired bones of the pectoral girdle), furcula (wishbone), wing, and pygostyle (a specialized bone in birds which is formed by a number of fused tail vertebrae) that are indistinguishable from those of modern birds. Therefore, they represent the most skilled fliers of that time. They probably also represent the earliest known endothermic birds based on studies of their flight, paleophysiology, and paleohistology.

In addition to *Hesperornis* and *Ichthyornis*, there exist some other ornithurines in the Late Cretaceous such as *Patagopteryx* from Argentina, a flightless bird living on the ground, and *Apsaravis* from Mongolia that preserved some enantiornithine features. One of the most important Late Cretaceous discoveries is *Vegavis*, a bird that can be referred to Anseriformes from the late Late Cretaceous of Antarctica. This finding provides for the first time definitive evidence for the hypothesis that extant bird radiation occurred before the Cretaceous/Tertiary boundary.

Conclusions. New discoveries of birds in the last two decades have significantly increased our understanding of the diversity of birds in the Cretaceous. The bird radiation in the Cretaceous, the first ever in avian history, was explosive, with the sudden appearance of over 20 different orders and families of birds with a very diverse morphology, flight capability, body size, locomotion, and dietary adaptation. Ornithurine birds first appeared and radiated in the Early Cretaceous. They finally gave rise to living orders of birds by the end of the Late Cretaceous.

For background information *see* ARCHAEORNITHES; AVES; CRETACEOUS; DINOSAURIA; PALEONTOLOGY in the McGraw-Hill Encyclopedia of Science & Technology.   Zhonghe Zhou

Bibliography. L. M. Chiappe and L. M. Witmer, *Mesozoic Birds: Above the Heads of Dinosaurs*, University of California Press, Berkeley, 2002; J. A. Clarke et al., First definitive fossil evidence for the extant avian radiation in the Cretaceous, *Nature*, 433:305–308, 2005; A. Feduccia, *The Origin and Evolution of Birds*, 2d ed., Yale University Press, New Haven, 1999; Z. Zhou, The origin and early evolution of birds: Discoveries, disputes, and perspectives from fossil evidence, *Naturwissenschaften*, 91:455–471, 2004.

Dark energy

Dark energy comprises the majority of the energy of the universe and is responsible for the accelerating cosmic expansion. Its name derives from the inference that it is nonluminous. The nature of dark energy is speculative. Leading theories propose that dark energy is a static, cosmological constant consisting of quantum zero-point energy, and a dynamical condensate of a new, low-mass particle. Alternatively, it has been proposed that the dark energy phenomena are due to a change in the form of gravitation on cosmological length scales. Confirmation of any one of these ideas would have a profound impact on physics. Dark energy appears to be distinct from dark matter, nonluminous particles which make up the majority of the mass of galaxies and galaxy clusters. Determining the nature of dark energy is widely regarded as one of the most important problems in physics and astronomy. *See* DARK MATTER.

Observational evidence. The empirical evidence for dark energy is indirect. Yet the evidence is manifest in what is, perhaps, the single most remarkable feature of the universe—the cosmic expansion. According to Albert Einstein's general theory of relativity, a cornerstone of the standard big bang cosmology, the rate of cosmic expansion is determined by the average energy density of all forms of matter and radiation in the universe. Measurements of the expansion, discovered by Edwin P. Hubble in 1929 and refined by the National Aeronautics and Space Administration's Hubble Space Telescope Key Project, therefore determine a critical energy density, $\rho_{crit} = 0.88\ (\pm 0.19) \times 10^{-9}$ J/m^3 (68% confidence level), to which all forms of matter and radiation must sum. For comparison, the energy density of water is nearly 30 orders of magnitude larger.

The first piece of evidence for the existence of dark energy arises from measurements of the gravitational mass density of galaxies and galaxy clusters. Diverse measurements accumulated over many decades, from Fritz Zwicky's 1933 observations to recent galaxy-redshift surveys, indicate that the fractional energy density in the form of atoms and even dark matter falls far short of the critical density, with $\Omega_m \equiv \rho_m/\rho_{crit} \approx 0.3$. The fraction in baryons, referring to all atoms or nuclear material, is a mere $\Omega_b \approx 0.04$, implying that the nature of $\sim$96% of the energy of the universe (26% dark matter and 70% dark energy) is unknown.

The second piece of evidence derives from measurements of the mean curvature of space. Just as mass and energy distort space-time, a mean curvature of space itself contributes an equivalent energy density. A space that is sufficiently curved so as to explain the dark energy must thereby possess a geometry in which initially parallel light rays traversing cosmological distances are distorted and eventually diverge. But detailed measurements of the cosmic microwave background radiation, most recently by NASA's *Wilkinson Microwave Anisotropy Probe*, have eliminated this possibility in confirming that the mean curvature of space is negligible. The $\sim$70% gap in energy must be due to something other than the curvature or dark and baryonic matter. *See* SPACE FLIGHT.

The third piece of evidence, which provides the strongest clue to the nature of the dark energy, derives from the 1998 discovery of the accelerating cosmic expansion. Cosmological distance-redshift mea-surements of the scale of the universe as a function of look-back time reveal that the rate of expansion has been increasing during the last several billion years. In general relativity, the acceleration of a test particle due to the gravitational field of a fluid sphere is proportional to $\rho + 3p$ of the fluid, where ρ is the fluid density and p is the pressure. In the cosmic context, acceleration implies that the universe is under tension, like a fluid with a negative effective pressure. (An example is the stress parallel to a uniform electric or magnetic field.) Because no other cosmologically abundant forms of energy support a sufficiently negative pressure across cosmic scales, the dark energy must carry the cosmic pressure responsible for the acceleration. The strength of the pressure is characterized by an equation-of-state parameter, w, defined as the ratio of the homogeneous pressure to the energy density of the dark energy, $w \equiv p/\rho$. Formally, acceleration requires $w < -1/3(1 - \Omega_m)$, with current measurements giving $-1.2 \leq w \leq -0.8$ (95% confidence level).

These are the main lines of evidence for dark energy. Numerous other observations and measurements—including the age of the universe, the variation of distances with redshift, and distortions of the cosmic microwave background radiation by the accelerated expansion—are consistent with the interpretation that some form of dark energy with negative pressure is responsible for the majority of the energy of the cosmos.

Theoretical solutions. The leading hypothesis is that the dark energy is a cosmological constant (Λ), a uniform sea of positive energy and negative pressure with $w = -1$. A detailed comparison between theoretical predictions and observations finds that a cosmological constant supplying $\sim$70% of the critical energy density, or $\Omega_m \approx 0.3$, fits current data. In the absence of further clues, a cosmological constant provides the most economical description of dark energy.

The cosmological constant was introduced in 1917 by Einstein as a mathematical device in his ground-breaking application of general relativity to the problem of cosmology. A physical explanation of its origin was offered in 1967 by Yakov Zeldovich, who demonstrated that the vacuum of quantum-mechanical particles is equivalent to a cosmological constant. However, the predicted amplitude is at least 60 orders of magnitude too large—a mismatch regarded as one of the most enigmatic problems in theoretical physics. It is speculated that a complete understanding of the cosmological constant requires a theory of quantum gravity.

The need to adjust or finely tune the cosmological constant to a value so many orders of magnitude below the predicted value has prompted speculation that perhaps $\Lambda = 0$ or that it does not contribute to the gravitational field equations, and some other mechanism is responsible for the accelerated expansion.

A widely studied possibility is that the dark energy is dynamical, consisting of a condensate of a new species of particles of very small mass. In the simplest version of this theory, these particles have mass $m \approx 10^{-68}$ kg and are identified as excitations of a cosmic field which oscillates with wavelength $\lambda \approx 10^{23}$ km, just within the size of the observable universe. This form of dark energy is also referred to as quintessence, as distinct from other fields and forms of matter or radiation. In contrast to a cosmological constant, the equation of state of quintessence is time varying, with $w > -1$, and its energy density and pressure fluctuate in response to gravitational field variations spanning length scales larger than λ. Through measurements of such time variation or fluctuations it is hoped to distinguish among competing theories of dark energy.

The theory of quintessence dark energy has two significant, although unproven, precedents. First, the dark energy phenomenon is very similar to primordial inflation, an epoch of accelerated expansion conjectured to have taken place in the early universe. According to inflation theory, the mechanism that drives the expansion is a cosmic field identical to quintessence except scaled up in mass and energy to match the conditions in the early universe. Second, the quintessence field is very similar to the axion particle, a dark matter candidate particle, originally proposed to resolve a fine-tuning problem in quantum chromodynamics. Despite these motivations for quintessence dark energy, any such theory also faces the challenge that it must explain why the new particle does not readily interact with known particles and mediate a new force, and why the new particles are so light compared to known particles.

More speculative theories of dynamical dark energy with equation of state $w < -1$ have been proposed. One such example, called phantom dark energy for a similarity to so-called ghost fields in quantum field theory, is also a cosmic field but with very different properties from quintessence. The energy density of the phantom field grows with time, rather than decays as for quintessence. The viability of this theory has implications for quantum gravitational processes.

Numerous other theories of dynamical dark energy have been proposed. Most can be characterized simply by an equation of state of the quintessence ($w > -1$) or phantom ($w < -1$) type. In all cases, the magnitude of the cosmic acceleration is determined by the combination $-[1 + 3w(1 - \Omega_m)]/2$ so that, for equal matter density, quintessence gives the weakest acceleration and a phantom gives the strongest. Important clues to the nature of dark energy can be gained from measurements of the equation of state and any possible time variation.

Gravity. It has been proposed that the dark energy phenomena are due to a change in the form of gravitation on cosmological scales, as alternative to a cosmological constant or quintessence. Einstein's general relativity is well-tested within the solar system, and underpins the successes of the big bang cosmology. However, a modification of the gravitational field equations required to balance the expansion against the energy density and pressure of the cosmological fluid could obviate the need for the above-described theories of dark energy.

New theories of gravitation which introduce extra gravitational degrees of freedom under the rubric of scalar-tensor gravity, or which incorporate the effects of additional dimensions of space-time based on string theory or other fundamental theories of physics, are a subject of intense study. Most such theories proposed to explain the dark energy phenomena include a new cosmic field which is similar to quintessence but also modulates the strength of gravitation, thereby resulting in variability of Newton's constant, the amount of space-time curvature produced per unit mass or energy, and the degree to which the curvature produced by a massive body exceeds the sum of the curvatures due to its individual constituents. Most simple extensions of general relativity are heavily constrained by solar system observations in such a way that viable cosmological models are essentially indistinguishable from general relativity with a cosmological constant or quintessence dark energy.

Fate of the universe. The discovery of the cosmic acceleration suggests that the future evolution and fate of the universe will be determined by the nature of the dark energy. Previously, the main speculation was whether the big bang would terminate in a big crunch, as the universe collapsed under its own weight. Now a big chill appears as a likely fate: In a universe containing a cosmological constant or quintessence dark energy which maintains the inequality $-1 \leq w < -1/3(1 - \Omega_m)$, the expansion will continue to accelerate forever. Only gravitationally bound structures today will survive into the distant future. Furthermore, the acceleration will prevent any communication across distances greater than $\sim 10^{23}$ km due to the appearance of a cosmological event horizon, and the universe will grow cold and rarefied. But in a universe containing phantom dark energy which maintains the inequality $w < -1$, the future would appear to be a big rip: the catastrophic end to the universe. In this case, the expansion would undergo a runaway process resulting in the gravitational disruption of clusters, galaxies, and then both gravitationally and nongravitationally bound objects on smaller and smaller scales. The Milky Way would be torn apart ~ 50 million years before the big rip. The universe would be shredded to its fundamental constituents, unless quantum gravity effects intervene, and space-time itself would end ~ 50 billion years in the future.

Prospective observations and experiments. The current challenge to cosmology is to tighten the evidence for dark energy, refine measurements of the cosmic expansion so as to narrow the bounds on the equation of state w, and continue to test the framework of general relativity and big bang cosmology. The main impact of the dark energy is on the rate of cosmic expansion, so that new clues to the physics of dark energy may be obtained from measurements of distance, volume, and age versus redshift; the evolution and abundance of galaxies and clusters; and the distortion of the cosmic microwave background radiation. Numerous experiments and observatories dedicated to investigating dark energy are planned.

For background information *see* BIG BANG THE-ORY; COSMIC BACKGROUND RADIATION; COSMOL-OGY; HUBBLE CONSTANT; INFLATIONARY UNIVERSE COSMOLOGY; QUANTUM GRAVITATION; RELATIVITY; SUPERSTRING THEORY in the McGraw-Hill Encyclo-pedia of Science & Technology.   Robert Caldwell

Bibliography. R. R. Caldwell, Dark energy, *Phys. World*, 17(5):37–42, 2004; W. L. Freedman and M. S. Turner, Measuring and understanding the uni-verse, *Rev. Mod. Phys.*, 75:1433–1447, 2003; J. P. Ostriker and P. J. Steinhardt, The quintessential universe, *Sci. Amer.*, 284(1):46–53, 2001; P. J. E. Peebles and B. Ratra, The cosmological constant and dark energy, *Rev. Mod. Phys.*, 75:559–606, 2003.

Dark matter

Dark matter consists of particles or objects that exert a gravitational force but do not emit any detectable light. Dark matter is the dominant form of matter in our Galaxy. Astronomers have detected the pres-ence of dark matter through its gravitational effects and have shown that dark matter is not composed of ordinary atoms. Particle physicists have suggested several plausible candidates for dark matter; planned experiments are capable of detecting these new par-ticles.

Astronomical evidence for dark matter. A variety of astronomical observations imply that dark matter is ubiquitous; it is detected in dwarf galaxies, in spiral galaxies, and in elliptical galaxies. It is the dominant form of matter in galaxy clusters and leaves clear sig-natures in the large-scale distribution of galaxies and in the microwave background. Astronomers infer the presence of matter through its gravitational effects. Since they have not been able to detect any light di-rectly associated with this matter, they have labeled it "dark matter."

Dark matter is also sometimes called the "missing matter." This is a misnomer since astronomers de-tect the mass but they are unable to detect the light associated with the matter.

Galaxy motions in clusters. Many galaxies are bound to-gether in galaxy clusters. These galaxy clusters con-tain hundreds and sometimes thousands of galax-ies. These galaxies are moving with very high relative velocities. Typical velocities are roughly 1000 km/s (600 mi/s), and some galaxies that ap-pear to be bound to the cluster move at velocities as high as 2000 km/s (1200 mi/s). Within an individual galaxy, stars and gas move much slower with typical velocities of only 200 km/s (120 mi/s).

As early as the 1930s, F. Zwicky recognized that clusters must be very massive to have a strong enough gravitational field to keep these high-velocity galaxies bound to the cluster. Zwicky noted that the inferred mass was much larger than the mass in stars. The result was so surprising that most of his col-leagues ignored his conclusion. Current estimates

Fig. 1. Sloan Digital Sky Survey image of the Perseus galaxy cluster. Galaxies in the cluster are moving relative to each other at velocities of thousands of kilometers per second. Since the cluster appears to be gravitationally bound, this implies that the cluster has a mass of 10^{11} solar masses, roughly 100 times the mass contained in stars. (*Robert Lupton; SDSS Collaboration*)

suggest that stars comprise only 1% of the mass of the cluster (**Fig. 1**).

Galaxy rotation curves. Most galaxies, including our own Milky Way, are spiral or disk galaxies, where cold atomic and molecular gas (primarily hydrogen) settles into a rotating disk. Much of the molecular gas is found in spiral arms. These spiral arms are active sites of star formation.

The gas and stars in disk galaxies move on nearly circular orbits. The Sun is moving on a nearly circu-lar orbit around the center of the Milky Way Galaxy. To keep material moving on a circular orbit, centrifu-gal acceleration must balance gravitational accelera-tion. This balance can be expressed by the equation below, where V is the velocity of the gas or stars,

$$\frac{V^2}{R} = \frac{GM}{R^2}$$

R is the distance from the center of the galaxy, G is Newton's constant, and $M(R)$ is the mass contained within radius R.

Astronomers can measure the velocity of gas and stars in other galaxies through the Doppler effect. In the 1960s and 1970s, Vera Rubin pioneered opti-cal observations of the motions of stars near the vi-sual edges of galaxies. She found that the stars were moving rapidly. This large velocity implied the pres-ence of hundreds of millions of solar masses of matter where little light was detected. Radio observations of the motions of neutral gas beyond the visual edge of spiral galaxies confirmed this remarkable result: Most of the mass in a galaxy is not in its disk of young stars, but rather in a halo composed of dark matter (**Fig. 2**).

Hot gas in elliptical galaxies and clusters. X-ray satellites have detected copious amounts of hot gas in ellipti-cal galaxies and in galaxy clusters. Since the thermal

Fig. 2. Sloan Digital Sky Survey image of galaxy M101. Measurements of the velocities in the disk of the galaxy imply the presence of enormous amounts of mass out toward the edge of the disk. (*Robert Lupton; SDSS Collaboration*)

pressure in this 1–10 million degree gas balances the gravitational field of the galaxy, measurements of the density and temperature profiles of the gas can be used to directly measure the mass distribution in these galaxies. These measurements reveal that the dark matter problem is ubiquitous: Stars can account for only a small fraction of the mass in elliptical galaxies. Observations of x-ray gas in clusters con-

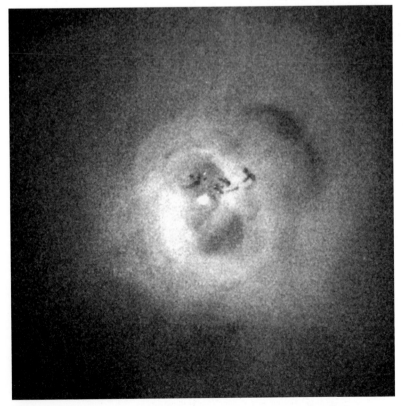

Fig. 3. *Chandra X-ray Observatory* image of hot gas in the Perseus galaxy cluster, indicating that stars account for only a small fraction of the mass of the cluster. (*NASA/Chandra X-ray Center/Institute of Astronomy/A. Fabian et al.*)

firm Zwicky's inference that stars account for only a small fraction of the mass in a cluster (**Fig. 3**).

Gravitational lensing. John Wheeler has remarked that "General relativity is simple: Matter tells space how to curve; the curvature of space tells matter how to move." Since mass curves space and distorts the path of light, images of galaxies that are behind rich clusters can be used to infer the distribution of mass in clusters. The masses in clusters distort background galaxies into arcs whose orientations are very sensitive to the distribution of matter in the cluster. Observations of gravitational lenses detect the presence of significant amounts of dark matter, and are consistent with the x-ray and optical observations of the same clusters (**Fig. 4**).

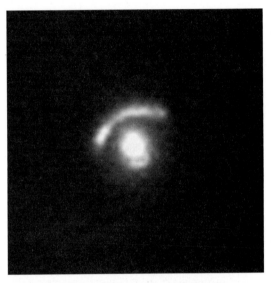

Fig. 4. *Hubble Space Telescope* image of a lensed background source in the form of an extended arc about an elliptical lensing galaxy. By measuring the distortions in the background, the mass of the foreground lensing galaxy can be inferred. (*Kavan Ratnatunga, Carnegie Mellon University; NASA*)

Novel nature of dark matter. The astronomical evidence discussed above shows that most of the mass in galaxies is not in the form of luminous stars and is consistent with a wide range of plausible candidates. Most astronomers assumed that the dark matter was similar to the "ordinary stuff" that makes up stars: protons, neutrons, and electrons. Cosmologists often refer to this "ordinary stuff" as baryonic matter. (Unfortunately, cosmologists use the term "baryonic matter" to refer to protons, neutrons, and electrons, while particle physicists, more correctly, refer to only protons and neutrons as baryonic matter.) Astronomers have searched for various possible forms of ordinary matter, including gas, dust, and low-mass stars, but they have been unable to detect it.

Cosmological observations can measure the density of atoms in the early universe. These observations imply that baryonic matter can account for only one-sixth of the mass in galaxies. These observations imply that most of the mass in galaxies is in the form of some novel and not yet identified form of matter.

Gas. Gas can be detected through either emission or absorption. Astronomers have measured the mass in atomic gas through observations of the 21-cm line and molecular gas through millimeter observations of carbon monoxide (CO) clouds. These observations suggest that only 1% of the mass of the Milky Way Galaxy is in cold gas in the galactic disk.

X-ray satellites can detect 1–100 million degree gas through its thermal bremsstrahlung emission. These observations reveal that most of the baryonic matter in rich clusters is not in stars but in hot gas. In some rich clusters, there is ten times as much mass in hot gas as in stars. While significant for the evolution of the cluster, only 10–20% of the mass in a cluster is in hot gas. Dark matter accounts for most of the remaining 80–90%.

Low-mass stars. While massive stars can be easily detected through their light, very low mass stars and planet-size objects are so dim that they can evade detection by optical telescopes. Astronomers, however, can detect these stars through their gravitational effects.

When a low-mass star passes in front of a distant star, it serves as a gravitational lens. This lensing brightens the image of the background star. Using large-area cameras capable of monitoring millions of stars, astronomers have searched for and detected these effects as low-mass stars pass in front of stars in the Large Magellanic Clouds and in front of the Milky Way bulge. While these observations have detected hundreds of low-mass stars through their gravitational effects, the number of events are consistent with models of the Milky Way Galaxy and imply that there are not enough low-mass stars to account for the dark matter.

Deuterium and helium. During the first minutes of the big bang, all of the neutrons in the universe combined with protons to make deuterium (one proton and one neutron). Two deuterium nuclei then collided to make helium (two protons and two neutrons). Most of the helium and deuterium in the universe was produced during the first few minutes of the big bang, and thus deuterium and helium can be said to be fossils from that time. The cosmological abundance of deuterium depends on the density of baryons (protons and neutrons) in the early universe: the more baryons, the lower relative abundance of deuterium. Astronomers use measurements of the ratio of the relative number of deuterium nuclei to provide a direct estimate of the density of atoms in the universe.

These observations imply that ordinary matter makes up only 4% of the energy density of the universe. These observations suggest that dark matter is not composed of protons and neutrons (or of nuclei made of protons and neutrons).

Microwave background fluctuations. Measurements of microwave background fluctuations are an important probe of the composition of the universe. The tiny variations in temperature seen by the *Wilkinson Microwave Anisotropy Probe* (and by ground and balloon-based microwave experiments) were produced during the first moments of the big bang. The statistical properties of these fluctuations depend on the composition of the universe: the more atoms, the smoother the shape of the fluctuations; the more matter, the higher the relative amplitude of small-scale fluctuations. The current observations confirm that ordinary matter makes up only 4% of the energy density of the universe and also imply that dark matter comprises 25% of the energy density of the universe.

Dark matter candidates. The big bang theory implies that the early universe was very hot and very dense, the perfect environment for creating particles through collisions. The densities and temperatures in the first microsecond of the big bang exceed the energies achieved in even the most powerful particle accelerators. Many cosmologists suspect that the dark matter is composed of some yet undiscovered fundamental particle that was produced in copious numbers during the first moments of the big bang.

Supersymmetry. The most popular dark matter candidate is the neutralino, a new particle posited by the theory of supersymmetry. Supersymmetry is an extension of the standard model of physics and is a vital element in almost all attempts to unify the forces of nature and in attempts to connect gravity and quantum mechanics.

Modern particle physics is based on the symmetries observed in nature. For example, all particles have antiparticles. The positron is the antiparticle of the electron. The antiproton is the antiparticle of proton. While seemingly bizarre, antimatter has not only been produced in the laboratory and detected in space, but now plays a major role in medical imaging (PET scans).

Particle physicists have speculated on the existence of a new, not yet confirmed symmetry: supersymmetry. Supersymmetry implies that particles such as electrons and positrons have partners called selectrons and spositrons. These superparticles have the opposite statistical properties to ordinary matter. Selectrons behave like photons, while the supersymmetric partner of the photon, the photino, behaves quantum mechanically like an electron. This new symmetry has many aesthetic attractions and helps to explain the relative strength of the fundamental forces of nature. These new particles interact weakly with ordinary matter and have so far escaped detection. (They also may not exist.)

Supersymmetry implies the existence of a new stable particle: the neutralino. The neutralino would have been produced in abundance during the first moments of the big bang. During the first nanosecond of the big bang, there were as many neutralinos as photons in the universe. While the number of neutralinos was reduced as they annihilated into ordinary matter, the current best estimates of their predicted residual abundance imply that they could be the dark matter.

The primary scientific goal of the new Large Hadron Collider (LHC) at CERN in Switzerland, scheduled for start-up in 2007, is to detect the experimental signatures of supersymmetry. By colliding nuclei at velocities very close to the speed of light, the

Fig. 5. CERN Axion Solar Telescope (CAST), which uses a 9-tesla, 10-m (33-ft) prototype dipole magnet for CERN's Large Hadron Collider, enhanced by the use of a focusing x-ray telescope, to search for axions from the Sun's core. (*CERN*)

LHC aims to recreate some of the conditions in the big bang. The LHC may be able to provide direct evidence for the existence of supersymmetry and may even be able to produce dark matter particles in the laboratory.

Deep-underground experiments may also detect dark matter. If the neutralino is the dark matter, then hundreds of millions of these particles are streaming through our bodies every second. Because the neutralino interacts so weakly with ordinary matter, our bodies (and our detectors) are nearly transparent. These particles, however, do have rare weak interactions. In a kilogram detector, the dark matter particles may have a few collisions per day. These collisions are difficult to detect since each collision deposits only 10^{-3} (10^{-15} erg) of energy. Physicists have built a number of very sensitive low-background experiments capable of directly detecting neutralinos. Because the experiments need to be shielded from cosmic rays and other terrestrial sources of background signal, they must operate in deep-underground mines and tunnels.

Gamma-ray satellites may also be able to indirectly detect the neutralino. While the neutralinos interact weakly, occasionally one neutralino collides with another neutralino. This collision produces a shower of particles and radiation. Several on-going and planned experiments are looking for this dark matter annihilation signal.

Axions. The axion is another hypothetical particle, which was invented to explain one of the symmetries seen in the strong nuclear interaction. If this explanation is correct, then axions would have been produced copiously during the first microsecond of the big bang. While axions interact very weakly with ordinary matter, they can be converted into ordinary photons in the presence of a very strong magnetic field. Experimentalists are using very strong fields to search for the axion (**Fig. 5**).

Black holes. If black holes were produced in large numbers during the first moments of the big bang, then they would be plausible candidates for the dark matter. Currently, there are no viable scenarios for producing large numbers of black holes in the early universe. However, since there are significant uncertainties in our understanding of physics during the first microsecond of the big bang, some astronomers consider black holes to be a viable dark matter candidate.

If the dark matter were composed of black holes and the black holes were significantly more massive than the Sun, then the black holes could be detected through their gravitational effects on galaxies and globular clusters, or through gravitational microlensing (discussed above in connection with the detection of low-mass stars). However, low-mass black holes could easily evade any current (and planned) observational searches.

Failure of general relativity. Many astronomers have speculated that the discrepancy between the mass seen in stars and gas and the mass inferred by applying general relativity (and Newtonian gravity, its low-velocity simplification) to astronomical observations is not the signature of some new exotic particle, but instead is the observational signature of the breakdown of general relativity. Classical Newtonian physics works well in daily life: Quantum mechanics is usually important only at very small scales and general relativity is only important on very large scales and near massive objects such as black holes and neutron stars. Perhaps the dark matter problem is the observational signature of the failure of general relativity. The discovery of dark energy strengthens the motivation for considering these alternative models.

Modified Newtonian dynamics (MOND), developed by Mordehai Milgrom and Jacob Bekenstein, is the most carefully examined alternative gravitational theory. It has the advantage of explaining galaxy rotation curves without resorting to dark matter. However, MOND has difficulties fitting the data implying the existence of dark matter in clusters and is inconsistent with WMAP's observations of microwave background fluctuations.

For background information *see* BIG BANG THEORY; BLACK HOLE; COSMIC BACKGROUND

RADIATION; COSMOLOGY; GALAXY, EXTERNAL; GRAV-
ITATIONAL LENS; INTERSTELLAR MATTER; MILKY
WAY GALAXY; PARTICLE ACCELERATOR; RELATIVITY;
SUPERSYMMETRY; UNIVERSE; WEAKLY INTERACTING
MASSIVE PARTICLE (WIMP); X-RAY ASTRONOMY in the
McGraw-Hill Encyclopedia of Science & Technology.
David N. Spergel

Deep brain stimulation

Depression refers to a set of prevalent, extremely
debilitating disorders that can be characterized by
a triad of symptoms: extreme anhedonia (complete loss of pleasure from previously pleasur-
able activities), depressed mood, and low energy.
Other cognitive symptoms (for example, pessimistic
thoughts, feelings of guilt, low self-esteem, and sui-
cidal ideations) and somatic symptoms (for example, sleep and psychomotor disturbances, and food-
intake and body-weight dysregulation) are also often
present. Unipolar major depression (unipolar refers
to the presence of one pole or one extreme of mood,
namely depressed mood) is the leading cause of disability worldwide and is associated with a high mortality due to suicide and increases in mortality in
comorbid somatic disorders such as cardiovascular
disorders or cancer.

Modern medication treatments, in conjunction
with certain methods of psychotherapy, are effective
at alleviating depressive symptomatology in most
patients. However, these treatments do not work
for all patients. A sizable minority of patients does
not respond. Indeed, 17–21% of patients suffering
from major depression have a poor outcome after
2 years, and 8–13% have a poor outcome even after
5 years of treatment. These patients thus do not respond to any known treatment combination, including electroconvulsive therapy, and are referred to as
treatment-resistant patients. This underserved population has had little hope of recovering from this
disease.

Psychotropic drugs work by altering neurochem-
istry to a large extent in widespread regions of
the brain, many of which may be unrelated to depression. It might be that more focused, targeted
treatment approaches that modulate specific networks in the brain will prove a more effective approach to help treatment-resistant patients. In other
words, whereas existing depression treatments approach this disease as a general brain dysfunction,
a more complete and appropriate treatment will
arise from thinking of depression as a dysfunction of specific brain networks that mediate mood
and reward signals, in particular, the cortical-limbic-
thalamic-striatal network. This conceptualization
leads to novel ideas about targeted neuromodula-
tory treatments that have been researched in the last
decade.

Use of deep brain stimulation. The use of neurosurgical interventions for psychiatric disorders dates
back to the origins of neurosurgery itself. Targeted
irreversible destruction of brain tissue—more or less

guided by hypotheses on the neurophysiology of the
disorders to be treated—was used to modulate networks and behaviors. Deep brain stimulation (DBS)
is a well-established procedure that refers to stereo-
tactic placement of uni- or bilateral electrodes in a
given brain region, with the electrodes connected
to a neurostimulator implanted under the skin of the
chest. Chronic high-frequency (150–185 Hz) stimulation likely leads to a decrease of neural transmission through inactivation of voltage-dependent
ion channels. Electrical stimulation achieves this by
producing neuronal deactivation either by direct disruption of neuronal activity or by increasing gamma-
aminobutyric acid (GABA)–mediated inhibitory neu-
rotransmission.

So far, DBS has been approved by the U.S. Food and
Drug Administration for the control of severe forms
of tremor in Parkinson's disease, essential tremor,
and primary dystonia (disorder or lack of muscle
tonicity) via stimulation of the internal globus pal-
lidus and the subthalamic nucleus. The most effective target to reduce tremor in Parkinson's disease
is the subthalamic nucleus, although other targets
such as thalamic and pallidal sites are stimulated. Currently, DBS is also being investigated for use in cluster
headache (via stimulation of the ipsilateral ventro-
posterior hypothalamus), obsessive-compulsive disorder, and major depressive disorder.

Advantages. Advantages of DBS versus ablative
neurosurgical procedures—such as anterior capsulo-
tomy, anterior cingulotomy, subcaudate tractotomy,
or limbic leucotomy (essentially a combination of
the latter two procedures)—are the lower risk, the
reversibility, the ability to adjust stimulation for the
individual patient, and the possibility to investigate
its effectiveness in blinded protocols (separate investigation of on-off cycles).

Affective symptoms induced. Cognitive and neuropsychiatric symptoms such as apathy, anxiety, im-
pulsivity, mood lability, and depression are often seen
early in the course of Parkinson's disease, an observation which contributes to the concept that major
depression and Parkinson's disease might be associated with dysfunctions in overlapping pathways in
the basal ganglia. On the other hand, DBS for Parkinson's disease has the potential to induce a wide range
of affective symptoms depending on the brain area
stimulated. Stimulation of areas in or around (sub-
stantia nigra, zona incerta) the superior thalamic
nucleus has been associated with acute and delayed
induction of depressive symptoms, including suici-
dality (there are reports of patients committing suicide, of hypomania, and of cognitive disturbances).
Other investigators, however, also reported on dramatic enhancement of mood after subthalamic nucleus stimulation. The theories why such divergent
affective symptoms can be induced by DBS are sev-
eralfold:

1. *Electrode placement.* The subthalamic nucleus
is believed to regulate not only motor circuits but also
associative and limbic-cortical-subcortical circuits involved in mood regulation. Since affective symptoms
occur in most cases (75%) of Parkinson's disease

1–3 months after surgery, they are most likely due to adaptive changes of brain circuits.

2. *Reduction of antidopaminergic medication.* DBS often allows for dramatic reduction of dopaminergic medication, which in turn may lead to insufficient dopaminergic stimulation of mesolimbic and mesocortical projections, resulting in lower mood.

3. *Progression of psychiatric disorder.* Depression is very common in Parkinson's disease, and some clinicians suggest that increased control of motor function may unmask subclinical depression.

Currently it is not clear whether affective symptoms following DBS for Parkinson's disease are due to direct or indirect effects of DBS. However, because the targets of DBS stimulation used in Parkinson's disease are involved not only in motor control but also in mood regulation, a direct mood-modulating effect of DBS is extremely likely.

Putative treatment for neuropsychiatric disorders. In integrating findings from functional neuroimaging studies on treatment effects of different antidepressant modalities, depression can probably be best conceptualized to be a multidimensional, systems-level disorder affecting limbic-cortical pathways. The subgenual cingulate region (Brodmann area cg25) has received particular interest as a possible target since it plays an important role in acute stimulus-induced sadness, it is metabolically overactive in treatment-resistant depression, and clinical improvement of depression after pharmacotherapy with selective serotonin reuptake inhibitors (SSRIs) has been associated with pre- to posttreatment decreases in its regional metabolic activity. H. Mayberg et al. demonstrated that chronic high-frequency electrical stimulation with DBS to this area led to significant clinical response in four of the six study patients with sustained improvement through 6 months to the study endpoint.

Another promising approach for possible treatment of refractory (resistant to treatment) major depression with DBS is currently being investigated by an international study. The target of this investigation is the ventral portion of the anterior limb of the internal capsule and in the surrounding ventral striatum, an area containing converging connections of networks linking prefrontal cortex with subcortical, paralimbic, and limbic regions strongly implicated in the neurobiology of depression. DBS of this target in patients with obsessive-compulsive disorder led to significant elevation of mood in some patients.

Concerns in psychiatry. Although promising, the use of DBS for the putative treatment of major depression raises several concerns that are not associated or less problematic with other treatments being researched. First, there is a broad range of available treatments for chronic and severe major depression, and clinical experience shows that some patients even with severe and chronic forms of the disorder may finally respond after months or years of creative treatment. Second, DBS is associated with rare but severe risks, such as bleeding into the brain, infection related to the implants, or hemiparesis (muscle weakness on one side of the body). Third, as discussed

above, the use of neurosurgery for psychiatric disorders has a long but somewhat tainted history. The use of surgery on the brain—even when fully reversible and adjustable—in extremely desperate patients suffering from a potentially deadly mental disorder requires special ethical consideration. These reasons necessitate a careful multidisciplinary and systematic assessment of treatment refractoriness in the consideration of patient eligibility. The strict adherence to the inclusion criteria approved by an ethical board should be confirmed for each patient by a review board independent from the research group, or at least by an independent psychiatrist. Early on, consensus guidelines have to be established by psychiatrists, neurosurgeons, and ethicists together, continuously weighing potential risks and benefits of the procedure as new data become available.

Although the elegant work by H. Mayberg et al. can certainly be seen as a proof of the principle that reducing depression-associated hyperactivity in a certain brain region (cg25) is associated with symptomatic improvement, professionals are far from embracing DBS as a new clinical opportunity. Further studies should use controlled designs, explore different stimulation sites and parameters within known depression networks, and follow up patients for longer time periods to determine the necessary length of treatment and possible cessation effects. Following this path, we might not only find new treatment modalities for very severe, chronic, and treatment-refractory forms of major depression, but also learn more about the underlying neurobiology of one of the most disabling psychiatric disorders—knowledge which might ultimately lead to the development of treatment modalities with an even more favorable risk-benefit profile. In the meantime, DBS in the treatment of major depression will take place within the constraints of carefully planned clinical trials with the highest ethical standards possible.

Historically, neurosurgical interventions for psychiatric disorders were radically destructive crude interventions, indiscriminantely applied, often undertaken with the aim to better control difficult patients, and neither hypothesis-guided nor with the effects systematically assessed. The refinement of neurosurgical methods, the results from translational research on the pathways implicated in depression, the standard of clinical research in psychiatry, and the application of ethical research guidelines have changed this perspective completely. Neurosurgery might have the potential to help those patients with the severest forms of major depression.

For background information *see* AFFECTIVE DISORDERS; BRAIN; ELECTROENCEPHALOGRAPHY; MOTOR SYSTEMS; NERVOUS SYSTEM (VERTEBRATE); PARKINSON'S DISEASE in the McGraw-Hill Encyclopedia of Science & Technology.   Thomas E. Schlaepfer

Bibliography. J. L. Abelson et al., Deep brain stimulation for refractory obsessive-compulsive disorder, *Biol. Psychiat.*, 57(5):510–516, 2005; S. Breit, J. B. Schulz, and A. L. Benabid, Deep brain stimulation, *Cell Tissue Res.*, 318(1):275–288, 2004; B. D.

Greenberg et al., Neurosurgery for intractable obsessive-compulsive disorder and depression: Critical issues, *Neurosurg. Clin. N. Amer.*, 14(2):199–212, 2003; W. M. Grill, A. N. Snyder, and S. Miocinovic, Deep brain stimulation creates an informational lesion of the stimulated nucleus, *Neuroreport*, 15(7):1137–1140, 2004; P. E. Holtzheimer III and C. B. Nemeroff, Advances in the treatment of depression, *NeuroRx*, 3(1):42–56, 2006; K. Kumar, C. Toth, and R. K. Nath, Deep brain stimulation for intractable pain: A 15-year experience, *Neurosurgery*, 40(4):736–746, discussion 746–747, 1997; H. Mayberg et al., Deep brain stimulation for treatment-resistant depression, *Neuron*, 45(5):651–660, 2005; T. E. Schlaepfer and K. Lieb, Deep brain stimulation for treatment of refractory depression, *Lancet*, 366(9495):1420–1422, 2005; J. Schoenen et al., Hypothalamic stimulation in chronic cluster headache: A pilot study of efficacy and mode of action, *Brain*, 128(pt. 4):940–947, 2005.

Deep-sea siphonophore

One of the classic images of bioluminescence (light produced by a chemical reaction that originates in an organism) is that of an anglerfish, dangling its glowing lure. However, of all the luminous organisms in the sea, only a few are thought to use light in this way. Fewer still have been known to make red light. Recently, a siphonophore, specifically an angler jellyfish (**Fig. 1**), has been discovered which seems to do both: it uses small red lures to attract prey.

Definition. Siphonophores (**Fig. 2**) are animals in the phylum Cnidaria, a diverse group which includes the corals (Anthozoa) and the familiar "true" jellyfish (Scyphozoa). This phylum is characterized by the possession of stinging cells (nematocysts or cnidae). Siphonophores fall within the cnidarian class Hydro-

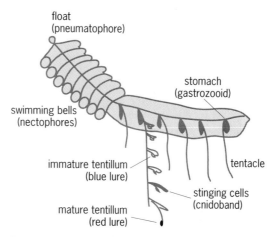

Fig. 2. Parts of a siphonophore. Tentacles and their side branches trail down alongside the feeding polyps. At the anterior end, swimming bells help the colony move and deploy the tentacles. In *Erenna*, the side branches develop sequentially along their length. The farthest end, which is also the oldest, has mature lures which contain a fluorescent red coating over their core of white bioluminescent material.

zoa, which includes hydroids, small anemone-like polyps that grow on rocks, as well as the small and transparent jellies called hydromedusae.

Siphonophores are not colonies of single-celled organisms, nor do they come together to form a colony. They are, however, considered colonial by some definitions (that is, they are composed of many physiologically integrated zooids, where the zooids are all attached to each other rather than living independently). It is easiest to consider them as a single super-organism, which grows by budding off specialized polyps and medusae. (Coloniality comes in when you consider the polyps as individuals.) Each type of polyp has a particular function, including feeding, buoyancy, propulsion, and reproduction. The Portuguese man-of-war is a familiar example of a siphonophore, but it is not a typical representative of the group as it lacks many of the polyp types. Most planktonic forms (living in the water column) have one or more swimming bells and grow in the form of a long chain.

Feeding. Of special interest here are the feeding polyps, known as gastrozooids. Siphonophores may have a dozen or more of these feeding buds, arrayed along the length of their central stem. In the genus *Erenna*, each gastrozooid has a tentacle which branches off it at the base. This tentacle, in turn, has side branches called tentilla, which hold the stinging cells.

Most siphonophores catch prey by putting out their long tentacles and waiting for something to bump into their tentilla. These "nets" can be very effective: in the longest species, which can be 40 m (130 ft), the tentacles can hang down 2 m (6.5 ft) sweeping an area of 80 m^2 (860 ft^2).

Not all siphonophores operate in this general way. Using deep-diving remotely operated vehicles from the Monterey Bay Aquarium Research Institute, the author and coworkers found a new species in the genus *Erenna*. Siphonophores like *Erenna*,

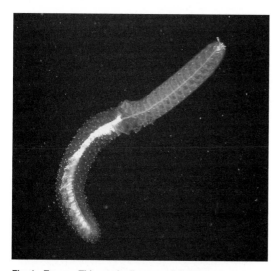

Fig. 1. *Erenna.* This newly discovered deep-sea siphonophore is about 45 cm (18 in.) long. The upper half of the colony consists of swimming bells that pulse like jellyfish to keep the colony moving through the water. The lower half carries dozens of pale white stinging tentacles, which are used to capture small deep-sea fishes.

especially the new species in question, do not seem to follow this same passive mode of operation. When seen from a submersible, this species holds its tentacles close to its body. On the face of it, this would not seem to be an efficient feeding strategy.

What makes this species more unusual is that it is known to feed only on fish, and it lives more than 1600 m (5250 ft) deep, where fish and other organisms are relatively scarce. Thus, without a net, how does it catch its food?

Stinging cells and lures. A clue is found in the arrangement and "behavior" of the tentacle side branches. These tentilla include a large battery of stinging cells attached to a central stalk. The battery of stinging cells, known as a cnidoband, can contain a thousand tiny poison-filled harpoon-like cells. Once triggered, the cnidoband fires off all its harpoons at once in a painful and paralyzing explosion (**Figs. 3** and **4**).

At the end of the tentillum that supports the cnidoband is a red bulb suspended by a flexible transparent stalk. The bulb is filled with white material and surrounded by a coating of fluorescent material. The white material is calcium-activated protoprotein, a bioluminescent material found in many Cnidaria. It would normally release blue-green light when stimulated. However, based on the fluorescence emission and excitation spectra of the surrounding red substance, blue light emitted at the center of the lure would be expected to excite the fluor and produce orange-red light.

This wavelength of light is very unusual in the ocean for two reasons. One is that it doesn't penetrate very far: meters or several centimeters rather than tens of meters. In addition, because red light

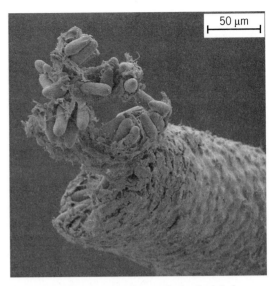

Fig. 4. Scanning electron micrograph of a disrupted tentillum shows the stinging cells (nematocysts) as individual capsules. These are normally packed into a dense white band, to which the lures are attached. When triggered, the nematocysts fire simultaneously to paralyze prey.

is rare, it is not surprising that the eyes of deep-sea organisms have only rarely been found to be sensitive to red light. Thus, long-wavelength light is not a signal that will be seen by many organisms, and not over very long distances.

The lures are not just capable of producing light, but the transparent stalk also contracts rapidly to flick the lure. This behavior is eerily similar to the motion of many deep-sea copepods. Often they are seen to do a "hop and sink" behavior, where they alternate between rapid jumps and slow sinking behavior. If a copepod-eating fish were able to see the motion of the lure, it would almost certainly be a tempting target.

Purpose of luminescence. For a jellyfish to have a lure is unusual enough; only anglerfish and a few squid have been suspected to use luminescence as a lure in the sea. However, because so few fish see red light, there is additional resistance to the idea of a jelly using long-wavelength light in its luminous lure. It would seem to be more effective to simply use blue light, rather than add the extra touch of fluorescence.

It may be useful to imagine this from the point of view of the deep-sea fish which are the potential prey. The most likely target species for this deep-living siphonophore are small (a few centimeters long) bristlemouth fish in the family Gonostomatidae. Because the deep sea is so vast—about 10 times larger than the next-largest habitable volume—the most common of these fish, *Cyclothone*, is probably the most common vertebrate on the entire planet. Despite its abundance, its visual sensitivities have never been measured. The eyes are extremely small and too fragile for many deep-sea trawling operations. Thus, we do not know what colors of light this fish might see. If any fish were to "slip through the cracks" of our scientific knowledge and have a

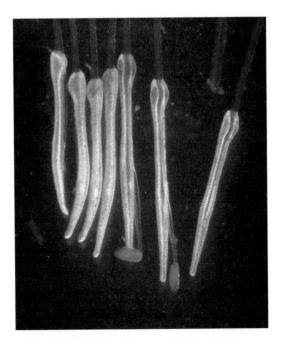

Fig. 3. This photograph shows a tentacle from *Erenna*, along with its tentilla. Each tentillum contains thousands of stinging cells. The red lures are on separate stalks, which contract rapidly, causing the lures to flick like swimming copepods (a typical food of small midwater fishes).

hidden ability to see red light, then *Cyclothone* would be a great candidate.

Even if we accept that this ability might go undetected the question of why a deep-sea fish would need to detect red light remains. One possibility is that it is using the ability to detect prey. Red fluorescent material, especially derived from plant chlorophyll, is common in the ocean, including the deep sea. Particles called marine snow, even at depths of more than 2000 m (6560 ft), will still produce red fluorescence when illuminated with blue light because of the algal cells they have accumulated. With a yellow filter in their eye, or visual pigments able to distinguish short- and long-wavelength light, a fish would see anything red as a conspicuous target over short distances. This strategy might be an effective way to find food for a tiny fish. Light to excite fluorescence would have to come from ambient blue light (at shallower depths) or from bioluminescence itself.

Conclusions. The use of a dim red lure in the deep sea is almost impossible to observe undisturbed. Any illumination would obviously affect the visual environment, and a low-light camera would be hard-pressed to find these rare siphonophores, let alone document a feeding event. Nonetheless, the most conservative interpretation of the evidence seems to be that this jelly dangles red lures to attract its prey. If more specimens can be obtained, there are many avenues for further study of this special function of bioluminescence.

For background information *see* ADAPTATION (BIOLOGY); BIOLUMINESCENCE; COELENTERATA; DEEP-SEA FAUNA; FLUORESCENCE; HYDROZOA; MARINE ECOLOGY; SIPHONOPHORA in the McGraw-Hill Encyclopedia of Science & Technology..

Steven H. D. Haddock

Bibliography. W. J. Broad, Cousin of the jellyfish spurs fresh theories on seeing red, *New York Times*, July 12, 2005; R. H. Douglas and J. C. Partridge, On the visual pigments of deep-sea fish, *J. Fish Biol.*, 50:68–85, 1997; R. H. Douglas, J. C. Partridge, and N. J. Marshall, The eyes of deep-sea fish, I: Lens pigmentation, tapeta and visual pigments, *Prog. Ret. Eye Res.*, 17:597–636, 1998; C. W. Dunn, P. R. Pugh, and S. H. D. Haddock, Molecular phylogenetics of the Siphonophora (Cnidaria), with implications for the evolution of functional specialization, *Systemat. Biol.*, 54:916–935, 2005; S. H. D. Haddock et al., Bioluminescent and red-fluorescent lures in a deep-sea siphonophore, *Science*, 309:263, 2005; P. R. Pugh, A review of the genus *Erenna* Bedot, 1904 (Siphonophora, Physonectae), *Bull. Nat. Hist. Mus. (Zoo. Ser.)*, 67:169–182, 1904.

Designing for and mitigating earthquakes

Earthquakes cause loss and suffering every year in large parts of the world. Most of this loss and suffering is due to lack of, or inadequate, structural design for the forces that earthquakes impose on buildings and other structures. Earthquakes impose

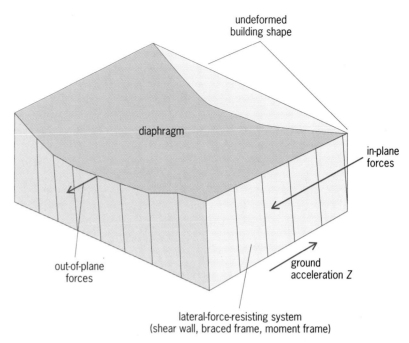

Fig. 1. Building deformation due to lateral seismic forces.

vertical and lateral forces on structures, due to the inertia of the structure. While vertical forces are significant, most structures have substantial capacity to resist vertical forces, since their primary structural design is for the force of gravity. Therefore, the primary problem for most structures is force in the horizontal, or lateral, direction, which tends to subject buildings to large horizontal distortion (**Fig. 1**). When these distortions get large, the damage can be catastrophic. Therefore, most buildings are designed with lateral-force-resisting systems (or seismic systems) to resist the effects of earthquake forces. In many cases, seismic systems make a building stiffer against horizontal forces and thus reduce the amount of relative lateral movement and consequently the damage. Seismic systems are usually designed to resist forces that result from horizontal ground motion, as well as from vertical ground motion.

When strong earthquake shaking occurs, a building is thrown mostly from side to side, as well as up and down. That is, while the ground is violently moving from side to side, taking the building foundation with it, the building structure tends to stay at rest, similar to a passenger standing on a bus that accelerates quickly. Once the building starts moving, it tends to continue in the same direction, but the ground moves back in the opposite direction (as if the bus driver first accelerated quickly, then suddenly braked). Thus, the building gets thrown back and forth by the motion of the ground, with some parts of the building lagging behind the foundation movement and then moving in the opposite direction. The force F that an upper floor level or roof level of the building should successfully resist is related to its mass m and its acceleration a, according to Newton's law, $F = ma$. The heavier the building, the more the force is exerted. Therefore, a tall, heavy,

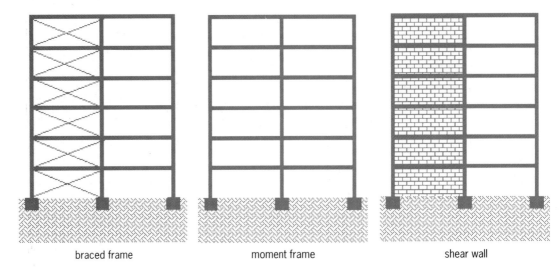

braced frame   moment frame   shear wall

Fig. 2. Earthquake lateral-force-resisting systems.

and reinforced concrete building will be subject to more force than a lightweight, one-story, wood-frame house, given the same acceleration. The combined action of seismic systems along the width and length of a building can typically resist earthquake motion from any direction. The earthquake-resisting systems in modern buildings take many forms (**Fig. 2**).

In moment-resisting steel frames, the connections between the beams and the columns are designed to resist the rotation of the column relative to the beam. Thus, the beam and the column work together and resist lateral movement and lateral displacement by bending. Steel frames sometimes include diagonal bracing configurations, such as single diagonal braces, cross bracing, and K-bracing, which resist forces through tension and compression in the braces. Steel buildings are sometimes constructed with moment-resistant frames in one direction and braced frames in the other.

In concrete structures, shear walls are sometimes used to provide lateral resistance in the plane of the wall, in addition to moment-resisting frames. Ideally, these shear walls are continuous walls extending from the foundation to the roof of the building. They can be exterior walls or interior walls. They are interconnected with the rest of the concrete frame and thus resist the horizontal motion of one floor relative to another. Shear walls are most often reinforced concrete walls but can also be reinforced masonry walls constructed of bricks or concrete blocks with steel reinforcing.

Damage can be structural and nonstructural, both of which can be hazardous to building occupants. Structural damage means degradation of the building's structural support systems (that is, vertical- and lateral-force-resisting systems), such as the building frames and walls. Nonstructural damage refers to any damage that does not affect the integrity of the structural support systems. Examples of nonstructural damage are chimneys collapsing, windows breaking, or ceilings falling. The type of damage is a complex issue that depends on the structural type and age of the building, its configuration, construction materi-

als, the site conditions, the proximity of the building to neighboring buildings, and the type of nonstructural elements.

Design. Traditionally, engineers have designed structures to resist earthquake forces, intending the strength of structural members to be equal to or greater than the seismic demand placed on them (Fig. 2). In earthquake analysis and design, a structure is commonly assumed to remains elastic (that is, the design neglects the effects of damage), although inelastic analyses which take into account material nonlinearity are increasingly being used in the practice. The procedures for earthquake design are required by building codes and are provided in the same building codes or related documentation. In the United States, the primary source for seismic design for buildings is ASCE 7 (2005), in the European Union the primary source is Euro code 8 (1998), and in Japan AIJ (1994) and PWRI (1998). An excellent source for codes for most countries is IAEE (2004), while IAEE (1986) is a guide for seismic design and construction of nonengineered buildings.

For ordinary design, the actual earthquake forces are reduced via a response modification factor, which varies depending on the assumed inherent ductility of the structure, based on its lateral-force-resisting-system's material (wood, steel, concrete, etc.) and system (moment frame, braced, shear wall, etc.). These reduced forces, with appropriate safety factors, are combined with gravity and other loads for the design of the overall structure. Pseudostatic or linear dynamic analytical methods are used to determine member forces for design. Pseudostatic methods are based on the structure's natural period (that is, first mode of vibration) and are appropriate only for simple structures. Linear dynamic methods account for the first and higher modes, which are usually more quickly performed in the frequency domain using techniques based on the fast Fourier transform. For larger and more important structures, nonlinear dynamic analyses are employed, typically in the time domain. In a nonlinear dynamic analysis, the structure is subjected to earthquake acceleration time histories (actual or synthesized records, scaled to match the site hazard), and member response into the inelastic range is taken into account, including P-Δ effects (that is, the increase in overturning moment due to the structure's weight P times its lateral deflection Δ). Dynamic analysis has become common due to the advent of more powerful computers and specialized software—ETABS, SAP2000, ANSYS, STAAD, and LARSA are some of the structural analysis packages more commonly employed today.

During the 1980s and 1990s, new approaches to seismic design emerged which involved modifying the structural response to reduce earthquake loads to more tolerable levels. These included base isolation, supplemental damping, and active control. Base isolation involves placing special components called isolators within the structure, which are relatively flexible in the lateral direction, yet can sustain the vertical load. However, the isolators are not always at the base, so that the technique is more properly termed structural isolation. When the earthquake causes ground motions beneath the structure, the isolators allow the structure to respond much more slowly than it would without them, resulting in lower seismic demand on the structure. Isolators may be laminated steel and high quality rubber pads, sometimes incorporating lead or other energy-absorbing materials (**Fig. 3**), or parabolic dish-shaped base plates which rely on the structure's own weight trying to "climb" the sloping sides of the "dish" to counteract the lateral force of the earthquake. Other methods include supplemental damping and active control. The most recent development in earthquake engineering is performance-based design, in which the expected structural damage due to the maximum expected earthquake is quantified. If the expected damage is unacceptably high, the owner and the engineer may agree on a design in excess of the building code requirements.

Mitigation. Mitigating earthquakes is a much broader field than simply designing for earthquakes. Mitigation is a several-stage planning and management process which involves (1) estimating the potential losses that earthquakes might cause; (2) deciding if the potential losses are acceptable or not; (3) if not, examining alternative loss reduction techniques, which involves identifying the effectiveness of each alternative in reducing losses and the associated cost; (4) setting some criteria for deciding which alternative is the most effective, such as benefit-cost, lives saved, least regret, or other paradigms, and applying the criteria to select the most effective package of alternatives; (5) developing a design and implementation program for the package of alternatives; and (6) implementing the program. A crucial and often difficult aspect of mitigation is finding the political will and resources to examine the risk of earthquakes and then to implement the program. Mitigation alternatives can be broadly grouped into four categories: structural, locational, operational, and risk transfer. Structural mitigation generally involves resisting or avoiding earthquake forces via hardware solutions. Locational mitigation typically

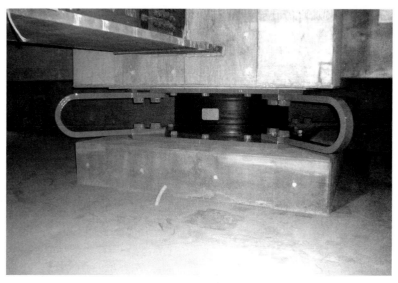

Fig. 3. Rubber isolator under concrete-encased column (note steel girder framing-in on the upper left). The "paper clip" steel bars are hysteretic dampers. As the building above the rubber isolator moves under earthquake motions, the steel bars flex, absorbing energy. (*Courtesy of C. Scawthorn*)

avoids earthquake effects via alternative land uses. Operational mitigation refers to emergency planning and related measures that respond to earthquake effects to reduce the impacts to acceptable levels. Risk transfer implies not reducing the loss in an absolute sense but only in a relative sense, where the loss is shared with others, typically via insurance but sometimes in other ways to reduce the loss to an acceptable level.

For background information *see* ARCHITECTURAL ENGINEERING; BUILDINGS; EARTHQUAKE; LOADS, DYNAMIC; SEISMOLOGY; STRUCTURAL ANALYSIS; STRUCTURAL DESIGN; STRUCTURE (ENGINEERING) in the McGraw-Hill Encyclopedia of Science & Technology.

Charles Scawthorn

Bibliography. *AIJ (1994) Design Guidelines for Earthquake Resistant Reinforced Concrete Buildings Based on Ultimate Strength Concept*, Architectural Institute of Japan, Tokyo, 1994; *ASCE 7-05 (2006) Minimum Design Loads for Buildings and Other Structures*, ASCE/SEI Standard 7-05, American Society of Civil Engineers, Reston, 2006; B. A. Bolt, *Earthquakes*, W. H. Freeman, San Francisco, 1993; W. F. Chen and C. Scawthorn (eds.), *Earthquake Engineering Handbook*, CRC Press, 2002; A. K. Chopra, *Dynamics of Structures*, Prentice Hall, 1995; *Eurocode 8 (1998)* [Part 1 covers general rules, seismic actions, and rules for buildings, Part 2 bridges, Part 3 the strengthening and repair of buildings, Part 4 silos, tanks and pipelines, Part 5 foundations, retaining structures, and pipelines, Part 6 towers, masts and chimneys]; *IAEE Guidelines for Earthquake Resistant Non-Engineered Construction*, International Association of Earthquake Engineering, Tokyo, 1986; *IAEE (2004) Earthquake Resistant Regulations: A World List*, International Association of Earthquake Engineering, Tokyo, 1992; S. L. Kramer, *Geotechnical Earthquake Engineering*, Prentice Hall, 1995; F. Naiem,

Seismic Design Handbook, 2d ed., Springer, 2001; *PWRI (1998) Design Specifications of Highway Bridges—Part V: Seismic Design*, Earthquake Engineering Division, Earthquake Disaster Prevention Research Center, PWRI, Tsukuba-shi, Japan; C. R. Scawthorn, J. M. Eidinger, and A. J. Schiff (eds.), *Fire Following Earthquake*, TCLEE Monogr. 26, American Society of Civil Engineers, Reston, 2005.

Digital geological mapping

Geological maps provide a summary of the distribution, character, age, and geological history of the rocks that underlie all land areas. As interpretive documents rather than simply presentations of data, they are essential tools for mineral and energy resource exploration, land-use planning and urban development, and assessing geological hazards such as landslides and earthquakes. Geological maps are an important information resource in modern society. Accordingly, their creation and interpretation are fundamental components of geoscience education programs worldwide, and numerous textbooks are available that provide introductions to these activities.

Today, the digital mapping process captures more data more effectively. It presents geological maps in ways that can be more effectively queried and analyzed, both individually and together with other geological, environmental, and cultural datasets. In addition, they can be readily accessed on the Internet. Consequently, geological maps are more relevant and useful to more people in more fields than ever before.

Analog mapping. Until recently, geological fieldwork was done manually (**Fig. 1**). The geologist and his or her assistant walked a traverse determined previously by examining aerial photographs to identify where the bedrock outcrops were most plentiful, and the terrain was safe to navigate. Using the aerial photos and a detailed topographic map, the field crew followed a preset course, navigating by dead reckoning or by pace and compass. The outcrop localities were sketched onto a field map and annotated. Then, field notes were recorded in a notebook, identifying the basic, critical aspects of the rock types present and the orientations of any characteristic features such as sedimentary bedding or compositional layering, deformation fabrics, major fractures, and so on. The traces of important regional-scale features, such as boundaries between rock formations (geological contacts) and structural trends, were also sketched on the field map. Additional notes might include interpretations and hypotheses to help focus report writing back in the office. Also, sketches were made and photos taken to illustrate critical relationships. As the field program progressed and more traverses documented the positions and orientations of the mappable features, a map of the project area was developed.

After completion of the field program, the geological interpretations were reviewed, the geological contacts—inferred to extend through areas between the outcrops—were revised, as necessary, with the help of the aerial photos and other information sources such as geophysical or geochemical data, if available. Fossil collections were sent for expert identification to guide the geologist's interpretation of the rock units. Finally, the field maps were redrawn more rigorously. After critical reviews identified and corrected errors, the resulting manuscript map was submitted for drafting by specially trained technicians, who created a professional product by scribing the line work onto a specially coated film, which was then photographically reproduced. After checking for errors, color separations were made and

Fig. 1. Analog data capture, 1976. (*Photo courtesy of J. F. Psutka*)

the product was offset-printed. Altogether, the compiling, revising, checking, scribing, correcting, and printing took months to complete.

Digital map making. Over the past few years, digital processes have changed all aspects of the map-making process. These changes have been driven by interest in geographic information system (GIS) technologies, the development of new technologies that have made GIS work flows more cost-effective and user-friendly, increased demand for digital formats, the advantages of faster production, and especially the Internet, which has created unprecedented access to digital data.

The digital revolution began slowly. As early as the 1970s, field notes were captured on forms designed for easy retyping into databases that would facilitate the selection, analysis, and evaluation of a suite of geological properties such as rock types and diagnostic fossils. By the late 1980s, relational databases were being linked to graphics software so that selected properties could be readily displayed according to their spatial distribution, creating an integrated digital mapping and display system. By the early 1990s, such integrated systems were still uncommon; however, map production had been largely transformed into a digital process. Thus, manuscript maps were digitized; polygons, lines, and points were created using computer-assisted drafting software; and the finished products were plotted on-demand in time frames of days instead of months.

In the past 5 years, advances in handheld computer technology, development of new mapping software suited to these products, and the increasing numbers of computer-literate professionals in the field have revolutionized the data collection process. Today, the mapping process from start to finish is largely digital. Starting at the outcrop, locality information is collected using a global positioning system (GPS) receiver in conjunction with topographic maps and aerial photos. Field notes are largely captured digitally and organized in a database in the field. To accomplish this, some geologists use a handheld computer to enter the most basic data using pick lists; others use printed data-entry forms, which are processed in the base camp. The interpreted positions of large-scale features, such as geological contacts, faults, and folds, may also be digitally captured on the outcrop, although many mappers still use paper maps and notebooks as well. By the time the fieldwork is completed and the crew has returned to the office, the database is already populated and a first draft of the map is largely complete.

Data organization. The most important aspect of this digital revolution is not how the data are collected and processed or how quickly the information is published, but how digital mapping has forced a fundamental rethinking of how data are organized, presented, and interpreted. In the analog process, little thought was given to data organization, because far more data were collected than could be adequately displayed on a map and "searchability" was a purely visual process. Maps were two-dimensional representations (on one layer)

of a three-dimensional ground surface. The differences between lines, points, and polygons were implicitly understood but were not particularly relevant.

In analog mapping, the principal issue for the mapper was the level of certainty. The most certain information, and therefore the most valuable and durable, was what could be carefully recorded from direct observation at the outcrop. Everything else was interpretive to a greater or lesser degree. Thus, geological contacts, or structural trends, were estimated (interpolated) between outcrops or beneath bogs. Even the identification of map units was interpretive. The level of certainty was, and is, symbolized on maps by distinctive line styles. Typically, solid lines represent a high level of certainty, with dashed and dotted lines representing decreasing levels of certainty.

In digital mapping, even before any data are collected, organizing the data is the critical first step in the digital mapping process, including how the data are intended to be used, which data are the most critical, and which properties are dependent on others. These and many similar issues must be addressed in order to develop a workable database. As GIS products, maps can be uploaded to an institutional database used in concert with other geological datasets, as well as many other types of data. This has hugely increased the utility of geological maps for a broad spectrum of applications, including traditional uses such as resource exploration, land-use planning, landslide hazards, and construction aggregates, as well as for an increasing number of applications that were previously not widespread such as vegetation mapping, wildlife habitat, environmental impact assessments, and tourism. However, if many datasets are to be used together, it is important to ensure that the definitions, data structure, and attributes are applied consistently from one dataset to the next. In any natural science and particularly in geology, most definitions are nuanced by wide natural variability. Consequently, considerable effort has been taken to develop consistent definitions and database structures. This ensures that a query applied to an amalgamated dataset, derived from several individual datasets, will capture all of the relevant information.

Data display. Although paper maps represent all of the information (that will fit legibly) on a single layer, this is not the most efficient way to present data in a GIS. Primarily, this is because GIS software provides valuable computational and analytical functions that are most effective if the data are appropriately organized. Accordingly, once the data are collected and satisfactorily organized within a database, an appropriate method of graphically organizing and displaying them must be devised in the GIS. First, the data are divided into geometric types. At any given scale of inquiry, points are all the features that can be symbolized by a dot, star, or asterisk, and lines are features symbolized by lines of various styles (such as faults and geological contacts). Polygons represent features that enclose an area on the map. Rasters (bitmaps) are images that cannot be edited, but are commonly very useful for reference and analysis when other

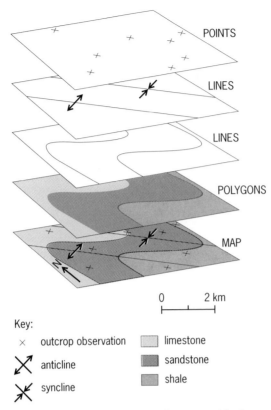

POINTS

LINES

LINES

POLYGONS

MAP

0   2 km

Key:

× outcrop observation

⤢ anticline

⤦ syncline

limestone

sandstone

shale

Fig. 2. Schematic representation of how geometric classes (points, lines, and polygons) are separated into multiple independent themes in GIS products, greatly simplifying data management.

data are overlaid. Examples include satellite or aerial photography, and geophysical features such as gravity anomalies or geothermal (earth temperature) gradients. Each geometric type is represented by one or more layers in the GIS. Thus, a GIS dataset presents information in many layers (themes), which are characterized by distinct attributes (**Fig. 2**). The number and content of the layers are discretionary and dependent on many factors, such as the variety and complexity of the data, maximizing the efficiency of periodic updates, the potential for extracting and reusing data subsets in other contexts, and integration with other datasets. Finally, because the GIS contains multiple layers and the database contains a detailed description of the attributes for every point, line, and polygon on the map, a digital GIS product contains far more of the collected data than a simple paper map possibly could.

For background information *see* AERIAL PHOTOGRAPHY; CARTOGRAPHY; COMPUTER GRAPHICS; DATABASE MANAGEMENT SYSTEM; GEOGRAPHIC INFORMATION SYSTEMS; INTERNET; MAP DESIGN; MAP REPRODUCTION; SATELLITE NAVIGATION SYSTEMS; TOPOGRAPHIC SURVEYING AND MAPPING; VEGETATION AND ECOSYSTEM MAPPING in the McGraw-Hill Encyclopedia of Science & Technology.   Larry S. Lane

Bibliography. J. W. Barnes and R. J. Lisle, *Basic Geological Mapping*, 4th ed., Wiley, 2004; B. Brodaric, Field data capture and manipulation using GSC Fieldlog v3.0, in *Proceedings of a Workshop on Digital Mapping Techniques: Methods for Geologic Map Data Capture, Management, and Publication*, U.S. Geological Survey Open-file Report 97–269, 1997; D. Viljoen, Topological and thematic layering of geological map information: Improving efficiency of digital data capture and management, in *Proceedings of a Workshop on Digital Mapping Techniques: Methods for Geologic Map Data Capture, Management, and Publication*, U.S. Geological Survey Open-file Report 97–269, 1997.

DNA repair

Every cell in our body contains genetic information for its proper functioning. This information is present in the long stretch of nucleotides—adenosine (A), cytosine (C), guanosine (G), and thymidine (T)—that form our DNA (deoxyribonucleic acid). Several hundreds to thousands of these A, C, G, and T nucleotides contain the information for one gene, and our DNA contains more than 30,000 genes. Each of these genes has to be intact for proper functioning. However, the integrity of our DNA is threatened by many different agents, such as sunlight, chemicals, and even the oxygen in the air. If damaged DNA were left unrepaired, the cells of our body would quickly lose essential gene functions and life would not be possible. However, all living organisms contain a multitude of enzymes that counteract such damage and restore the proper DNA sequence. Failure of these DNA repair functions can lead to various diseases, such as cancer and developmental disorders. In recent years, considerable progress has been made in understanding the various DNA repair pathways. We are now on the verge of an era in which this knowledge can be exploited to develop more effective and specific anticancer drugs.

Pathways. The large variety of possible DNA lesions necessitates a number of DNA repair pathways (see **table**). DNA lesions can be divided into two categories: one type of damage affects only one strand of the DNA double helix, and the other type affects both strands.

Damaged nucleotides in one strand of the DNA are fixed by either base excision repair or nucleotide excision repair. These complicated repair pathways recognize the affected nucleotide, cut it out of the DNA, and restore the original sequence, making use of the intact strand of the DNA double helix.

DNA mismatch repair functions in normal DNA replication, the process that duplicates the cellular DNA before cell division takes place. DNA replication is quite accurate, but the enzyme that carries out DNA synthesis makes a mistake once in every 10 million nucleotides. Mismatch repair reduces these mistakes (mutations) a hundredfold. This is quite important since mutations can cause cancer or hereditary diseases.

DNA double-strand breaks can be mended by two different repair pathways. One of these is homologous recombination, which functions mainly during DNA replication. This repair pathway makes use of the information in the newly synthesized intact DNA

Overview of DNA damage, repair pathways, and related diseases		
Type of damage	Repair pathway	Disease connected to repair defect
Oxidative damage; single-strand breaks	Base excision repair	Neurodegenerative diseases
UV damage; (bulky) chemicals	Nucleotide excision repair	Xeroderma pigmentosum; Cockayne's syndrome; trichothyodystrophy
Mismatches	Mismatch repair	Hereditary non-polyposis colon cancer
Double-strand breaks	Nonhomologous end joining	Severe combined immunodeficiency
Double-strand breaks	Homologous recombination	Hereditary breast cancer
Interstrand cross-links	Fanconi's anemia pathway; homologous recombination; nucleotide excision repair	Fanconi's anemia

strand to restore the broken strand. The other repair pathway is nonhomologous end joining, which functions in all cells. It repairs the break without using a template and can therefore result in mutations at the site of the original break. It is extremely important to repair double-strand breaks, since large pieces of chromosomes, containing a large number of genes, may be lost if these lesions are left unrepaired.

DNA interstrand cross-links are particularly harmful, since they hamper essential cellular processes. Although the repair mechanisms that handle these lesions are only partially understood, several cross-link repair genes have been identified over the past few years.

Repair defects and disease. Defects in DNA repair may give rise to a variety of cancer predisposition syndromes and developmental disorders (see table). Classical examples of such cancer predisposition syndromes are xeroderma pigmentosum and hereditary non-polyposis colon cancer (HNPCC). Xeroderma pigmentosum is characterized by a high incidence of ultraviolet light–induced skin cancers and is caused by defects in the nucleotide excision repair machinery. HNPCC patients harbor a mutated mismatch repair gene, which may cause colon tumors. Various other cancer predisposition syndromes have now been characterized, notably hereditary breast cancer (Brca) caused by mutations in the Brca1 or Brca2 genes. Normal cells from HNPCC and hereditary breast cancer patients carry one intact and one nonfunctional copy of the gene. In the tumor, the good copy is lost, resulting in cells that have reduced mismatch repair (HNPCC) or double-strand break repair (Brca1 or Brca2). These cells accumulate additional changes in their DNA much faster than normal cells, which increases their risk of turning into a cancer cell.

DNA repair defects can also lead to various developmental disorders. For example, a defective immune system may be the result of the inability to carry out nonhomologous end joining. This repair pathway is required for formation of the genes that recognize invading microorganisms, which is necessary for combating infections. Another example of a developmental disorder is Fanconi's anemia, which is associated with a defect in removal of interstrand cross-links from the DNA. In addition to the common feature of a reduced capacity to make blood cells, these patients can show a variety of other defects, such as missing fingers or toes. These abnormalities

may be explained by the loss of certain progenitor cells in early development as a consequence of their reduced repair capacity.

Another common complication of repair dysfunction is neurodegenerative disease. Nucleotide excision repair defects can cause mild to very severe neurodegeneration. Defects in repair of single-strand or double-strand breaks can also result in neurological problems, ranging from microcephaly (the condition of having an abnormally small head) to progressive ataxia (lack of muscular coordination). This shows that neurons are particularly sensitive to persisting DNA lesions, possibly because of the limited regenerative capacity of neuronal tissue.

Neurodegeneration is the most obvious sign of premature aging in patients suffering from various DNA repair syndromes. However, it is not the only link between reduced repair capacity and aging. The generation of mice lacking one or more DNA repair systems has contributed greatly to our understanding of the link between DNA damage, aging, and cancer. Some of the more severe DNA repair defects, such as the deficiency of the nucleotide excision repair protein ERCC1, recapitulate many aspects of normal aging at a highly accelerated pace, suggesting that normal aging may at least partially be explained by accumulation of mutations.

Thus, reduced DNA repair capacity results in accumulation of unrepaired DNA lesions or mutations. This may give rise to cell death, which is probably the major cause of aging, or to activation of oncogenes and inactivation of tumor suppressor genes, which can lead to cancer. The balance between cell death and mutation accumulation has to be maintained to balance the rate of aging and the chance to develop cancer.

Interaction with other cellular processes. DNA damage not only triggers DNA repair but also regulates other cellular processes, such as cell cycle progression and programmed cell death (**Fig. 1**). Cell cycle checkpoints prevent cells with damaged DNA from going into the next cell cycle phase. Cells accumulate much higher levels of mutations or chromosomal aberrations in the absence of cell cycle checkpoints, which may lead to cancer or cell death. Multicellular organisms may benefit from killing a (potentially dangerous) mutated cell, and heavily damaged cells can initiate programmed cell death (apoptosis). Damaged cells can somehow measure the level of genomic injury and then decide whether they should

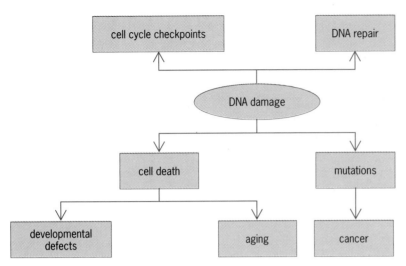

Fig. 1. DNA damage and its possible consequences.

die. Little is known about this decision point, but it is clearly important for cells to balance the risk of cancer formation and premature aging.

Another example of such a balancing act between cancer and aging is UV-induced DNA damage. This DNA lesion can be repaired by nucleotide excision repair, but during DNA replication the DNA polymerase sometimes runs into a lesion before repair has taken place. In that case, specialized DNA polymerases can take over the job and incorporate nucleotides across the damaged site. This translesion synthesis process contributes to cell survival at the expense of increased mutation levels.

Repair defects in tumor cells: use for anticancer therapy. Most tumor cells have acquired a defect in their DNA damage response. This observation indicates that rational tumor therapies that target these defects hold great promise. The first example of such a rational treatment has been described recently for hereditary breast cancer that is caused by mutations in the Brca1 or Brca2 genes. Cells carrying a mutation in either of these genes show defects in DNA double-strand break repair by homologous recombination. Therapeutic approaches make use of the fact that the tumor cells carry only the mutated copy, whereas cells in the surrounding tissues carry one mutated and one functional copy of the gene. Thus, only tumor cells have a repair defect. Replication through

a region that contains a single-strand break will result in a loose DNA end that needs homologous recombination for its repair (**Fig. 2**). Therefore, a treatment that would cause single-strand breaks to persist longer should be quite deleterious to Brca1- or Brca2-deficient (tumor) cells. Indeed, Brca-deficient cells were found to be extremely sensitive to inhibition of an enzyme that is required for efficient single-strand break repair, poly-ADP-ribose polymerase (PARP). The first clinical trials have started to investigate the possible use of PARP inhibitors for breast cancer treatment.

As outlined above, most tumor cells have acquired DNA repair defects. In principle, these defects may be targeted in a way that is similar to the PARP inhibitor treatment of Brca-deficient breast cancer cells. In most cases the surrounding cells will not have the same repair defects as the tumor cells, and therefore they should be much more resistant to the treatment. It is to be expected that novel cancer treatment strategies will be developed when more detailed knowledge of the various DNA repair pathways and their interactions with other cellular processes is acquired. Such treatments should be much more specific for tumor cells than conventional chemotherapy, and are therefore expected to have much less severe side effects on healthy cells. Genomics and proteomics approaches will greatly facilitate the detailed molecular diagnosis of the exact defects in tumors, which raises hopes for truly tailor-made medication.

For background information *see* AGING; CANCER (MEDICINE); DEOXYRIBONUCLEIC ACID (DNA); GENETICS; MUTATION; NUCLEOTIDE; TUMOR in the McGraw-Hill Encyclopedia of Science & Technology.

Dik C. van Gent

Bibliography. J. H. J. Hoeijmakers, Genome maintenance mechanisms for preventing cancer, *Nature*, 411:366–374, 2001; S. Madhusudan and M. R. Middleton, The emerging role of DNA repair proteins as predictive, prognostic and therapeutic targets in cancer, *Cancer Treat. Rev.*, 31:603–617, 2005; J. R. Mitchell, J. H. J. Hoeijmakers, and L. J. Niederhofer, Divide and conquer: Nucleotide excision repair battles cancer and ageing, *Curr. Opin. Cell Biol.*, 15:232–240, 2003; M. O'Driscoll and P. Jeggo, The role of double-strand break repair—insights from human genetics, *Nat. Rev. Gene.*, 7:45–54, 2006; D. C. van Gent, J. H. J. Hoeijmakers, and R. Kanaar, Chromosomal stability and the DNA double-stranded break connection, *Nat. Rev. Genet.*, 2:196–206, 2001.

Earliest tools

In the Late Pliocene (2.6 million years ago), during the transition of the apelike australopithecines to the earliest representatives of the genus *Homo*, one of the most significant elements in the human evolutionary process appeared: the use of stone tools. The earliest archeologically documented technology consisted of unmodified cobblestones (hammerstones),

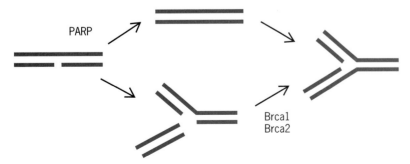

Fig. 2. Consequences of single-strand breaks in DNA when a DNA polymerase duplicates the DNA. Inhibition of poly-ADP-ribose polymerase (PARP) prevents efficient repair of the single-strand break, which can then be converted into a double-strand break. Repair of this DNA lesion requires Brca1 and Brca2.

Fig. 1. Stone tools from the oldest archeological site in the world: OGS7 at Gona (Ethiopia). (*a*) Refitting of flakes to a core. There are a variety of artifacts, including (*b*) a chopper and (*c, d*) cores in which centripetal flaking is documented for the first time. (*Courtesy of S. Semaw*)

modified cobblestones (cores), and the pieces detached from them (variously called debitage, flakes, and debris). Cobblestones were flaked in various ways. Flaking occurred along one edge (resulting in choppers and discoids) or along various edges (resulting in polyhedrons and subspheroids) [**Fig. 1**]. The pieces detached from these cobbles usually consist of fragments with one striking platform (that is, area of impact) and a sharp-edge perimeter (**Fig. 2**). This industry was called Oldowan based on the eponymous site in Olduvai Gorge (Tanzania), dating to 1.8 million years ago, where it was first described by Mary Leakey. However, the oldest evidence of stone tools can be found in Gona in the region of Afar (Ethiopia), dating to 2.6 million years ago.

Some cores and flakes were used as tools. In the beginning, as can be observed at Gona, most of the cobbles were used as cores to produce flakes. However, some were also initially used as hammerstones, as suggested by pitting on the end opposite the flaked area. It was previously assumed that stone tools were developed for animal butchery, but most archeological sites earlier than 2 million years old lacked any evidence of this. Recently, the discovery of cut-marked bones at Gona and in the nearby area of Bouri in Ethiopia has yielded evidence of stone tools used for butchery as early as 2.6 million years ago. In Bouri, no stone tools were found in association with the cut-marked bones, and sources of raw material are several kilometers away. This suggests that hominids

were transporting stone tools over much longer distances than reported for chimpanzees, which also use and transport stone tools. Planning, therefore, must also have been more advanced than is seen in chimpanzees.

Technology. Technological analysis of the earliest stone tools suggests that Pliocene hominids differed from chimpanzees in planning and flaking skills. Therefore, they represent the earliest documented evidence of behavioral complexity beyond that observed in extant apes in the human evolutionary process. In order to obtain a usable flake, it is necessary to select a point in the striking platform neither too close nor too far from the edge of the core, so that angular fragments, usually devoid of sharp edges, are not produced. The hammerstone must hit the striking platform with a certain force and at a certain angle, which varies with the size of the core and the hammerstone. Experiments with chimpanzees suggest that this degree of precision is highly unusual. Most of their flaking takes place in the form of multiple strikes on the core surface, usually near the core's edge. This frequently results in small and useless debris fragments or in core fragmentation. Once chimpanzees obtain a flake, they do not rotate the core to produce a new one; rather, they tend to abandon the core or to continue flaking a different part of it.

Pliocene hominids were more sophisticated than their chimpanzee ancestors. They selected raw

Fig. 2. Functional flakes (that is, large detached pieces with sharp edges) from OGS7 at Gona (Ethiopia). (*Courtesy of S. Semaw*)

materials according to the flaking and functional properties. Most of the Gona archeological sites were created a few meters away from the source of raw material, in contrast with the Bouri cut-marked bones. Several artifact assemblages from Gona were compared with random cobble samples taken from associated conglomerates (natural occurrences of cobblestones) that record raw material availability in the region at the time of occupation. The results prove that hominids were selecting specific fine-grained materials over others. Some of these fine-grained materials have a limited representation in the conglomerates. The selection of raw materials together with their regular transport over long distances clearly differentiates Pliocene hominids from chimpanzees.

Hominids also had a good understanding of the principles of conchoidal fracture. An example of conchoidal fracture is a bullet hitting a window or a drop of rain landing in a puddle: energy moves away from the epicenter of the strike in concentric waves. When a hammerstone strikes a core, these waves are often visible on both the flake and its negative scar. Very few failed strikes have been recorded at these early sites. Most blows were aimed at the correct striking platforms to detach functional flakes. In addition, hominids rotated the core to exploit the negative scars of previous flakes, thereby sequentially flaking the core. As noted above, this has not been documented in chimpanzees.

Functionality. There is evidence that the first stone tools were used for butchering animal carcasses. However, this does not imply that animal butchery was their main function. Among all sites dated to more than 2 million years old, less than two dozen cut-marked bones have been found. In most of these early sites, bones are not present at all; in these sites, pitting on cores suggests that pounding activities (for example, of plant foods) might have taken place. This is suggested by a recent analysis of the 1.9–1.3-million-year-old sites at Olduvai Gorge, where the most widely represented activity documented is heavy-duty pounding not related to carcass exploitation.

As important as the emergence of stone tools is the appearance of the sites at which they are found. While chimpanzees move linearly from one eating place to the next, hominids began to select specific loci (central places) in the landscape to which they would bring raw material and transform it into tools. Some raw materials and tools were further transported away from these spots. Sometimes, hominids also brought bones to the same sites. This selection of specific places in which certain activities are performed, and radial movement to and from these locations to points on the landscape, is still seen among living hunter-gatherers.

Original tool makers. Scholars still debate which hominid ancestor is responsible for creating the earliest stone tools. At Bouri, the cut-marked bones are associated with *Australopithecus garhi*. However, the brain size of this hominid is the same as that of other australopithecines and is not much larger than that of apes. There is no indication in the fossil record that this hominid had neurophysiological skills different from those of earlier apes. Fragmentary remains of early *Homo* dating to 2.4–2.3 million years ago have been retrieved at Hadar (Ethiopia) and West Turkana (Kenya). Until the fossil record is more fully reconstructed, it is difficult to assign the earliest stone tools to a specific hominid. However, considering that these stone tools are part of a new behavioral repertoire that included creation of central places, a substantial amount of planning, and the beginning of animal butchery, stone tools should be associated with a hominid that shows significant physiological differences from earlier hominids, as can be clearly observed in the genus *Homo* 2 million years ago, and that probably emerged earlier.

Bone tools. Making stone tools has been discussed as the earliest technology; bone tools appear later in the archeological record. Claims of early bone tools no older than 1.8 million years have been made at Sterkfontein (Member 5) and Swartkrans (Members 1–3) in South Africa, based on the type of polishing and abrasion shown on the ends of these bones. It was first argued that this polishing and abrasion resulted from digging tubers and later digging into termite mounds, suggesting an early presence of insectivory in the hominid diet. However, these claims should be viewed cautiously for the following reasons: (1) In most of the fossil contexts where these "tools" have been found, there is either lack of associated evidence for human activity (that is, stone tools or modified bone) or the evidence is very minimal. (2) A recent analysis of Swartkrans has shown that coarse-grained sediment compression and movement has created a large number of naturally abraded bone specimens. (3) Several of these purported tools are too small to be successfully manipulated in any useful way.

Conclusions. Archeologists have reconstructed how the earliest stone tools were created. They have also discovered that sometimes these tools were used for butchering animals. The location of the cut and percussion marks created on bones by the use of stone flakes has allowed the identification of several butchery activities: skinning, dismembering, filleting, and demarrowing. The distribution of cut marks created through defleshing has also been used to suggest that as early as 2.6 million years ago hominids might have had primary access to fleshed carcasses (in contrast to secondary access, that is, scavenging).

However, there is still a long way to go toward understanding the range of activities for which stone tools might have been used. We do not yet know if animal butchery at that time was an important or marginal part of hominid activities, that is, the extent to which meat formed part of the diet. Finally, the behavioral meaning of most early archeological sites where these stones tools are found continues to elude us.

For background information *see* EARLY MODERN HUMANS; FOSSIL HUMANS; NEOLITHIC; PALEOLITHIC; PREHISTORIC TECHNOLOGY in the McGraw-Hill Encyclopedia of Science & Technology.

M. Domínguez-Rodrigo

Bibliography. A. Delagnes and H. Roche, Late Pliocene hominid knapping skills: The case of Lokalalei 2C, West Turkana, Kenya, *J. Human Evol.*, 48:435–472, 2005; T. Plummer, Flaked stones and old bones: Biological and cultural evolution at the dawn of technology, *Yearbook of Physical Anthropology*, 47:177–204, 2004; K. Schick and N. Toth, *Making Silent Stones Speak: Human Evolution and the Dawn of Technology*, Simon & Schuster, New York, 1993; S. Semaw, The world's oldest stone artifacts from Gona, Ethiopia: Their implications for understanding stone technology and patterns of human evolution between 2.6-2.5 million years ago, *J. Archaeol. Sci.*, 27:1197-1214, 2000; S. Semaw et al., 2.5 million-year-old stone tools from Gona, Ethiopia, *Nature*, 385:333-338, 1997.

Ecosystem valuation

Food, water, shelter, fuel, transportation, recreation, scenic amenities, waste recycling, biodiversity, and climate stabilization are just a few examples of the goods and services that are supported by ecosystems (that is, functional systems that include the organisms of natural communities together with their environment). Human life would be impossible without nature's services. However, at the scale of individual ecosystems, people make decisions every day that affect the health and sustainability of ecosystems and that reflect the people's values for ecosystems and their services.

Some decisions are clearly connected to their effects on ecosystems—for example, the decision to fill a wetland for construction. However, most decisions are not clearly connected to their effects on ecosystems. Furthermore, many seemingly small choices made by individuals can add up to huge cumulative and possibly catastrophic changes to the ecosystems that provide human life support. Choices that affect ecosystems are complicated by the fact that prices do not reflect the full social costs or benefits. Because markets do not necessarily provide adequate incentives for maintaining healthy and sustainable ecosystems, collective actions are required.

Conservation and management of ecosystems invariably require difficult trade-offs. Such trade-offs are always based on human values, whether explicitly or implicitly stated. Using the methods of ecosystem valuation, economists attempt to explicitly measure and present people's values for the services of ecosystems.

Market failures and ecosystems. Because markets fail to provide adequate incentives, people's choices do not generally account for their effects on the health of ecosystems. Three causes of market failure are particularly relevant to ecosystems and their services—public goods, externalities, and open-access resources.

Many ecosystem services, such as a pleasing view, are public goods that can be enjoyed by any number of people without affecting other people's use or enjoyment. Although people value public goods, no individual has an incentive to protect or maintain these goods.

Numerous human actions result in unintended side effects, or externalities, that affect ecosystems. For example, industrial facilities may discharge heated or polluted water into an adjacent water body, altering the aquatic or marine habitat and affecting fish populations. If commercial fishers or recreational anglers then catch less of the fish they seek, the industrial facility has imposed what is known as a negative externality on the fishers and anglers, resulting in uncompensated losses.

Finally, if property rights for natural resources are not clearly defined, there is no incentive to conserve them. For example, unregulated fisheries are an open-access resource—anyone who wants to harvest fish can do so. Because no one owns open-access resources, they are likely to be depleted to unsustainable levels.

In the cases of public goods, externalities, and open-access resources, individuals or businesses do not consider the costs they impose on others through their actions. Further, they have no incentive to take corrective actions without some form of government intervention or collective action.

Description of values. Ecosystem valuation measures the importance of ecosystem services to people in terms of economic value. Economic value is measured as the most someone is willing to give up in other goods and services in order to obtain a particular good, service, or desirable state of the world. This is often referred to as willingness to pay. Economic value is sometimes measured as the minimum a person would accept to give up a good or service, or willingness to accept. In a market economy, dollars are a universally accepted measure of economic value. However, ecosystem values may be measured without assigning dollar values, through indicators of relative value.

It is often incorrectly assumed that a good's market price is equivalent to its economic value. However, many people are willing to pay more than the market price for a good, and hence the economic value of the good exceeds the market price. In the case of individual consumers, consumer surplus is a measure of economic value that is equal to the maximum willingness to pay minus the amount actually paid.

Producers of goods also receive economic benefits, based on the profits they make when selling a good. If producers receive a higher price than the minimum price they would sell their output for, they receive a benefit from the sale, referred to as the producer surplus. When measuring economic benefits of a policy or action, economists measure the total net economic benefit. This is the sum of consumer surplus plus producer surplus, less any costs associated with the policy or action.

Economists measure the value of ecosystem services to people by estimating the amount that people are willing to pay to preserve, restore, or enhance the services. However, this is not always straightforward for a variety of reasons. While some services of ecosystems, such as fish or lumber, are bought and sold in markets, many ecosystem services, such as a day of wildlife viewing or a view of the ocean, are not traded in markets and thus have no established price. Additionally, people value many ecosystem services that they do not use or experience personally.

Ecosystem values include market values, such as the values for food, lumber, or fish, and nonmarket values, the values for ecosystem goods and services that are not directly bought and sold. Two categories of nonmarket values are use values and nonuse values. Use value is defined as the value derived from the actual use of a good or service through such activities as hunting, fishing, birdwatching, or hiking. Use values may also include indirect uses. For example, a wilderness area provides direct use values to the people who visit the area. Other people might enjoy watching a television show about the area and its wildlife, thus receiving indirect use values. People may also receive indirect use values from an input that helps to produce something else that people use directly. For example, the lower organisms on the aquatic food chain provide indirect use values to recreational anglers who catch the fish that eat those organisms. Option value is the value that people place on having the option to enjoy something in the future, although they may not currently use it. It is sometimes categorized as a nonuse value.

Nonuse values, also referred to as passive use values, are values that are not associated with actual use or even the option to use a good or service. Existence value is the value that people place on simply knowing that something exists, even if they will never see it or use it. Other types of nonuse values include altruistic values, which arise from concerns for other people's enjoyment, and bequest values, which are based on concern for the well-being of future generations. Nonuse values may also include intrinsic values, which are based on the belief that some ecological services are valuable regardless of human values. However, while intrinsic values may be important to debates over protecting ecosystems, they are by definition not measurable as economic values.

A single person may benefit in more than one way from the same ecosystem. Thus, total economic value is measured as the sum of all relevant use and nonuse values for a good or service.

Estimation methods. In order to estimate ecosystem values, economists must collaborate with natural scientists to understand the links between human systems and ecosystems. Natural scientists must first evaluate how human actions affect the structure and functioning of ecosystems, and must express these changes in terms of changes in ecosystem goods and services that people value. Economists then apply the methods of ecosystem valuation.

Some ecosystem products, such as fish or wood, are traded in markets. Thus, their values can be estimated using the market price method, by estimating consumer and producer surplus, as with any other market good. Other ecosystem services, such as clean water, are used as inputs in production, and their value may be measured using the productivity method, by estimating their contribution to the value of the final good.

Many ecosystem services, including esthetic views or many recreational experiences, may not be directly bought and sold in markets. However, the prices that people are willing to pay in markets for related goods can be used to estimate the values of these goods. Thus, people reveal their values for these services through their purchases or other decisions. Based on statistical analysis of people's choices, economists use revealed preference methods to estimate values for such ecosystem services. Revealed preference approaches include hedonic pricing methods and travel cost methods.

For example, people often pay a higher price for a home with an exceptional view or adjacent to open space. Economists use the hedonic pricing method to estimate how such environmental amenities affect home prices, and thus to estimate the value of the amenities to local residents.

Often, people take the time to travel to a special spot for outdoor recreation, such as fishing, hiking, or wildlife viewing. Economists use travel cost methods to estimate the value of a recreational experience based on statistical analysis of the time and money spent to travel to a certain location.

Many ecosystem services are not traded in markets and are not closely related to any marketed goods. Thus, people cannot reveal what they are willing to pay through their market purchases or other actions. In these cases, economists use stated preference methods, based on surveys, to estimate value. The contingent valuation method asks people what they are willing to pay for changes in ecosystem services, based on a hypothetical scenario. Conjoint analysis, or the contingent choice method, asks people to make trade-offs among, or to rank, different alternatives, described in terms of their attributes. Stated preference methods are currently the only generally accepted methods available for measuring nonuse values.

Finally, economists may infer values for ecosystem services by observing what people are willing to pay to avoid the adverse effects that would occur if these services were lost, or to replace the lost services. These methods include the averting behavior method, the damage cost avoided method, the replacement cost method, and the substitute cost method. For example, wetlands often provide protection from floodwaters. The amount that people pay to avoid flood damage in areas similar to those protected by the wetlands can be used to estimate willingness to pay for the flood protection services of the wetland. Strictly speaking, these methods mea-sure costs rather than benefits, so they do not measure economic values but provide an indication of the magnitude of actual willingness to pay.

Summary. Ecosystem valuation is an often controversial approach that economists use to evaluate the costs and benefits of human actions that affect the health and sustainability of ecosystems. Ecosystem valuation provides information that can help individuals, businesses, activists, and policy makers fully understand the ramifications of choices in terms of their costs and benefits to individuals, institutions, and society as a whole. This information may be used in various ways: to help regulators adjust prices and other market signals so that market incentives promote choices that more fully reflect costs and benefits to people and ecosystems; to help policy makers compare alternative policy choices for preserving, restoring, or enhancing ecosystems; to guide government in crafting laws that regulate activities that affect ecosystems, and in imposing penalties for damages to ecosystems; and to educate individuals, businesses, and other organizations about the full costs and benefits of their actions.

For background information *see* ADAPTIVE MANAGEMENT; CONSERVATION OF RESOURCES; ECOLOGY, APPLIED; ECOSYSTEM; ENVIRONMENTAL MANAGEMENT; RESTORATION ECOLOGY; SYSTEMS ECOLOGY in the McGraw-Hill Encyclopedia of Science & Technology.   Marisa J. Mazzotta

Bibliography. P. Champ, K. J. Boyle, and T. C. Brown (eds.), *A Primer on Nonmarket Valuation*, Kluwer, Boston, 2003; G. C. Daily (ed.), *Nature's Services: Societal Dependence on Natural Ecosystems*, Island Press, Washington, DC, 1997; J. Gowdy and S. O'Hara, *Economic Theory for Environmentalists*, St. Lucie Press, Delray Beach, FL, 1995; G. Heal, *Nature and the Marketplace: Capturing the Value of Ecosystem Services*, Island Press, Washington, DC, 2000; National Research Council, *Valuing Ecosystem Services: Toward Better Environmental Decision-making*, National Academies Press, Washington, DC, 2004.

Ectomycorrhizal symbiosis

Ectomycorrhizal (ECM) symbiosis is an obligate association formed between the roots of many plant species and a diverse range of soil fungi. The fungi and the finest, terminal root tips interact to form joint organs called ectomycorrhizas (**Fig. 1**). The fungi exchange soil-derived nutrients for carbohydrates from the host plant. Nutrient uptake into the host is enhanced by the efficient spatial exploitation of the soil by the fungi and via access to nutrients, particularly organic sources, which would not be directly available to the plant. The percentage of root tips converted to ectomycorrhizas is usually close to 100%, which means that the belowground absorptive surfaces of most ECM plants are effectively isolated from the soil environment by a 30–80-micrometer-wide layer of fungal tissue (the mantle; **Fig. 2**). Nutrients and water taken up by the trees and any carbon

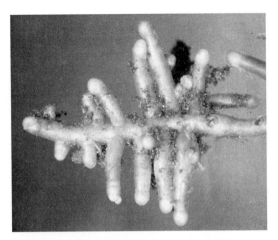

Fig. 1. Branched ectomycorrhizal cluster formed between Norway spruce (*Picea abies*) and the fungus *Russula ochroleuca*. A pigmented fungal mantle covers the root surface.

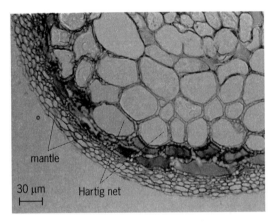

Fig. 2. Cross section through an ectomycorrhizal tip formed between Norway spruce and *Russula ochroleuca*.

which enters the soil environment from the tree roots must pass through this structure. Consequently, by occupying the interface between the soil and the absorptive root surface, the ECM symbiosis can have a considerable influence not only on the nutrition of the host trees but also upon the soil environment in which they grow.

Within the root, hyphae (filaments composing the mycelium of a fungus) of the ECM fungi ramify between the outer root cells to form a complex structure called the Hartig net, which provides a large surface area of contact between the fungus and the host, allowing efficient transfer of metabolites. External to the root, fungal hyphae grow out into the soil from the surface of the mantle. These form the extraradical mycelium, which is considered to be the primary site for nutrient and water uptake. Hyphal growth may vary from a small number of hyphae growing out a few millimeters (Fig. 1) to highly developed, extensive mycelial systems that occupy large volumes of soil. Recent studies that have used molecular markers to localize ECM mycelium in soil have found evidence for the spatial separation of ECM fungal species in different soil layers and substrates.

Fungi. It is believed that ectomycorrhizal fungi evolved on several occasions from a diverse range of free-living saprotrophic fungi (that is, those that obtain all their nutrients and carbon from the breakdown of organic material). Surprisingly, it has been suggested that the switch from a free-living to a mycorrhizal life style may be reversible, with some existing saprotrophic fungi having evolved from former ECM fungi. A multiple origin of the mycorrhizal habit would help explain the very diverse range of ECM fungi that can be found today. The great majority of ECM fungi are Homobasidiomycetes, with Ascomycetes usually considered to make up only a few percent of the global diversity. However, recent work suggests that the importance of Ascomycetes as ECM formers may have been underestimated. In addition, there are increasing reports of members of the Heterobasidiomycetes forming ectomycorrhizas with a wide range of host plants. The use of molecular techniques to directly identify the fungi on ectomycorrhizal tips has greatly increased our knowledge of the taxonomic range of fungi capable of forming ectomycorrhizas. There are no good recent estimates of the total number of ECM fungal species worldwide, but there may be about 7000–10,000.

Most ECM fungi produce macroscopic structures, fruit bodies, which bear sexually produced spores. The autumn displays of fungal fruit bodies in temperate and boreal forests are primarily produced by ECM fungi. These structures are rich sources of nutrients and are an important source of food for both small (for example, slugs and snails) and large animals (for example, deer). A significant number of ECM fungi also produce fruit bodies belowground (for example, truffles), and spore dispersal from these structures relies upon consumption by animals which excavate the fruit bodies.

Plants. Around 8000 species, about 3%, of seed plants form ectomycorrhizas, but this relatively small number of plant species constitutes the dominant component of forest and woodland ecosystems over much of the Earth. The ectomycorrhizal symbiosis is, therefore, of enormous ecological and economic importance. Most ECM plants are trees, but some herbaceous plants and sedges (for example, *Polygonum*, *Kobresia* spp.) also form ectomycorrhizas. Traditionally, the ECM symbiosis has been considered primarily as an adaptation for nutrient capture in boreal and temperate forests. Trees in the plant families characteristic of these regions (for example, Fagaceae, Betulaceae, Salicaceae, Myrtaceae, Leptospermoideae, Pinaceae) invariably form ectomycorrhizas. However, arctic and alpine habitats (*Dryas, Salix*) and Mediterranean vegetation (*Pinus, Cistus, Arbutus, Arctostaphylos*) may also contain plant species that support species-rich ECM fungal communities. It is increasingly clear that the ECM symbiosis is also very important in many tropical forests.

The Dipterocarpaceae, a diverse ECM plant family with over 500 species, has a center of diversity in southeastern Asia, where dipterocarps make up 80% of the canopy trees and up to 40% of the understorey. The family is an important source of tropical hardwood timber. Dipterocarps also occur in South America, South Africa, and Madagascar, suggesting that the ECM habit was present more than 130 million years ago (Ma), that is, prior to the separation of South America from Africa. In the tropical rainforest and savannah woodlands of Africa, many trees within the pea family (the legumes) also form ectomycorrhizas. Some of these (for example, *Tetraberlinia and Julbernardia*) form groves or extensive monodominant stands in the rainforests of the Guineo-Congolian basin, while others (for example, *Isoberlinia*) are important in the miombo and savannah woodlands of east and south-central Africa. Another leguminous tree, *Dicymbe*, also forms monodominant stands on poor soils throughout the Guyana shield of South America. The genus *Eucalyptus* also forms ectomycorrhizas and is widespread throughout temperate and subtropical forest ecosystems in Australia.

Interactions with other soil organisms. Traditionally, saprotrophic fungi and ECM fungi have been regarded as two distinct functional groups. However, over the past two decades the distinction between the two groups has become less well defined. It is now clear that some ECM fungi have the potential to degrade and assimilate nitrogen and phosphorus from organic molecules in plant, microbial, and animal detritus. This may occur either directly via the production of catabolic enzymes or indirectly via combative interactions with saprotrophic fungi. The ECM symbiosis evolved in soils already occupied by fungivorous microarthropods (for example, soil mites), and it seems likely that defense against grazing was an important evolutionary pressure in the development of the ECM structures seen today. In some groups of ECM fungi (for example, the genera *Lactarius* and *Russula*), the mycelium and the mantle contain chemical deterrents; in other groups, structural defense structures include dense coverings of thick-walled, melanized spikes on the mantle surface.

Facultative fungi. It has been claimed that some ectomycorrhizal fungal species are capable of obtaining sufficient carbon from the breakdown of organic material to complete their life cycles in the absence of a host plant; that is, they are facultative symbionts. These claims are often based upon the appearance of fruiting bodies of ectomycorrhizal ECM fungi many meters away from the nearest host plant. However, the roots of large trees can extend more than 20 m (66 ft) from the stem base, and this may account for fruit bodies that appear in fields or parklands that are edged by large ECM hosts. The appearance of fruit bodies in vegetation apparently devoid of potential ECM hosts can also be misleading. Although the majority of ECM hosts are large conspicuous trees, a number of small woody or herbaceous plants also form ectomycorrhizas. Dwarf *Salix*, *Dryas* spp., and *Polygonum viviparum* are often very inconspicuous components of short turf vegetation, but they can support a wide variety of ECM fungi. Surprisingly, recent studies have also found an ECM fungus colonizing the root system of some species of *Carex*, a plant genus usually considered to be nonmycorrhizal.

Some ECM fungi can be grown in pure culture on artificial media containing high concentrations of soluble carbon and may even produce small fruit bodies under these conditions. This phenomenon has been interpreted as an indication that the fungi could be capable of a free existence in nature. It is hard to envisage, though, a situation in nature where the fungi would have unhindered access to such carbon-rich conditions for a sufficient period of time to complete their life cycle. It has been established that some ECM fungi can produce a number of extracellular enzymes that degrade a range of organic macromolecules. However, there is no evidence that any ECM fungus can obtain sufficient nutrients and carbon via these enzymes to exist in the absence of a plant host. This assertion is backed by the very close link between carbon supply from the host plant and growth of the fungus: when this link is broken by either decapitation of the host or severing of the phloem transport down the stem of the host, there is an almost immediate cessation of fungal growth and a 50% drop in soil reduction in soil respiration. These observations also highlight the importance of ECM fungi as a major pathway by which carbon enters the soil.

For background information *see* FUNGAL ECOLOGY; FUNGI; MYCORRHIZAE; RHIZOSPHERE; ROOT (BOTANY); SOIL CHEMISTRY; SOIL ECOLOGY in the McGraw-Hill Encyclopedia of Science & Technology.

Andy F. S. Taylor

Bibliography. J. R. Leake and D. J. Read, Mycorrhizal fungi in terrestrial ecosystems, in *The Mycota IV Environmental and Microbial Relationships*, ed. by D. Wicklow and B. Söderström, pp. 281–301, Springer-Verlag, Berlin, 1997; T. P. McGonigle, The significance of grazing on fungi in nutrient cycling, *Can. J. Bot.*, 73(suppl.1): S1370–S1376, 1995; S. E. Smith and D. J. Read, *Mycorrhizal Symbiosis*, Academic Press, San Diego, 1997; M. G. A. Van der Heijden and I. R. Sanders, *Mycorrhizal Ecology*, Ecological Studies 157, Springer-Verlag, Berlin, 2002.

Field-programmable gate arrays

Field-programmable gate arrays are very large scale integrated (VLSI) circuits that can be electrically programmed to implement complex digital logic. By combining many of the advantages of custom chips with the programmability of microprocessors, they are useful in a wide range of electronics applications. The name derives from their structure—a grid of programmable logic elements (gate array) and interconnection wires that can be configured at an engineer's workbench, or even inside a consumer electronic device (programmed in the field), as opposed to other technologies that must be preconfigured during chip fabrication.

Design. Field-programmable gate arrays generally consist of a large number of programmable logic blocks embedded in a flexible routing structure (**Fig. 1**). The logic blocks implement the logic

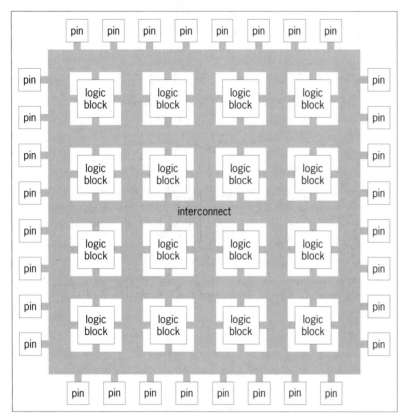

Fig. 1. Basic structure of a typical field-programmable gate array.

Look-up tables are usually coupled with an optional D flip-flop (Fig. 2). All D flip-flops in the field-programmable gate array receive a periodic signal called a "clock," which transitions millions to hundreds of millions of cycles a second. Every time the clock changes from true to false, the D flip-flop's input D is captured and sent on to the output Q. In this way, the global clock signal synchronizes the operation of all memory elements in the system, and allows the design to remember and react to past inputs and behaviors.

The muxiplexor (Fig. 2) picks between the outputs of the look-up table and the flip-flop. If the programming bit P stores a 0 (false) value, the look-up table's output is sent to the output of the logic block; while if the programming bit is 1 (true), the D flip-flop's output is sent. In this way, the logic block can be programmed to compute a clocked or unclocked function.

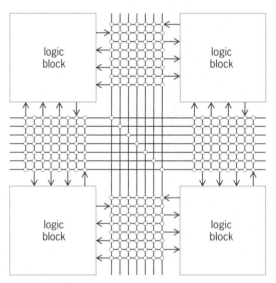

Fig. 3. Abstract representation of the interconnect structure for a typical field-programmable gate array. The circles are programmable connections between crossing wires.

functions and flip-flops (simple memory elements that allow the system to remember past inputs and behaviors), while the routing structure wires these elements together into a complete system (**Figs. 2 and 3**). Typically, the logic blocks contain look-up tables, small 1-bit-wide memories that store the desired output for any possible combination of their inputs. To configure a look-up table to compute the function "return true if both A and B are true, or C is true," the table would have false for the three cases where C is false and at least one of A and B are false (ABC = {000, 010, 100}), and true for the other five cases. In a similar manner, any 3-input combinational Boolean function can be implemented in a 3-input look-up table.

The routing structure provides fairly flexible wiring connections between the outputs and inputs of the logic blocks on the chip. This allows the relatively simple logic blocks to be combined to support a complex computation, perhaps computing a function equivalent to a million or more basic 2-bit Boolean functions. The routing structure also connects to the chip input and output pins. The routing wiring pattern is generally fixed once the field-programmable gate array is programmed, instead of the dynamically changing communication pattern found in a typical computer network.

Modern field-programmable gate arrays often augment these basic elements with more complex units. Logic blocks have elements to speed up additions and subtractions of large values, while other parts of the array replace the logic blocks with more special-purpose elements. These might include multipliers, memories, and even complete microprocessors.

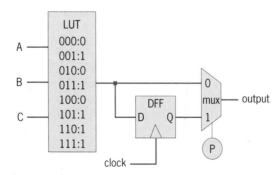

Fig. 2. Abstract representation of a logic block in a typical field-programmable gate array. LUT = look-up table. DFF = D flip-flop (input D; output Q). Mux = muxiplexor (programming bit P).

Also, the input/output (I/O) pins often include the special hardware needed to support high-speed communication protocols to other chips or communication networks.

Programming. All of these elements within the field-programmable gate array are electrically programmable. The chip contains huge numbers of programming bits, individual elements that optionally turn on or off different features in the array. For example, each entry in a look-up table will be a bit programmable to represent either a true or a false. These programming bits might be implemented using antifuse, SRAM (static random-access memory), or EEPROM/Flash (electrically erasable programmable read-only memory/flash memory) technology. Antifuse programming points are normally disconnected wires in the array that can be "blown" by a high voltage to make a permanent electrical connection. SRAM are memory bits like the cells in a computer memory; they can hold a value for as long as power is maintained to the chip, but can also be changed at will. EEPROM/Flash is a hybrid of the two, based on isolated capacitors that can hold a value without power being applied, yet can be reprogrammed via high voltages.

The basic programming technology of a field-programmable gate array affects the way that it can be used. Chips based upon SRAM technology generally must have some external memory to hold the configurations, since they lose their values when the power is turned off; antifuse and EEPROM/Flash do not. Antifuse devices can be programmed only once; SRAM or EEPROM/Flash devices can be reconfigured to fix problems or change configurations. Because of this reconfigurability, SRAM-based and EEPROM/Flash-based FPGAs will be referred to as reconfigurable field-programmable gate array here. Overall, SRAM-based devices are generally the most successful commercially, since they do not need special fabrication techniques and thus can use the latest chip technologies.

To configure a field-programmable gate array for a specific circuit, a designer typically starts with code written in a hardware description language such as Verilog or VHDL. These are languages similar to C or Ada but are optimized for specifying hardware designs. This code is then converted to basic logic gates, such as ANDs and ORs, by a logic synthesis tool. A "technology-mapper" takes these logic gates and groups them into chunks that fit in the look-up tables in the field-programmable gate array. Finally, these look-up-table size groupings are assigned into specific logic blocks and routed together by the "placement" and "routing" tools, respectively. At the end of this process, a file with the assignments to the field-programmable gate array's configuration bits (known as a bitstream) is generated. When this file is downloaded into a field-programmable gate array, it is configured to implement the user's design. Each of these steps is highly automated by software tools.

Applications. Field-programmable gate arrays can be viewed as occupying a middle ground between programmable processors and custom chips. Compared to custom chips developed for a specific application (application-specific integrated circuits, or ASICs), field-programmable gate arrays are generally slower, are less power-efficient, and have a smaller logic capacity. However, an ASIC requires a much greater design time and a large up-front cost to set up fabrication for the first chip. Furthermore, unlike reconfigurable field-programmable gate arrays, ASICs cannot be changed after fabrication to handle bug fixes or functionality upgrades.

Compared to programmable processors, such as microprocessors and digital signal processors, field-programmable gate arrays can be faster, denser, and more power-efficient. However, field-programmable gate arrays are only more efficient than processors on applications involving long repetitive tasks, such as multimedia processing and networking. Also, designing for field-programmable gate arrays is more complex than for microprocessors, and field-programmable gate arrays usually cannot support running multiple applications simultaneously (multitasking).

Although a microprocessor-based solution is usually preferred over an implementation with either an ASIC or a field-programmable gate array, some applications have speed or power requirements that make programmable processors unsuitable. Because of their relative strengths versus ASICs, field-programmable gate arrays are typically used in the following applications.

Low- to medium-volume hardware. Because ASICs have very high initial costs that must be amortized over each system sold, custom hardware may be affordable only for products that will sell a large number of units. And while in the past field-programmable gate arrays were able only to implement small parts of an application, their significant increase in capacity have allowed field-programmable gate arrays to now support complete systems in a single chip.

Fast time-to-market. Even for applications with enough sales volume to justify the initial costs of an ASIC, the design time may be too long. Field-programmable gate arrays, which require less design time and have no extra delays for custom fabrication, can help get a product to market much faster than an ASIC.

Rapid prototyping. When a designer is developing a new circuit, it is often crucial to create a working prototype of the system for testing. This prototype will need to be changed multiple times as bugs and limitations are found. While ASICs cannot be changed, the reprogrammability of reconfigurable field-programmable gate arrays makes them ideal for prototyping.

Upgradability. For some applications, it is clear the circuitry will change after the system is sold. This includes devices supporting evolving protocols, complex designs susceptible to bugs, and applications where new features might be added over time. Reconfigurable field-programmable gate arrays can easily handle such systems, with the logic changed simply by providing a new bitstream.

Multimode systems. Some systems will need to support many different applications. For example, a single cellphone design might need to handle multiple protocols, depending on the country and the cellphone network provider. A field-programmable gate array can implement each protocol simply by providing it with the appropriate configuration file.

Run-time reconfiguration. A more aggressive use of a field-programmable gate arrays reconfigurability is to change the loaded configuration rapidly during the operation of the system. A field-programmable gate array otherwise is too small to implement a large computation but can do so if that computation is broken down into multiple, smaller pieces. This significantly increases the virtual capacity of the field-programmable gate array. ASICs would have to implement all of the configurations simultaneously.

Field-programmable gate arrays were introduced commercially in 1985, then became a multibillion-dollar industry. Due to their reprogrammability, ease of use, and low cost of hardware and support software, SRAM-based field-programmable gate arrays are one of the best hardware technologies for the novice or hobbyist.

For background information *see* INTEGRATED CIRCUITS; MICROPROCESSOR; PROGRAMMING LANGUAGES; SEMICONDUCTOR MEMORIES in the McGraw-Hill Encyclopedia of Science & Technology.

Scott A. Hauck

Bibliography. S. Brown and Z. Vranesic, *Fundamentals of Digital Logic with Verilog Design*, McGraw-Hill, 2002; K. Compton and S. Hauck, Reconfigurable computing: A survey of systems and software, *ACM Comput. Surveys*, 34(2):171–210, June 2002; S. Hauck, The roles of FPGAs in reprogrammable systems, *Proc. IEEE*, 86(4):615–639, April 1998; J. Villasenor and W. H. Mangione-Smith, Configurable computing, *Sci. Amer.*, pp. 66–71, June 1997.

Fingerprint identification

Fingerprint identification is the world's most widely used and trusted method of biometric identification. Fingerprint identification exploits the variability of the patterns formed by papillary ridges (the corrugated ridges that swirl around and across the fingertips). Historically, the two primary uses of fingerprint identification have been for criminal record-keeping and for forensic identification. Fingerprints are also used for civilian identification purposes, such as controlling access to entitlements (government benefits), keeping motor vehicle records, and immigration control. Due to recent innovations in fingerprints scanners, fingerprints are increasingly used for civil identification and security purposes.

History of forensic identification. Impressions of the papillary ridges, transferred through a medium such as ink, were used as crude signatures since ancient times, especially in Asia. Western scientists noted papillary ridges beginning in the seventeenth century. Fingerprint identification in the modern sense did not begin until the late nineteenth century, when

Western criminal justice systems began to demand methods of indexing criminal records according to physical attributes, rather than names. This demand was stimulated by reformist jurisprudence, which held that offenders should be punished differentially according to the length of their criminal records. This was not practical if offenders could "reset" their criminal records simply by inventing an alias.

The first new technology applied to this problem was photography, beginning in the 1840s. But photographic images were not easily amenable to indexing. In the 1880s, Alphonse Bertillon, a Paris police official, developed a method of indexing criminal records according to 11 anthropometric measurements (bodily measurements). The Bertillon system was quite successful and widely disseminated. The **table** summarizes fingerprint identification advances since 1856.

The idea of using impressions of the papillary ridges to systematically identify individuals was proposed almost simultaneously around 1880 by two Britons living abroad: William Herschel, a colonial administrator in India, and Henry Faulds, a Scottish physician doing missionary work in Tokyo.

Although fingerprint identification is known today primarily as a system of criminal identification, the first modern use of it was for civil identification. Herschel used fingerprints to verify the identity of Indian pensioners. The major innovations in fingerprint identification, however, resulted from efforts to adapt papillary ridges to the needs of the criminal justice administration—to match individuals in custody to existing criminal records using bodily attributes,

Fingerprint identification advances	
Year	Event
1856	John Maloy identifies bloody print, Albany, New York
1858	William Herschel takes hand print, Hooghly, Bengal, India
1877	Thomas Taylor publishes suggestion of using "markings on the palms of the hands and the tips of the fingers," U.S.
1880	Henry Faulds and William Herschel publish in *Nature*
1883	Thumb printing proposed and rejected for identification of Chinese immigrants, California
1883	Bertillon system adopted, France
1892	Francis Galton devises arch-loop-whorl classification scheme
1892	Juan Vucetich identifies bloody print, Argentina
1893	Juan Vucetich and Edward Henry, Azizul Haque, and Chandra Bose devise ten-print classification systems
1897	Edward Henry identifies bloody print, Bhutan
1901	Britain adopts fingerprinting
1902	Henry DeForest introduces first civil fingerprint to U.S., New York City
1902	Latent print cases, Britain and France
1903	James Parke institutes first criminal fingerprint file in U.S., Albany, New York
1904	John Ferrier disseminates Henry system at World's Fair, St. Louis
1905	Deptford murder trial, Britain
1905	First use of latent prints in U.S., New York
1911	First U.S. legal precedent upholding use of fingerprint evidence, People v. Jennings, Illinois

rather than names, which could be falsified. This required developing a system of indexing fingerprint records according to fingerprint pattern types, much as Bertillon had done for anthropometric records.

The first crucial step was taken by Francis Galton, who proposed classifying all fingerprint patterns into three basic types: arches, loops, and whorls. Edward Henry, Azizul Haque, and Chandra Bose, working in India, added methods for subdividing loops and whorls, through processes of ridge counting and ridge tracing, and developed a system for filing fingerprint records according to the pattern types of all 10 fingers. Juan Vucetich, a police official in La Plata, Argentina, simultaneously developed another fingerprint classification system. "Ten-print" classification systems, as they were known, allowed police officials to retrieve criminal records based on nothing more than the fingerprint patterns of an individual in custody. Theoretically, this rendered the use of aliases obsolete. This development allowed criminal justice officials to have greater confidence in the accuracy of their criminal records, enabling them to calibrate punishments according to the length of the offender's criminal history.

Fingerprint identification supplanted anthropometric identification only slowly, due in part to investments that law enforcement agencies had already made in building databases of anthropometric records, and in part to concerns about its scientific reliability. Only by the 1920s was the predominance of fingerprinting assured, primarily based on the perceived lower costs and training requirements associated with recording fingerprint data.

Forensic fingerprint evidence. The second major use of fingerprint identification is for forensics—to link an individual to a crime scene by attributing an impression found at the scene to his or her hand. The prints found at crime scenes are often called latent prints (because they are left in oily secretions and require a medium, such as powder, in order to be "developed" or made visible) or, more precisely, marks (because it is not necessarily known whether they are in fact representations of papillary ridges at all). Less common are patent prints (that need not be developed because they are left in a visible medium, such as blood) and plastic prints (such as those left in wet paint).

Forensic fingerprint identification was originally only an accidental side benefit of ten-print identification systems. In the 1890s, when prints were found at crime scenes, the police records clerks in charge of maintaining fingerprint records were asked to determine whether a certain suspect had made the crime scene print. These clerks reasoned that they could conclude that the same finger produced a latent print and an inked print known to have come from the suspect by finding corresponding minutiae, as Galton called them. Today these are called friction ridge details; they are essentially endings and bifurcations of papillary ridges. Forensic identifications were effected in Albany, New York, in the late 1850s, Tokyo around 1880, Argentina in 1892, India in 1897, France and Britain in 1902.

Criminal defendants opposed the new evidence. One early argument held that fingerprint evidence should just be placed before the jury, rather than being interpreted by an expert. Indeed, Galton predicted that fingerprint expert witnesses would only be necessary until the novelty of the technique wore off. As it happened, courts supported the right of fingerprint experts to interpret latent print evidence, and the experts never did fade away.

Despite some initial misgivings, courts quickly accepted and enshrined into law fingerprint examiners' claims that fingerprint identification was a science, that fingerprint identification was reliable, and that it had been used for years without error. The courts' conclusions were largely based on the fallacious but intuitively appealing logic that if all human papillary ridge patterns were different from one another, then fingerprint identification must be absolutely reliable. In 1892, Galton had attempted to model the individuality of complete fingertip patterns, estimating the likelihood of any one finger having an exact duplicate as 1 in 64 billion. Assuming 16 billion fingerprints in the world, he then calculated the probability of a set of exact duplicates existing as 1 in 4.

Galton's calculations had little relevance to the value of latent print identifications. Latent prints are typically smaller in area than a complete fingertip and suffer from degradation of information through blurring, artifacts, substrate noise, and so on. Moreover, the probability of exact duplication of the pattern of the papillary ridges is not the same as the probability of misattribution of a small distorted fragmented impression of a small section of the papillary ridges contained on a fingertip. This point was articulated by Faulds in the 1905 Deptford Murder Trial in Britain. Faulds expressed confidence in ten-print identification, but argued that identification from small latent print fragments, which he called smudges, were another matter altogether.

Faulds's criticisms persuaded few, and fingerprint identification enjoyed rapid acceptance by the courts and soon acquired a reputation as the most trustworthy of forensic techniques. Indeed, the term "fingerprinting" became almost synonymous with infallibility and unique identification.

Probably because of this rapid legal acceptance, no one ever conducted a comprehensive study of the rarity of configurations of ridge details, of the accuracy rate of latent print attributions, or of the number of corresponding ridge details necessary to effectively eliminate all other fingertips of potential donors of the latent print.

Current debate. For nearly a century, fingerprint identification was seldom questioned and was apparently trusted, even by those expected to doubt it (such as defense attorneys), as well as in popular culture. In the 1990s, however, the assumption of the infallibility of fingerprint identification began to be questioned. One reason was the rise of forensic DNA typing. DNA analysts accompanied DNA "matches" with calculations of the rarity of the matching profile in given populations. These rarities were strenuously debated both in the scientific community and the courts. It came as somewhat of a surprise, therefore, to realize that no one had ever made such calculations for fingerprint matches. Instead, fingerprint examiners simply assumed that the frequency of the features found in any given latent print containing "sufficient" detail was 1.

Another reason for the growing skepticism was a 1993 United States Supreme Court case, Daubert v. Merrell Dow Pharmaceuticals, which mandated that judges ensure the reliability of expert evidence put before juries in trials. Criminal defendants argued that no study demonstrated the reliability of latent print evidence. Proponents of fingerprint evidence were unable to identify any such study. Instead, they argued that the long-standing use of fingerprint evidence in criminal justice proceedings constituted a de facto study.

Although courts accepted this argument initially, they eventually realized that the ground truth is not known in criminal proceedings. Nonetheless, United States courts have thus far unanimously upheld the legal admissibility of fingerprint evidence.

Although most commentators continue to evince great confidence in the ultimate accuracy of fingerprint identification, its accuracy remains unmeasured and understudied. Fingerprint identification continues to more closely resemble a subjective clinical judgment than an objective scientific measurement. Latent print examiners continue to use intuition to decide when the amount of consistent detail between a latent print and a print of known origin is sufficient to eliminate all other possible donors and support a conclusion of individualization. The current debate about fingerprint evidence has generated increased attention to these problems from the scientific community, and new research may help resolve some of these vexing questions.

For background information *see* CRIMINALISTICS; DEOXYRIBONUCLEIC ACID; EPIDERMAL RIDGES; FINGERPRINT; INTEGUMENTARY PATTERNS; SKIN in the McGraw-Hill Encyclopedia of Science & Technology.

Simon A. Cole

Bibliography. D. R. Ashbaugh, *Quantitative-Qualitative Friction Ridge Analysis: An Introduction to Basic and Advanced Ridgeology*, CRC Press, Boca Raton, FL, 1999; C. Champod et al., *Fingerprints and Other Ridge Skin Impressions*, CRC Press, Boca Raton, FL, 2004; S. A. Cole, *Suspect Identities: A History of Fingerprinting and Criminal Identification*, Harvard University Press, Cambridge, 2001; H. C. Lee and R. E. Gaensslen, *Advances in Fingerprint Technology*, CRC Press, Boca Raton, FL, 2001; C. Sengoopta, *Imprint of the Raj: How Fingerprinting Was Born in Colonial India*, Macmillan, London, 2003.

Florigen

Florigen is a hypothetical stimulus that controls when a plant will flower in response to the changes in day length associated with the different seasons.

Flowering occurs when shoot apical meristems—small regions of prolific cell division in the buds of the main shoot and lateral branches—stop producing vegetative tissues (stems and leaves) and instead produce floral tissues (a process called evocation). The transition to flowering is fundamental to the production of offspring and the survival of the species, and must be coordinated with seasonal change to (1) synchronize flowering within the species to promote cross-pollination, (2) harmonize flowering with the life cycles and behaviors of pollinators such as insects, and (3) coordinate flowering with the seasons to ensure that progeny (seeds) are sufficiently mature to survive unfavorable conditions, such as winter or hot, dry summers. The timing of this transition is also a concern for agriculture since seeds and fruits must be harvested within the growing season of the region.

Many plants determine the season by measuring changes in photoperiod—the light/dark cycles that occur over a 24-hour period (that is, in the Northern Hemisphere, plants perceive the long days of spring and summer, and the short days of autumn and winter). Leaves perceive photoperiod, but the bud produces the flower, indicating that a stimulus is transmitted from leaf to bud. Mikhail Chailakhyan proposed in the 1930s that this stimulus is a hormonelike substance and called it florigen (Latin, *flora*, "flower"; and Greek, *genno*, "to beget"). Efforts to clearly identify florigen have failed, however, earning it the title of the Holy Grail of plant biology. Several candidates have been proposed, and current efforts suggest a macromolecule such as ribonucleic acid (RNA) or protein.

Relation to photoperiod. Any discussion of florigen requires an understanding of how photoperiod affects plants. The first detailed studies on flowering in response to photoperiod were performed in the 1910s by Henry Allard and Wightman Garner at the U.S. Department of Agriculture in Beltsville, Maryland. They observed that a new hybrid of tobacco (*Nicotiana tabacum*), Maryland Mammoth, grew to 5 m (16.4 ft) without flowering during the summer, but flowered profusely when less than 1 m (3.3 ft) tall if grown in the greenhouse during the winter. They also documented that a given cultivar of soybean (*Glycine max*) would flower at roughly the same time in the summer, irrespective of when planting occurred in the spring. After testing numerous environmental factors, they concluded that the tobacco and the soybean bloomed when days were shorter than a critical length. They also found that other plants, such as spring barley and spinach, flowered when the days were longer than a critical length. From these experiments, plants that flower when days are shorter than a certain critical period are called short-day plants, and those that flower when the days are longer than a certain critical period are long-day plants (we now know that the dark period is the actual stimulus so that short-day plants are induced to flower by long nights). Plants that do not respond to photoperiod are day-neutral plants. This response to day or night length was called photoperiodism by Allard and Garner, and comparable phenomena have been discovered in many plants and animals.

Soon after the discovery of photoperiodism, it was found that leaves, not the bud, perceive the photoperiodic signal, implying that a signal is transmitted through the plant to induce reproductive growth. Chailakhyan tested this by grafting stems of flowering plants onto plants that were not flowering, and maintaining the chimera under noninductive conditions. Flowering was observed in the noninduced stems, indicating that a signal (florigen) had crossed the graft junction. Grafts between different species suggested that florigen is active across a broad range of species irrespective of whether the plants are short-day, long-day, or day-neutral. For example, reciprocal grafts between short-day tobacco and long-day black henbane (*Hyoscamus niger*) caused the black henbane to flower if the tobacco had been maintained under short-day conditions, and conversely the tobacco would flower if the black henbane had been maintained under long days. Many similar experiments suggested that florigen is the same compound, or at least induces the same physiological response, in most angiosperms (plants that flower and form fruits with seeds). Further experiments showed that florigen requires a living-tissue union to move between two graft partners, and moves with the products of photosynthesis, indicating that movement is in the phloem.

Attempts to define the chemical nature of florigen were largely unsuccessful, probably due to limitations in the approaches used. It was assumed that florigen was a small hormonelike molecule, and hormones were typically assessed by applying extracts to a plant and recording the response. When extracts from flowering plants were applied to noninduced plants, results were ambiguous and not fully reproducible. A limitation of these experiments is that the active compound, once applied to the leaf, may not enter the cells and move through the phloem to the bud. In this regard, modern studies that utilize transgenic plants to express specific genes and generate specific compounds in defined tissue are a significant advance (see below).

The failure to isolate florigen led to skepticism toward its existence, and several alternative hypotheses were put forward. Principal among these are (1) the nutrient diversion hypothesis, by Roy Sachs and Wesley Hackett, that proposes that an increase in the flow of nutrients to the apical meristem stimulates the transition to flowering; and (2) the model of multifactorial control, by Georges Bernier and colleagues, which proposes that flowering results when a number of factors including promoters, inhibitors, phytohormones, and nutrients are present in the buds at the appropriate time and in the appropriate concentrations. Although numerous physiological and genetic experiments support both hypotheses, the transition to flowering is influenced by several stimuli in addition to photoperiod. Both of these alternative hypotheses tend to group these different pathways together. When photoperiod is considered

in isolation, the concept of florigen as a single signal transmitted from leaves to buds retains its appeal.

Genetics and molecular biology. During the last 20 or so years, advances in genetics and molecular biology have provided new tools for identifying florigen, and a small member of the mustard family (Brassicaea), *Arabidopsis thaliana*, has emerged as an exceptionally useful model plant. Arabidopsis is a facultative long-day plant, meaning that the transition to flowering is accelerated in long days, but it will eventually flower under short-day conditions after an extended vegetative phase. Although research with other plants has contributed to the field, the most significant advances, and those that are discussed here, are from Arabidopsis.

A series of mutants that flower late under long days but are identical to wild type under short days define a genetic pathway that promotes flowering in long-day photoperiods. One of these genes is *CONSTANS* (*CO*), which appears to coordinate external light stimuli with the plant's internal circadian rhythms (cyclic patterns of activity occurring on a 24-hour basis that are maintained in the absence of external stimuli). Consistent with a role in promoting flowering in long days, *CO* messenger RNA (mRNA) accumulates in long days relative to short days, and oscillates with a circadian rhythm. Under long-day conditions *CO* mRNA begins accumulating during the light period; in contrast, in short days accumulation begins during the night period. These patterns of *CO* expression in relation to light stimuli may be a mechanism by which plants coordinate flowering with day length. Several methods show that *CO* is naturally expressed in vascular tissues of leaves and stems, as well as in meristemic regions, and recent experiments with transgenic plants demonstrate that *CO* expression exclusively in the veins of leaves is sufficient to generate a phloem mobile signal that promotes evocation. Grafting experiments with these plants show that the signal is graft-transmissible. Based on these results, it is clear that *CO* participates in generating florigen.

CO encodes a transcription factor that promotes the expression of another gene called *FT* (*FLOWERING LOCUS T*). *FT* expression also follows a circadian rhythm, with maximal expression 20 h after dawn. *FT* expression localizes to vascular tissues of leaves and stems, and overexpression of *FT* from a constitutive promoter compensates for mutations in *CO*. Therefore, *FT* is situated between *CO* and florigen in the photoperiod pathway. *FT* encodes a 20-kilodalton protein similar to phosphatidylethanolamine-binding proteins, but its biochemical function in plants is not known.

Although florigen was assumed to be a small hormonelike molecule, Laurel Mezitt and William Lucas of the University of California at Davis proposed in 1996 that it may be a macromolecule, such as protein or RNA (RNA was proposed later). These ideas inspired others to look for proteins and RNA in the phloem that may have a role in long-distance signaling. When expressed exclusively in the phloem of mature leaves, *CO* promotes the transition to

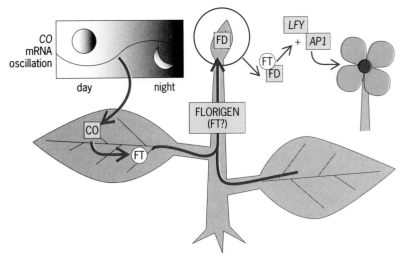

Florigen coordinates photoperiod perception in leaves with the transition to flowering in buds. (Left) In leaves, *CO* expression oscillates as a circadian rhythm and accumulates with appropriate day length. *CO* induces the expression of *FT* in leaf vascular tissues. **(Center)** Florigen, possibly as FT protein or RNA, migrates in the phloem from leaves to buds, where FT protein interacts with FD protein. **(Right)** FD together with FT induces the expression of meristem-identity genes, *LEAFY* (*LFY*) and *APETALA1* (*AP1*), and the meristem stops producing vegetative tissues and instead produces floral tissues.

flowering as described above, but the protein localizes predominantly, if not exclusively, to the nucleus, and neither protein nor RNA crosses graft junctions in detectable amounts. There is, however, compelling evidence that FT, in the form of protein or RNA, may travel from leaves to meristems to promote flowering (see **illustration**). To promote flowering, *FT* requires the activity of another gene, *FD*, which encodes a transcription factor. FD and FT proteins interact directly and colocalize to the nucleus when in the same cells. *FD*, however, is expressed at the shoot apex, whereas *FT* is expressed in leaf vascular tissues. The movement of *FT* RNA from leaves to the shoot apex is documented, implicating it as a component of the florigen signal. Presumably, transported *FT* mRNA is translated into functional protein in the bud.

Together, FT and FD induce the expression of *LEAFY* (*LFY*) and *APETALA1* (*AP1*) in the bud. *LFY* and *AP1* are known as meristem-identity genes since their expression heralds the conversion of vegetative mersitems to floral meristems. Buds expressing these genes will produce flowers instead of leaves.

Despite recent progress, a word of caution is warranted. Translation of *FT* RNA into active protein in recipient shoot apices is not demonstrated, and movement of the FT protein out of leaves was not sought. Furthermore, it cannot be excluded that *FT* expression in leaves produces an unidentified signal that comigrates in the phloem to induce flowering. Thus, the search for florigen may not be over yet.

For background information *see* FLOWER; PHOTOPERIODISM; PLANT GROWTH; PLANT HORMONES; PLANT PROPAGATION; PLANT (REPRODUCTION) in the McGraw-Hill Encyclopedia of Science & Technology.

Brian G. Ayre

Bibliography. M. A. Blazquez, The right time and place for making flowers, *Science*, 309:1024–1025, 2005; T. J. Lough and W. J. Lucas, Integrative plant biology: Role of phloem long-distance macromolecular trafficking, *Annu. Rev. Plant Biol.*, 57:203–232, 2006; A. Mouradov, F. Cremer, and G. Coupland, Control of flowering time: Interacting pathways as a basis for diversity, *Plant Cell*, 14(suppl.):S111–S130, 2002; F. B. Salisbury and C. W. Ross, *Plant Physiology*, 4th ed., Wadsworth, Belmont, CA, 1991; L. Taiz and E. Zeiger, *Plant Physiology*, 4th ed., Sinauer Associates, Sunderland, MA, 2006.

Forensic linguistics

Forensic linguistics applies the theories, constructs, and analytical methods of linguistics to questions that arise in civil, criminal, and security investigations and adjudication. Understanding linguistics as a social science is essential, because almost all of linguistics has been applied to forensically significant questions.

Social science. Linguistics is the social science concerned with all aspects of language. It has many subfields. Primary subfields of linguistics focus on the structural principles underlying human language, while secondary subfields focus on language in relation to other human behaviors. Phonetics and phonology study the system of sounds in human language. Morphology studies the minimal units of sounds that convey meanings and how these minimal units are combined to form words. For example, the sound represented by the English letter "s" is phonetically described by its acoustic properties of hissiness and continuance, phonologically described by its contrast to other English sounds such as "z," "t," and "n," and morphologically described as an indicator of plurality and possession on some nouns like "aunts" and "aunt's" and of singularity on present-tense verbs like "laughs." Syntax studies the ways in which words can combine into phrases, and phrases can combine into sentences, and the dependency relations between phrases that can affect word order and referential meaning. Semantics focuses on word, phrasal, and sentential meanings. For example, in the sentence "Jack asked Mary to go to the store," the dependency relation between "ask," "Mary," and "to go" determines the meaning such that "Mary," not "Jack," is going to the store if Jack's request is granted. Semantics and syntax are both involved in resolving or interpreting ambiguity in language. Pragmatics focuses on language use in conversation and other forms of discourse. Language change studies the ways in which language evolves over time, and is closely associated with child language acquisition. Another aspect of child language acquisition focuses on the genetic disposition toward language. Psycholinguistics studies how the human cognitive system processes language at all levels of structural analysis and at all levels of human development. Language acquisition by adults focuses on the cognitive system as well as the best practices in pedagogy. Neurolinguistics focuses on the human neuroanatomy and neuronal functioning related to the acquisition, genetics, and disintegration of linguistic ability. Sociolinguistics examines how human social systems use language to define boundaries and interaction. Dialectology focuses on the identification of dialects related to regional boundaries, while sociolinguistics focuses on the identification of linguistic patterns constrained by social, class, racial, and gender boundaries. Computational linguistics develops tools through which computers can be used to perform linguistic tasks for humans, such as information retrieval, machine translation, and language analysis.

Applied linguistics. Essentially, forensic linguistics applies linguistics to forensic issues. Forensic linguistics draws upon multiple fields in answering questions of legal significance which are related to language structure and use. The **table** illustrates the kinds of forensically significant questions on which linguistics has been or could be used.

Two examples which demonstrate clearly the utility of forensic linguistics are speaker identification and author identification. Research in both speaker and author identification applies basic linguistic analysis so that the identification becomes a classification procedure; that is, the questioned speech or text is classified or identified according to a statistical model built on the quantitative linguistic features extracted from samples of known speakers or authors. Current research in both speaker and author identification demonstrates that these problems must be solved by including linguistic features from various levels of linguistic structure and use. Speaker identification accuracy is not just a matter of identifying shared

phones, just as author identification is not just a matter of identifying shared words. Instead, the current research shows that accuracy for identifying speaker or author improves when different kinds of features are fused.

Current research in speaker identification and author identification is demonstrating high accuracy rates when the computational linguistics fusional approach is used. In speaker identification, R. D. Rodman and coworkers of North Carolina State University report high accuracies, while D. Reynolds and coworkers at MIT's Lincoln Laboratories report accuracies of 99.8%. In author identification, C. E. Chaski at the Institute for Linguistic Evidence reports average accuracies of 95%, with some experiments as high as 98% that have as few as 100 sentences per author. J. Li and coworkers at the University of Arizona report accuracies as high as 99% with 30 e-mail documents per author. This core work in speaker and author identification in academia and research institutes demonstrates how forensic linguistics as an applied science can provide reliable evidence to law enforcement for investigative purposes and, under the right conditions, admissible evidence for adjudication.

Misconceptions. Forensic linguistics is not the analysis of anything having to do with language and law because such an approach does not draw upon the theories, constructs, and methodology of linguistic science and does draw upon commonly held misconceptions about language. We all have ideas about language. It is commonly held that some languages

Linguistics needed for forensic questions	
Forensic questions	Linguistics
Who spoke this message or statement? What geographic area is the speaker from? What social class is the speaker from? What language is being spoken? Is this statement deceptive?	Phonetics, phonology, morphology, dialectology, sociolinguistics, computational linguistics
Who wrote this text or statement? What geographic area is the writer from? What social class is the writer from? What language is being written? Is this text deceptive?	Morphology, syntax, pragmatics, sociolinguistics, computational linguistics
Are there alternate interpretations of this text? Is the worldview of this document consistent with the worldview of the plaintiff or defendant?	Semantics
Is this a real threat letter or a hoax? Is this a real suicide note or a hoax? What is the best way to present this case to a jury or a judge?	Semantics, pragmatics
Is this document authentically from the particular time from which it purports to be?	Language change
Can the child legitimately testify? Is this kind of testimony consistent with a child of this age or could the testimony have been coached?	Child language
How comprehensible is this text, in general, and specifically for this plaintiff or defendant?	Psycholinguistics
Is this language behavior consistent with linguistic disorders as claimed? Could a person with this specific disorder produce this specific text or speech?	Neurolinguistics
Does this text or speech represent someone associated with a specific class or gang or cell or criminal activity?	Sociolinguistics

are more difficult than others, or that some phrases are bad grammar while others are correct. It is also commonly held that people can reliably recognize voices in speech and in print. These ideas about language, though intuitively plausible, are not tenable in the light of what we know from linguistics about language structure, human processing capabilities, and language use.

In speaker identification, a nonlinguist method for identifying the voice on a tape is known as voiceprinting. Voiceprinting uses a visual inspection of a sound spectrogram, which plots sound-wave frequency on the vertical axis and time on the horizontal axis. Introduced in 1962, voiceprinting was championed by the Federal Bureau of Investigation and renounced by leading phoneticians. Voiceprinting has survived and even won admissibility into trial evidence as late as 1999, although many jurisdictions have excluded voiceprint testimony for years. Another nonlinguist method for identifying authorship of a written document is known as forensic stylistics. Forensic stylistics uses a visual inspection of the documents for page and textual formatting, spelling errors, grammatical errors, and such. The method has received mixed admissibility into trial evidence since the early 1900s. Under recent rules of evidence, testimony based on forensic stylistics has either been excluded from trial or admitted without the expert being allowed to state an actual opinion regarding authorship.

The primary difference between forensic linguistics and nonlinguist methods is the scientific approach. In forensic linguistics, the scientific method requires hypothesis testing and a litigation-independent testing of the method for its accuracy. These tests are performed with robust controls regarding data quantity, data sources, and analytical objectivity. If the practitioners of nonlinguist methods were to emulate this kind of research agenda for testing and validation of their methods, their analyses would be taken much more seriously by both linguists and lawyers.

For background information *see* CRIMINALISTICS; LINGUISTICS; PHONETICS; PSYCHOLINGUISTICS; SOUND; SPEECH in the McGraw-Hill Encyclopedia of Science & Technology.   Carole E. Chaski

Bibliography. C. E. Chaski, Who's at the keyboard? Authorship attribution in digital evidence investigations, *Int. J. Digital Evidence*, Spring 2005; B. S. Howald, Comparative and non-comparative forensic linguistic analysis techniques: Methodologies for negotiating the interface of linguistics and evidentiary jurisprudence in the American judiciary, *Univ. Detroit Mercy Law Rev.*, vol. 83, issue 3, Spring 2006; J. N. Levi, *Language and Law: A Bibliographic Guide to Social Science Research in the USA*, American Bar Association, 1994; J. Li, R. Zheng, and H. Chen, From fingerprint to writeprint, *Commun. ACM*, 49(4):76–82, April 2006; D. Reynolds et al., SuperSID project: Exploiting high-level information for high-accuracy speaker recognition, *ICASSP*, 2003; R. D. Rodman et al., Forensic speaker identification based on spectral moments, *Forensic Linguistics: Int. J. Speech, Lang. Law*, vol. 9, 2002; L. M. Solan and P. M. Tiersma, *Speaking of Crime: The Language of Criminal Justice*, University of Chicago Press, 2005.

Fungal bioconversion

Humans throughout history have created environments conducive to the natural production of chemicals by organisms, or have produced chemicals to achieve the various marvels that shape the world. Beginning with the earliest production of alcohol by fermentation of grape juice, all the way to the formation of complex industrial polymers that surround us today, chemicals have been exploited to make various processes possible, easier, or more enjoyable. Household and industrial cleaners, pesticides, herbicides, cosmetics, dyes, paints, fuel additives, lubricants, plastics, and many other chemical entities have molded a world unrecognizable without them.

During the manufacture of these products, many other chemicals and large quantities of energy are used, requiring the burning of fossil fuels to produce the energy necessary for manufacturing or the disposal of chemical hazardous waste. Historically, the excessive use of fossil fuels and hazardous chemicals (often dumped in rivers or abandoned brownfields) has not been of concern, and in most cases was the economically competitive decision to make. In recent decades, however, there has been increasing concern raised by the environmental movement, as well as growing evidence of global warming, that has fostered the need to question the unforeseen "ends" to our "means." This increasing concern has led humans away from the concept of "out of sight, out of mind" toward a more genuinely global responsibility for the compounds that humans produce, involving ideals such as the precautionary principle.

The precautionary principle elevates consideration of the end point of a chemical into the decision making of its production, use, and disposal, ideally leading to its reuse, recycling, or complete mineralization. Since achieving these ends would typically require more chemicals (followed by more treatment) or other uneconomical methods, many industrial researchers are turning to the biological world to do the work for them, in most cases more efficiently. According to the law of microbial infallibility, if a substance exists, there is a microbe that can or will degrade it. However, finding that microbe may be quite a challenge. Bioprospecting, sparked by the discovery and development of penicillin from the fungus *Penicillium chrysogenum*, is the search for commercially valuable biochemical and genetic resources in plants, animals, and microorganisms. This is a field of research that is increasingly popular, particularly in regard to enzymes.

Enzymes. Produced by all organisms, enzymes are proteins that assist chemical reactions by establishing the essential environment for the necessary bonds to be broken or formed. Each enzyme is structured to function in the breakdown (catabolism) or buildup (anabolism) of a specific type of bond or molecule, creating the need for several enzymes to digest or construct a wide variety of molecules or even a single complex molecule. Over time, a species may evolve to efficiently produce the enzymes required to exploit the chemicals by which it is surrounded or to which it is exposed. Bioprospecting identifies enzymes that can be applied to economically eliminate or transform unwanted chemicals or pollutants into useful and/or harmless by-products. Fortunately, this form of bioconversion or bioremediation is proving to be an ideal ingenuity to solve several industrial needs.

Bioconversion-bioremediation. Initial uses of bioconversion or bioremediation involved only wastewater treatment, but since the 1960s more effort has turned to the breakdown of industrial wastes, pesticides, and petroleum-based products, and even to creation of food additives. Industrial wastes such as effluent from textile factories containing azo, diazo, and other active dyes, effluent from lumber treatment facilities containing dibenzo-*p*-dioxins, and pentachlorophenol (PCPs) have all been demonstrated to be susceptible to bioconversion into innocuous by-products using microbial enzymes. Trichloro-ethylene (TCE) or substituted aromatics like toluene, benzoate, or phenol present in coal tars and preservatives have also been found to be degradable. High concentrations of explosives like trinitrotoluene (TNT), solvents like polycyclic aromatic hydrocarbons (PAHs), and many more have recently been shown to be susceptible to enzymatic conversion.

When pesticide and herbicide residues like dichlorodiphenyltrichloroethane (DDT), isoxaflutole, or atrazine are added to the list, and conversions of toxic metal and radioactive compounds to less harmful forms are included, the catalog of biodegradable contaminants becomes quite impressive. Initially, these wastes and contaminants have been bioremediated in the lab and the environment, but recently many industries have tried employing biodegradation even before the compounds become environmental contaminants, largely for public-relations, legal, and economic reasons.

Especially in the textile industry, many microbes have been tested in attempts to reduce, eliminate, or convert harmful chemicals necessary in production and processing. Three major areas of application include the decolorization of dye effluents, and enzyme treatment of both natural and manufactured fibers. Microbes can also assist by increasing the efficiency of mechanical or otherwise lengthy processes during production to decrease chemical use or high energy consumption, thus reducing the overall environmental impact of a given industry. The breakdown of dyes using various liquid cultures of fungi, enzyme extracts from cultures of fungi, and strains of the bacterium *Pseudomonas* have allowed for effluent to be recycled during industrial procedures or discharged without the typical expensive chemical treatments.

Relying on chemical applications can also create hazardous by-products requiring further treatment, leading to increased costs. The natural fiber treatments use enzyme extracts from microbes in scouring, retting, seed coat removal, and finishing of the clothing fibers. Often the use of these enzymes can result in an improvement in the quality of the fibers. Manufactured fiber research using serine proteases for finishing polyester fibers has also shown a reduction in effluent pollution, leading to greater cost reduction. Even the production of stonewashed jeans had assistance from enzymes produced by a species of *Trichoderma* fungus, instead of using harsh bleaches. The realization that manipulating microbes could decrease environmental as well as monetary impact has led researchers of wood decay fungi to apply their enzymatic systems to several chemical problems.

Wood decay fungi. When bioprospecting for organisms to degrade persistent pollutants and toxic industrial chemicals, wood decay fungi have several advantages over generic bacteria or even other fungi. The powerful nature of their enzymatic arsenal becomes obvious when learning more about what these fungi live on in the environment. Remember that wood is mainly composed of two chemicals: cellulose, which is white, and lignin, which is brown. Wood decay fungi are classified on the basis of which of these two is still present in degraded wood (**Fig. 1**). Brown rot fungi can degrade cellulose, leaving the brown lignin. White rot fungi, in their simplest form, degrade lignin and leave the white cellulose behind. There are also "simultaneous white rotters" that can degrade both cellulose and lignin, but in these cases the less abundant lignin is degraded first, leaving white cellulose. The ability of white rot fungi to break down lignin, one of nature's largest and most complex compounds (often referred to as nature's cement), makes the task of digesting DDT look like easy.

White rot fungi are able to degrade such physically and chemically diverse organic compounds because of both their inherent avenues of metabolism and the nonspecificity of their lignin-degrading enzymatic systems. The enzymes, called ligninases, utilize a free radical–based mechanism, where unlike typical microbial enzymes the structure of the substrate (food molecule) is not required to fit a specific site on the enzyme. Ligninases use hydrogen peroxide and other chemicals to generate highly active free radical molecules that break down, and generate other free radicals when they collide with other molecules. It is the explosive nature of this mechanism that has led to the term "enzymatic combustion." In addition, all fungal enzymes, called exoenzymes, are excreted outside the fungal cell wall, allowing the enzymes to reach substrates otherwise not available to ingestion.

The knowledge of white rot fungal enzymatic systems has blazed a path for this inquisitive fascination to become environmentally beneficial. Initial research was in the papermaking industry in a process known as biopulping (discussed below) and was completely based on utilizing the fungi's ability to degrade lignin. However, new research is finding that

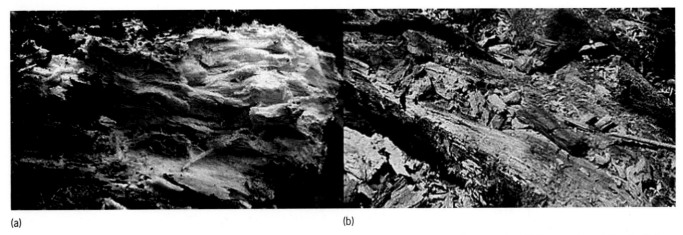

(a)   (b)

Fig. 1. Fungal action on wood. (*a*) White rot fungi digest lignin and leave cellulose behind. (*b*) Brown rot fungi digest cellulose and leave lignin behind.

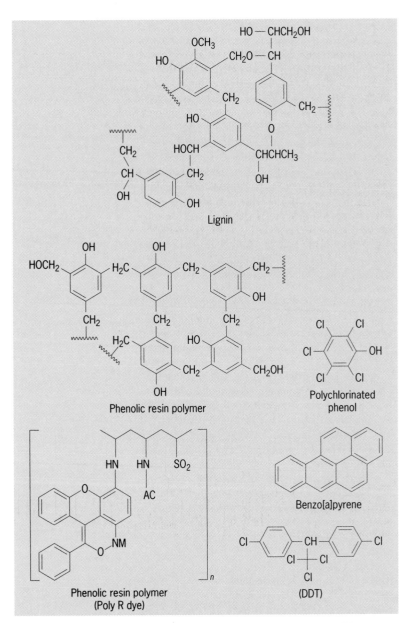

Fig. 2. Chemical structure of lignin compared with the structure of some recalcitrant molecules. (*Adapted from J. A. Field et al., 1993*)

many wood decay fungi can be extremely beneficial in converting otherwise problematic agricultural and lignocellulosic wastes into economical commodities like biofuels, fragrances, food additives, and animal feedstock. A large list of recalcitrant chemicals that are degradable by white rot fungi is given in the **table**. **Figure 2** shows the structures of some of these chemicals compared to lignin.

Recent research conducted in our lab in La Crosse, Wisconsin (A. C. Gusse et al., 2006), has demonstrated the ability of a white rot fungus to degrade phenolic resin, a previously nonbiodegradable industrial polymer (plastic-like) that is not commercially recycled. These phenolic resin polymers, first known as Bakelite, glue layers of plywood together, provide the binding matrix to particleboard, or laminate the surface of Formica™ countertops. They are also used in constructing rotary telephone casings, bowling balls, toilet seats, motor casings, and many other everyday products. The annual production of phenolic resin is 2.2 million tons (2 million metric tons) per year. Formerly considered to be nonbiodegradable and contributing greatly to landfills, phenolic resin polymers can be degraded by *Phanerochaete chrysosporium*, as our recent research has demonstrated, through three lines of independent evidence.

Biopulping. A more studied use for white rot fungi has been in biopulping. Fungi were first used in the industrial breakdown of wood, or biopulping, as far back as the 1950s. Biopulping is the process of using white rot fungi to delignify wood chips as a pretreatment to pulping during the papermaking process. Currently, chemical and mechanical treatments, alone or in combination, are used to remove the structurally steadfast and complex lignin molecules from the cellulose and hemicellulose fibers needed for papermaking. Chemical treatments generate waste effluent that must be treated, therefore increasing costs. Mechanical treatment plants alone can produce up to 90% higher yields, are less polluting, and are more economical to construct. However, they produce low-quality pulps that are limited in their uses and require demanding

Environmental pollutants degraded by white rot fungi*	
Compound	Fungal species†
Polycyclic aromatic compounds	
Anthracene	Pc, Tv, Ba, Ni
Benzo[a]pyrene	Pc, Tv, Ba, Ni
Chrysene	Pc
Fluorene	Pc
Phenanthrene	Pc, Tv, Cl
Pyrene	Pc
Chlorinated aromatic compounds	
Chloroguaiacols	Pc
4-Chloroaniline	Pc
Di- & trichlorophenol	Pc
Di- & tetrachlorodibenzo-*p*-dioxin	Pc
2,4,5-Trichlorophenoxyacetic acid	Pc
Pentachlorophenol (PCP)	Pc, Ps, Po, Th, Cs
Polychlorinated biphenyls (PCBs)	
Aroclor 1254	Pc, Tv, Pb, Fg
Tetrachlorobiphenyl	Pc
Pesticides	
Chlordane	Pc
DDT	Pc
Lindane	Pc
Toxaphene	Pc
Dyes	
Azo dyes	Pc
Crystal violet	Pc
Polymeric dyes	Pc
Munitions	
Di- & trinitrotoluene (DNT and TNT)	Pc
Cyclotrimethylenetrinitoamine (HMX)	Pc
Cyclotetramethylenetetranitramine (RDX)	Pc
Phenolic resin polymers	Pc
Others	
Cyanides	Pc
Azide	Pc
Aminotriazole	Pc
Carbon tetrachloride	Pc

*Compiled from D. P. Barr and S. D. Aust, 1994, J. A. Field et al., 1993, and A. C. Gusse et al., 2006.
†Ba, *Bjerkandera adjusta*; Cl, *Chrysosporium lignosum*; Cs, *Ceriporiopsis subvermispora*; Fg, *Funalia gallica*; Pb, *Phlebia brevispora*; Pc, *Phanerochaete chrysosporium*; Po, *Pleurotus ostreatus*; Ps, *Phanerochaete sordida*; Th, *Trametes hirsuta*; Tv, *Trametes versicolor*; Ni = new isolates (species unknown).

amounts of energy. Chemimechanical treatments can improve pulp quality and strength, but again produce chemical waste, decrease overall yield, and obtain no significant reduction in energy use. The hope has been to reduce these burdens by substituting fungi that are naturally adapted to performing this task, thereby eliminating the need for outside chemical usage.

Since its initial description in 1974, notable background knowledge has been gained on degradative mechanisms used by the fungus *Phanerochaete chrysosporium* (see http://TomVolkFungi.net), but any attempts to scale up the process beyond the lab have had little success. The experiments focused on fungal screening methods, inoculum preparation, bioreactor evaluation, mechanical refining, handsheet preparation, and their effects on strength and quality of the end product. This research resulted in hundreds of different biopulping runs that would show, from 1987 to 1992, how this process could reduce pulping energy by 40%, and could create a return on investment between 14.9 and 100%, depending on the bioreactor method that was chosen.

Unfortunately cutbacks in the pulp and paper industry and a lack of immediately applicable large-scale processes resulted in dissolution of biopulping efforts. The final research demonstrated that biopulping would be able to handle the demanded 200–2000 tons/day (182–1820 metric tons/day) in commercial setups. This final research provided an annual return of 62%, and has resulted in four patents, with five additional applications pending. These results, especially the industrial-sized tests, indicate that biopulping could be commercially implemented with *P. chrysosporium*, but to date no actual paper manufacturers are capitalizing on these benefits.

Outlook. The "lowly" fungi have developed this remarkable cadre of enzymes and are of importance to society. Besides those fungi described herein, other species of fungi have been shown to degrade all sorts of waste products—and even materials like stonewashed jeans. Although we have discussed mainly the white rot fungi, several volumes could be written about the thousands of other fungal species that can degrade thousands of different compounds. The fungi have developed many diverse and powerful enzymes, only relatively few of which are currently exploited in industrial processes. Research on the manipulation of fungal enzymes and enzymatic combustion for human use will undoubtedly lead to innovative applications for many decades to come.

For background information *see* BIODEGRADATION; ENZYME; FUNGAL BIOTECHNOLOGY; FUNGI; LIGNIN; PAPER; WOOD CHEMICALS in the McGraw-Hill Encyclopedia of Science & Technology.

Adam C. Gusse; Thomas J. Volk

Bibliography. D. P. Barr and S. D. Aust, Mechanisms white rot fungi use to degrade pollutants, *Environ. Sci. Tech.*, 28(2):78A–87A, 1994; J. A. Field et al., Screening for lignolytic fungi applicable to the biodegradation of xenobiotics, *TIBTECHS*, 11:44–49, 1993; A. C. Gusse, P. A. Miller, and T. J. Volk, White rot fungi demonstrate first biodegradation of phenolic resin, *Environ. Sci. Tech.*, 40:4196–4199; I. R. Hardin, D. E. Akin, and S. J. Wilson (eds.), *Advances in Biotechnology for Textile Processing*, Department of Textiles, Merchandising and Interiors, University of Georgia, Athens, 2002; T. K. Kirk et al., *Biopulping: A Glimpse of the Future?*, U.S. Department of Agriculture Forest Service, Forest Products Laboratory Res. Pap. FPL-RP-523, Madison, WI, 1993; R. E. Parales et al., Biodegradation, biotransformation, and biocatalysis (B3), *Appl. Environ. Microbiol.*, 68(10):4699–4709, 2002; D. VanderZwaag, The precautionary principle in environmental law and policy: Elusive rhetoric and first embraces, *J. Environ. Law Pract.*, 8:355–358, 1999; T. J. Volk, http://TomVolkFungi.net.

Genetic code

Proteins are essential constituents of all living cells. The myriad functions that proteins play are largely determined by the order in which different amino acids form the polypeptide chain, which is determined by the gene sequence encoding the protein. Transcription of a gene reproduces its information as a messenger ribonucleic acid (mRNA) used to make the protein during translation by the ribosome. The genetic code describes the pairing rules dictating to the ribosome which amino acids are encoded by which nucleotide triplet (codon) of the mRNA. Key to this process are transfer RNAs (tRNAs), each of which matches one amino acid with the right codon. Each amino acid has one or more dedicated tRNAs, attached by a cognate (partner) enzyme called an aminoacyl-tRNA synthetase (AARS). Twenty of these enzymes were known, one for each amino acid. The classic textbook account of translation therefore portrayed the genetic code as hardwired to encode 20 amino acids. However, it is now clear that the paradigm of a 20 amino-acid genetic code is not sufficient to describe processes in either nature or the laboratory. Two more novel genetically encoded amino acids, selenocysteine and pyrrolysine, have been found in some organisms. Further, as many as 30 unnatural amino acids have been added artificially to the genetic code of organisms.

Amino acids 21 and 22. Selenocysteine and pyrrolysine are very rare amino acids, in part explaining why they were not identified until 1987 and 2002, respectively, as genetically encoded. Hundreds of other rare amino acids exist in proteins but are modifications of existing amino acids made after translation. Selenocysteine and pyrrolysine are therefore sometimes called the 21st and 22d amino acids to denote their special status.

Selenocysteine is the major biological form of selenium, an essential trace element for humans. Although relatively rare in proteins, it is widespread and found in many organisms from the three domains of life (the Bacteria, Archaea, and Eukarya), including humans. Selenocysteine is similar to cysteine, one of the 20 amino acids, the only difference being that selenium atom replaces sulfur (**Fig. 1**). Many selenoproteins are enzymes catalyzing oxidation-reduction reactions. Well-studied examples are the mammalian glutathione peroxidases that remove hydroperoxides and protect cells from oxidative damage. Selenocysteine forms part of the catalytic site of many selenium-containing enzymes that are thought to carry out reactions faster, or under more diverse conditions, than their counterparts that contain cysteine rather than selenocysteine.

L-Pyrrolysine resembles the genetically encoding amino acid lysine, except that it is modified by the addition of a pyrroline ring (a ring of four carbon atoms and one nitrogen atom) [Fig. 1]. While selenocysteine is rare in proteins but found in many organisms, pyrrolysine appears rarely in proteins and is rare in organisms. As pyrrolysine was only recently discovered, it may yet prove to be more widespread. Genetic encoding of pyrrolysine has thus far been found only in certain methane-forming archaeal microbes (relatives of the genus *Methanosarcina*) and one type of Bacteria (*Desulfitobacterium hafniense*). Pyrrolysine is part of the catalytic site of methylamine methyltransferases, enzymes that begin the process

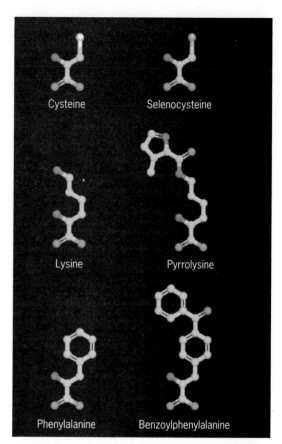

Fig. 1. Old and new members of the genetic code. Three of the original set of 20 genetically encoded amino acids (left side) are compared with the most recently discovered natural additions to the genetic code, selenocysteine and pyrrolysine, along with an example of an unnatural addition to the code, benzoylphenylalanine. Cysteine is analogous to selenocysteine, except that sulfur has been replaced by selenium. Pyrrolysine resembles lysine except for the addition of a chemically reactive five-membered ring, called a pyrroline ring. Benzoylphenylalanine resembles phenylalanine, a common amino acid, except for the addition of a photochemically active group. Benzoylphenylalanine was added to the code by mutagenesis of an AARS specific for tyrosine. Each of the new members of the code brings unusual chemistry previously lacking in other genetically encoded amino acids. The common natural amino acids are included for comparison only, and no precursor relationship to the new code additions should be inferred.

of making methane from methylamines formed from decaying biomass.

Genetic encoding of selenocysteine and pyrrolysine. For an amino acid to enter the genetic code, it must have (1) a codon assigned to it, (2) a tRNA that carries the amino acid to the ribosome and pairs the amino acid to its codon, and (3) an AARS that specifically attaches the tRNA to the amino acid. Moreover, the aminoacyl-tRNA (amino acid attached to the tRNA) must bind to elongation factors that participate in translation at the ribosome.

As all 64 possible codons are already assigned as the 20 amino acids, or as signals to stop translation, a codon must be reassigned (or recoded) for an additional amino acid to join genetic codes. UGA and UAG are normally stop codons that signal the ribosome to stop making protein, but in the right mRNA and right organisms they also mean selenocysteine (UGA) or pyrrolysine (UAG). Recoding UGA as a sense codon (a codon specifying an amino acid) requires an additional signal in the mRNA, a selenocysteine insertion sequence (SECIS) that forms a stem-and-loop structure downstream of the UGA codon. SECIS binds proteins that assist the ribosome in translating UGA as a selenocysteine codon, rather than signaling a stop. Similarly, a stem-loop found following UAG has been suggested to be a pyrrolysine insertion element (PYLIS), and may play a role encouraging pyrrolysine insertion into protein, rather than terminating translation.

Fulfilling further prerequisites of entrance to the genetic code, pyrrolysine and selenocysteine have their own dedicated tRNA species. For selenocysteine, the UGA-decoding tRNA (tRNA^{Sec}) is first attached to serine by the same AARS that cells use when serine is put into proteins (**Fig. 2**). However, serine attached to tRNA^{Sec} reacts with selenophosphate to form selenocysteinyl-tRNA^{Sec}. Unlike selenocysteine, pyrrolysine is made free in the cytoplasm by an unknown pathway that involves the PylB, PylC, and PylD proteins (**Fig. 3**). Pyrrolysine is attached to the UAG-decoding tRNA^{Pyl} by the PylS protein, an AARS specific for pyrrolysine. PylS was the first AARS found for an amino acid that is not one of the common 20 amino acids. The 21st and 22d amino acids therefore have very different strategies for entering the genetic code. Unlike all other natural genetically encoded amino acids, selenocysteine lacks its own AARS, and so is made indirectly on tRNA^{Sec}. Pyrrolysine is more similar to the original 20 amino acids, being preformed before attachment to tRNA.

Having been made, selenocysteinyl-tRNA^{Sec} and pyrrolysyl-tRNA^{Pyl} are taken to the ribosome by elongation factors. In Bacteria, selenocysteine attached to tRNA^{Sec} is bound to selenocysteine-specific elongation factor SelB (Fig. 2). In Eukarya, the selenocysteinyl-specific elongation factor mSelB (also called EF$_{Sec}$) binds SECIS indirectly via another SECIS binding protein, SBP2. The use of specialized

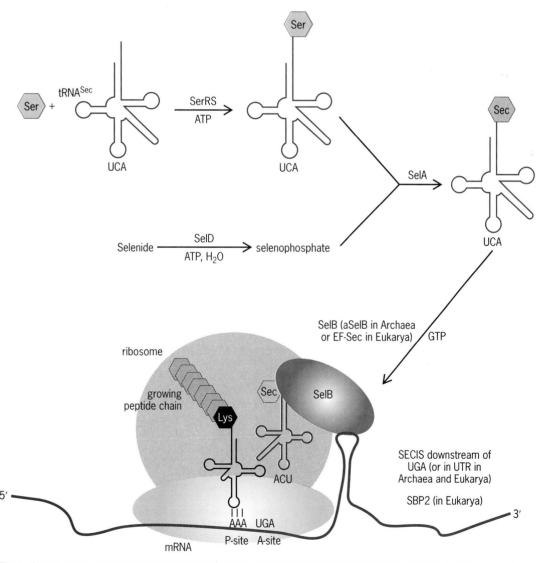

Fig. 2. Genetic encoding of selenocysteine. The reactions catalyzed by the bacterial enzymes are indicated. The means by which selenocysteinyl-tRNA is made in eukaryotes is less clear. Once made, selenocysteinyl-tRNA is bound to an elongation factor (SelB and homologs) that rejects all other aminoacyl-tRNA species. Binding of the specific elongation factor to SECIS helps selenocysteinyl-tRNA to compete with a release factor that would otherwise bind to the UGA codon and cause protein synthesis to stop, resulting in UGA translation as selenocysteine.

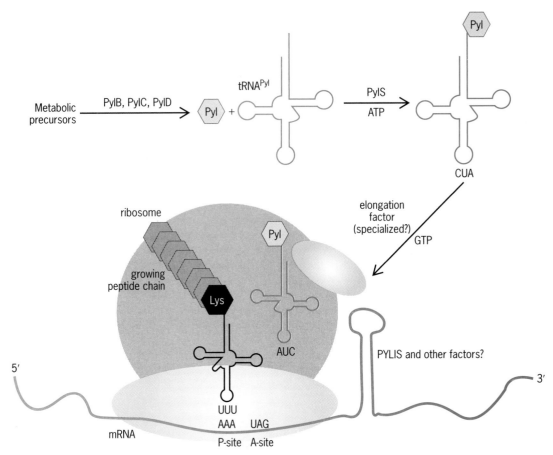

Fig. 3. Genetic encoding of pyrrolysine. The path of pyrrolysine biosynthesis is unknown, except for the involvement of the indicated enzymes. Once made, pyrrolysine is the substrate of PylS, a pyrrolysyl-tRNA synthetase. The pyrrolysyl-tRNA thus made may be carried to the ribosome by the normal host elongation factor, but a specialized factor has not been ruled out. The PYLIS element may help increase the efficiency of UAG translation as pyrrolysine, as opposed to UAG acting as a translation stop.

elongation factors binding to SECIS elements discourages translation factors from recognizing UGA as a stop codon. The location of SECIS in the untranslated part (UTR) of the mRNA in Eukarya and Archaea imparts the ability to insert more than one selenocysteine into a single protein; in contrast, Bacteria have SECIS close by UGA, and so can insert only one selenocysteine per protein chain. UGA encoding of selenocysteine is thought to be a fairly inefficient process, and UGA still functions often as a stop codon with selenoprotein mRNA.

It is unknown if pyrrolysyl-tRNA^{Pyl} requires a pyrrolysine-specific elongation factor, but the *Escherichia coli* elongation factor Ef-Tu (used for the 20 common amino acids) will bind pyrrolysyl-tRNA. This property allowed expansion of the genetic code of *E. coli* to include supplied pyrrolysine with only the genes for PylS and tRNA^{Pyl} added to the cell. In further contrast to selenocysteine, UAG encoding as pyrrolysine appears efficient. How pyrrolysyl-tRNA bound to its elongation factor and PYLIS achieve the high efficiency of UAG translation as pyrrolysine is unknown.

Artificial expansion of the code. Both selenocysteine and pyrrolysine have novel chemical properties not found in the first 20 amino acids, leading to their use in key enzymes of metabolism. This suggests

evolutionary pressures that might have driven the addition of these amino acids to the genetic code. Similarly, it is possible to artificially put pressure on cells in the laboratory to add unnatural amino acids to their code. Such unnatural amino acids are incorporated at sites throughout a protein and possess groups that are fluorescent, glycosylated, or chemically reactive. Artificial additions to the genetic code promise a boon to research, permitting novel precise ways to study how proteins function, as well as producing proteins with enhanced or new functions to serve biotechnology and medicine.

In order to add an amino acid to the genetic code, an orthogonal tRNA and AARS cognate pair must be introduced into the host cell whose genetic code will be changed. An orthogonal tRNA/AARS pair is foreign to the host cell and will interact minimally with other AARS or tRNA in the cell, but is still capable of interacting with the host ribosome. The orthogonal pair's natural amino acid specificity will be changed to a new unnatural amino acid, often a derivative of the natural amino acid originally recognized by the AARS (Fig. 1). The orthogonal tRNA's codon will be changed to a rarely used stop or sense codon, or even a 4-nucleotide (rather than 3-nucleotide) codon. UAG is often selected as the new codon. Many organisms tolerate change of this rare stop codon to a sense

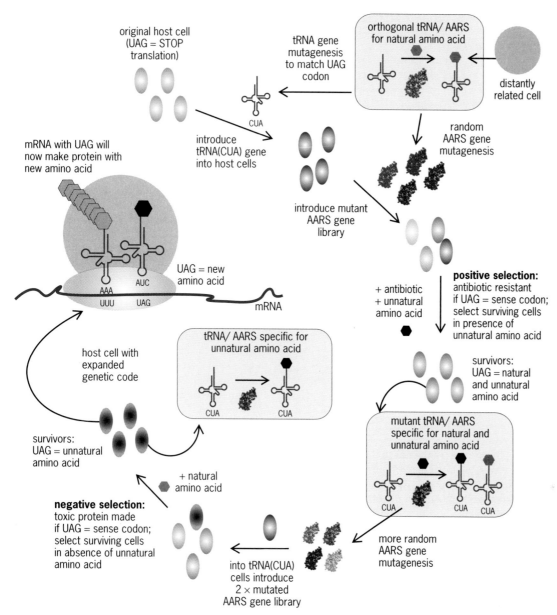

original host cell
(UAG = STOP
translation)

tRNA gene
mutagenesis
to match UAG
codon

orthogonal tRNA/ AARS
for natural amino acid

distantly
related cell

random
AARS gene
mutagenesis

mRNA with UAG will
now make protein with
new amino acid

introduce
tRNA(CUA) gene
into host cells

introduce mutant
AARS gene
library

UAG = new
amino acid

AAA   AUC
UUU   UAG

mRNA

positive selection:
antibiotic resistant
if UAG = sense codon;
select surviving cells
in presence of
unnatural amino acid

+ antibiotic
+ unnatural
amino acid

survivors:
UAG = natural
and unnatural
amino acid

host cell with
expanded
genetic code

tRNA/ AARS specific for
unnatural amino acid

CUA   CUA

mutant tRNA/ AARS
specific for natural and
unnatural amino acid

CUA   CUA   CUA

survivors:
UAG = unnatural
amino acid

+ natural
amino acid

negative selection:
toxic protein made
if UAG = sense codon;
select surviving cells
in absence of unnatural
amino acid

into tRNA(CUA)
cells introduce
2 × mutated
AARS gene library

more random
AARS gene
mutagenesis

Fig. 4. Adding a synthetic amino acid to the genetic code. An orthogonal pair (AARS and its cognate tRNA) that normally serves to bring a natural amino acid to the ribosome is converted to serve an unnatural amino acid by mutagenesis and selection. In this example, the tRNA is first mutated to recognize UAG rather than its former codon (tRNA$_{CUA}$). The AARS gene goes through two rounds of mutagenesis that create libraries of cells which each contain a slightly different mutant AARS enzyme. The mutants are screened by positive selection for UAG translation in the presence of the unnatural amino acid; then following another round of mutation, by negative selection against UAG translation with only the natural amino acid. Each type of selection relies on a gene whose mRNA contains a UAG codon that must be translated to make the protein. For positive selection, the translation of UAG in the gene results in a protein that allows the cell to survive treatment with antibiotics. For negative selection, translation of UAG codon without the unnatural amino acid (and with the natural amino acid) allows production of a protein that degrades cellular RNA. Finally, the gene selected by these cycles of mutation and selection makes an AARS, attaching only the unnatural amino acid to tRNA$_{CUA}$. The unnatural aminoacyl-tRNA$_{CUA}$ then enters protein synthesis at the ribosome.

codon, even in the absence of additional signals in mRNA. The tRNA must be mutated to match the new codon, and cycles of mutagenesis and selection of the AARS and tRNA are undertaken (**Fig.** 4). The tRNA is mutated so that it is not recognized by the host AARS, and the AARS is mutated so that it attaches only the unnatural amino acid (not the natural amino acid) to only its cognate tRNA. The cells containing the modified orthogonal tRNA/AARS will now specifically use the desired codon to encode the new amino acid, leaving the organism with an expanded genetic code.

Similar approaches have now been applied to not only *E. coli* but also eukaryotes like yeast and even mammalian cells in culture. The latter achievements have led to predictions that expansion of the genetic code of multicellular organisms will soon be possible.

For background information *see* AMINO ACIDS; BIOCHEMISTRY; GENE; GENETIC CODE; GENETIC

ENGINEERING; NUCLEOTIDE; PROTEIN; RIBONUCLEIC ACID (RNA) in the McGraw-Hill Encyclopedia of Science & Technology.

Anirban Mahapatra; Joseph A. Krzycki

Bibliography. S. K. Blight et al., Direct charging of tRNA(CUA) with pyrrolysine in vitro and in vivo, *Nature*, 431:333–335, 2004; A. Böck et al., Selenoprotein synthesis: An expansion of the genetic code, *Trends Biochem. Sci.*, 16:463–467, 1991; D. Driscoll and P. R. Copeland, Mechanism and regulation of selenoprotein synthesis, *Annu. Rev. Nutr.*, 23:17–40, 2003; B. Hao et al., A new UAG-encoded residue in the structure of a methanogen methyltransferase, *Science*, 296:1462–1466, 2002; T. L. Hendrickson, V. de Crécy-Lagard, and P. Schimmel, Incorporation of nonnatural amino acids into proteins, *Annu. Rev. Biochem.*, 73:147–176, 2004; G. Srinivasan, C. M. James, and J. A. Krzycki, Pyrrolysine encoded by UAG in Archaea: Charging of a UAG-decoding specialized tRNA, *Science*, 296:1459–1462, 2002; L. Wang, J. Xie, and P. G. Schultz, Expanding the genetic code, *Annu. Rev. Biophys. Biomol. Struc.*, 35:225–249, 2006.

Giant squid

The giant squid, *Architeuthis*, is renowned as the largest invertebrate in the world. The largest squid so far recorded, which measured 55 ft (16.8 m) in total length (from tip of the fins to tip of the longest tentacle), was found stranded on a Newfoundland seashore in 1878. Considerable effort to view this elusive creature in its deep-sea habitat has been expended, but until the morning of September 30, 2004, no one had ever reported an observation of a live giant squid in the wild. On that morning, images of a live giant squid in its natural environment were captured by an underwater digital camera and depth recorder system in the deep sea off the Ogasawara Islands in the western North Pacific. This is the first evidence showing the behavior and biological characteristics of a live giant squid in the wild.

Previous records. During the midnineteenth to early 20th century, 19 nominal species of *Architeuthis* were reported from oceanic localities around the world. Most of them were so inadequately described and poorly understood that the systematics of *Architeuthis* are still misleading. Today *A. dux* in the northern Atlantic, *A. sanctipauli* in the Southern Hemisphere, and *A. japonica* in the northern Pacific are suggested as valid species, although some systematists consider that all may be synonymous with *A. dux*, the first described species of the genus. A few sightings and strandings of dead and/or moribund giant squids have been reported, especially from seashores of Europe, Newfoundland, New Zealand, South Africa, and Japan. This indicates that the giant squids are widely distributed primarily in the subarctic, subantarctic, and temperate waters of the world oceans. They do not live in the colder waters of the Arctic and Antarctic oceans. Giant squids

have also been reported from stomach contents of sperm whales harpooned during 1960–1980, prior to the worldwide prohibition of commercial whaling. Studies on the feeding habits of sperm whales, regardless of the different areas and localities where they were captured, revealed that most of their prey consisted of unfamiliar, large deep-sea squids, including *Architeuthis*. The giant squids were generally few in number in the stomachs; however, when estimating prey composition in volume, they became a quantitatively important prey for sperm whales caught off northwestern Africa, New Zealand, and South Africa. Judging from sperm whale distribution and feeding behavior, the giant squid is estimated to live in the deep sea at about 500–2000 m (1640–6560 ft) and is especially abundant in the waters off the Madeira Islands in the Atlantic and off northwestern New Zealand in the Pacific.

Characteristics. Although the giant squid lives in the deep sea, its external appearance is almost the same as those of common squids living in shallower waters, irrespective of the giant squid's huge size (**Fig. 1**). It has a long cylindrical, massive mantle with relatively small heart-shaped fins, a squarish head with large round eyes, and eight long arms with two rows of suckers. The most dramatic characteristic of the giant squid is the pair of extremely long tentacles which generally make up to two-thirds of the total length. Giant squids are unique among cephalopods as they can hold the long tentacle shafts together with a series of small suckers and corresponding knobs along the tentacles, enabling the shafts to be "zipped" together. This results in a single shaft bearing a pair of tentacle clubs in a clawlike arrangement at the tip. The giant squids do not have bioluminescent organs (photophores), which are common in small- to medium-sized deep-sea squids. As with many deep-sea squids, the giant squid incorporates minute pouches of ammonium solution within its flesh to provide neutral buoyancy. Live animal orientation and hunting techniques by giant squids had previously been unknown, although many authors presumed them to be sluggish predators.

Observation in the wild. An individual giant squid attacked the lower squid bait of the camera system at 900 m (2950 ft) over a sea-floor depth of 1200 m (3937 ft) off Chichi-Jima Island (26°57.3′N, 142°16.8′E) at 09:15 a.m. on September 30, 2004. The system contained a digital camera, timer, strobe, depth sensor, data logger, and depth-activated switch. The camera captured JPEG images of around 150 KB every 30 seconds for 4–5 hours. The system weighed about 3 kg (6.6 lb) in the air and 1.5 kg (3.3 lb) in the water. The system was attached to the end of a vertical long-line which was 1000 m (3280 ft) in length.

The first image (**Fig. 2**, top) captured by the digital camera showed that the giant squid approached the bait horizontally and wrapped the bait in a ball with its two long tentacles, with the other arms spreading widely. In the bottom image taken 30 seconds later, the squid was disappearing from view as it actively swam away from the camera system. The giant squid

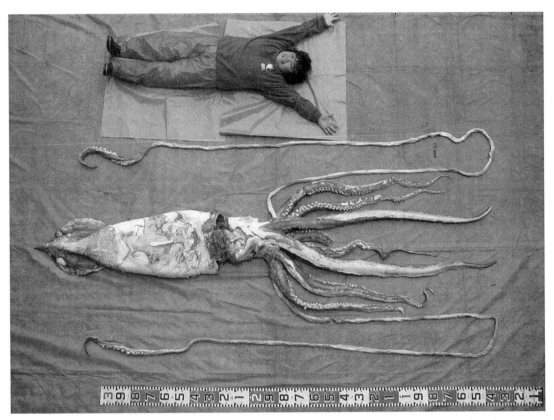

Fig. 1. A giant squid *Architeuthis* sp., fished up off Okinawa, Japan, on December 23, 2002: 141 cm (55.5 in.) mantle length. The Ryukyu woman is shown for size comparison.

appeared again after a few minutes because it became snagged on the squid jig (a fishing device jerked up and down or drawn through the water) by the club of one of these long tentacles. For the next 80 minutes, the squid repeatedly approached the line, spreading its arms widely or enveloping the line as it tried to detach from the jig. During this period, the camera system was drawn upward by the squid and/or current from a depth of 900 m (2950 ft) to 600 m (1968 ft). Over the subsequent 3 hours, the squid and system slowly returned to the planned deployment depth of 1000 m (3280 ft). During this period, the camera captured only the line and tentacle club at the corner of the camera frame. For the last hour, the line was out of the frame, suggesting that the squid was tired and could not pull the line strongly enough to turn the camera toward it. Finally the line came into the camera frame again when the giant squid started to swim forcefully and detach itself from the system. Four hours and 13 minutes after becoming snagged, the tentacle broke and the squid escaped from the jig.

The severed tentacle remained attached to the line and was subsequently retrieved from the camera system. The recovered section of tentacle was still functioning, with the large suckers of the tentacle club repeatedly gripping the boat deck and any offered fingers. The tentacle portion was 5.5 m (18 ft) long and the club length was 72 cm (28 in.), with the largest sucker being 28 mm (1.1 in.) in diameter. Based on calculations between club length, sucker diameter, and mantle length, the size of the squid was esti-

mated to be 1.6–1.7 m (5.2–5.6 ft) in mantle length and approximately 4.7 m (15.4 ft) from the tip of the fins to the tip of the arms. Combined with the retrieved tentacle, the squid should be more than 8 m (26 ft) in total length.

New findings. The photographs indicate that the giant squid was hunting at 900 m (2950 ft) during the day. Sperm whales feed at this depth during the day and at 400–500 m (1312–1640 ft) at night. Giant squids probably rise in the water column at night to feed in these shallower depths. Based on the images, giant squids are probably much more active predators than previously suggested and appear to attack their prey from a horizontal orientation. The giant squids can retract their long tentacles once a prey has been captured. The tentacles apparently coil into an irregular ball in much the same way that pythons rapidly envelop their prey within coils of their body immediately after striking. The long tentacles are clearly not weak fishing lines dangled below the body. The tentacle resisted for more than 4 hours the pull of the camera system in the ocean current and that of its own swimming (finning and/or jetting). The giant squid had the swimming ability and power to pull the camera system against the current for several hours and enough strength to break off the tentacle by itself after 4 hours of grappling. This encounter has given us important knowledge on the habitat and behavior of the giant squid, although there is still much to learn about these spectacular animals.

For background information *see* CEPHALOPODA;

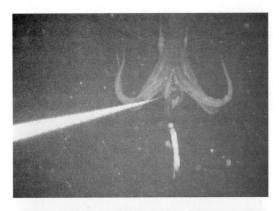

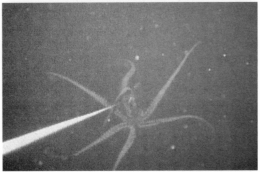

Fig. 2. First image (top) of the giant squid in the wild, and an image showing arms spreading (bottom). (*Courtesy of the Proceedings of the Royal Society: T. Kubodera and K. Mori, First-ever observations of a live giant squid in the wild, Proc. R. Soc. Lond. (Biol. Sci.), 272(1581): 2583–2586, 2005*)

DEEP-SEA FAUNA; MARINE ECOLOGY; SQUID; TEUTHOIDEA in the McGraw-Hill Encyclopedia of Science & Technology.   Tsunemi Kubodera

Bibliography. F. A. Aldrich, Some aspects of the systematics and biology of squid of the genus *Architeuthis* based on a study of specimens from Newfoundland waters, *Bull. Mar. Sci.*, 49: 457–481, 1991; T. Kubodera, *Studies on Systematic and Phylogeny of Giant Squid, Architeuthis, around Japanese Waters: Report of JSPS Grant-in-Aid for Scientific Research, 2001-2003* (13660197), 1–15 (in Japanese), 2004; C. F. E. Roper and K. J. Booss, The giant squid, *Sci. Amer.*, 246:96–105, 1982.

Gold-catalyzed reactions

The activation of multiple bonds is one of the central reactivity patterns in transition-metal catalysis. Selectivity is a crucial issue in terms of the connection of the atoms (chemoselectivity) and the steric arrangement of the atoms (stereoselectivity) in the product. If out of several possible products only one is formed, transition-metal catalysis will contribute to the responsible use of resources in the sense of green chemistry. A high catalytic activity is important, too. The discovery of new direct chemical transformations, which previously could only be achieved by multistep synthesis, is highly innovative.

The periodic table of the chemical elements offers only 82 nonradioactive elements as basic building blocks for chemistry, and gold is one of them.

While gold has been investigated in stoichiometric reactions with the same intensity as every other chemical element, only a few papers have been published in the field of catalysis, especially for homogeneous catalysis. There has been widespread prejudice among scientists that gold is too expensive and too unreactive to be a catalyst. However, catalysis metals, such as rhodium and platinum, are more expensive than gold. And based on the huge amount of gold produced to date, the price of gold is more stable than the price of platinum or rhodium. In addition, the ligand necessary to convert a metal ion to the catalytically active complexes often is more expensive than the metal itself.

There were sporadic earlier reports that such gold complexes showed catalytic activity. In 2000, A. S. K. Hashmi and coworkers reported new gold-catalyzed carbon-carbon bond-forming reactions. This initiated a kind of gold rush in homogeneous transition-metal catalysis, as other researchers began to take a closer look at gold. Since 2005, the number of publications has exploded in the field of gold-catalyzed organic reactions for the synthesis of both bulk and fine chemicals.

Most gold catalysts are based on gold(III) complexes with halide, nitrogen, or oxygen ligands, gold(I) complexes with halide or phosphane ligands, or gold nanoparticles.

Nucleophilic additions to carbon-carbon multiple bonds. Figure 1*a* shows the reaction pattern that has been most widely studied. The reaction is initiated by the coordination of the carbon-carbon multiple bond to the gold catalyst, which activates it for the addition of a nucleophile (Nu). The fact that gold has a high affinity for carbon is very helpful for organic synthesis, with alkynes being the most popular substrates. The more reactive allenes are used less often, while alkenes are increasingly popular. The nucleophilic groups can be introduced by intramolecular (including ring expansion) or intermolecular means, with many examples using carbon, nitrogen, oxygen, sulfur, and chloride nucleophiles having been reported. The process is highly diastereoselective, with anti addition observed in most cases. Often, similar reactions can be catalyzed by other metal complexes, but the activity of the gold complexes has proven to be significantly higher.

The most innovative reactions were proceeded by gold carbenoids (carbene complexes), which ultimately led to interesting cyclopropanation reactions. Two major reactions that lead to these intermediates are known: the reactions of propargyl esters and the reactions of enynes, which contain double and triple bonds (Fig. 1*b*).

Activation of terminal alkynes. The C-H bond of a terminal alkyne can be activated by gold catalysts, too. A number of three-component reactions which assemble propargylic amines from an aldehyde, an amine, and a terminal alkyne have been described.

Hydrogenations. Hydrogen can be added to alkynes, allenes, or alkenes. Compared to the many well-known transition-metal-catalyzed

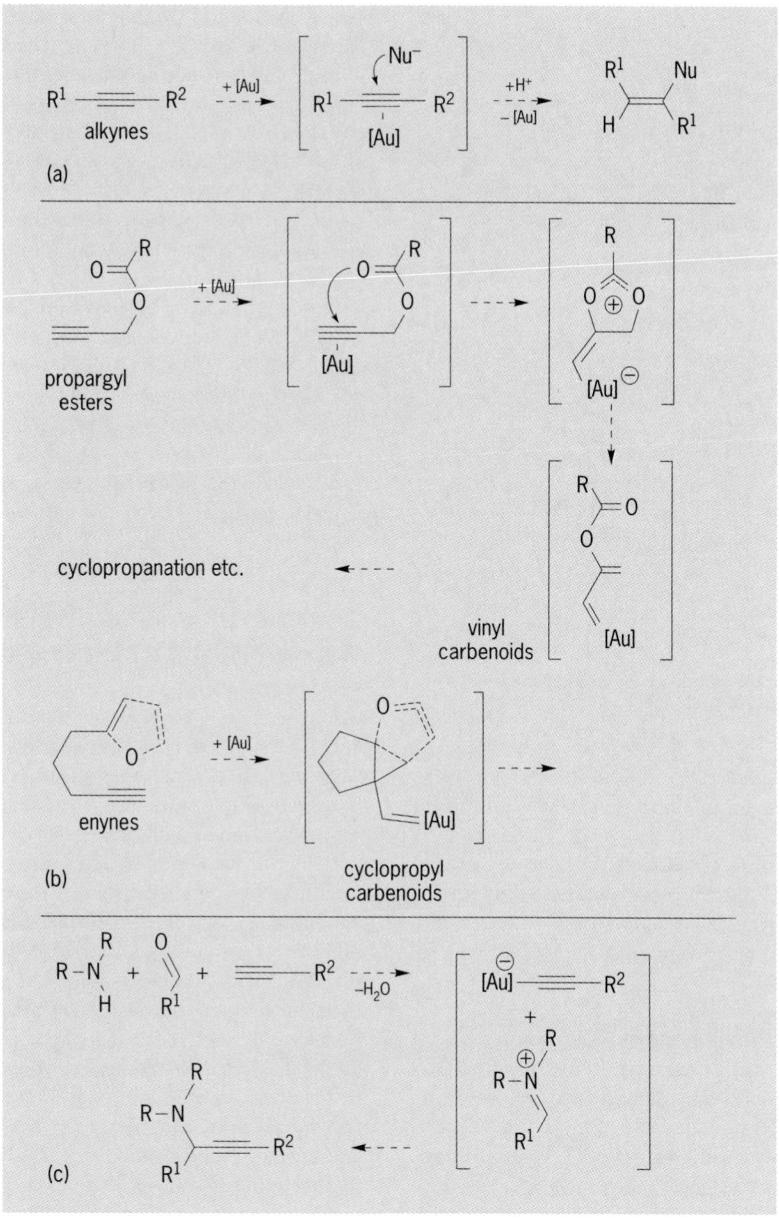

Fig. 1. Activation of unsaturated organic molecules by gold complexes [Au]. (*a*) General activation for the attack of a nucleophile. (*b*) Formation of gold carbenoid intermediates. (*c*) Activation of terminal C-H bonds.

hydrogenation reactions, gold-catalyzed hydrogenations offer interesting advantages. One of the strengths of heterogeneous gold catalysts is the ability to selectively (but not specifically) reduce α,β-unsaturated aldehydes at the C=O double bond (**Fig. 2a**). Catalytic asymmetric hydrogenations with soluble chiral gold complexes to enantiomerically pure products were possible, and in certain cases better enantiomeric excesses were obtained than with the corresponding platinum and iridium complexes. Mechanistically, it is interesting that such reactions can proceed via gold(III) monohydride complexes which form via dissociation of dihydrogen to hydride (cooridnated to the gold) and a proton (transferred to the substrate) in a heterolytic mode, avoiding a change of the oxidation state of gold in the sense of an oxidative addition (Fig. 2*b*).

Related processes that were recently developed include highly chemoselective dehydrogenative silylation of alcohols with triethylsilane which show a high functional group tolerance, diborylations, the synthesis of distannane (organotin) compounds, and dehydrogenative aromatization, for example, in a pyridine synthesis.

Selective oxidations. Unlike hydrogenations for which no uncatalyzed "background" reaction is known, oxidation with molecular oxygen (the ultimate green oxidant) always suffers from the problem of uncatalyzed, usually unselective background reactions. Here the focus has been on two reaction types: epoxidation and selective oxidation of alcohols to aldehydes.

In epoxidation reactions with heterogeneous gold catalysts, the competing allylic oxidation can be

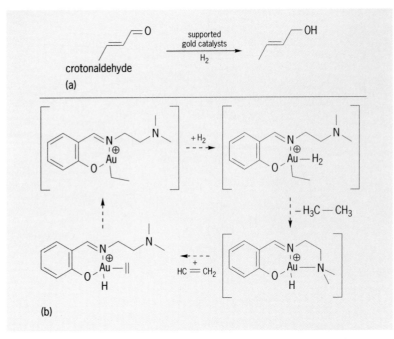

Fig. 2. Gold-catalyzed hydrogenations. (*a*) Selective reduction of crotonaldehyde. (*b*) Activation of the dihydrogen molecule without changing the oxidation state of gold.

suppressed, achieving up to 80% selectivity for the epoxides produced in these industrially important processes (**Fig. 3a**).

The oxidation of alcohols by gold nanoparticles produced exciting results with carbohydrates, as a positionally selective oxidation of glucose to gluconic acid with impressive catalyst activity was reported (Fig. 3b).

The selective oxidation of alcohols using air or oxygen and homogeneous gold complexes or gold particles on a heterogeneous support has been reported, with outstanding yields and selectivities.

Outlook. One of the greatest challenges with immense commercial significance is the selective activation of C-H bonds in alkanes. Using cyclohexane as a typical test substrate, some promising results with heterogeneous catalysts have been reported. Further development is needed, however, since the

highly symmetric cyclohexane does not show positional selectivity. The same is true for results with soluble catalysts and methane as the substrate.

For background information *see* ALKENE; ALKYNE; ASYMMETRIC SYNTHESIS; CATALYSIS; GOLD; HETEROGENEOUS CATALYSIS; HYDROGENATION; ORGANIC SYNTHESIS; OXIDATION-REDUCTION; REACTIVE INTERMEDIATES; STEREOCHEMISTRY in the McGraw-Hill Encyclopedia of Science & Technology.

A. Stephen K. Hashmi

Bibliography. A. S. K. Hashmi et al., A new gold-catalyzed C-C bond formation, *Angew. Chem. Int. Ed.*, 39:2285, 2000; A. S. K. Hashmi et al., Highly selective gold-catalyzed arene synthesis, *J. Amer. Chem. Soc.*, 122:11553, 2000; A. S. K. Hashmi and G. Hutchings, *Angew. Chem. Int. Ed.*, vol. 45, in press, 2006; S. Ma et al., Gold-catalyzed cyclization of enynes, *Angew. Chem. Int. Ed.*, 45:203, 2006.

High-throughput materials chemistry

The speed and efficiency of research aimed at the discovery and optimization of catalysts, polymers, and other materials have been significantly improved over the last several years through the application of high-throughput research and development techniques. Automated laboratory tools for the rapid or parallel preparation, characterization, and testing of new materials, and specialized software enable the simultaneous design of multiple experiments, automate the hardware, and facilitate the evaluation of the large and numerous data sets generated. The high-throughput approach is especially applicable to problems where there is a large parameter space to be investigated and where the outcome is the result of an unpredictable interdependence among the variables of the parameter set. Homogeneous and heterogeneous catalysis fit this description. It is therefore not surprising that research groups are reporting high-throughput approaches to catalyst discovery. Polymer science and electronic materials are other examples. The number of experiments that can be done using state-of-the-art high-throughput techniques is more than an order of magnitude greater than was possible only a few years ago using conventional research techniques. A high-throughput program can yield 50,000 experiments per year, compared to 500 to 1000 experiments using traditional methods.

The drivers for these changes, especially in industry, include the need to reduce the commercialization time for new and optimized materials and processes, increase the probability of technical success made possible by far greater numbers of experiments, achieve better intellectual property protection via the newly available thoroughness with which an area can be explored, shorten/execute more research projects per unit time, and increase organizational efficiency and transparency because of improved data storage, access, analysis, and sharing. Many of these advantages had already been

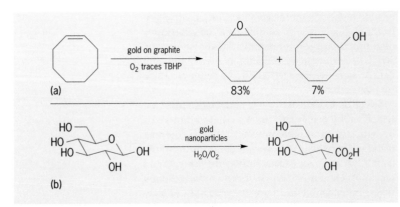

Fig. 3. Selective oxidations of different organic substrates. (*a*) Heterogeneous gold-catalyzed epoxidation reaction [TBHP = tert-butyl hydroperoxide]. (*b*) Gold-nanoparticle-catalyzed oxidation of glucose to gluconic acid.

realized in the pharmaceutical industry, where long developmental times and high research costs have forced the development of high-throughput approaches to accelerate the drug discovery process.

Hierarchical workflow. High-throughput research, development, and scale-up workflows leading to a final commercialized material can be divided into primary, secondary, and tertiary screening stages (**Fig. 1**).

Primary screening. Primary screening usually consists of very high throughput synthesis and testing activities done on small-scale samples, often using unconventional reactor designs. Most often, this phase focuses on discovering new materials rather than improving existing materials, with the objective being to screen a large and diverse set of material types for the desired catalytic reaction and physical properties. Discoveries, termed "hits," are then taken to the secondary screening phase. It is critical that the primary screening results correlate with the real properties of the material. This can usually be accomplished by screening for qualitative trends and by means of relative performance rankings, generated by using a simplified analogue of the real process parameter. For example, the exothermicity of a polymerization reaction, measured by infrared thermography, would be a primary screen for the activity of a set of olefin polymerization catalysts, conventionally measured by directly quantifying the olefin consumption rate. In another example, the viscosity and density of a formulation, such as a lubricant, can be screened by measuring the frequency response of a miniature mechanical resonator immersed in it. The primary screen must be designed to minimize both false positive and, especially, false negative results. The former can be eliminated in the secondary screen, but the latter means missing potentially good materials. Throughput for a primary screen can reach 1000 samples per day or more.

Secondary screening. Secondary screening is used to confirm and optimize primary hit materials or to optimize known materials. In this step, many of the primary screening hits, including the false positives, are eliminated. In contrast to the primary screen, the synthesis procedure, form of the material (such as formulation, macroscopic shape, and so on), reactor geometry, physical property testing devices, and process analytics are designed to represent closely those obtained from the real bench- or pilot-scale tests. The data quality and precision here should be high since the goal is to observe small improvements in performance as a function of material modification. In many cases, secondary screening technology has evolved to the point where the synthesis and screening of hundreds of materials per week is possible, with data quality equivalent to that obtained using conventional laboratory technology. In some instances, large pilot-plant studies can proceed directly from secondary screening, without tertiary screening.

Tertiary screening. Optimized hits, termed "leads," are usually taken to tertiary screening to generate commercial development candidates. The equipment

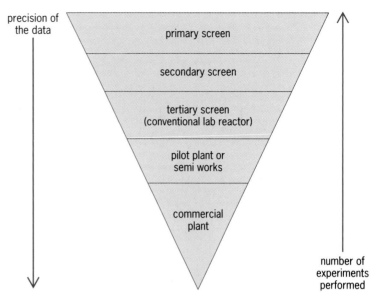

Fig. 1. Stages in catalyst discovery, scale-up, and commercialization.

used are usually commercial reactors or small pilot plants, in the case of catalyst testing. The material should be in its commercial form (for example, the pellet size and shape or overall formulation). Tertiary screening of a heterogeneously catalyzed reaction might be done, for example, by using conventional fixed-bed reactors with full reactant and product detection and a full mass balance. In some cases, this work can be done in a properly designed secondary screen. This is advantageous in that the cost and time required for multiple pilot tests on a large scale is prohibitive.

Software tools. A key component of any high-throughput workflow is specialized software that enables and integrates experimental design, material library design, automation of characterization and testing results, calculations, and data mining and visualization. An example of a commercial software package is Symyx® Technologies' Library Studio®, a graphics-based application for the rapid specification of up to hundreds of experiments (**Fig. 2**). All the necessary calculations for the synthesis of the specified materials are done by the software, a task that could take many hours if performed manually. The resulting experimental designs are saved to a database and later correlated to the testing results after the library is screened. Software control (automation) of the advanced synthesis and screening hardware is also essential to an efficient workflow. Software applications allow researchers to construct protocols for experimental execution (for example, reactor control) graphically in a flow-chart format. The data generated by a protocol during execution is stored in a server and linked to the library design. Visualization software then allows one to query the data and to view it in tables or graphs, spectra, chromatograms, images, and library summaries.

Hardware tools. High-throughput materials synthesis and screening equipment is quite diverse, given the

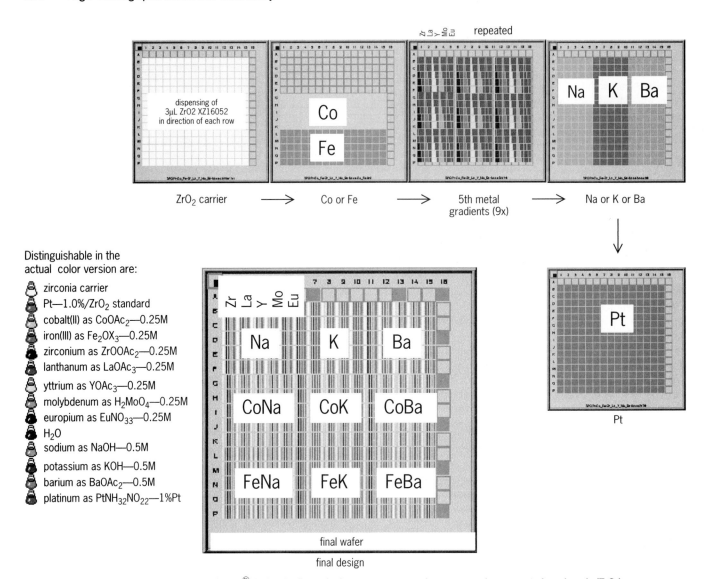

ZrO₂ carrier ⟶ Co or Fe ⟶ 5th metal gradients (9x) ⟶ Na or K or Ba

Distinguishable in the actual color version are:

- zirconia carrier
- Pt—1.0%/ZrO₂ standard
- cobalt(II) as CoOAc₂—0.25M
- iron(III) as Fe₂OX₃—0.25M
- zirconium as ZrOOAc₂—0.25M
- lanthanum as LaOAc₃—0.25M
- yttrium as YOAc₃—0.25M
- molybdenum as H₂MoO₄—0.25M
- europium as EuNO₃₃—0.25M
- H₂O
- sodium as NaOH—0.5M
- potassium as KOH—0.5M
- barium as BaOAc₂—0.5M
- platinum as PtNH₃₂NO₂₂—1%Pt

Pt

final wafer

final design

Fig. 2. Library Studio® design for five-point heterogeneous catalyst quaternaries supported on zirconia (ZrO₂).

variety of applications in materials research. Here we describe some primary and secondary synthesis and screening equipment used in the field of heterogeneous catalysis as illustrative examples.

For primary screening, catalyst arrays can be prepared and tested on two-dimensional substrates. **Figure 3** shows the automated procedures for preparing and testing 256-element catalyst libraries (16 × 16 arrays, about 1 mg each) on quartz wafers. Catalyst supports, such as silica or alumina, are first dispensed onto the wafers as slurries and, after drying, impregnated with premixed catalyst precursor solutions typically consisting of water-soluble metal salts. Support-free metal-oxide catalysts can be prepared by directly applying the precursors to the wafer and drying. The wafers are heat-treated in ovens and then tested for catalytic activity in primary screening reactors. Other primary synthesis methods, such as split pool, which are not based on spatially addressable libraries, have also been described in the literature.

Larger quantities of bulk catalysts (up to about 1 g) are used for secondary screening in heterogeneous catalysis because the testing reactors are generally fixed-bed units. The synthesis methods are similar to those used in larger-scale conventional research, but at a much smaller scale, and are done in parallel 50 to 100 at a time. Some of the challenges in these syntheses, especially when making unsupported catalysts, are sizing the catalyst particles reproducibly for the secondary-reactor fixed-bed dimensions, as well as solids-processing steps, such as grinding, which are often important synthetic variables.

Screening reactor designs in gas-solid heterogeneous catalysis range from scaled-down conventional laboratory reactors that share some common components such as feed systems and detectors, to parallel units that allow the evaluation of hundreds of catalysts simultaneously. Symyx researchers have developed a reactor system for primary screening based on a microfabricated microfluidic flow distribution and equalization device (a 16 × 16-catalyst array) and

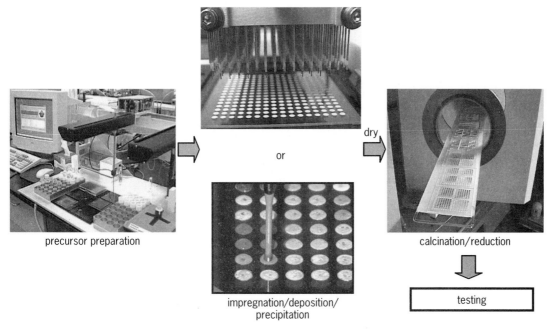

Fig. 3. Example of a primary synthesis workflow. Catalysts can be prepared by impregnation, deposition/evaporation, or precipitation.

an optical detection methodology that allows parallel reaction and detection (**Fig. 4**). Microreactors are formed by stacking the catalyst wafer and a gas distribution wafer. Each isolated reactant stream contacts a 2-mm-diameter by 0.2-mm-deep well containing the catalyst. After reaction, all the 256 product streams flow simultaneously out of the reactors onto an absorbent detection plate, where the reaction products are trapped. The absorbent array is removed from the reactor and sprayed with a dye solution that selectively interacts with the reaction products, causing a change in the absorption spectrum, or

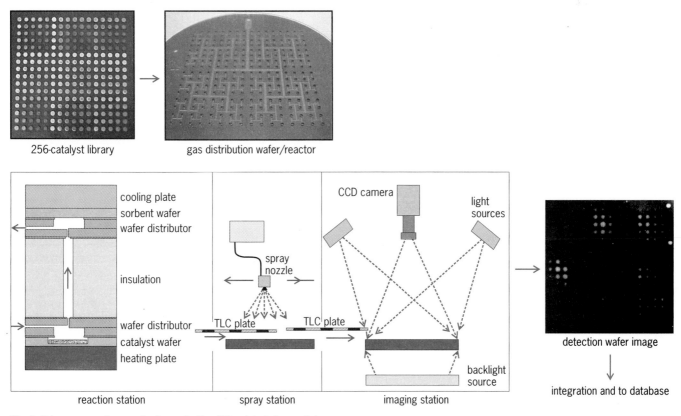

Fig. 4. Primary screening reactor for evaluating 256 catalysts in parallel.

enhanced or bleached fluorescence. The developed absorbent array is then imaged using a charge-coupled-device (CCD) camera, and the intensities are integrated and the data stored. This reactor has been used to screen catalysts for a variety of reactions, including the partial oxidation of ethane to acetic acid and the ammoxidation of propane to acrylonitrile.

Secondary screening is done in parallel fixed-bed reactors sized for catalyst loadings of 15–500 mg. A Symyx system, for example, is made up of 48 individual reactor beds arranged in six modules of eight reactor wells each. Six channels are analyzed in parallel, and effluent stream selection valves feed the analytical hardware, in this case a gas chromatograph. Reactant feeds are generated by vaporizing liquid flows from pumps or using gas feeds metered through mass-flow controllers. The feed is distributed equally among the 48 channels by using flow restrictors, ensuring equal feed gas residence time in each reactor vessel containing the catalyst. During the screening process a number of parameters, such as the temperature, pressure, and flow rate, are varied and monitored. The data are stored in a database for later analysis.

For background information *see* CATALYSIS; COMBINATORIAL CHEMISTRY; COMBINATORIAL SYNTHESIS; MATERIALS SCIENCE; POLYMER; SOFTWARE in the McGraw-Hill Encyclopedia of Science & Technology.

Anthony F. Volpe, Jr.; W. Henry Weinberg

Bibliography. A. Hagemeyer, P. Strasser, A. F. Volpe, Jr. (eds.), *High Throughput Screening in Chemical Catalysis: Technologies, Strategies and Applications*, Wiley VCH Verlag GmbH, Weinheim, 2004; H. Koinuma and I. Takeuchi, Combinatorial solid-state chemistry of inorganic materials, *Nat. Mater.*, 3(7):429–438, 2004; J. Margitfalvi (guest ed.) and R. B. van Breemen (ed.), Special Issue: Combinatorial Catalysis, *Combinatorial Chemistry & High Throughput Screening*, 2006; V. Murphy, High-throughput organometallic chemistry: Chemical approaches, experimental methods and screening techniques, in R. H. Crabtree and D.M.P. Mingos (eds.), *Comprehensive Organometallic Chemistry III*, vol. 1, ch. 10, 2006; U.S. Schubert and E. J. Amis (guest eds.) and I. Meisel (ed.), Special Issue: Combinatorial Material Research and High Throughput Experimentation in Polymer and Material Research, *Macromolecular Rapid Communications*, vol. 25, 2004.

Hybridization and plant speciation

The role of hybridization (crossing or mating of two plants) in plant speciation (species evolution) has fascinated generations of botanists and evolutionary biologists. Significant progress in understanding this topic has been made recently by cross-disciplinary studies combining quantitative trait locus (QTL) mapping, comparative genomics, ecological fieldwork, chip-based gene expression analysis, and genetic analysis of specific candidate adaptive traits and the genes that encode them. The combination of these approaches has allowed evolutionary biologists to revisit long-standing hypotheses and controversies regarding the role of hybridization in plant evolution, and conceptual developments fueled by this work are beginning to transform our view of how plant species are formed and maintained.

The potential roles of hybridization in plant speciation may comprise several different mechanisms (see **illustration**). Hybridization between existing species may lead to hybrid speciation. The term "homoploid hybrid speciation" is used if the hybridizing progenitors are of the same ploidy (the number of complete chromosome sets in a nucleus), whereas the term "allopolyploid speciation" refers to speciation that involves genome doubling. However, hybridization may also contribute to speciation via other mechanisms, such as the reinforcement of reproductive barriers between diverging populations. Also, continued hybridization (introgression) among related species may effectively enrich the gene pool of groups of taxa undergoing explosive bursts of speciation, a process also known as adaptive radiation. This article reviews recent progress in understanding the various mechanisms through which hybridization may contribute to plant speciation, highlighting new or previously underexplored concepts. The focus is on recent developments in homoploid and allopolyploid hybrid speciation.

Homoploid hybrid speciation. Hybridization has often been viewed in the past as evolutionary noise—a local phenomenon with only transient effects—but recent work indicates that it may often be a potent evolutionary force, creating opportunities for adaptive evolution. This possibility has been demonstrated not only in the case of allopolyploid speciation but also for homoploid hybrid speciation. The latter phenomenon has been demonstrated most convincingly for three natural diploid (having two complete chromosome pairs in a nucleus) hybrid species of wild annual sunflowers (*Helianthus* spp.), and the list of other potential and plausible examples in the plant kingdom has been growing steadily.

In sunflowers, three natural diploid hybrid species are known to stem from crosses between two parental sunflowers, *H. annuus* and *H. petiolaris*, with wide geographic distributions across North America. The three hybrid species are adapted to extreme habitats that are not tolerated by either of their parents: sand dunes (*H. anomalus*), desert floors (*H. deserticola*), and salt marshes (*H. paradoxus*). The hybrid origin of these three taxa became apparent from classical systematics and more recent molecular phylogenetics studies of the genus. Genetic map–based comparisons among the genomes of three experimentally synthesized hybrid lineages and one natural hybrid species (*H. anomalus*) revealed that the genomic composition of synthesized and ancient hybrids was concordant, which indicated an important role for selection in hybrid species formation.

Recent experimental work on *Helianthus* hybrid species has addressed the role of ecological selection in homoploid hybrid speciation in this group. In doing so, this work has also provided convincing evidence that hybridization per se can facilitate the evolution of ecological divergence and even major ecological transitions. Phenotypic selection studies with synthetic hybrid crosses in the greenhouse and the field have facilitated what may be seen as a replication of the earliest steps of the hybrid speciation process. These experiments allowed researchers to estimate the strength of ecological selection on extreme (transgressive) phenotypic traits in hybrids, and on the quantitative trait loci (chromosome segments) controlling these traits. This revealed that selection acting on the relevant chromosome segments was strong enough to counteract the homogenizing effect of gene flow, which is a prerequisite for speciation in the absence of geographic isolation. The genetic architecture of adaptive traits also was consistent with the complementary gene action hypothesis; that is, the origin of new adaptive phenotypes in sunflowers is facilitated by the reshuffling of genetic contributions (alleles) in hybrids with opposing effects in each parental species. In effect, contributions from both parental species are required to generate new adaptive phenotypes in hybrids via a process coined transgressive segregation. This hypothesis was strongly supported by results from a comparative genomics study involving all three sunflower hybrid species, their parents, and synthetic early generation hybrids. That study clearly showed that the same combinations of parental chromosome segments required to generate extreme phenotypes in synthetic hybrids also occurred in the ancient hybrids. Thus, hybridization contributed to major ecological transitions in this plant group.

Allopolyploid speciation. Allopolyploid speciation has also attracted renewed interest in recent years. It has long been known that new polyploid species can arise by hybridization between existing species and that new phenotypic and ecological characteristics can result from genome doubling. The nature of genetic changes following polyploidization, however, is only beginning to be understood, with a particular focus of current interest being changes in gene expression and "gene silencing" following polyploid formation. Studies on synthetic lines of the allotetraploid (having four sets of chromosomes and one or more sets derived from a different species) *Arabidopsis suecica*, a derivative of the model plant *A. thaliana* and its relative *A. arenosa*, for instance, revealed that approximately 1% of the transcriptome (the complete set of RNA transcripts produced by the genome at any one time) was silenced following polyploidization. Similarly, 5% of the genes examined in allopolyploid populations of *Tragopogon miscellus* have been silenced upon genome doubling. Both genetic changes (changes in the DNA sequence itself) and epigenetic changes (for example, localized methylation of the DNA, or so-called interference through RNA molecules) can lead to gene silencing

in polyploids. In addition, subfunctionalization may occur; that is, different copies (genetic loci) of a duplicated gene may be silenced in different organs or tissues of a polyploid plant. This phenomenon was observed in allotetraploid *Gossypium* (cotton). The development of chip-based analysis of gene expression, for example, in genera such as *Senecio*, holds enormous promise for understanding the impact of genetic changes following allopolyploid hybrid speciation.

Gene silencing, however, is just one of several possible genetic changes in allopolyploids. Other, perhaps even more enigmatic or drastic changes include intergenomic cross talk and genomic downsizing. Intergenomic cross talk refers to the movement of genetic material from one parental genome to another after the genomic merger. This is best exemplified by the colonization of the D genome of allotetraploid cotton by repetitive DNA elements of its A genome. Genomic downsizing, on the other hand, refers to a process of genome contraction in polyploids. This process is little understood but has been clearly documented via a general decrease in physical genome size (measured through flow cytometry) with increasing ploidy in the plant kingdom.

Clearly, genetic changes in allopolyploids pose many open questions for geneticists. However, numerous other questions remain to be addressed as well, such as the question as to how newly arisen allopolyploids are able to escape competition from their usually much more abundant diploid ancestors. One factor that may increase the chances of survival for newly arisen polyploids is ecological divergence. Alternatively, long-distance dispersal of a pollen grain onto a nearby sister species may lead to polyploid formation in the immediate presence of just *one* of its progenitors. This may make it easier for the newly formed polyploid to escape competition and may thus facilitate its establishment. The role of this process in polyploid formation, however, has not been evaluated experimentally yet.

Other potential roles of hybridization. Recent progress has also been made in understanding two contentious topics, the role of plant hybridization in speciation through reinforcement and the role of hybridization during adaptive radiation (see illustration). The reinforcement hypothesis includes the notion that selection against interspecific matings may lead to even stronger barriers to gene exchange between diverging populations and may thus contribute to speciation. Convincing evidence for this scenario in plants is rare, but a recent study of the grass *Anthoxanthum odoratum* in a long-term ecological experiment in the United Kingdom indicates that such prezygotic selection may indeed lead to population divergence.

Two hypotheses have recently been put forward, or developed further, to discuss the potential role of hybridization during adaptive radiation. The hybrid swarm hypothesis assumes that hybridization between radiating taxa will instantaneously elevate heritable variation in ecologically relevant traits.

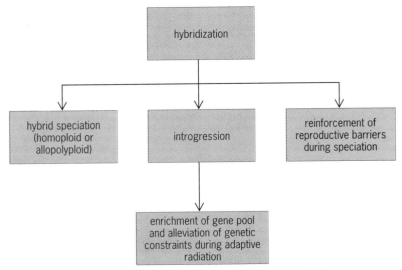

Three mechanisms through which hybridization may lead to plant speciation.

As a consequence, increased phenotypic variation will be available for natural selection to act upon. Thus, hybridization upon invasion of new environments should facilitate rapid adaptive radiation. The syngameon hypothesis predicts that new variation created by continued hybridization (introgression) between radiating taxa will alleviate genetic constraints imposed when natural selection has exhausted genetic variation at ecologically relevant genes. Thus, hybridization will maintain and prolong the radiation process.

Clearly, a "red thread" through recent work on plant hybridization is the emerging view that hybridization between divergent populations or species may sometimes be a potent force in adaptive evolution and a stimulus for speciation at the diploid level, in addition to its widely accepted role in allopolyploid speciation. Recent developments in this field are enabled by rapid developments in genomic and postgenomic plant science. This allows scientists to connect research on plant speciation in a whole-organism context with research to understand its exact molecular underpinnings. Hybridization and gene flow between divergent populations may have more profound and diverse evolutionary consequences than previously thought.

For background information *see* BREEDING (PLANT); CHROMOSOME; GENE; GENETICS; PLANT EVOLUTION; POLYPLOIDY; SPECIATION in the McGraw-Hill Encyclopedia of Science & Technology.

Christian Lexer

Bibliography. J. A. Coyne and H. A. Orr, *Speciation*, Sinauer Associates, Sunderland, MA, 2004; V. Grant, *Plant Speciation*, Columbia University Press, New York, 1981; M. J. Hegarty et al., Development of anonymous cDNA microarrays to study changes to the *Senecio* floral transcriptome during hybrid speciation, *Mol. Ecol.*, 14:2493–2510, 2005; L. H. Rieseberg et al., Major ecological transitions in wild sunflowers facilitated by hybridization, *Science*, 301:1211–1216, 2003; O. Seehausen, Hybridization and adaptive radiation, *Trends Ecol. Evol.*, 19:199–207, 2004.

Hypoxia-inducible factor

It is said that one can survive for 3 weeks without food, 3 days without water, 3 hours without warmth, but only 3 minutes without oxygen. Although these numbers may not be entirely accurate, it is undeniable that oxygen is essential. Oxygen is so important in mammals and other animals that intake through breathing occurs without conscious thought. At the cellular level, oxygen is the final electron acceptor in the mitochondrial electron transport chain, facilitating the efficient generation of energy in the form of ATP (adenosine triphosphate). Unlike some other intermediates in this pathway, cells are unable to store and retrieve molecular oxygen. Consequently, cells are often affected by varying levels of hypoxia, a situation where the cell's supply of oxygen is insufficient for its metabolic requirements. Low environmental oxygen availability (for example, at high altitude), underdeveloped vasculature (for example, in growing embryos), various diseases which result in restricted blood flow, and vigorous exercise are examples where cellular energy homeostasis may be adversely affected. To compensate for the lack of oxygen-storing capacity, cells are equipped with exquisitely sensitive oxygen-sensing machinery, which in response to hypoxia triggers a battery of adaptive responses to minimize damage to cells or the organism as a whole. Among the most acute of these responses (within seconds) is the rapid release of transmitters from oxygen-sensitive ion channels, a process which mediates neurological responses such as hyperventilation. A more chronic response to hypoxia (minutes to hours) is the regulated expression of a program of specific genes. A variety of transcription factors are involved in this genomic response, most importantly the hypoxia-inducible factors (HIFs). Approximately 100 direct target genes of the HIFs have been reported, and the collective activation of these target genes facilitates adaptation to hypoxia via an increase in oxygen-independent energy production (largely through glycolysis), improved oxygen transport and delivery, and careful modulation of cell proliferation and apoptosis (programmed cell death) [**Fig. 1**].

Specific factors. The HIF transcription factors first came to the attention of researchers through their ability to be specifically activated by hypoxia. The two best-characterized, HIF-1 and HIF-2, are heterodimeric transcription factors, comprising one alpha and one beta subunit. The beta subunit, referred to as HIF-1β or Arnt (aryl hydrocarbon receptor nuclear translocator), is common to both transcription factor complexes, while the alpha subunits are the products of the paralogous genes (that is, they are derived from the duplication of an ancestral gene), *HIF-1α* or *HIF-2α*. A third HIF-α paralog, *HIF-3α*, exists, but preliminary analysis suggests that it serves to repress activity of HIF-1 and HIF-2, and its overall role in the hypoxic response remains unclear. Collectively, the HIF subunits are part of a large family of transcription factors containing a basic helix-loop-helix/Per-Arnt-Sim homology (bHLH-PAS) structural motif. In addition to playing a role in the HIF transcription factor complex, Arnt partners with many other members of the bHLH-PAS transcription factor family. It is therefore unsurprising that the hypoxic response of the HIFs is regulated exclusively by the alpha subunits. HIF-1α and HIF-2α have very similar domain structures, consisting of the N-terminal bHLH-PAS motif (which is responsible for dimerization and DNA binding), two transactivation motifs (the NAD and CAD, referring to their relative N- and C-terminal positions), and a central oxygen-dependent degradation domain (ODD) [**Fig. 2**]. As discussed below, the functions of the two HIFs are not redundant; however, both alpha subunits are subjected to similar basic regulatory mechanisms.

Oxygen-dependent regulation of HIF-α subunits. As the transcription factors primarily responsible for upregulation of genes involved in adaptation to hypoxia, much research has focused on the mechanism by which the HIFs are "switched on" by hypoxia. Early observations of HIF-1α activity indicated that hypoxia has two major effects. First, low oxygen causes the rapid and extensive stabilization of HIF-1α protein (normally undetectable in normoxia, that is, the normal oxygen state), thereby permitting HIF-1α protein to translocate to the nucleus, partner with Arnt, and bind its target genes. Additionally, hypoxia also substantially upregulates the transactivation capacity of the HIF-1α CAD, resulting in greater target gene activation. In the last 5 years, considerable progress has been made in understanding the basis of this regulation.

Oxygen-dependent hydroxylation. Three separate HIF prolyl hydroxylases (PHD 1–3) and a single asparaginyl hydroxylase, FIH-1 (factor inhibiting HIF-1), are responsible for hydroxylating HIF-α subunits. Each hydroxylase belongs to a large family of oxygen-, iron-, and 2-oxoglutarate-dependent oxygenases. Critically, these enzymes directly use molecular oxygen as a cosubstrate; hence the efficiency of hydroxylation is modulated by physiologically relevant changes in oxygen tension. Such a mode of action immediately suggests a mechanism for the oxygen sensitivity of HIF-α, and has led to designation of the hydroxylases as the likely primary oxygen sensors for the HIF pathway.

In normoxia, three highly conserved residues are hydroxylated in an oxygen-dependent manner in both HIF-1α and HIF-2α: two proline residues in the ODD (Pro402 and Pro564 in human HIF-1α) and a single asparagine in the CAD (Asn803). Each modified proline forms a structure which acts to recruit the von Hippel-Lindau (VHL) E3 ubiquitin ligase, resulting in ubiquitination of HIF-α, which tags the protein for rapid proteasomal degradation. However, in hypoxic conditions the reduced oxygen availability decreases hydroxylation, thus inhibiting binding of VHL, and allowing HIF-α to escape destruction and accumulate (**Fig. 3**). The asparagine residue in the HIF CAD is also hydroxylated in normoxic conditions, which prevents recruitment of transcriptional coactivator proteins, leading to repression of the CAD. This repression is reversed by hypoxia, again

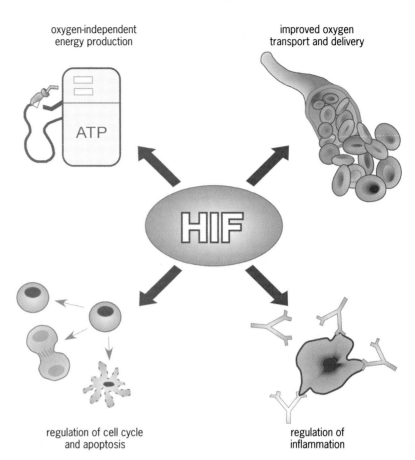

Fig. 1. Major physiological processes mediated by HIFs. Transcriptional upregulation of direct target genes by HIF assists cellular adaptation to hypoxia.

due to the reduction in available oxygen, allowing efficient coactivator recruitment and maximal upregulation of HIF target genes.

Reactive oxygen species. Reactive oxygen species (ROS) [that is, oxygen-containing molecules that may be free radicals, such as superoxide, hydrogen peroxide, and hydroxyl radical] have also been implicated in the regulation of HIF activity. Cells tightly regulate the production of ROS as they are involved in cellular signaling, and excessive levels cause damage via oxidation of numerous cellular components. It has been reported that during hypoxia cellular ROS levels are altered, and this change can influence mechanisms of HIF regulation, including the activity of the HIF hydroxylases. It has also been suggested that ROS, rather than molecular oxygen, are the direct mediators of the cellular hypoxic response, although this remains controversial. It is clear, however, that these effects vary both within and between cell types, and further study is required to fully define the mechanisms involved.

Oxygen-independent activation. In addition to hypoxic stress, a variety of other stimuli have the capacity to activate HIF and switch on certain subsets of HIF target genes in normoxia. Prominent examples include growth factors, such as insulin, which utilize HIF's ability to control production of glycolytic enzymes to assist in maintaining homeostasis in carbohydrate metabolism. HIF's role as a mediator of

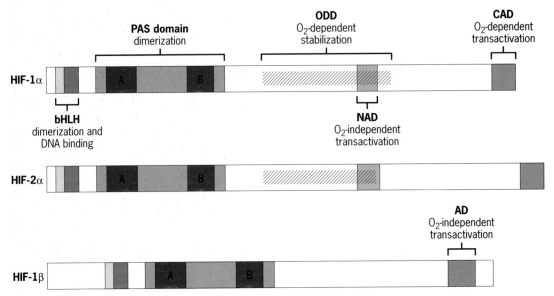

Fig. 2. **Domain structure of HIF transcription factor subunits.** Basic structure of the mammalian HIF-1α, HIF-2α, and HIF-1β proteins, showing conserved functional domains including basic helix-loop-helix (bHLH), Per-Arnt-Sim homology (PAS), oxygen-dependent degradation (ODD), and N-terminal transactivation (NAD) and C-terminal transactivation (CAD) domains.

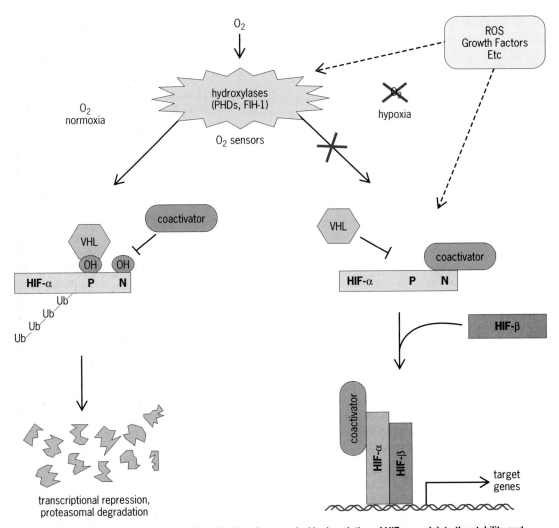

Fig. 3. **Oxygen-dependent regulation of HIF-α.** Prolyl and asparaginyl hydroxylation of HIF-α modulate the stability and transactivation capacity of HIF-α in an oxygen-dependent manner, and therefore overall activity of HIF target genes.

the immune response has also become apparent, and normoxic activation of HIF by cytokines such as interleukin-1 and tumor necrosis factor-alpha can facilitate upregulation of a variety of markers of inflammation. It is highly likely that many of these stimuli also work in concert with hypoxia to fine-tune HIF regulation, and it will be essential to determine the contribution of each signaling pathway involved to fully understand these cellular responses.

HIF-1α versus HIF-2α. Among the many known HIF target genes, a significant proportion are exclusively HIF-1 targets, relatively few are exclusively HIF-2 targets, and the remainder can be regulated by both HIFs. Evidence exists to demonstrate that the functions of HIF-1 and HIF-2 are far from redundant. The expression patterns of both alpha subunits differ significantly, as HIF-1α is ubiquitously expressed, whereas HIF-2α (also known as endothelial PAS domain protein, EPAS) has a more restricted expression pattern, with highest levels in vascular endothelial cells. Furthermore, the efficacy of hypoxic activation of HIF-1 seems to be significantly higher than that of HIF-2. By far the most convincing observations for nonredundancy, however, come from gene deletion studies. The targeted disruption of *HIF-1α* or *HIF-2α* genes in mice results in substantially different developmental phenotypes. *HIF-1α*–deficient mice die at an early embryo stage, largely due to heart defects and poor blood vessel formation. In contrast, *HIF-2α*–deficient mice display bradycardia (slow heart rate), disrupted catecholamine synthesis, vascular disorganization, and lung defects. These studies clearly show that neither HIF-α can compensate for loss of the other, and much research remains to be done to understand the exact physiological role of each transcription factor.

Therapeutic potential. Exposure of cells to mild hypoxia is common during normal growth, development, and metabolic fluctuation of cells and organisms. However, hypoxia can also cause or contribute to many serious diseases that are increasingly prevalent in today's society, such as anemia or ischemic vascular, cardiac, or cerebral disease. Ischemia is the restriction of blood flow and can cause significant damage due to deprivation of oxygen and other nutrients. Given its prominent role in promoting the growth of blood vessels and cells, and generally responding to oxygen deficiency, HIF is a prime candidate for therapeutic manipulation. Elucidation of the prominent roles of the hydroxylase enzymes has opened the way for enzyme-specific inhibitors to be developed to facilitate increased HIF activity and minimize the damage associated with ischemia. Indeed, preliminary analysis of the effects of broad-specificity hydroxylase inhibitors in laboratory models of ischemic disease has produced promising results.

The role of HIF in disease has also become apparent through analysis of the gene expression profiles of malignant tumors. Unlike normal tissue, cancerous tissue can grow and divide very rapidly, generally outpacing the growth of new blood vessels, thus creating an oxygen- and nutrient-deprived microenvi-ronment. Aberrant activation of HIF-1α or HIF-2α by these cells serves to counter these detrimental conditions and to promote tumor growth, and hence elevated HIF levels are often indicative of a poor prognosis. Once again, therapeutic targeting of the HIF pathway may prove beneficial by restricting tumor progression.

The discovery and characterization of the HIF transcription factors has had a major impact on our understanding of this important genomic response to hypoxia. This exciting groundwork still requires more basic knowledge, but is already providing invaluable tools and approaches to understand, diagnose, and combat some of our most challenging and detrimental diseases.

For background information *see* BIOLOGICAL OXIDATION; HYPOXIA; OXYGEN; RESPIRATION; RESPIRATORY SYSTEM; RESPIRATORY SYSTEM DISORDERS in the McGraw-Hill Encyclopedia of Science & Technology.   Daniel Peet; Rachel J. Hampton-Smith

Bibliography. J. I. Bardos and M. Ashcroft, Negative and positive regulation of HIF-1: A complex network, *Biochim. Biophys. Acta*, 1755(2):107–120, 2005; C. P. Bracken, M. L. Whitelaw, and D. J. Peet, The hypoxia-inducible factors: Key transcriptional regulators of hypoxic response, *Cell. Mol. Life Sci.*, 60:1376–1393, 2003; K. S. Hewitson and C. J. Schofield, The HIF pathway as a therapeutic target, *Drug Discovery Today*, 9(16):704–711, 2004; S. Lahiri, N. Prabhakar, and G. L. Semenza (eds.), *Oxygen Sensing: Responses and Adaptation to Hypoxia*, Marcel Dekker, New York, 2003; C. J. Schofield and P. J. Ratcliffe, Signalling hypoxia by HIF hydroxylases, *Biochem. Biophys. Res. Commun.*, 338(1):617–626, 2005; G. L. Semenza, Targeting HIF-1 for cancer therapy, *Nat. Rev. Cancer*, 3(10):721–732, 2003.

Ichnology

Ichnology is the study of animal-sediment relationships and encompasses the examination of both modern and ancient structures. Trace fossils are biologically produced structures that include tracks, trails, burrows, borings, fecal pellets, and other traces made by organisms. Owing to their nature, trace fossils can be considered as both paleontological and sedimentological entities, thereby bridging the gap between two of the main subdivisions in sedimentary geology.

The contributions of ichnology to sedimentary geology are considerable, including knowledge relating to: (1) the production of sediment by boring organisms; (2) the consolidation of sediment by suspension feeders; (3) the alteration of grains by sediment-ingesting organisms; (4) the destruction of sedimentary fabrics and sedimentary structures; (5) the construction of new fabrics and sedimentary structures; (6) the initial history of lithification; (7) the interpretation of depositional environments; (8) the delineation of facies and facies successions; (9) rates of deposition; (10) substrate coherence

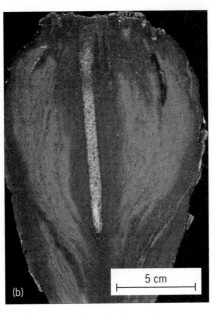

Fig. 1. Zoological affinities of trace fossils. (*a*) Structure produced by the modern terebellid polychaete *Cirriformis* sp. that produces a *Rosselia*-like burrow. (*b*) Trace fossil *Rosselia socialis*, Upper Cretaceous of Alberta.

and stability; (11) aeration of water and sediments; (12) the enhancement of porosity/permeability in reservoir units; and (13) the amounts of sediment deposited or eroded.

Sedimentary geologists are well aware that credible interpretations of depositional environments cannot depend upon data from a single discipline, whether it be sedimentology, stratigraphy, paleontology, geochemistry or any of a host of others. Instead, environmental interpretations are made much stronger when all available evidence is utilized. Ichnology, the study of animal-sediment interrelationships, combines elements of most of these disciplines and, like petroleum geology, is more a blending of related disciplines than a singularly unique entity.

Conceptual framework. The importance of ichnology to the fields of stratigraphy, paleontology, and sedimentology stems from the fact that trace fossils display the following characteristics: (1) long temporal range, which greatly facilitates paleoecological comparisons of rocks differing in age; (2) narrow facies range, which reflects similar responses by trace-making organisms to given sets of paleoecological parameters; (3) no secondary displacement—trace fossils are closely related to the environment in which they were produced; (4) occurrence in otherwise unfossiliferous rocks—trace fossils are commonly enhanced by the very diagenetic processes that can obliterate body fossils; (5) creation by soft-bodied biota—many trace fossils are formed by the activities of organisms that generally are not preserved because they lack hard parts; (6) the same individual or species of organism may produce different structures corresponding to different behavior patterns; and (7) identical structures may be produced by the activity of systematically different trace-making organisms, where behavior is similar.

Such characteristics make trace fossils very useful in facies analyses, including reconstruction of individual paleoecological factors, sedimentary dynamics, and the documentation of local and regional facies changes.

Behavioral classification. Perhaps the most important facet of ichnology is the behavioral interpretation of trace fossils. Fundamental behavioral (or ethological) patterns are dictated and modified not only by genetic preadaptations but also by prevailing environmental parameters. Eight basic categories of behavior have been recognized; resting traces, locomotion traces, dwelling traces, grazing traces, feeding burrows, farming systems, predation traces, and escape traces. The basic ethological categories evolved early and have generally persisted throughout the Phanerozoic. Although individual trace-makers have evolved, basic benthic behavior has not.

One of the most frustrating, albeit most fascinating, aspects of ichnology is the attempt to establish the zoological affinities of specific ichnofossils (**Fig. 1**). This is because ichnofossils reflect the behavior of animals, and only to a small extent reflect their anatomy or morphology. The result is that more than one genus or species of trace fossil may have been constructed by a single species of animal; or conversely, different species of animals may have made identical species or genera of trace fossils. Therefore, each occurrence of a given trace fossil must be treated on an individual basis, and sweeping generalizations on their zoological affinities should be avoided. In most cases, attributing a particular trace to a particular soft-bodied organism depends on a uniformitarian approach and other indirect lines of evidence.

Ichnofacies concept. The essence of trace fossil research involves the grouping of characteristic ichnofossils into recurring ichnofacies. Eleven recurring ichnofacies have been recognized, each named for a representative ichnogenus: *Scoyenia, Mermia, Coprinisphaera, Trypanites, Teredolites, Glossifungites, Psilonichnus, Skolithos, Cruziana, Zoophycos,* and *Nereites.* These trace fossil associations reflect adaptations of trace-making organisms to numerous environmental factors such as substrate consistency, food supply, hydrodynamic energy, salinity, and oxygen levels. Traces in nonmarine and brackish marine settings are in need of further study. The marine soft-ground ichnofacies are distributed according to numerous environmental parameters; traces in the firmground, woodground, and hardground ichnofacies are distributed on the basis of substrate type and consistency (**Fig. 2**).

These archetypal models, particularly the marine ones, have proven to be valuable indicators of general environmental conditions. For instance, primary physical sedimentary structures of fluvial point bars (sediment ridges) may be strikingly like those of estuarine point bars. However, the biogenic structures are very different in the two settings. Equally important is the long temporal duration of most kinds of trace fossils. These basic benthic behavioral patterns are more nearly like stable ecologic niches than

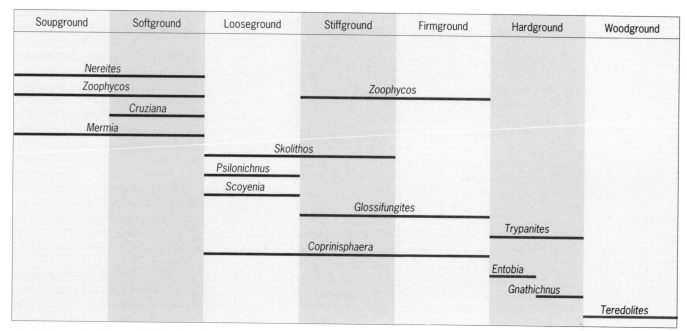

Fig. 2. Distribution of ichnofacies based on the type of substrate.

individualistic records of particular animal species. Many different animal species, over long intervals of geologic time, may be expected to exploit it, and their preserved traces are strikingly similar and have equivalent significance.

The purpose for recognizing these recurrent ichnofacies (**Fig. 3**) must not be overlooked. Interpreted in terms of the original trace fossil assemblage, these ichnofacies are merely archetypal facies models with which the local ichnofacies may be compared. The archetypes are intended to supplement, not supplant, local ichnofacies designations, some of which are quite distinctive. The basic consideration rests not with such inanimate backdrops as water depth or distance from shore, or some particular tectonic or physiographic setting, but with innate, dynamic controlling factors as substrate consistency, hydraulic energy, rates of deposition, turbidity,

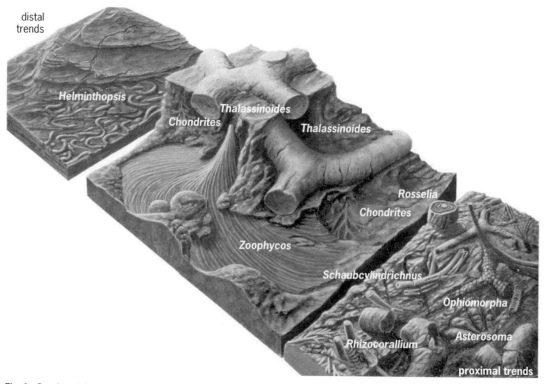

Fig. 3. *Cruziana* **ichnofacies that is typical of offshore marine environments. (***Modified from S. G. Pemberton et al., 2001***)**

oxygen and salinity levels, toxic substances, the quality and quantity of available food, and the ecologic or ichnologic prowess of trace-makers themselves.

Paleoenvironmental significance of trace fossils. The concept of functional morphology, a basic premise employed by ecologists and paleoecologists in environmental reconstruction, is equally applicable to ichnology. Trace fossils are unique in that they represent not only the morphology and behavior of the trace-making organism but also the physical characteristics of the substrate; they are closely linked to the environmental conditions prevailing at the time of their construction.

The application of ichnology to paleoenvironmental analysis goes far beyond the mere establishment of archetypal ichnofacies. For instance, shallow-water, coastal marine environments comprise a multitude of sedimentological regimes which are subject to fluctuations in many physical and ecological parameters. In order to fully comprehend the depositional history of such zones in the rock record, it is imperative to have some reliable means of differentiating subtle changes in these parameters. Detailed investigations of many of these coastal marine zones in Georgia (United States) have shown the value of using biogenic sedimentary structures (in concert with physical sedimentary structures) in delineating them. For example, the shoreface consists of a seaward-sloping sediment wedge extending from the low-tide mark, generally to the fair-weather (minimum) wave base, corresponding to approximately 10–20 m (33–66 ft) of water depth. The shoreface setting is dominated by wave energy and shows a pronounced basinward fining, as a result of decreasing wave interaction with the substrate in a seaward direction. The shoreface is classically divided into three subenvironments (from seaward to landward)—the lower, the middle and the upper shoreface—and grades distally into offshore units and landward into foreshore deposits. There are several excellent outcrop examples of Cretaceous shoreface deposits in the Western Interior Seaway. The integration of the sedimentology and ichnology within these deposits affords the opportunity to characterize the facies and facies successions, and to explain the observed facies variability.

Ichnology can supply a wealth of environmental information that cannot be obtained in any other way and that should not be ignored. Its potential usefulness is accentuated when fully integrated with other (chemical, physical, and biological) lines of evidence.

Reservoir ichnology. Trace fossil research in hydrocarbon reservoir rocks is usually confined to exploration geology; however, recent research is showing that ichnology has significant applications in

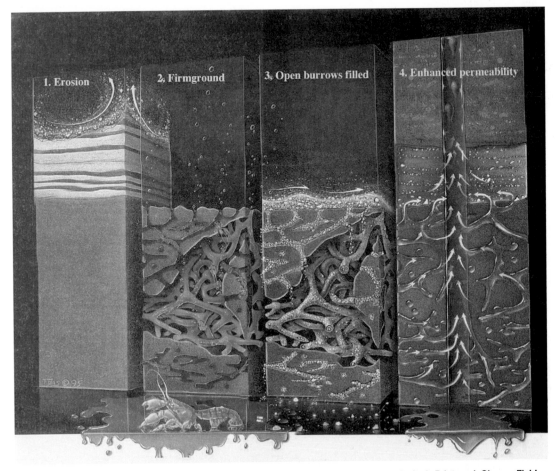

Fig. 4. Burrow-enhanced permeability and development of superpermeability in the Jurassic Arab-D interval, Ghawar Field, Saudi Arabia. (*Modified from S. G. Pemberton and M. K. Gingras, 2005*)

production geology. Applications of ichnofossil research to reservoir rocks revolve around the textural modifications a trace fossil might impose on flow media (**Fig. 4**). Any measurable textural modification invariably results in modification of the permeability fabric, commonly evoking a dual porosity-permeability system.

The dual porosity-permeability system has notable implications where applied to resource quality in porous media. Conceptually, these fabrics have some of the flow characteristics that fracture systems exhibit. Several of the considerations pertinent to fluid production from sedimentary rocks are: (1) diffusion into and out of the unburrowed media affects the rate at which a fluid or contaminant might be removed from the rock; (2) fluids in higher-permeability conduits can be cut off from the main flow channels, reducing the overall recoverable resource in a reservoir; (3) the net permeability and porosity are a nonlinear function of the matrix and burrow properties; and (4) flow networks might control later diagenetic processes, such as cement (mineral) precipitation and dissolution. Although studies that relate ichnology to reservoir permeability fabrics are only now becoming popular, a small body of literature already exists.

For background information *see* DEPOSITIONAL SYSTEMS AND ENVIRONMENTS; DIAGENESIS; FACIES (GEOLOGY); FLUVIAL SEDIMENTS; FOSSIL; GEOLOGIC TIME SCALE; MARINE SEDIMENTS; OIL AND GAS FIELD EXPLOITATION; PALEOECOLOGY; PALEONTOLOGY; PETROLEUM GEOLOGY; SEDIMENTOLOGY; STRATIGRAPHY; TRACE FOSSILS in the McGraw-Hill Encyclopedia of Science & Technology.   S. George Pemberton

Bibliography. R. G. Bromley, *Trace Fossils, Biology and Taphonomy*, 2d ed., Unwin Hyman, 1996; A. A. Ekdale, R. G. Bromley, and S. G. Pemberton, *Ichnology: The Use of Trace Fossils in Sedimentology and Stratigraphy*, Society of Economic Paleontologists and Mineralogists, Short Course Notes 15, 1984; R. W. Frey and A. Seilacher, Uniformity in marine invertebrate ichnology, *Lethaia*, 13:183–207, 1980; S. G. Pemberton, J. A. MacEachern, and R. W. Frey, Trace fossil facies models: Environmental and allostratigraphic significance, *Facies Models: Response to Sea Level Change*, Geological Association of Canada, 1992; S. G. Pemberton et al., *Ichnology and Sedimentology of Shallow and Marginal Marine Systems: Ben Nevis and Avalon Reservoirs, Jeanne D'Arc Basin*, Geological Association of Canada Short Course Notes 15, 2001; S. G. Pemberton, and M. K. Gingras, Classification and characterizations of biogenically enhanced permeability, *Amer. Ass. Petrol. Geol. Bull.*, 89:1493–1517, 2005.

Inkjet printing

Inkjet printing refers to a family of related technologies in which airborne droplets of ink are electronically guided to form images. The roots of this technology reach as far back as 1867, when William Thomson (later, Lord Kelvin) acquired a patent for "Receiving or Recording Instruments for Electrical Telegraphers." The system used electrostatic forces to control the release of ink drops onto paper to record telegraph messages. In 1951, Siemens patented the first continuous-stream inkjet printer, but it achieved little commercial success. Later patents by Winston, Ascoli, Sweet, Hertz, Elmquist, and others led to commercially successful printers from A. B. Dick, Mead, and Toshiba in the 1960s.

Inkjet printing emerged in the 1970s as one of the competing digital printing technologies able to produce hard copy documents directly from digital files without the need for intermediate photomechanical processes. IBM promoted the technology after introducing the 4640 inkjet printer in 1976. Today, inkjet printing encompasses an array of technologies that fulfill an ever-larger number of graphic applications. Inkjet printing is currently the most versatile printing method available. It dominates both the wide-format and the desktop printer markets, and has become the digital proofing method of choice for other printing processes. Inkjet printing is also the principal photofinishing method for making prints from digitally captured images. Currently inkjet technology is not used for the mass production of printed products because it is relatively slow compared to other processes for reproducing the same image thousands of times.

The inherent advantages of inkjet technology are that it is a noncontact printing method using simple and inexpensive printers which are compatible with a wide variety of inks and other fluids. The inkjet process involves printing thin ink film layers and overprinting as needed to produce multiple gray levels, allowing varying quality in the prints produced. Quality can be enhanced with the use of specially designed coated papers that limit the feathering and penetration of the ink. The inkjet process also has the advantage of printing directly from digital files without intermediate image carriers. The data that drives the printer can be changed on the fly, allowing each successive print to be unique. All types of paper and board can be used as a substrate for inkjet printing, as well as fabrics, plastics, ceramics, metals, and glass.

In the 1970s, inkjet technology began to be used for the in-line printing of address labels onto magazines as they were being produced. This application of inkjet printing using web offset lithography is still in use. There were also a number of other moderately successful inkjet document printing systems in the 1970s. Widespread markets for inkjet printing did not emerge, however, until the development of the drop-on-demand (DOD) inkjet methods in the mid-1980s. With the introduction of the DOD methods, particularly the innovation of the disposable print head used in the process, inkjet printing came to dominate the desktop printer market. As multicolor inkjet printers gained quality and resolution, inexpensive desktop printers began to deliver near-photographic color quality. Inkjet technology is commonly used today to make prints from digital images.

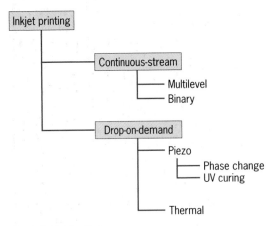

Fig. 1. Inkjet family tree.

In the early 1990s, multicolor, large-format DOD inkjet printers replaced previous systems for the production of large-sized color graphics. This market has grown rapidly over the past 15 years due in part to the improved quality and economy of the inkjet printing process.

In addition to the graphic media applications, inkjet printing is increasingly adapted for use in the manufacture of other products. In 2004, for instance, a modified inkjet printer produced the world's first ultrathin 20-layer circuit board. The inkjet process is also used in the manufacture of LCDs and other flat screen displays. Ironically, screen printing production, a competitive process in some markets, often uses inkjet printers to print stencil material directly onto printing frames. Inkjet technology is even used for biomedical research to lay down successive layers of live cells onto gelatin substrates. The cell layers then grow to form complex structures such as blood vessels.

Continuous-stream inkjet printing. The family of inkjet printing methods is broadly divided into continuous-stream and drop-on-demand methods as seen in **Fig. 1**. The DOD methods are most popular, but continuous-stream inkjet is preferred for some applications.

Continuous-stream inkjet methods involve the production of a unbroken stream of droplets, only some of which contribute to the formation of the image. The printer then captures and recycles unused drops for subsequent use.

The continuous-stream inkjet format can itself be subdivided into multilevel deflection and binary deflection methods. **Figure 2** shows schematic diagrams of the two types of continuous-stream inkjet printing techniques. In both systems a nozzle emits a stream of equally sized, equally spaced ink droplets, produced by an ultrasonic vibration induced in the pressurized ink stream as it flows through the nozzle.

Continuous-stream inkjet printing forms images by placing electrostatic charges on individual drops of ink to control their trajectories—some receive a charge, others do not. To allow this, the ink must be capable of holding an electrostatic charge. Generally, the drops of ink receive the charge as they exit the printing nozzle. They fly past a deflection plate carrying a strong charge of the opposite type. This deflects the charged droplets from their flight paths but leaves the trajectories of the uncharged droplets unchanged.

Multilevel inkjet printing. Multilevel inkjet printing deflects droplets through a range of trajectories, thereby directing them to various placements on the paper or other substrate. The degree of deflection depends on the strength of the charge applied at the point of droplet isolation.

Usually, droplets are deflected in only one direction, but some designs use bidirectional deflection. Multilevel deflection systems use a single nozzle to print an element about 0.10 in. high. The nozzle moves across and down the substrate to cover the entire print area. The process tends to be slow, a limitation that can be overcome in part with the use of multinozzle print heads.

Binary continuous-stream inkjet printing. Binary continuous-stream inkjet printing uses only one charge level. Uncharged droplets hit the substrate and print while charged droplets are deflected away from the substrate and into the ink recycling system.

A notable development in binary continuous inkjet was the invention of the multinozzle array. In the late 1960s, Mead introduced the DIJIT (Direct Imaging for Jet Ink Transfer) process, in which a multinozzle array fired downward to produce images on a moving web of paper. The manufacture of the nozzle array on a silicon wafer, with charging electronics formed directly on the silicon, made the technology possible. The key to the success of such multinozzle binary systems was the use of a common ink reservoir, agitated by a single ultrasonic vibration, which fed all of the nozzles in the array. This caused

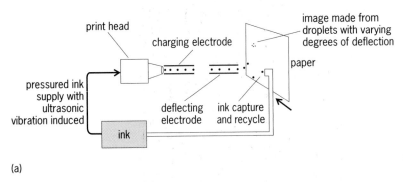

(a)

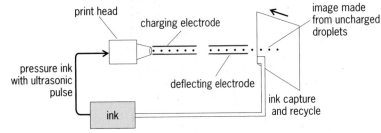

(b)

Fig. 2. Two types of continuous-stream inkjet printing techniques. (a) Multilevel. (b) Binary.

the synchronous emission of droplets from all the nozzles.

Despite its marginal image quality, the high speed of this printer made it useful for certain applications. Mead sold the DIJIT division to Kodak in 1983, which spun it off as Diconix. Scitex later acquired Diconix. Today some high-speed document printers still employ this downward-firing multinozzle array technology. Some experts in the field believe it will be the technology of choice for producing books on demand in the future.

Drop-on-demand inkjet printing. Although continuous-stream inkjet processing fulfilled the production needs of some applications, the process had drawbacks related to the ink charging, deflection, and recycling functions. Only semiconductive inks could be used, for example. Air turbulence, created by the simultaneous flight of many drops with various trajectories, also made accurate deflection very difficult. And the inclusion of components to control ink capture, filtering, and remixing (to replenish lost solvents) required additional space and increased the complexity of the systems. Moreover, in continuous inkjet printers, the distance between the nozzle and the substrate must be great enough to accommodate the ink charging and droplet deflection activities. To place the drops across this distance, however, nozzle exit velocities must reach 12 m per second.

DOD inkjet processes were developed to sidestep these inherent drawbacks. The essential characteristic of DOD systems is that drops are emitted only as needed. There is no need for ink charging, deflection, or recycling. The nozzles are typically positioned very close to the substrate, and the systems require low nozzle velocities.

Today two technologies compete for dominance of the DOD inkjet market: the piezo and the thermal systems. These two methods differ in the way in which the droplets are forced from the nozzle. **Figure 3** shows a schematic diagram of the systems.

Piezo DOD inkjet printers were first commercialized in 1975. This method relies on the electromechanical properties of piezoelectric crystals which flex or twist under electrical impulses. This conversion from electrical to mechanical energy momentarily pressurizes the ink chamber, forcing droplets of ink to be emitted. Early piezo DOD inkjets suffered from nozzle clogging and so failed to displace the popular and reliable dot matrix printers in use at the time. During the 1980s, piezo inkjet systems greatly improved in quality and reliability. At the same time, developers created the thermal DOD inkjet process. Piezo DOD systems are addressable at higher frequencies than thermal printers, however, and they can use a much wider variety of fluids.

Solid inkjet printing, introduced in the mid-1980s, is a variation of the piezo inkjet process. This process uses phase-change inks and heat-resistant piezo DOD inkjet nozzles. Phase-change inks, used in other forms of printing since the early 1950s, are solid

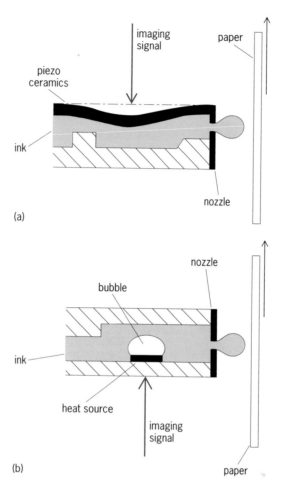

(a)

(b)

Fig. 3. Cross sections of (a) piezo and (b) thermal DOD inkjet systems.

at room temperature. The inks convert to a liquid state when heated for printing and then solidify very rapidly on the cold substrate. The ink remains on the surface of the paper rather than soaking in, which produces vibrant colors. The inks, often wax-based, are resistant to water; however, they abrade easily and tend to have poor transparency.

Thermal inkjet printing resembles the piezo printing process in basic design, except that it is the rapid heating of small volumes of ink that stimulates the fluid-pressure impulses which produce the droplet. The heat is generated by a thin-film resistor attached to the print head. This resistor heats rapidly when current is applied, vaporizing the ink in closest proximity to it and emitting ink droplets from the nozzle.

This process occurs very rapidly, allowing printers to operate at over 15 kHz. Since the print heads could be manufactured inexpensively, they were made to be disposable units. It was the introduction of thermal DOD inkjet printers with the convenience and reliability of disposable cartridges—coupled with their production of high-quality prints—that swept the office printer market in the late 1980s.

Electrostatic inkjet printing. Electrostatic inkjet technologies are variants of the DOD process in which electrostatic forces guide the flight of ink droplets. Control impulses cause the release of ink drops from a nozzle. An electrostatic field between the print head and the paper then causes the ink to migrate to the paper. The electrostatic forces can shape the ink drops as well, creating drop sizes smaller in diameter than the nozzles which emitted them. Some electrostatic systems use heating at the nozzle rims to expel drops. Other systems use electrical impulses. Ink mist printers, another electrostatic technology, focuses an ultrasonic wave on the nozzle outlets to produce a fine ink mist.

Appeal of inkjet systems. Over the past 30 years, inkjet printing has grown to dominate several printer markets, finding several industrial applications. The appeal of inkjet as a printing process lies in its extreme simplicity. Although based on complex engineering, inkjet printers are small, low-cost, self-contained devices that are easily integrated into other manufacturing lines.

In fact, inkjet printing is the most compact technology for transferring an image to a substrate. With inkjet printing, a multicolor print head can be incorporated into a single unit that scans across the print area. The imaging is directly controlled by digital information with no intermediate image carriers. The imaging is noncontact, allowing a wide variety of potential substrates. An astonishing array of inks and other fluid can be used for printing, allowing a large variety of potential applications. Another important appeal of inkjet printing is the high image quality, especially in multicolor prints. Inkjet prints of photographic images rival the quality of photographic prints.

Applications for inkjet printing. The number of printing applications using inkjet printing continues to increase. Among these new applications are systems for digital proofing. These specialty printers mimic other processes. Today color proof production employs inkjet technology more than any other.

Inkjet desktop printers are the most commonly used devices for the home office or small business applications, outselling laser printers by 12 to 1. This is especially true where color output is needed.

Specialty inkjet photographic printers are popular for producing photographic prints from digitally captured images. These printers use six or more colors and typically print with resolutions of 600 dpi (dots per inch) or greater.

The development of DOD inkjet systems in the mid-1980s enabled the development of wide-format inkjet printers that printed on rolls of material in full color. These printers were first marketed in 1990, and by 1998 inkjet technology had an 80% share of the large-format graphics market. Inkjet printing created new markets for large-format printing by offering photolike quality at short turnaround times and relatively low cost. Today many different types of inkjet vie for slices of the large-format market. Together, their popularity dwarfs the other processes that compete for this market.

Two recent trends in large-format inkjet processes include the development of flatbed printer designs and the use of UV (ultraviolet light) curing inks. Flatbed presses create images on materials such as glass, metal, fabric, foam core, and virtually any other flat surface. UV-cured inks are frequently used for outdoor applications because, once cured, the ink is waterproof and scratch-resistant. UV curing also circumvents any problems that arise during ink drying because these inks set immediately upon exposure to a UV energy source.

High-productivity inkjet systems have been introduced for books-on-demand printing. The average print head today contains more than 300 nozzles and can print images at resolutions above 1200 × 1200 dpi at 14,000 dots per second.

Textile producers have begun exploiting the potential applications for inkjet printing on woven and knitted materials. Products ranging from customized designer fabrics to trade show banners are now routinely printed on large-format multicolor inkjet printers. The inkjet textile market has recently been expanding as multicolor flatbed inkjet T-shirt printers have come into use starting in 1998. These printers can transfer high-resolution vibrant color graphics onto fabrics in customized designs. However, they cannot match the production speeds of screen-printing, nor can inkjet print many of the specialty inks used in T-shirt design today. Nonetheless, inkjet processes require neither makeready (final preparation) nor cleanup, and each T-shirt can be customized.

In 2005, German researchers used inkjet technology in the production of microlenses and microvessels. In this process, the inkjet deposits drops of solvent onto a polymer substrate. The solvent dissolves the substrate surface, leaving a small cavity when the solvent evaporates. The cavities then can be used as microreaction vessels for chemistry applications or as plano-concave lenses.

In summary, inkjet printing is a remarkably versatile process with a growing number of applications that promises to be a major force in the production of graphic products in the years ahead.

For background information *see* ELECTROSTATICS; INK; PAPER; PRINTING; ULTRAVIOLET RADIATION in the McGraw-Hill Encyclopedia of Science & Technology.   Anthony Stanton

Bibliography. W. Benedetti, Textile printing market exploding, *Focus on Imaging*, pp. 10–11, August 27, 2003; D. Burgess, Ink-jet technique produces microvessels and microlenses, *Photonics Spectra*, pp. 22–24, May 2005; C. D. Cullen, Inkjet proofing systems: The proof is in the proof, *High Volume Printing*, pp. 40–45, August 2003; M. Flippin, Wide-format inkjets: Inspiring new trends in signage and display graphics, *Screen Printing*, pp. 44–48, April 2004; P. Henry, Inkjet claims a place in high-volume printing, *High Volume Printing*, pp. 40–45, October 2004; H. Kipphan (ed.), *Handbook of Print Media*, Springer-Verlag, 2001; S. Partridge, Inkjets enter industrial markets: What it means for screen printers, *Screen Printing*, pp. 42–50, June 2003.

Insulator (gene)

Eukaryotic chromosomes are divided into independent domains with distinct chromatin structures (euchromatin and heterochromatin). Most genes reside within euchromatic domains that are associated with irregular nucleosomal arrays enriched in histones (basic proteins of cell nuclei) that are hyperacetylated and methylated at lysine 4 of histone H3. Such nucleosomal modifications establish chromatin structures that are generally permissive to transcription factor association and gene transcription. Within a given euchromatic domain, neighboring genes often display different expression patterns, indicating that distinct functional domains exist within large euchromatic regions. Transcriptionally active euchromatic domains are interspersed with heterochromatic regions that are gene-poor and organized into regular arrays of hypoacetylated nucleosomes that are methylated at lysine 9 of histone H3. Heterochromatic histone modifications generate a platform for interaction with proteins that spread in *cis* direction along the chromosome and limit transcription factor access to target sequences. A class of deoxyribonucleic acid (DNA) elements known as insulators defines the junctions between structural and functional chromatin domains.

Properties and classes. Insulators are characterized by two properties. First, insulators prevent enhancers from activating transcription when placed between these regulatory elements and a promoter. This position-dependent blocking activity does not cause inactivation of the enhancer, but disrupts processes that are required for communication between an enhancer and promoter. Second, insulators protect against positive and negative chromosomal position effects that result when transgenes (recombinant genes that are reintroduced into the genome) are inserted randomly within eukaryotic genomes (see **illustration**). Mechanisms involved in protection from chromosomal position effects may involve the direct blocking of regulatory elements or prevention of the spread of silencing complexes. Insulators are located throughout eukaryotic genomes, indicating that these elements play a conserved role in defining independent domains of gene function.

Several classes of insulators have been identified. Some insulators possess only enhancer-blocking activity (called enhancer blockers), some have only silencer-blocking activity (called barriers), while some have both properties. These observations indicate that insulators assemble multiprotein complexes to confer regulatory autonomy. This concept is illustrated by consideration of the chicken ß-globin 5′ hypersensitive site 4 (HS4) insulator. This 250-base-pair insulator separates the ß-globin domain from an upstream region of heterochromatin and has both enhancer-blocking and barrier functions. Deoxyribonuclease I (DNase I) footprinting studies revealed that the 5′ HS4 insulator contains five protein-binding sites (FI to FV). Two proteins have been identified that bind these regions. CCCTC-binding factor (CTCF) binds FII and is necessary and

sufficient for enhancer blocking. Upstream stimulatory factor (USF) protein binds FIV and provides barrier activity. These findings highlight the composite nature of the 5′ HS4 insulator and imply that the DNA-bound proteins bring distinct functions that cooperate to confer isolation from enhancers and silencers. In contrast to 5′ HS4, the Drosophila *gypsy* insulator binds a single protein, Suppressor of Hairy-wing [Su(Hw)], that establishes both enhancer-blocking and barrier function. In this case, the Su(Hw) protein recruits at least two proteins, Modifier(mdg4)67.2 [Mod67.2] and Centrosomal Protein (CP) 190, to form an insulator complex. It is possible that these recruited proteins bring distinct activities to the insulator to provide regulatory isolation, similar to each DNA-binding protein of the 5′ HS4 insulator.

Insulator proteins fall into two general classes. One class is associated with the ability to modify histones. For example, the barrier function of the 5′ HS4 USF protein involves recruitment of histone acetyltransferases (HATs) and histone methyltransferases (HMTs) to counteract the propagation of chromosomal silencing. A second class of insulator proteins directs nuclear positioning of the underlying DNA sequences. These proteins associate with nuclear substructures such as the nuclear matrix, the nuclear pore, or the nucleolus to form looped domains in the chromatin fiber. Investigations of Su(Hw) and CTCF suggest that these proteins fall into this latter category. The Su(Hw) protein directs interactions between *gypsy* insulator proteins bound at remote

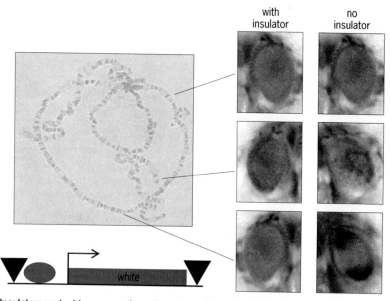

Insulators protect transgenes from chromosomal position effects. Production of transgenic animals can depend upon the ectopic integration of transgenes randomly into a eukaryotic genome. In this example, a transgene carrying the Drosophila *white* gene (oval and rectangle) flanked by *gypsy* insulators (triangles) was integrated at three distinct genomic locations (shown by arrows pointing to the *Drosophila* polytene chromosomes). The *white* transgene encodes a protein required for eye pigmentation, with high levels of *white* expression leading to the production of red eyes. Transgenic flies with insulator function display a uniform pigmentation at all three locations (with insulator). A loss of insulator function (no insulator) caused by crossing the transgenes into genetic backgrounds mutant for the *gypsy* insulator proteins revealed that in two genomic locations the transgenes were integrated into repressive chromatin. Without the insulator, a spread of silent chromatin occurs into the unprotected *white* transgene, causing decreased *white* transcription and a loss of eye pigmentation.

sites, as well as factors associated with the nuclear periphery, whereas CTCF interacts with nucleophosmin to tether the underlying sequences at the nucleolus.

Insulators have been identified in many ways. These include studies of transcriptional regulation, changes in chromatin structure, or the presence of binding sites for insulator proteins. The variety of DNA sequence requirements and protein components involved in insulator function suggest that these elements have specialized roles that are optimized for each genomic location. This postulate is supported by findings that insulator elements are not interchangeable, such that replacement of one insulator for another has been found to cause aberrant gene expression.

"Regulated" regulatory elements. Insulators do not establish static blocks of enhancer and silencer action. Several studies have identified mechanisms where insulator activity was modulated. One mechanism involves posttranslation modification of the insulator-binding proteins. For example, CTCF can be conjugated with poly-adenosine diphosphate (ADP)-ribose. Poly(ADP-ribosyl)ation does not alter the DNA binding properties of CTCF, but is believed to alter insulator properties by affecting the spectrum of proteins that CTCF interacts with. In addition, the Mod67.2 and CP190 proteins of the *gypsy* insulator can be modified by addition of the small ubiquitin-like modifier (SUMO) protein. Sumoylation of these proteins disrupts nuclear positioning of the *gypsy* insulators, causing a loss of insulator function. A second mechanism of insulator regulation involves controlling the DNA association of the insulator-binding protein. This has been observed for CTCF binding to the H19 insulator. This insulator is located in the imprinting control region (ICR) positioned between the mammalian *Igf2* and *H19* genes. CTCF binds the unmethylated H19 insulator on the maternally inherited allele, forming an insulator that blocks enhancer activation of the *Igf2* promoter. In contrast, methylation of H19 on the paternally inherited allele prevents the CTCF binding, allowing enhancer activation of *Igf2*. Finally, insulator function may depend upon proteins that accumulate in a tissue-restricted manner. The Drosophila *Fab-7* insulator that affects transcriptional regulation in a tissue-specific manner demonstrates this mode. Taken together, these findings imply that the organization of chromatin domains is dynamic, being regulated by changes in insulator function.

Mechanisms of function. The mechanisms used by insulators to establish regulatory autonomy are not well understood. Two general models have been proposed, called structural and transcriptional models. Neither model accounts for all observations of insulation function. Given the variety of insulators identified, multiple mechanisms might be employed.

The prevailing model of insulator function proposes that insulators define independent structural domains that confine the activity of transcriptional regulatory elements. In this model, specialized nucleoprotein complexes are assembled on insulators that associate with other insulator complexes or nuclear substructures to separate chromosomes into topologically distinct loop domains. Loop domains are proposed to form higher-order chromatin structures that interfere with the reception of signals from enhancers and silencers that reside outside the loop domain, resulting in regulatory isolation of a promoter inside the loop domain. Observations that insulator proteins tether chromosomal regions to nuclear substructures, such as the nucleolus, have provided support for this model.

An alternative transcriptional model suggests that insulators act directly on processes involved in the transduction of the enhancer or silencer signal to a promoter. According to this model, insulators might assemble protein complexes that resemble complexes bound next to promoters that are responsible for capturing transcriptional signals. As insulator effects depend upon insertion between the transcriptional regulatory element and promoter, interactions between a "capture complex" and enhancer or silencer would be nonproductive, and would divert these regulatory elements away from the target promoter. Support for this model comes from findings that some promoters when taken out of context are insulators, indicating that promoter regions can disrupt enhancer and silencer action.

Human disease. Insulators constrain transcriptional regulatory interactions, implying that these elements may be essential for proper maintenance of gene expression throughout development. The finding that the loss of insulator function is associated with human disease supports this postulate. Myotonic dystrophy is a dominantly inherited disease that encompasses symptoms of myotonic myopathy, cataracts, and cardiac conduction defects. The most common form is associated with expansion of a CTG repeat in the 3′ noncoding region of the *DMPK* gene that results in accumulation of a toxic repeat-containing ribonucleic acid (RNA). Unaffected individuals have fewer than 38 CTG repeats, with muscular dystrophy individuals containing more than 100. Even larger CTG expansions (>100) cause a more severe congenital form of muscular dystrophy. The mechanism for the increased disease severity has been found to result from DNA methylation associated with the large CTG expansions. This DNA methylation includes a CTCF-dependent insulator that separates *DMPK* from the nearby downstream *SIX5* gene. It has been proposed that in congenital muscular dystrophy CTCF binding to the insulator is lost, which permits enhancers from the *SIX5* gene to activate the *DMPK* promoter and expand the expression domain of the toxic *DMPK* messenger RNA. As CTCF represents a major vertebrate insulator protein, it is possible that CTG expansion at other loci might interfere with insulator function.

Properties of insulators suggest that these elements have great potential for gene therapy. In such approaches, retroviruses direct stable integration of transgenes into the genome to provide long-term gene expression. However, barriers to the success of these approaches exist. Insertional mutagenesis, as shown in the development of a leukemia-like disorder in the *SCID-X1* trial, is one challenge. A second difficulty is the variability in, or even lack of, prolonged expression of the therapeutic gene. Inclusion of insulators within gene transfer vectors holds promise to alleviate these unwanted effects by preventing inappropriate regulatory interactions.

Conclusions. Insulators are genomic elements that play a conserved role in organizing chromosomes into independent functional domains to maintain transcriptional autonomy. These elements represent a diverse class of sequences that act by multiple molecular mechanisms. A better understanding of insulator function will advance gene therapy strategies and the treatment of human diseases.

For background information *see* CHROMOSOME; DEOXYRIBONUCLEIC ACID (DNA); GENE; GENE ACTION; NUCLEOPROTEIN; NUCLEOSOME in the McGraw-Hill Encyclopedia of Science & Technology.

Xingguo Li; Pamela K. Geyer

Bibliography. G. N. Filippova et al., CTCF-binding sites flank CTG/CAG repeats and form a methylation-sensitive insulator at the DM1 locus, *Nat. Genet.*, 28:335–343, 2001; E. J. Kuhn and P. K. Geyer, Genomic insulators: Connecting properties to mechanism, *Curr. Opin. Cell. Biol.*, 15:259–265, 2003; F. Recillas-Targa, V. Valadez-Graham, and C. M. Farrell, Prospects and implications of using chromatin insulators in gene therapy and transgenesis, *BioEssays*, 26:796–807, 2004; A. G. West, M. Gaszner, and G. Felsenfeld, Insulators: Many functions, many mechanisms, *Genes Dev.*, 16:271–288, 2002; H. Zhao and A. Dean, Organizing the genome: Enhancers and insulators. *Biochem. Cell Biol.*, 83:516–524, 2005.

Intelligence, race, and genetics

The explosion of genetic research within the last 10–15 years has brought the concept of race back into prominence. One might think that, because the concept of race originated as a social proxy for the description of biological differences, at least biologists studying race would agree on its definition. However, this is not the case.

One view is that socially defined racial differentiation is most pronounced, and even is discontinuous, when it is evaluated on the basis of residence on a particular continent. A second view is that there is continuity in genetic variation across socially defined races. In this view, various races are not distinct, but a single lineage with a shared evolutionary fate. Hence, in this view there is no biological value at all in the concept of race.

In considering these views, it is important to understand that there is some agreement. Even within these extreme views, researchers agree that, although human populations might differ dramatically in terms of proportions or frequencies of alternative forms of genes (allelic variants), they do not differ in the kinds of genes they possess.

Genes and intelligence. Although attempts have been made to find connections between the variation in genes and intelligence, no genes for intelligence have been conclusively identified. The IQ (Intelligence Quotient) QTL project—aimed at identifying quantitative trait loci (QTL) contributing to genetic variation in intelligence—has attempted to establish QTLs associated with intelligence. But to date whatever positive findings have emerged have either failed to replicate or produced weak signals that have not yet been attempted to be replicated in independent samples. In addition, there are a number of recent reports of results of screens of the whole genome and exploration of specific candidate genes that produced interesting preliminary results, but these findings have not been convincingly supported. Of course, the future may bring conclusive identifications: we just do not know yet. Given that we cannot identify genes for intelligence, we certainly cannot define genetic bases of racial differences.

Genes and race. In a study using a set of protein markers (blood groups, serum proteins, and red blood cell enzymes) as indirect indicators of genetic differences between populations, Richard Lewontin estimated that roughly 85% of the genetic variance occurs between any two individuals within any socially identified racial groups, roughly 9% occurs among different populations with a socially identified race, and only the remaining 6–7% occurs between socially identified races. Other researchers arrived at the same conclusions using more powerful data sets obtained with more technologically advanced methodologies or through simulation analyses.

Richard Nisbett reviewed published studies investigating sources of differences in cognitive abilities between white and black individuals. These studies have directly sought to investigate genetic and environmental effects on intelligence. For example, one design has been to look at black children adopted by white parents. Of seven published studies, six supported primarily environmental interpretations of group differences and only one study did not. The results of this one study, that of Sandra Scarr and Richard Weinberg, are equivocal. What the Scarr and Weinberg study did show is that IQs of adopted children are more similar to those of their biological mothers than to those of their adopted mothers. Less clear are the racial implications of their findings. Moreover, there is evidence that heritability estimates vary across populations.

Concept of race. One would hope that, because the concept of race was originally, if falsely, conceived as a concept to signify "large" qualitative biological differences between groups of people, the strongest support for the concept of race would originate from biological and genetic data. Does it?

Modern *Homo sapiens* evolved once in Africa about 200,000 years ago and then spread throughout the rest of the world and simultaneously diversified starting about 50,000 to 100,000 years ago. During that spreading, modern humans supplanted now-archaic humanlike populations identifiable as having spread outside Africa at an earlier time, such as Neanderthals. The gradual geographic expansion of modern humans resulted in gradual changes in gene frequencis for the millions of sites in DNA with variation, accumulating along the paths of expansion.

Today in the United States, there are populations from very different parts of that geographically continuous spectrum of allele frequencies. Those distinct allele frequencies do not mean that different races exist, only that different parts of a continuum have been sampled. An analogy is the distinction between the colors blue, yellow, and red as samples from a continuous spectrum of light. Those colors have meaning only because the spectral sensitivities of the photoreceptors in our eyes and the neurological circuits interpreting the signals interact with a label arbitrarily imposed on some narrow range of wavelengths from a continuous spectrum.

There is no question that populations, defined geographically, demonstrate dramatic variability in allele frequencies, not only for the several million normal polymorphisms not associated with causing genetic disorders but also for many disease-related genetic alleles (variants). A diverse pattern of these genetic alleles can be readily seen in ALFRED, the ALlele FREquency Database (http://alfred.med.yale.med). The issue is not whether this variation is present or not; the issue is whether explaining this variation should occur at the levels of populations per se (for example, Lapps, Chuvash, Nyanja, or Corsicans), continents, or alleged races. Based on our review of the literature, variation that seems to be meaningful and transferable into helpful public health or educational policies is primarily at the level of specific populations and secondarily at the level of large geographic regions of ancestry. Global socially constructed categories such as race do not appear to be useful proxies for genetic features.

Social labels. When biological and behavioral markers of socially defined races are investigated, the studies primarily or even exclusively rely on participants' self-reporting of socially defined racial, ethnic, and cultural groups. Many studies use social labels such as Asian-American, African-American, Chinese, or Hispanic, implicitly ignoring the fact that these labels generalize across substantial amounts of cultural, linguistic, and biological diversity. Even ignoring the substantial variation within each of these large classifications, there is no basis, except for certain social-cultural traits, for grouping these individuals.

Moreover, self-naming of social labels might change, depending on past and present social surroundings of the surveyed participants. For example, during the Soviet era, many emigrating Soviet Jews referred to themselves as Jewish by ethnicity, but upon their arrival in Israel or the United States they referred to themselves as Russians. In the United States, indeed, Judaism is not viewed as an ethnicity but as a religion. Similarly, individuals who met the classifications of "Coloured" established by the apartheid government of South Africa would have probably self-identified themselves as black in the United States. Thus, because most medical and psychological research on racial differences is based on self-defined racial or ethnic categories and there is substantial evidence questioning the accuracy of these self-classifications, the validity of racial and ethnic differences as commonly investigated is questionable.

People will probably always label themselves and others, regardless of what scientists find. The problem is not the use of social labeling per se, but the confusion of it with biological labeling. Further, it is especially problematical when scientists contribute to this confusion by using social labels in a way that suggests they are somehow biological.

Conclusions. The important message here is that the division lines between racial and ethnic groups are highly fluid and that most genetic variation exists *within* all social groups—not *between* them. Recent studies based on hundreds of genetic polymorphisms confirm earlier studies such as that by Lewontin and show that only 11–23% of observed genetic variation is due to differences among populations and is mostly attributable to differences in allele frequencies, not all-or-nothing genetic differences. In fact, most common genetic variants exist in almost all populations. The overwhelming majority of the variation occurs among individuals with different genotypes within each population. One study found even less variation among population, but highly polymorphic multiallelic markers were studied and they may have been biased toward high heterozygosity (that is, the two chromosomes of an individual having different alleles) in many different populations, thereby minimizing the between-population variation. Clearly, when common polymorphisms are studied, there is only a minority of the genetic variation that occurs among populations. Variants that are restricted to only a few populations in one part of the world are almost never common even in those populations. The few examples of high-frequency, region-specific alleles are notable because of their rarity. The lactase (LCT) variant for adult lactose tolerance, for example, became common in northern Europe because of a unique history of selective advantage.

In sum, at this time there is no convincing evidence that any differences that may exist in average intelligence across groups can be traced to genetic structures or mechanisms.

For background information *see* ALLELE; GENE; HUMAN BIOLOGICAL VARIATION; HUMAN GENETICS; INTELLIGENCE; POLYMORPHISM (GENETICS) in the McGraw-Hill Encyclopedia of Science & Technology.

Robert J. Sternberg; Elena L. Grigorenko; Kenneth K. Kidd

Bibliography. T. Bersaglieri et al., Genetic signatures of strong recent positive selection at the lactase gene, *Amer. J. Human Genet.*, 74:1111–1120, 2004; K. K. Kidd et al., Understanding human DNA sequence variation, *J. Heredity*, 95:406–420, 2004; R. L. Lamason et al., SLC24A5, a putative cation exchanger, affects pigmentation in zebrafish and humans, *Science*, 310(5755):1782–1785, 2005; M. V. Osier et al., A global perspective on genetic variation at the ADH genes reveals unusual patterns of linkage disequilibrium and diversity, *Amer. J. Human Genet.*, 71:84–99, 2002; D. Posthuma and E. J. C. de Geus, Progress in the molecular genetic study of intelligence, *Curr. Directions Psychol. Sci.*, in press; S. A. Tishkoff and K. K. Kidd, Implications of biogeography of human populations for "race" and medicine, *Nat. Genet.*, 36(11, suppl.):S21–S27, 2004.

Intelligent search engines

The increased use of Web resources has created a need for more efficient and useful search methods. The current mechanisms for assisting the search and retrieval process are quite limited, mainly because they lack access to documents' semantics and because of the underlying difficulties in providing suitable search patterns.

Recent advances in intelligent search suggest that these limitations can be partially overcome by providing search engines with more intelligence and with the user's underlying knowledge. In this sense, intelligence is seen as the ability of systems to interact with users by natural language dialog so that the engine can learn user profiles and likes. User behavior suggests that feedback in terms of natural dialog interactions can play a key role in decreasing information overload and getting accurate search results.

Smarter search engines. Intelligent searching agents have been developed to assist information retrieval systems. Agents can utilize spider technology used by Web search engines, but in new ways. Usually these tools are robots that are trained by the user to search the Web for specific information. The agent can be personalized so that it can build individual profiles or precise information needs. An intelligent agent can also be autonomous, so that it is capable of making judgments about the likely relevance of the material on its own.

To guide the Web search process, one promising method is to discover user preferences and needs by either extracting deep knowledge from what users are looking for or interactively generating explanatory requests to focus users on their interests. Although some research has been done using natural language processing (NLP) technology to capture users' profiles, it has only been in very restricted domains that use general-purpose electronic linguistic resources to act on their requirements. In particular, using techniques for automatically generating natural language (NL) sentences allows the system to produce a useful dialog with the user and guide her or his preferences.

Designers of natural language generation (NLG) systems have strongly focused on generating natural language text and its contents at the discourse level, where complex tasks such as discourse planning play a key role in generating effective texts. When dialog processing involves managing dialog interactions (user-system), NLG systems are capable of capturing underlying knowledge, such as conversational turns (interactions), to provide replies according to the user's knowledge and goals, to react to mistakes, and to deal with unexpected reactions from the user.

Natural language feedback. In recent years, a few approaches to intelligent Web search using natural language processing technology have emerged, mainly designed as question-answering systems. These address the problem of using linguistic processing on different levels to retrieve documents containing specific paragraphs in which target natural language queries are answered. So far, there is no dialog, and the effort is centered on obtaining accurate paragraphs (within documents), instead of capturing a user's preferences.

As a part of a major interactive searching system, the model is based on task-dependent discourse and dialog analysis capabilities.

Figure 1 shows a model for intelligent searching and filtering using natural language feedback. The operation starts with natural language queries provided by a user (that is, general queries, general responses, feedback, and confirmation) and then passes them on to the discourse-processing phase, which generates the corresponding interaction turns (natural language output), arriving at a specific search request. As the dialog continues, the system generates a refined query that is sent to a search agent.

Natural language generation for dialogs. The dialog generator is based on a number of stages that state the context, participants' knowledge (user and system), and goal of the interaction. It also consists of a set of modules for which input and output is delimited according to different stages of linguistic and nonlinguistic information extracted from the dialog.

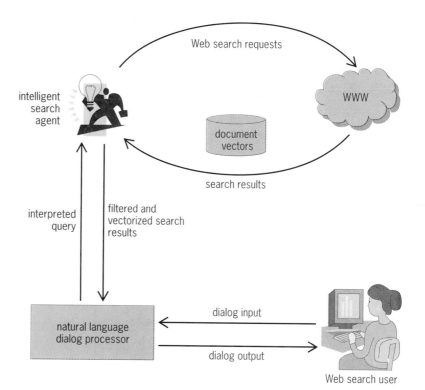

Fig. 1. Overall search-driven NL dialog agent.

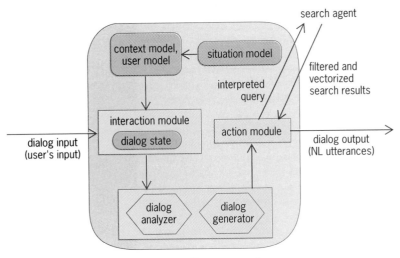

Fig. 2. Interactive natural language generation component.

This dialog-processing component is based on state-of-the-art linguistic models for discourse processing.

The natural language generation component in **Fig. 2** is capable of generating discourse outputs (that is, natural language utterances) from the results of a bibliographic Web search. This starts with the user's input (natural language query) and produces either an output consisting of a natural language conversation exchange to guide the dialog and focus the user, or a search request that is passed to the search agent. In order to understand its underlying workings, the model has been divided into components (Fig. 2).

Context model. The context model deals with information regarding the dialog's participants, that is, the user (who needs information from the Web) and the system (which performs the search). Here, the user model considers knowledge about the user with which the system interacts. The information regarding the communicative situation's characteristics in which the dialog is embedded is established in the situation model.

Natural language interaction module. The interaction module is based on cooperative principles. This involves a two-position exchange structure, such as question/answer, greeting/greeting, and so on. These exchange structures are subject to constraints on the system's conversation, regarding a two-way ability to transmit appropriate and understandable messages as confirmations.

Dialog analyzer. The dialog analyzer receives the user's query and analyzes the information to define the criteria that can address the system's response generation. In addition, recognition and interpretation are controlled by modules for semantic and pragmatics analysis, which process linguistic knowledge.

Dialog generator. The dialog generator takes the information obtained from the search agent and the dialog state, and generates a coherent utterance to the current dialog sequence.

Dialog begins by generating a kind of utterance, a query about information requested by the user. Next,

the system considers two possible generations: a specific query for communicating the situation (what topic do you want to search for?) and a general one on the context of the different kinds of information available on the Web (what kind of information do you need?).

As the dialog continues, the discourse generator produces its output (natural language sentences) based on search results, context information, and user feedback. In order to establish the starting point for the natural language generation process, high-level goals are identified.

Based on several samples obtained from experimental studies of users searching on the Web, basic initial criteria are extracted to restrict the natural language generation process, such as language, type of homepages, type of documents, and so on.

The natural language dialog generator can then produce two kinds of answers to explain the results of the search. One kind is for obtaining a more detailed specification of the user's query, for example, "Your query is too general, could you be more specific?" The other kind requires the user to state some feature of the topic, for example, "Which language do you prefer?" The discourse analyzer again performs the analysis of the user's specific answer in order for the search agent to perform a refined search. The search agent repeats the task, searching for the specific information on the topic in question.

At this point, the dialog analyzer processes the user's response in order for the generator to produce an output confirming or expressing the action being performed. Whenever a user's response is positive, the system will generate a sentence to give the user the opportunity to choose a new search topic. Otherwise, the dialog goes on. The overall process starts by establishing a top goal to build up the full structure in the sentence level.

Adaptive search agents. Unlike traditional search engines, the model for intelligent searching and filtering does not deliver all the information from Web search results to the user. Instead, the agent waits until it has sufficient knowledge about the user's feedback and goals, which has a positive effect in terms of information overloading. As the interaction continues, the agent refines the requests and filters the initial information obtained from the user's feedback, and then the search proceeds until a proper amount of information can be displayed (for example, 30 retrieved documents).

Using the knowledge obtained from user feedback, dialog samples, and current context information, the model detects the most frequent search patterns in vectorlike criteria, some of which involve the Web address of the page being selected, the author, the language, and so on.

Regardless of whether the search engine is fed with knowledge acquired from the user, the dialog, or the current intermediate search, the previously trained agent takes the matching vectors and performs the search request on the Web. When information extracted from the vectors and user feedback is not enough or unavailable, the agent makes

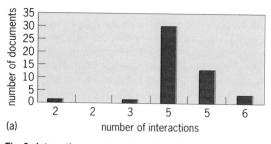 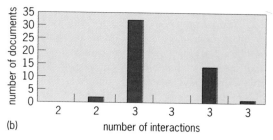

Fig. 3. Interactive experiment involving dialog interactions on (*a*) Java and (*b*) Animaniacs.

simple decisions by predicting the most likely actions to perform.

Intelligent search agents in action. A search model that uses intelligent-agent technology and natural language processing techniques was designed and implemented to investigate the extent to which information overloading and intelligent search capabilities can be effective for the Web search.

The results produced by a prototype system were defined in order to establish whether search-driven natural language dialog generation is possible. In addition, the approach exploits the benefits of using this kind of interaction to decrease information overloading as well as the time spent by users searching the Web. The following is an example of a dialog between the system (S) and a user (U):

S: What are you interested in?
U: I need basic information on linguistics.
S: Your query is too broad; could you please be more specific?
U: Fine.
S: There are twenty references about that topic. Do you want to check all of them?
U: Yes.
S: Do you need further information about that topic?
U: Yes.
S: I found commercial pages. What kind do you prefer?
U: I'd like to look at financial services.

The model was then assessed in terms of the average number of conversational turns in the dialog necessary to get a precise requirement, and information was filtered against the number of references/documents that matched these requirements. Initially, the set of possible candidate documents was more than 30,000, but the scope was reduced to 1000 or less.

Several experiments were done involving themes ranging from Java to Animaniacs (**Fig. 3**). In order to understand the analysis, each interaction is defined by one or more dialogs (exchanges) between a user and the system. Interactions for the experiment in Fig. 3 showed an increase in the number of documents matched as more than three turns are exchanged—this result does not come up by chance. For the same number of interactions (five), different results are shown mainly due to the adaptive way the dialog goes. That is, the context and kind of ques-

tions made by the agent are changing, depending on the situation and the document's contents. Different results were obtained for the same number of interactions because the kind of document searched for was changed as other features were restricted. A similar situation occurs as the dialog states a constraint regarding the language, in which case most of the original document references were not matched.

Experiments showed important drops in the results with a minimum of conversation turns due to constraints on the nature of the information finally delivered. The prototype search agent took into account previous issues, so there are some classes of high-level requests that are more likely to happen than others, depending on the context.

Overall, the current model, based on dialog interactions, shows promise as a novel and interesting work strategy to deal with specific information searching requirements. In addition, designing and implementing a natural language generation system easily can be adapted to tailored communicating situations.

For background information *see* ARTIFICIAL INTELLIGENCE; INFORMATION MANAGEMENT; INTERNET; LINGUISTICS; NATURAL LANGUAGE PROCESSING; WORLD WIDE WEB in the McGraw-Hill Encyclopedia of Science & Technology.

Anita Ferreira; John Atkinson

Bibliography. C. Holscher and G. Strube, Web search behavior of Internet experts and newbies, *9th International World Wide Web Conference*, Amsterdam, May 2000; B. Jansen and A. Spink, Real life, real users, and real needs: A study and analysis of user queries on the *Web, Info. Process. Manag.*, 36(2):207–227, 2000; D. Jurafsky and J. Martin, *An Introduction to Natural Language Processing, Computational Linguistics, and Speech Recognition*, Prentice Hall, 2000; A. Levy and D. Weld, Intelligent internet systems, *Artif. Intell.*, 11(8):1–14, 2000; E. Reiter and R. Dale, *Building Natural Language Generation Systems*, Cambridge University Press, 2000.

Interrogation and torture

In November 2001, three citizens from Tipton, England, were detained in northern Afghanistan and sent to Sherbegan Prison. Thereafter they were handed over to the U.S. military and transferred to Guantanamo Bay Naval Base in Cuba, where they remained for over 2 years. Following their release in 2004, the "Tipton Three" reported enduring what might be considered torture, including prolonged isolation, poor living conditions, deprivation of food and water, lack of medical treatment, sexual humiliation, drug inducement, and physical abuse. In addition, the three described hundreds of hours spent under interrogation, often placed in uncomfortable positions for long periods of time and questioned by persistent interrogators who claimed to have evidence of their involvement with the terrorist organization al-Qaeda. As a result of their harsh treatment, they eventually confessed to traveling to Afghanistan to fight a holy jihad and to having been present at a rally with al-Qaeda leader Osama bin Laden. The three were later returned to the United Kingdom and were able to prove to British authorities that they had no involvement with al-Qaeda—rather, their confessions had been elicited falsely as a result of the physical and psychological torture they had endured at Guantanamo. This article will review research on the effectiveness of physical and psychological torture in the context of interrogations, and discuss the psychological processes that lead to both true and false confessions.

Brief history of physical torture and psychological manipulation. Though shocking, the purported torture and coercive interrogation techniques recently evidenced at Guantanamo Bay and Abu Ghraib Prison (Iraq) have been recorded throughout history, including the trials and public confessions of Soviet citizens under Stalin's rule (mid-1900s) and the interrogation of U.S. military personnel by communist China (post-1949). Furthermore, organizations such as Human Rights Watch and Amnesty International regularly document instances of torture and coercive interrogation around the world. Standard torture tactics have included isolation and sensory deprivation, direct physical abuse and threats of death or harm, deprivation of food or sleep, sexual molestation or humiliation, confinement in "stress" positions for long periods of time, and inducements involving minor comfort items (bottled water, blanket, pillow, etc.).

Harsh interrogation tactics have also been employed within the U.S. criminal justice system. Although often more subtle than the extreme instances observed in military and political interrogations, "third degree" tactics were regularly employed by police investigators in the United States through the early twentieth century, including the use of physical abuse; incommunicado detention; deprivation of food, sleep, and medical attention; and explicit threats of harm to oneself or one's family. Reforms in the 1940s, including a Presidential commission's "Report on Lawlessness in Law Enforcement," eventually discouraged the use of physical coercion and led to the development of interrogation manuals that emphasized psychological manipulation. Such psychologically based interrogation techniques frequently involve presenting false evidence to a suspect that reinforces guilt ("we have a witness that identifies you as the murderer"), preventing a suspect from

ever denying his/her involvement in the crime ("we know that you did this, so there's no use in denying it"), maximizing the potential consequences of not cooperating with law enforcement ("if you don't cooperate with us, I'm sure the district attorney is going to place the death penalty on the table"), and attempts to minimize a suspect's potential involvement in the crime ("maybe it was an accident") or the perceived consequences associated with confessing to the crime ("I know the judge in your case, and if you confess I'd be happy to speak on your behalf").

Studies suggest that the use of both physical torture and psychological manipulation are rather effective in eliciting information from individuals (that is, true confessions); however, these methods will also compel individuals who are innocent to provide false confessions as a means of escaping the stresses of an interrogation. For example, social science researchers have found that psychologically based interrogation techniques employed by police in the United States lead to a significant number of false confessions and ultimately wrongful incarceration. One recent study documented 125 cases of proven false confession in the United States since the *Miranda v. Arizona* (1966) Supreme Court decision, while case studies suggest that 25% of individuals who had been wrongfully convicted (but were later exonerated by DNA evidence) were found guilty largely based upon coerced false confessions. Taken together, these studies suggest that physical torture and psychological manipulation fail to yield diagnostic intelligence; that is, these tactics are likely to yield information from both those who possess guilty knowledge and those who are innocent.

Factors associated with eliciting a confession. Social science researchers have systematically examined both the process of interrogation and the false confession phenomenon over the past several decades. This research has shown that several factors appear to be associated with the elicitation of confession evidence. First, with regard to the psychological processes leading to confession, decision-making theories suggest that individuals evaluate the available courses of action during an interrogation by considering their options ("should I confess or should I maintain my innocence?"), and weighing the likely consequences attached with those options ("what is likely to happen to me if I don't comply with the demands of my interrogator?"). If the benefits of providing information outweigh the costs of not providing that information, the individual is likely to provide a confession. As such, the use of torture tactics such as solitary confinement, physical abuse, and deprivation of food or sleep may force an individual to comply with the demands of an interrogator, while promises of leniency or release from imprisonment may provide a strong incentive for cooperation.

Second, physical torture and psychological manipulation can also more directly impair an individual's cognitive ability to reason and make decisions. Research shows that intense fear and anxiety can cause a person to become defensive, hypervigilant, and insecure regarding their environment. Similarly, fatigue and sleep deprivation significantly impair decision-making ability and promote susceptibility to social influence. These variables then lead to uncertainty and insecurity in what may happen beyond the interrogation, and ultimately promote feelings of isolation and dependency upon the interrogator. Once again, however, these techniques are likely to create a context for compliance by both those who possess critical information and those who do not.

Finally, research has demonstrated that some individuals are more vulnerable than others in the interrogation room. First, both field and laboratory studies indicate that children are more likely to confess during an interrogation when compared with adults. Second, individuals of low intelligence are often found to be more suggestible (that is, they have a tendency toward yielding to social influence) and less able to cope with the pressures of the interrogation room. Third, individuals who have never been interrogated before appear more susceptible to providing a confession when compared with those who have endured the pressures of interrogation previously. Finally, the psychological state of the individual at the time of interrogation has been shown to be associated with the likelihood of confession. In particular, individuals who suffer from mental illness (such as depression or anxiety disorders) appear to be more susceptible to the effects of interrogation, as well as individuals who may be under the influence of drugs or alcohol, or who may be in a state of detoxification.

Typology of false confession evidence. As discussed above, the use of physical torture and psychological manipulation appear quite effective in yielding confession statements during an interrogation. However, if the individual does not possess certain intelligence information or guilty knowledge, these tactics can lead the individual to provide false information that might appease the interrogator and alleviate the individual of the pressures of interrogation. Researchers have generally classified false confession statements into three categories. First, a voluntary false confession occurs when an individual provides false information without any torture or psychological coercion. An individual may provide a voluntary false confession due to his or her desire to protect someone else, for notoriety or attention, or due to an inability to distinguish reality from fantasy. Second, a coerced-compliant false confession generally involves an individual providing information for some immediate instrumental gain and in spite of his awareness that he is lying to the interrogator. Physical torture and psychological manipulation both increase the likelihood of coerced-compliant information—as the individual does not actually believe the false information that he is giving, but perceives the consequences of confessing as less damaging than the consequences of further torture. Finally, coerced-internalized false confessions occur when an individual falsely provides information and actually begins to believe that he or she is responsible for the criminal act. Individuals who provide coerced-internalized false confessions often are made to dis-

trust their memory for the event such that they are more willing to incorporate external suggestions. The long-term pressures of physical torture are also likely to promote the internalization of new beliefs and ideology, a tactic that proved effective in the political interrogation and indoctrination process employed by communist China.

Bottom line on torture and interrogation. In summary, social science research indicates that physical torture and psychological manipulation will successfully produce intelligence information and confession evidence. However, these methods are just as likely to produce information from those who possess guilty knowledge and those who are innocent. As a result, such harsh methods of torture and interrogation lack diagnostic value and, in the end, may prove rather costly to intelligence and criminal justice efforts to provide security and solve crimes, particularly with regard to financial and human resources that may be expended based upon false information. Furthermore, individuals who undergo extended periods of torture have been shown to exhibit considerable psychopathology and long-term effects of posttraumatic stress—a significant violation of human rights to those who are wrongfully imprisoned and interrogated.

For background information *see* BRAIN; COGNITION; EMOTION; LIE DETECTOR; NERVOUS SYSTEM (VERTEBRATE); PSYCHOLOGY in the McGraw-Hill Encyclopedia of Science & Technology.

Christian A. Meissner; Justin S. Albrechtsen

Bibliography. S. A. Drizin and R. A. Leo, The problem of false confessions in the post-DNA world, *North Carolina Law Rev.*, 82:891–1007, 2004; G. H. Gudjonsson, *The Psychology of Interrogations and Confessions*, Wiley, London, 2003; S. M. Kassin and G. H. Gudjonsson, The psychology of confessions: A review of the literature and issues, *Psychol. Sci. Public Interest*, 5:35–67, 2004; G. D. Lassiter (ed.), *Interrogations, Confessions, and Entrapment*, Kluwer Academic, New York, 2004; A. McCoy, *A Question of Torture: CIA Interrogation, from the Cold War to the War on Terror*, Metropolitan Books, New York, 2006.

Landslides

A landslide is the failure and movement of a mass of rock, sediment, soil, or artificial fill under the influence of gravity. Landslides range in volume from tens of cubic meters to tens of cubic kilometers, and they move at rates ranging from millimeters per year to more than 100 meters per second. Landslides are primarily associated with mountainous terrain but also occur in areas of low relief, for example, in roadway and building excavations, mine-waste piles, and river bluffs.

Landslides are an important natural hazard. On a global scale, they kill thousands of people each year and cause tens of billions of dollars in damage. One of the most deadly landslides on record occurred during an earthquake in Peru on May 31,

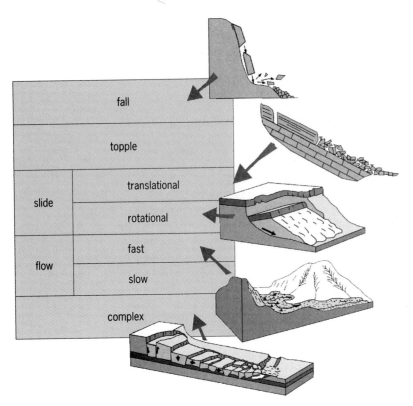

Fig. 1. Simplified classification of landslides.

1970. A streaming mass of blocky debris produced by a failure on a Western Andes mountain, Nevado Huascaran, killed 20,000 inhabitants of the towns of Yungay and Ranrahirca. Large landslides can also block rivers, creating impoundments and upstream flooding. Sudden failure of landslide dams may cause devastating downstream floods.

Types. Landslides are classified on the basis of their velocity, the type of movement, and the source material (**Fig. 1**). The presence or absence of water is important because it exerts a strong influence on the type of movement. The scheme shown in Fig. 1, although simplified, provides an understanding of the range and complexity of landslides.

Falls. Falls involve rolling and bouncing of rock, and less commonly sediment, from cliffs or down steep slopes. Initial failure occurs along steeply inclined fractures or other discontinuities in the rock

Fig. 2. Rockfall on Trans-Canada Highway near Yale, British Columbia. (*Duncan Wyllie*)

or sediment. This process is responsible for cones and aprons of talus, which are common landforms in mountains. Large blocks and boulders sometimes roll or bound beyond the foot of the talus slope, causing loss of life and property damage. Rockfall is also common along some roads and railways (**Fig. 2**). Rockfalls, although relatively small, are among the most costly of all types of landslides. In addition to economic losses due to traffic delays, the costs of scaling, blasting, and grouting threatening rock faces and removing debris from roads and rail lines are considerable.

Topples. Topples involve the forward rotation of rock or sediment about a pivot point under the influence of gravity. Movement occurs along steeply inclined fractures. Topples range from shallow movements to deep-seated displacements of large volumes of rock. The process operates almost imperceptibly, but a threshold of stability may be reached, at which time the material suddenly fails, producing a fall or a slide.

Slides. Slides are subdivided into translational and rotational types. Translational sliding takes place on planar or undulating surfaces. The slide mass commonly disintegrates as it moves downslope, but the fragments tend to retain their positions with respect to one another. Rotational sliding involves translation of rock or sediment along a concave-upward failure surface, producing slumps. Movement rates range from millimeters per hour to meters per second. Many slides are complex phenomena, involving both translational and rotational movements.

Flows. Flows are a large and varied group of landslides that share one similarity—the failed material moves in a fluidlike manner. In wet flows, rock fragments are partly supported by interstitial water. Debris flows are the most common type of wet flow; they consist of mixtures of water, rock fragments, and plant detritus that move down steep stream courses or ravines as slurries. Most debris flows are triggered by heavy rainfall. Those that move down the flanks of volcanoes during eruptions are called lahars. Mudflows are similar to debris flows, but the solid fraction consists of sand, silt, and clay, with little or no gravel or coarser material. Debris flows and mudflows can travel at speeds of up to a few tens of meters per second.

Flows of water-saturated sediment are also common on slopes underlain by permafrost in the Arctic. The failure plane is shallow, at the contact between the permanently frozen ground and the overlying active layer. These flows can occur on slopes as low as a few degrees and are especially common along river and coastal bluffs underlain by ice-rich sediments.

Sediment flows also happen in oceans and lakes, especially off deltas and at the heads of submarine canyons. Those that travel down submarine canyons into deep ocean waters are termed turbidity currents.

Rock avalanches (also known as sturzstroms) are large flows of fragmented rock that contain little or no water. They flow due to the energy released by particle interactions and particle comminution. Rock avalanches are the fastest of all landslides, in some cases achieving speeds of 100 m/s or more. They travel long distances where unimpeded by topography. Rock avalanches are rarer than other types of landslides, but they are important and of scientific interest because of their long travel distances and the destruction and death they cause. One famous example occurred in April 1903 and killed about 70 people in the mining town of Frank in Alberta, Canada (**Fig. 3**).

Many landslides do not fit comfortably into existing classification schemes. Prominent examples include sackung (from the German verb "to sag") and lateral spreads. Sackung is deep-seated downslope movement of large, internally broken rock masses, with no single, well-defined basal failure plane. Movement is manifested at the land surface by cracks, trenches, and scarps at mid and upper slope positions, and by bulging of the lower slope. Lateral spreading involves extension of a slab of earth material above a nearly flat shear plane. The moving slab may subside, rotate, disintegrate, or flow. Lateral spreading in silts and clays is commonly progressive—failure starts suddenly in a small area but spreads rapidly, ultimately affecting a much larger area (**Fig. 4**). Lateral spreading commonly results from liquefaction of a subsurface sand layer.

Many landslides, including most large ones, are complex, involving more than one type of movement. A rockslide may evolve into a debris flow by entraining water or saturated sediments along its path.

Causes and triggers. The cause of a landslide is the combination of external and internal factors that over time lead to failure. The main causes of landslides are material type and structure, steep topography, weathering, erosion by a stream, glacier or ocean waves, subsurface solution, depositional loading of the top of a slope, a change in climate, and human disturbance. A landslide can be triggered by a single event such as an earthquake, rainstorm, volcanic eruption, or freeze-thaw activity.

Most landslides triggered by earthquakes are rockfalls and small slides, but some very large landslides are seismically triggered. A common triggering mechanism of debris flows and rockfalls is intense rain. Rainwater infiltrates sediment and fractures in rock, raising the water pressures in these materials and inducing failure. Volcanic eruptions trigger lahars and, in some cases, huge flank collapses, such as happened at the onset of the cataclysmic eruption of Mount St. Helens in May 1980.

Many large landslides occur without known triggering events. In such cases, the slope has slowly deteriorated over centuries or millennia to the point that it failed of its own accord.

The water content of slope materials is an important factor in their stability. Slopes that are stable in arid environments may fail in humid ones, and some rock slopes in areas where temperatures frequently fluctuate above and below freezing are vulnerable to rockfall. Later in this century, climate change may

Fig. 3. Frank slide. This rock avalanche partially destroyed the town of Frank, Alberta, killing about 70 people. Note the long run-out of the blocky landslide debris. (*Geological Survey of Canada*)

alter the frequency and types of landslides in some areas.

Recognition and mitigation of hazards. The first step in landslide hazard mitigation is to identify and assess the hazard. It involves (1) determining the age and frequency of past landslides from geologic evidence and historical records and (2) assessing ground conditions where failure might occur. The second step is mitigation to minimize risk to people and property. Mitigation measures are of three types: (1) restrictions on land use, (2) monitoring and warning, and (3) corrective and defensive works.

Land-use restrictions. If a site is deemed too hazardous and the risk cannot be reduced to an acceptable level, development may be disallowed or restricted.

Fig. 4. Lemieux landslide. This 1993 landslide in the South Nation River valley, Ontario, Canada, is a typical "quick clay" failure. It was retrogressive and involved lateral spreading and flow on nearly flat bedding planes in fine-grained glaciomarine sediments. (*Geological Survey of Canada*)

Whether a risk is perceived as acceptable or unacceptable depends on social and economic factors. In general, less risk is tolerated in wealthy countries than in poor ones. In wealthy countries, knowledge of past landslides is generally taken into account during development. For this reason, land-use restrictions are more commonly accepted in these countries than in poor ones. The high level of risk that people in some poor countries accept contributes to the greater loss of life from landslides.

Monitoring and warning. Many large landslides are preceded by rockfalls, small slumps, the opening and widening of ground cracks, and tilting of trees. Observation or instrumentation of unstable slopes may provide warning of their imminent catastrophic failure.

Corrective and defensive works. Corrective and defensive measures include reforestation, control works, and protective structures. Careful management of forests reduces the likelihood of some types of landslides. Reforestation of logged or burned slopes, in combination with other corrective measures, may stabilize debris source areas. Control works, including dykes along stream channels, deflection dams and dykes, and debris retention basins, provide protection against debris flows. Steep rock slopes along highways and railways can be stabilized with retaining walls, anchored beams, rock bolts, bulkheads, toe buttresses, metal nets and fences, and ditches. Unstable slopes can be dewatered with tunnels or permeable pipes. Stabilization of large bedrock slides and sagging slopes requires extensive surface and subsurface drainage.

For background information *see* EARTHQUAKE; ENGINEERING GEOLOGY; LANDSLIDE; MASS WASTING; PERMAFROST; ROCK MECHANICS; SEISMIC RISK; SOIL; SOIL MECHANICS; TALUS; TURBIDITY CURRENT; VOLCANO in the McGraw-Hill Encyclopedia of Science & Technology.   John J. Clague

Bibliography. D. Cornforth, *Landslides in Practice: Investigation, Analysis, and Remedial/Preventative Options in Soil*, Wiley, 2005; J. E. Costa and G. F. Wieczorek (eds.), *Debris Flows/Avalanches: Process, Recognition, and Mitigation*, vol. 7, Geological Society of America, 1987; T. Glade, M. G. Anderson, and M. J. Crozier (eds.), *Landslide Hazard and Risk*, Wiley, Chicester, England, 2005; M.-L. Ibsen et al. (eds.), *Landslide Recognition: Identification, Movement and Causes*, Wiley, 1996; A. K. Turner and R. L. Schuster (eds.), *Transportation Research Safety Board*, Spec. Rep. 247, National Academy Press, Washington, DC, 1996.

Laonastes rodent and the Lazarus effect

Tropical southeastern Asia is an increasingly important center of biological diversity, with the discovery of plants and animals previously unknown to science. Among living mammals, for example, the number of newly recognized species has continued to grow during the last 15 years, now including at

Fig. 1. First photograph ever taken of *Laonastes*, the Laotian rock squirrel. This individual was caught, photographed, and released back into its rocky home in May 2006 by Dr. David Redfield, Florida State University.

least four artiodactyls, a bat, a primate, an insectivore, a lagomorph, and several rodents that have been discovered in this area. One recent addition from this biological hot spot is the living rodent *Laonastes aenigmamus*, described in 2005 by mammalogists working on a biodiversity survey in and around the Khammouan Limestone National Biodiversity Conservation Area in the central part of the Lao People's Democratic Republic (Laos). Current knowledge of *Laonastes*, known locally as the kha-nyou, is based primarily on dead specimens that were purchased in local markets and isolated bones collected from owl pellets in caves. Although information on living individuals is extremely limited, *Laonastes* seems to be nocturnal in its activity pattern and is thought to inhabit rocky terrain. It has a vaguely squirrellike appearance (**Fig. 1**) with an elongated head, long whiskers, black to grizzled pelage, a fringe of bristly hairs around the claws on the feet, and a long hairy tail. The limbs are nonspecialized, indicating that *Laonastes* employs a scampering mode of locomotion. Length of head and body averages about 26 cm (10.2 in.), with the tail adding another 14 cm (5.5 in.) to that length. Stomach contents from specimens purchased in markets were mostly plant remains, with inclusion of a few pieces of insects. Morphological and molecular differences of *Laonastes* from other rodents suggested to the original describers that it must represent a new rodent family, the Laonastidae. The morphological characters considered to set it apart from other rodents included the combination of the skull (**Fig. 2**) with an enlarged infraorbital foramen, lower jaw lacking a coronoid process and having the mandibular angle offset laterally (later shown to have been interpreted incorrectly), one premolar and three molars in each jaw, a transversely bilophodont pattern of the cheek teeth (that is, having two transverse ridges), and four roots on the lower molar teeth.

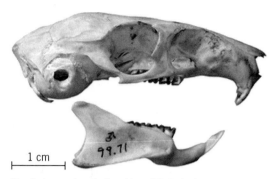

Fig. 2. *Laonastes* skull and jaw. (*Photo by L. Marivaux/Carnegie Museum of Natural History*)

While *Laonastes* is indeed different from other living rodents, the original description did not attempt to determine its possible antecedents among various groups of fossil rodents. This was left to subsequent paleontological investigation. The highly distinctive characteristics of *Laonastes* led to the determination that the origin of *Laonastes* could readily be traced to a known family of earlier Asian rodents, the Diatomyidae. The basic anatomical characters in the skull, jaws, teeth, and skeleton of *Laonastes* are not unique to a new family but are shared with members of this otherwise extinct rodent family, which is known from Oligocene and Miocene rocks in southern and eastern Asia. Accordingly, *Laonastes* is now regarded as a living diatomyid, with its closest known relative occurring more than 11 million years ago (Ma).

Fossil record of the Diatomyidae. Three fossil genera are currently recognized in the Diatomyidae: *Fallomus*, *Diatomys*, and *Willmus*. The geologically oldest, *Fallomus*, is known from isolated teeth and fragments of jaws from the Oligocene (about 25–30 Ma) of Pakistan, India, and Thailand and the early Miocene (about 21–23 Ma) of Pakistan and India. The geologically youngest is *Willmus*, known from only a few isolated teeth from the late Miocene (abouut 11 Ma) of Pakistan. *Diatomys* is a better-known and more widely distributed rodent, having Miocene records in Shandong and Jiangsu provinces of China, Kyushu Island of Japan, northern Pakistan, and the Lamphun district of Thailand.

Diatomys was originally described on the basis of two relatively complete skeletons of *D. shantungensis* from the early Miocene of Shandong Province in eastern China, now in the collections of the Institute of Vertebrate Paleontology and Paleoanthropology, Beijing. Both specimens are remarkably preserved in fine-grained, diatomaceous sediments and retain traces of whiskers and pelage. A third specimen (**Figs. 3** and **4**) was discovered in the same locality in 2005. All are laterally compressed to some degree by postmortem deformation, although the most recently discovered specimen is less distorted and shows more details of skull structure. The skeleton of *D. shantungensis*, which has a head and body length of about 25 cm (9.85 in.), is that of a generalized scampering rodent, lacking any obvious morphological adaptations for either leaping or burrowing. Although anatomy of the skull and jaw is somewhat more difficult to discern due to the lateral compression of the fossils, it is clear that *Diatomys* has a very large infraorbital foramen, showing that part of the major muscle for chewing, the masseter, passed anteriorly through it to originate on the side of the rostrum—the hystricomorphous condition. The mandible lacks a coronoid process and has a relatively low articular condyle. The masseteric fossa, marking the position of insertion of the masseter muscle on the lower jaw, extends forward to below the lower premolar. The angular process, at the posteroventral side of the jaw, is slightly inflected and in the same vertical plane as the incisor—the sciurognathous condition. The shaft of the enlarged lower incisor is relatively short, extending back to a line below the second lower molar. The microscopic structure of the incisor enamel, an important key to rodent relationships, was determined to be multiserial. The well-preserved dentition of these specimens shows that the cheek teeth in each jaw included one premolar and three molars and that each tooth had a simple, transversely bilophodont occlusal pattern. Another highly unusual characteristic is the presence of four roots on the lower molar teeth.

Fig. 3. *Diatomys shantungensis* fossil from China. (*Photo by Mark A. Klingler/Carnegie Museum of Natural History*)

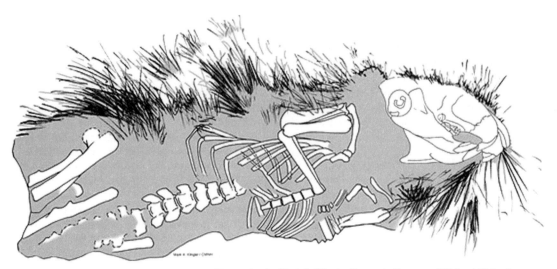

Fig. 4. *Diatomys shantungensis* fossil diagram. *(Illustration by Mark A. Klingler/Carnegie Museum of Natural History)*

Fossil diatomyids. One of the striking things about *Laonastes* is that it is virtually identical to the Miocene *Diatomys* in overall size and locomotor adaptations as well as in a suite of derived anatomical characters. The latter include the structure of the bilophodont cheek teeth, length of the lower incisor, details of the enamel structure of the incisor teeth, shape and muscle attachments of the jaws, articulation of the jaw, and skull. *Laonastes* was originally described as having an hystricognathous lower jaw, with the angle set lateral to the body of the jaw. Later investigation showed this interpretation to be incorrect, and the jaw structure, like that of *Diatomys*, has the angle in the same plane as the body of the mandible—the sciurognathous condition. Within the now known members of the Diatomyidae, *Fallomus* is more primitive than the others in having less well developed lophs (enamel ridges on molar teeth) and retaining more trace of cusps (pointed or rounded projections on the masticating surface of teeth) in the cheek teeth. Even *Fallomus*, however, shares the tendency toward bilophodonty of the cheek teeth, and has four rooted molars, the same type of incisor enamel and zygomasseteric structure as the other members of the family. The youngest known fossil diatomyid, *Willmus*, is very incompletely known but is larger and has higher crowns of the cheek teeth than in the geologically older members of the family.

Lazarus effect. The Lazarus effect is a phenomenon that is encountered in the fossil record. It refers to the reappearance of taxa following a lengthy temporal gap in the record, and is usually applied when there is a hiatus in time between extinct organisms. Discovery of a living example of taxa that were previously thought to be extinct, as occurs with the family Diatomyidae, is a very special case of the Lazarus effect, one that has only rarely been documented among mammals and other vertebrates. In most cases where fossil mammal taxa were subsequently discovered to be alive, the fossils in question have been Pleistocene in age and congeneric with their modern counterparts. For example, the Pleis-

tocene cricetid rodent *Blarinomys* and the canid *Speothos*, the bush dog, were first found as fossils and later discovered in the living fauna of South America. Similarly the Pleistocene peccary *Catagonus wagneri* (a wild piglike mammal) was discovered living in the Chaco of Paraguay. A longer gap exists in the record of the living South American marsupial *Dromiciops*, which was described as a didelphid marsupial but later was referred to the family Microbiotheriidae, otherwise known from the early and middle Tertiary of South America.

The case of *Laonastes*, with the reappearance of a family of mammals that was believed to have been extinct for more than 11 million years, is a striking case of the Lazarus effect. *Laonastes* has been resurrected in the extant biota after a temporal gap of roughly 11 million years. The living rodent occurs within the geographic range of its progenitors, but as currently understood is more limited in distribution. Further investigations, both paleontological and neontological, are required to establish where members of this interesting rodent family spent the millions of years intervening between its last appearance in the fossil record and its resurrection in the living biota of southeastern Asia. Diatomyids join tree shrews, flying lemurs, tarsiers, and some other mammals as examples of ancient and formerly wider-ranging mammalian taxa that are currently living with relictual distributions in southeastern Asia. Where diatomyids were in the intervening time, what climatic and habitat conditions they preferred, and why their range became restricted to southeastern Asia are questions that remain to be answered following the discovery of more fossils.

For background information *see* DENTITION; EXTINCTION (BIOLOGY); FOSSIL; RODENTIA; TAPHONOMY; TOOTH in the McGraw-Hill Encyclopedia of Science & Technology.　　　　　　Mary R. Dawson

Bibliography. M. R. Dawson et al., *Laonastes* and the "Lazarus effect" in Recent mammals, *Science*, 311:1456–1458, 2006; L. J. Flynn and M. E. Morgan, An unusual diatomyid rodent from an infrequently sampled late Miocene interval in the Siwaliks of Pakistan, *Palaeontologia Electronica*, 8.1.17A: 1-10, 2005; D. Jablonski, Causes and consequences of mass extinctions: A comparative approach, in D. K. Elliot (ed.), *Dynamics of Extinction*, Wiley, New York, 1986; P. D. Jenkins et al., Morphological and molecular investigations of a new family, genus and species of rodent (Mammalia: Rodentia: Hystricognatha) from Lao PDR, *Systemat. Biodivers.*, 2:419-454, 2005; L. Marivaux and J.-L. Welcomme, New diatomyid and baluchimyine rodents from the Oligocene of Pakistan (Bugti Hills, Balochistan): Systematic and paleobiogeographic implications, *J. Vert. Paleontol.*, 23:420-529, 2003.

Locating wind power

Wind is an important component of everyday weather, although unlike temperature and precipitation, we often pay little attention to wind outside of extreme events such as thunderstorms, tornadoes, or hurricanes. Wind is necessary for dispersing seeds and ventilating pollutants; wind speeds are considered in engineering design and construction; wind can strongly affect human comfort (for example, the wind chill); and, increasingly, wind is being tapped as a source of energy.

Wind primarily is a response to global, regional, and local temperature differences. Temperature differences affect atmospheric pressure, which causes the air to move from places with higher pressure toward places with lower pressure. Wind speed is directly related to the strength of these temperature and pressure differences. The larger the change (or gradient) of temperature and pressure over a given geographic distance, the faster the wind will blow.

Global wind patterns. The fastest wind speeds at the Earth's surface often are found beneath the midlatitude jet streams, ribbons of very fast winds that encircle the globe between about 30 and 60° north and south and approximately 33,000 ft (10,000 m) above the ground. These jet streams are located where temperatures and pressures change rapidly over short distances. Slower speeds typically occur in tropical and subtropical regions and in high latitudes, where temperatures vary little over large distances. These weak temperature (and thus pressure) changes lead to weak wind speeds (**Fig. 1**).

Seasonal and daily wind patterns. Just as temperature changes with seasons, so does wind speed. In general, midlatitude wind speeds are fastest in the spring and autumn, the times of year with the largest temperature differences between high and low latitudes. Winter also has fast wind speeds because the jet stream is farthest equatorward. In contrast, summertime wind speeds are comparatively weak, due both to smaller temperature differences and to the more poleward location of the jet stream.

Wind speed also varies with the time of day. Measurements at 10 m (33 ft) above the ground (a standard anemometer height) show that the fastest speeds usually occur in the mid to late afternoon and the slowest speeds occur during the nighttime hours. This daily variation is related to the corresponding variation in temperature. As the ground heats up during the day, the atmosphere becomes more turbulent and mixes the near-surface air with faster air that is farther above the ground. This turbulent mixing causes near-surface wind speeds to increase. As the ground cools off at night, turbulence decreases, and there is little mixing with the faster wind speeds above the ground. The result is that,

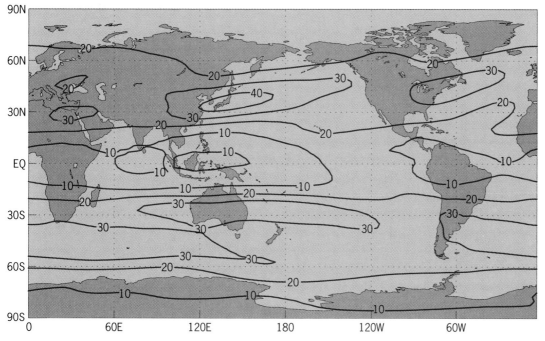

Fig. 1. Mean annual wind speed (m/s) for 1968–1996 at 250 millibars (about 33,000 ft or 10,000 m above the ground). (*NOAA-CIRES Climate Diagnostics Center, Boulder, Colorado*)

near the surface, nighttime speeds usually are lower than daytime speeds.

At higher heights, including typical wind turbine heights (50–90 m or 160–300 ft), turbulence has the opposite effect, producing slower winds during the day and faster winds at night. Daytime turbulence, which mixes faster wind speeds down toward the ground, also mixes slower speeds upward toward the location of the turbine. When turbulence decreases during the night, turbine-level wind speeds are faster because there is little mixing with slower winds near the surface.

Wind power varies with the cube of the wind speed, so small variations in speed can yield large changes in wind power. As a result, knowledge of the seasonal and diurnal variability of speed at a site is just as important as finding a site with fast mean wind speeds. Sites with the same mean annual wind speed but different patterns of variability can generate significantly different amounts of wind power (see **table**).

Variation of mean annual wind power at three sites with the same mean annual wind speed.

Site	Mean annual wind speed, m s^{-1}	Mean annual wind power, W m^{-2}
Culebra, Puerto Rico	6.3*	220
Tiana Beach, New York	6.3	285
San Gorgonio, California	6.3	365

*6.3 m s^{-1} is about 14 mi/h.
SOURCE: From D. L. Elliott et al., *Wind Energy Resource Atlas of the United States*, U.S. Department of Energy Report DOE/CH 10093–4, 1986.

Local effects on wind speed. A number of other factors affect the wind speed at a specific location. The main factors affecting speed at the local scale are topography, surface roughness, and land/water contrasts.

Topography is one of the strongest determinants of wind speed. As air passes over a mountain or

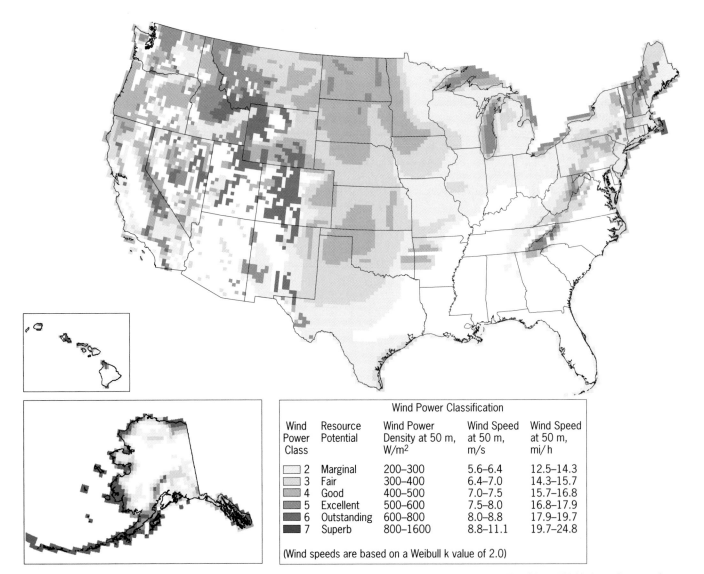

Wind Power Classification

Wind Power Class	Resource Potential	Wind Power Density at 50 m, W/m²	Wind Speed at 50 m, m/s	Wind Speed at 50 m, mi/h
2	Marginal	200–300	5.6–6.4	12.5–14.3
3	Fair	300–400	6.4–7.0	14.3–15.7
4	Good	400–500	7.0–7.5	15.7–16.8
5	Excellent	500–600	7.5–8.0	16.8–17.9
6	Outstanding	600–800	8.0–8.8	17.9–19.7
7	Superb	800–1600	8.8–11.1	19.7–24.8

(Wind speeds are based on a Weibull k value of 2.0)

Fig. 2. **United States wind resource map, showing the mean annual wind power at 50 m (about 160 ft) above the ground. Data from *Wind Energy Resources Atlas of the United States*, 1987. (*U.S. Department of Energy, National Renewable Energy Laboratory, Golden, Colorado*)**

Fig. 2. Carbonate pinnacles at the Lost City Hydrothermal Field. This photo was taken from the submersible *Alvin* during the 2003 Lost City expedition. The "Nature pinnacles" are located east of the Poseidon complex, and were some of the first chimneys imaged when the LCHF was discovered in 2000. These antlerlike formations are typical of many chimneys at the LCHF, where distinct slender pinnacles are welded together at their base. These delicate chimneys grow directly out of fissures in the serpentinite basement rock, emit fluids about 50°C (122°F), and are about 30 m (100 ft) tall.

beehive-shaped cones, and concave flanges that form on the sides of larger structures. The core of the field hosts the massive Poseidon complex, named after the actively venting chimney, which rises 60 m (200 ft) above the seafloor. The complex has a crown of four individual chimneys venting at about 60°C (140°F).

Radiocarbon ages of some of the actively venting chimneys are up to about 500 years old. The inactive chimneys span ages of hundreds of years to more than 25,000 years old. It is possible that the LCHF may be significantly older, and uranium-thorium dating is currently underway to examine this hypothesis.

Microbiology. Diffuse venting sites within the LCHF host microbial communities composed of Archaea and Eubacteria. The warm nutrient-rich fluids within the porous interiors of the chimneys support Archaea that utilize methane (either through methane oxidation or methanogenesis). Surprisingly, the Archaea are dominated by a single type of microbes related to a group of single-celled organisms know as Methanosarcinales, which are commonly found in methane seep and gas hydrate environments. In contrast, colonies of both Archaea and bacteria are found on the exterior of the chimneys that form delicate, filamentous strands about 1 cm long.

Macrobiology. Unlike the abundant animals found at black smokers, the macrofauna at the LCHF are generally sparse and small, with most organisms less than 1 cm in size. Gastropods, polychaetes, and amphipods are common. Less common are larger shrimp, crabs, sea urchins, and a variety of deep-sea corals. Large (about 1.5 m or 5 ft in length) wreckfish patrol the Atlantis Massif and are typically observed near the chimneys. Although most of the animal life at the LCHF is small and in low abundance, the diver-

sity is as high or higher than what is found at black smoker sites along the Mid-Atlantic Ridge.

Significance and future work. During the past 30 years, most hydrothermal vent research and exploration has been done along the narrow seafloor passage defined by the axial valley of the global mid-ocean ridge network. However, the discovery of the LCHF, nearly 15 km (9 mi) away from the spreading axis of the Mid-Atlantic Ridge, and the recent recognition that exposed mantle rocks are ubiquitous along slow- and ultra-slow-spreading ridges indicates that a much larger portion of the seafloor may host venting sites (and therefore life) than previously believed. The LCHF represents a novel environment for investigating the relationships between geological, biological, and hydrothermal processes in a system dominated by serpentinization reactions. Although the LCHF is the first of its kind to be explored, future studies will likely show that it is not unique to mid-ocean ridge environments and that similar systems have probably existed throughout much of Earth's history. *See* SLOW-SPREADING MID-OCEAN RIDGES.

For background information *see* CARBONATE MINERALS; EARTH INTERIOR; HYDROTHERMAL VENT; LIMESTONE; LITHOSPHERE; MARINE GEOLOGY; MASSIF; METHANOGENESIS (BACTERIA); MID-OCEANIC RIDGE; OLIVINE; ORE AND MINERAL DEPOSITS; PLATE TECTONICS; SERPENTINE; TRANSFORM FAULT in the McGraw-Hill Encyclopedia of Science & Technology.

Kristin A. Ludwig; Deborah S. Kelley

Bibliography. J. B. Corliss et al., Submarine thermal springs on the Galapagos Rift, *Science*, 203:1073–1083, 1979; G. L. Früh-Green et al., 30,000 years of hydrothermal activity at the Lost City vent field, *Science*, 301:495–498, 2003; D. S. Kelley et al., An off-axis hydrothermal vent field near the Mid-Atlantic Ridge at 30°N, *Nature* 412:145–149, 2000; D. S. Kelley et al., A serpentinite-hosted ecosystem: The Lost City Hydrothermal Field, *Science*, 307:1428–1434, 2005; K. A. Ludwig et al., Formation and evolution of carbonate chimneys at the Lost City Hydrothermal Field, *Geochim. Cosmochim. Acta*, 70:3625–3645, 2006; G. Proskurowski et al., Low temperature volatile production at the Lost City Hydrothermal Field: Evidence from a hydrogen stable isotope geothermometer, *Chemical Geology*, 229:331–343, 2006; M. O. Schrenk et al., Low archaeal diversity linked to subseafloor geochemical processes at the Lost City Hydrothermal Field, Mid-Atlantic Ridge, *Environ. Microbiol.*, 6(10):1086–1095, 2004; K. Von Damm, Controls on the chemistry and temporal variability of seafloor hydrothermal fluids, in S. E. Humphris et al. (eds.), *Seafloor Hydrothermal Systems: Physical, Chemical, Biological, and Geological Interactions*, pp. 222–247, American Geophysical Union, 1995.

Magnetic thin films

Magnetic thin films are composed of ferromagnetic or antiferromagnetic materials. The term "thin" is relative, but magnetic thin films generally have altered magnetic ordering or unique physical properties that distinguish them from their bulk counterparts as the result of the reduced thickness and the presence of interfaces. The thickness of magnetic thin films typically ranges from just one atomic layer to a few tens of nanometers.

The field of magnetic thin films originated in the 1880s with August Kundt, who studied the dispersion of light in thin iron and nickel films produced by electroplating. Modern research on magnetic thin films initially focused on such properties as magnetic domain structures and spin-wave excitations in single-layer films. With advances in deposition and characterization techniques, magnetic thin films have become the experimental platform for studying surface magnetism and for testing theoretical predictions from first-principle calculations. Furthermore, incorporating magnetic thin films into multilayered structures containing layers of different magnetic materials or even nonmagnetic materials has facilitated the discovery of a host of novel magnetic phenomena, including interlayer exchange coupling and giant magnetoresistance, both of which have fueled great advances in information storage technology.

Epitaxial growth. Most thin-film deposition techniques are suitable for creating magnetic thin films. In practice, physical vapor deposition methods such as evaporation, sputtering, and ablation are commonly used, because the high-vacuum or inert-gas environment helps to ensure material purity and interface quality. Epitaxy, the presence of atomic registry between the deposited film and the substrate, is an important issue in the fabrication of magnetic thin films. Repetitive multilayer structures where epitaxy is maintained throughout are termed superlattices. The strain induced by epitaxial growth provides a way to stabilize material phases that do not exist in bulk form under ambient conditions. For example, the tetragonal $SmFe_{12}$ phase, which does not otherwise exist without titanium doping, can be grown epitaxially on a W(100) surface. Controlling the crystallographic orientation of magnetic thin films via epitaxial growth makes it possible to tailor the magnetic anisotropy properties. In fact, in computer hard disks the crystallographic texture of the cobalt-based alloy recording media layer is engineered for high in-plane anisotropy and coercivity by using an appropriate underlayer such as (100)-textured chromium.

Finite-size effects. The magnetic exchange interaction, or coupling, originates from the overlap of the electron orbitals of magnetic ions. The exchange coupling keeps the directions of the magnetic ions correlated over a distance known as the correlation length. A magnetic system is considered ordered when all the magnetic ions contained within are correlated, that is, when the correlation length reaches the system size. For an infinitely large magnetic system, the coherence length becomes infinity at the critical temperature for magnetic ordering (T_c).

The reduced thickness (d) of magnetic thin films results in a suppression of T_c, given by the scaling law

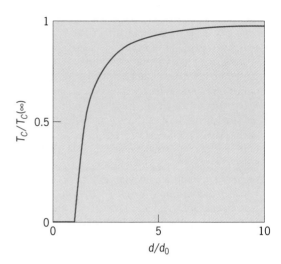

Plot of $T_c/T_c(\infty)$ versus d/d_0 for the case of a three-dimensional Ising system ($\nu = 0.64$).

in Eq. (1), where d_0 is a characteristic length, typi-

$$d/d_0 = [1 - T_c/T_c(\infty)]^{-\nu} \qquad (1)$$

cally a fraction to a few atomic layers, and $T_c(\infty)$ is the bulk critical temperature (see **illustration**). The value of the critical exponent ν depends on the universality class of the system. For epitaxial iron layers less than three atomic layers thick grown on Ag(100), ν is determined to be 1. This is coincident with the theoretically calculated value for a two-dimensional Ising system. Although a thin iron layer actually has a finite extent in the thickness direction, it behaves as a two-dimensional system. Other experimental reports on thicker ferromagnet and antiferromagnet films put ν at 0.6–0.7, consistent with that expected for three-dimensional Ising or Heisenberg systems.

A practical length scale for magnetic thin films is the characteristic distance over which a magnetic perturbation decays in a magnetic material. This is known as the exchange length, given by Eq. (2).

$$l_{ex} = (A/\mu_0 M^2)^{1/2} \qquad (2)$$

Here, A is the exchange constant that measures the energy of magnetic exchange coupling per unit length, M is the magnetization, and μ_0 is the permeability of free space. The exchange length l_{ex} is about 3 nm for ferromagnetic iron- or cobalt-based alloys. If a films thickness is less than l_{ex}, the magnetization is uniform across its entire thickness, and the film can be regarded as consisting of a single "giant" moment.

Interface effects. The large surface-to-volume ratio in thin-film structures makes surface and interface effects central to the many remarkable phenomena that occur in magnetic thin films. It was shown by Louis Néel that, at surfaces and interfaces of magnetic systems, the lowered symmetry modifies the magnetocrystalline contribution to the magnetic anisotropy, yielding the surface magnetic anisotropy. In monolayer-range iron films, it is this surface magnetic anisotropy that stabilizes the long-range ferromagnetic order. For (100)-oriented iron films, the favored magnetization direction, that is, the magnetic easy axis, is perpendicular to the film plane. In contrast, for Fe(211) thin films the surface anisotropy is in-plane, with the easy axis along the Fe[0$\bar{1}$1] direction. Another notable example where surface anisotropy dominates is that of cobalt/palladium and cobalt/platinum multilayers, where the large spin-orbit coupling at the palladium and platinum sites gives rise to very strong perpendicular magnetic anisotropy. This magnetic property, together with the large magnetooptical response of palladium and platinum, makes cobalt/palladium and cobalt/platinum multilayers suitable for perpendicular magnetooptical recording applications.

The interface between different kinds of magnetic materials hosts some fascinating phenomena due to the exchange coupling. The exchange-bias effect, discovered in 1956 by W. H. Meiklejohn and C. P. Bean in cobalt particles covered with a CoO surface layer, is the unidirectional pinning of a ferromagnetic material by an antiferromagnetic material. The unidirectional nature of exchange bias is useful in applications such as magnetic recording, where a reference magnetic direction is needed. The exchange-spring effect is similar, but originated from nanocomposite permanent magnet materials with a mixture of hard and soft magnetic grains. The exchange coupling of the soft magnet to the hard magnet makes the soft magnet more resistant to reversal in an opposing magnetic field, while the soft magnet generally increases the overall magnetization value. This scheme could lead to a new generation of ultrastrong permanent magnets. These interface effects are more appropriately explored, and often better exploited, in magnetic thin film structures, which offer the opportunity to adequately characterize and control the chemical states and the exchange interactions at the interface on an atomic scale, and to correlate them with the magnetic properties.

The giant magnetoresistance (GMR) effect is the large change in the electrical resistance of magnetic structures when a magnetic field aligns the electron-scattering magnetic elements. Although the giant magnetoresistance effect is not exclusively a property of layered structures, it was initially discovered in antiferromagnetically coupled iron/chromium multilayers and widely studied in magnetic thin films in layered structures. The giant magnetoresistance effect is due to the spin-dependent scattering of the conduction electrons at the ferromagnet/normal metal interfaces, and can be as large as 50% at room temperature. Prior to giant magnetoresistance, permalloy, with magnetoresistance of 1%, was used as the sensing element in magnetic recording heads. The availability of much more sensitive read heads based on the giant magnetoresistance effect has triggered an explosive increase in areal data density in magnetic recording.

Interlayer exchange coupling. Interlayer exchange coupling is the coupling between the magnetizations of two ferromagnetic layers separated by a nonferromagnetic spacer. The coupling is oscillatory, alternating between ferromagnetic and antiferromag-

netic, and decays in magnitude as the thickness of the spacer layer increases. Interlayer exchange coupling arises from the spin-dependent reflection of electrons at the two interfaces between the spacer and the ferromagnetic layers. The oscillation period ranges from 0.9 to 1.8 nm, and is governed by special features on the Fermi surface of the spacer material. While interlayer exchange coupling is bilinear (that is, the magnetization directions of the ferromagnetic layers are either parallel or antiparallel), fluctuations in the spacer layer thickness can lead to biquadratic coupling, a situation where the magnetization directions of the ferromagnetic layers are perpendicular to each other. Interlayer exchange coupling enabled the creation of an antiparallel arrangement of iron layers, leading to the discovery of the giant magnetoresistance effect, and it continues to be employed in magnetic structures and devices to tailor desired magnetic configurations.

Model systems based on thin-film structures. As noted above, when the thickness of a magnetic layer is less than the exchange length, the entire layer is equivalent to a giant magnetic moment. Thus, a multilayer structure containing many such layers can be considered as a one-dimensional chain of spins. This leads to the interesting possibility of "building" model magnetic systems out of magnetic multilayer structures. The interlayer exchange coupling provides a means to harness exchange interactions, while the surface supplies the magnetic anisotropy. In this fashion, all the parameters characterizing a magnetic system can be designed by tailoring the layer thickness or can be imparted during the film growth. Such experimental model systems are highly amenable to advanced characterization tools, provided by neutron and x-ray scattering. For example, an antiferromagnetically coupled Fe/Cr(211) superlattice is the exact model for an A-type antiferromagnet such as MnF_2. The surface spin-flop transition, theoretically predicted long ago, where the surface spins "flop" to be nearly perpendicular to an applied magnetic field, was first observed in this model system via surface-sensitive magnetooptical Kerr effect measurements, and the evolution of the surface spin-flop transition with magnetic field was followed via polarized neutron reflectometry. A "double superlattice" structure, where part of the iron/chromium superlattice can be made ferromagnetic and the other part antiferromagnetic by adjusting the chromium spacer thickness, serves as a model exchange bias system. This novel configuration was used to confirm the Meiklejohn-Bean exchange bias theory. This is especially significant because, although the theory accompanied the original exchange bias discovery, it had never before found quantitative verification within conventional exchange bias systems.

For background information *See* ARTIFICIALLY LAYERED STRUCTURES; CRITICAL PHENOMENA; CRYSTAL GROWTH; FERROMAGNETISM; ISING MODEL; MAGNETIC MATERIALS; MAGNETIC RECORDING; MAGNETIC THIN FILMS; MAGNETOOPTICS; MAGNETORESISTANCE in the McGraw-Hill Encyclopedia of Science & Technology.   J. Samuel Jiang; Samuel D. Bader

Bibliography. S. Chikazumi with S. H. Charap, *Physics of Magnetism*, Wiley, 1964; A. J. Freeman and S. D. Bader (eds.), *Magnetism Beyond 2000*, Elsevier, 2000, also published as *J. Magnetism Magnetic Mater.*, vol. 200, 1999; B. Heinrich and J. A. C. Bland (eds.), *Ultrathin Magnetic Structures*, 4 vols., Springer, 1994, 2005; D. L. Smith, *Thin-Film Deposition: Principles and Practice*, McGraw-Hill, 1995; S. X. Wang and A. M. Taratorin, *Magnetic Information Storage Technology*, Academic Press, 1999.

Melatonin

Derived from the essential amino acid tryptophan, melatonin was discovered in 1958. It is an indoleamine molecule that is found widely throughout nature (see structure). Until recently, melatonin was

thought to be produced exclusively by the pineal gland in most, if not all, vertebrate species, including mammals. It now ranks as one of the most phylogenetically old and fundamental biological signaling molecules present in organisms ranging through bacteria, unicellular organisms, plants, and animals. In humans, the pineal gland is a pea-sized, pineconeshaped structure located deep between the two halves of the brain. Until the mid-1960s, the pineal was generally thought to have no important physiological function(s); then it was discovered to be essential for the regulation of seasonal reproductive cycles in hamsters. Over the following decade, it became clear that melatonin, acting as a neurohormone from the pineal gland, was responsible for synchronizing the activity of the reproductive system with the time of year. In both day-active and night-active animals, melatonin is synthesized and secreted by the pineal during the night, whereas the daytime production of melatonin is virtually nil. The surge of melatonin during dark nights represents a biological timing signal that is internally driven by the activity of a central pacemaker located in the hypothalamus, another part of the brain near the pineal. The nocturnal rhythm of melatonin production persists even in the absence of an alternating light/dark cue, so that eventually the interval of time between nightly peaks of melatonin production tends to become somewhat longer (or shorter) than 24 hours. This type of internally generated rhythm that is no longer synchronized to an environmental light/dark cycle over a 24-h day is referred to as a free-running, circadian rhythm. The nocturnal melatonin signal provides time of day information to all the cells, tissues, and organs of the body, and is the most stable and reliable peripheral biomarker of the timing of the central circadian pacemaker. Since the duration of the melatonin surge defines the length of the biological night, the pineal gland acts not only as a clock but as a calendar by providing an organism with information about seasonal changes in day length. Melatonin is often referred to as the chemical expression or hormone of darkness. At either physiological or pharmacological concentrations, melatonin appears to be involved in far-flung physiological and pathophysiological processes, including the control of sleep, circadian rhythms, retinal physiology, seasonal reproductive cycles, cancer development and growth, immune activity, antioxidation and free radical scavenging, mitochondrial respiration, cardiovascular function, bone metabolism, intermediary metabolism, and gastrointestinal physiology.

Generation and regulation of the melatonin rhythm. The central circadian pacemaker, or master biological clock, is located in the hypothalamus and comprises a specialized group of neurons in the suprachiasmatic nuclei. The endogenous activity of this internal clock drives the complex molecular mechanisms responsible for the production of melatonin at night. In addition to the nocturnal rhythm of melatonin production, the central circadian pacemaker regulates other daily rhythms such as the sleep/wake cycle, body temperature, secretory patterns of hormones, food consumption, and metabolism. The circadian timing system facilitates an organism's adaptation to changes in its environment via the temporal coordination of these and other physiological functions; the rhythmic output of melatonin every night plays a pivotal role in this biological timing mechanism.

Photic (light) signals reaching the retina of the eye play an essential role in synchronizing this dynamic process. Light entering the eyes stimulates specialized photoreceptors in the retina from which neural signals are then sent not only to the primary visual system but also to nonvisual centers of the brain, particularly the suprachiasmatic nuclei. During the day, light input to the suprachiasmatic nuclei resets the central circadian pacemaker and thus synchronizes the internal near-24-h rhythms, including the melatonin rhythm, precisely to the environmental 24-h light/dark cycle. When an organism experiences a change in the timing of the ambient light/dark cycle, the central circadian pacemaker responds appropriately by advancing or delaying its own timing, which is referred to as a phase shift. Therefore, changes in an individual's exposure to light and darkness can induce changes in the phasing of the melatonin rhythm and other circadian rhythms, as are experienced during shift work, transcontinental jet travel, space flight, and other circadian disruptive activities. However, light present during darkness, provided that it is bright enough and of the proper wavelength, is detrimental to the circadian system and immediately turns off melatonin production. Although light is the most potent stimulus to the mechanisms governing internal time-keeping, much more light is actually required for circadian regulation than for vision.

Following its synthesis, melatonin is immediately secreted from the pineal directly into both the bloodstream and cerebrospinal fluid of the third ventricle (a cavity within the brain). The amount of melatonin produced during the night varies considerably from one individual to another, and this appears to be genetically determined. It appears to be greatest around the time just before puberty, with a steady decrease thereafter through middle and old age. Once released into the circulation, melatonin has a short half-life and is primarily metabolized in the liver, where it is converted by cytochrome P_{450} enzymes to its main metabolite, 6-sulfatoxymelatonin, which is excreted in the urine. Melatonin is also metabolized in a number of extrahepatic tissues, notably the brain, by both enzymatic and nonenzymatic mechanisms. Although the pineal gland is the primary source of melatonin release into the systemic circulation, there are a number of extrapineal sources of melatonin synthesis, including the retina, gastrointestinal tract, bone marrow, circulating lymphocytes, blood platelets, and skin; melatonin appears to exert a local action at these sites of production. Because of its unique molecular properties, melatonin has the capacity to pass in and out of all cellular and fluid compartments of the body with ease. In fact, the concentrations of melatonin detected in cells and tissues are orders of magnitude higher than those present in the blood circulation, indicating significant cellular/tissue uptake and storage of this molecule. Melatonin in other body fluids such as saliva, urine, ovarian follicular fluid, breast milk, and semen also exhibits a nocturnal rhythm. Due to its ability not only to produce but also take up and store melatonin, the gastrointestinal tract contains amounts of melatonin that far exceed the levels present in the pineal gland. Bile from the liver and gallbladder contains levels of melatonin that greatly exceed those in the general blood circulation.

Plants. Not only is melatonin present in pineal and extrapineal sites in animal species, but it is also found in abundance in many flowering and edible plants as well as in a large number of medicinal herbs. While concentrations of melatonin are quite low in some plants, many others (such as cherries, walnuts, and Chinese medicinal herbs) contain quantities that are orders of magnitude higher than those found in the blood of animal species at night. Many plants with high melatonin concentrations may serve as an excellent dietary source of this indoleamine. While the exact role that melatonin plays in plant species is unclear, it has been proposed that this compound may aid in protecting them from oxidative damage and hostile environmental conditions. Certain plant parts such as seeds are a particularly rich source of melatonin that may protect them from oxidative damage during dormancy and relatively dry states to help assure their eventual germination. In view of the ubiquitous distribution of melatonin among plants and animals, it has been proposed that melatonin be considered not only as a neurohormone but as a tissue factor, antioxidant nutrient, and per-

haps even a vitamin. In the United States, melatonin is not regulated by the Food and Drug Administration and is sold over the counter as a nutritional supplement.

Physiological and pharmacological actions. Physiological or pharmacological concentrations of melatonin exert an important regulatory influence over a broad range of physiological and pathophysiological processes in both lower animals and humans. Although current knowledge of the mechanisms by which melatonin modulates these diverse processes is incomplete, specific plasma membrane–associated melatonin receptors play a major role in mediating several of melatonin's actions at the cellular level. Melatonin receptors are expressed in diverse areas of the brain, including the suprachiasmatic nuclei as well as in several peripheral organs and tissues. Some physiological and pathophysiological situations may involve nuclear melatonin receptors, whereas in other biological contexts no specific receptors appear to be required for melatonin's actions.

Melatonin receptors mediate melatonin's ability to induce phase shifts in the firing of neurons making up the suprachiasmatic nuclei. This action lies at the heart of melatonin's chronobiotic action, namely its ability to cause phase shifts in overt circadian rhythms such as locomotor activity. Thus, an internal cue provided by the endogenous melatonin signal from the pineal coupled with an external cue provided by light ultimately synchronizes the central circadian pacemaker to the 24-h period of the rest/activity cycle. Physiological nocturnal melatonin concentrations appear to facilitate sleep and may accomplish this by suppressing the circadian wakefulness-generating program through a melatonin receptor-mediated mechanism. In the retina, locally synthesized melatonin, acting via melatonin receptors, is responsible for modulating certain aspects of retinal physiology.

Melatonin receptor-mediated inhibition of the hypothalamic-pituitary-gonadal axis suppresses reproductive function during the short days (long nights) of fall/winter in long-day seasonally breeding species such as hamsters. Conversely, in short-day breeding species such as sheep, the same short-day melatonin signal actually stimulates reproductive function during the fall/winter months. The seasonal expansion or contraction of the nighttime duration of circadian melatonin production and secretion provides certain species with critical information about day length, thus allowing them to adapt appropriately to alterations in seasonally limited resources necessary for their survival. This is particularly critical in reproductive physiology and behavior which requires that the birth of progeny be synchronized with the optimal time of year to ensure the survival of the species. Thus, it appears that a circadian/melatonin-based photoperiod timing mechanism interacts with a seasonal rhythm-based timing mechanism to ultimately govern seasonal cycles in physiology and behavior in several vertebrate species.

Other physiological and pathophysiological responses modulated by melatonin via melatonin receptor-mediated mechanisms include pituitary hormone release, testosterone production by the testes, cortisol secretion by the adrenal cortex, vascular tone, energy metabolism, fatty acid transport, immune activity, and cancer cell proliferation and tumor growth. Actions of melatonin that appear to involve nuclear receptors include stimulation of immune cells to produce biologically active substances called interleukins as well as inhibition of the proliferation of certain types of colon cancer cells.

There are a host of pharmacological actions of melatonin that apparently do not require melatonin receptors. One of the most important of these is melatonin's ability to directly scavenge and thus detoxify free radical and other reactive oxygen species. These molecules have the potential to cause extensive oxidative damage to lipids, proteins, and DNA in all cells, tissues, and organs throughout the body, which may be a root cause of many acute and chronic diseases associated with aging as well as the aging process itself. Melatonin can also protect the body against oxidative damage by bolstering the enzymatic machinery comprising other internal antioxidant defense systems possibly via mechanisms that may also involve membrane or nuclear melatonin receptors. Another important action of melatonin is its ability to maintain and improve the function of mitochondria in generating the energy molecule adenosine triphosphate (ATP) and prevent programmed cell death (apoptosis) in normal cells. There is some evidence that physiological nocturnal levels of melatonin may endow organisms with a degree of endogenous protection against oxidative stress.

Disease states and clinical applications. Melatonin has been tested in a number of preclinical animal models of and in-vitro systems mimicking human disease. For example, melatonin has been shown to be effective in protecting against the development or progression of gastrointestinal ulcers, ischemia/reperfusion injury in a variety of organs, inflammation, neurodegenerative disorders such as Alzheimer's and Parkinson's disease, diabetes, bone loss, and cancer development and growth. In the clinical setting, melatonin and/or bright light therapy has been used successfully to treat circadian phase disorders in sighted people such as advanced and delayed sleep syndromes, jet lag, shift work maladaptations, and winter depression. In blind people, melatonin is now the treatment of choice for correcting circadian rhythm disturbances. Melatonin promotes sleep and lowers core body temperature in humans, and it has been used to treat insomnia. Alterations in the magnitude or phasing of the endogenous, circadian melatonin rhythm have been described in patients with mood disorders (including major depression, bipolar affective, and seasonal affective disorders), cancer, and rheumatoid arthritis. New melatonin agonist drugs that work through melatonin receptors are currently on the market for the treatment of mood disorders and insomnia. In

some patients, melatonin may significantly improve sleep and circadian rhythm abnormalities associated with Alzheimer's disease and slow the progression of the disease itself. Additionally, melatonin appears to improve sleep physiology in parkinsonian patients and is effective in reducing blood pressure in hypertensive patients. Melatonin has also been successfully used to improve the clinical outcome, and in some cases prevent death, in newborns suffering from septic shock.

The nocturnal, circadian melatonin signal is a newly identified inhibitory link between the central circadian pacemaker in the brain and the processes governing the evolution of cancer in humans. Bright light at night, which suppresses the cancer-inhibitory melatonin signal, represents a new cancer risk factor that may account for the increased incidence of breast cancer in night-shift workers. Clinical trials indicate that nighttime melatonin supplementation may offer a promising new approach in the treatment of advanced stage malignancies and reduction in the toxicity of chemotherapy and/or radiation. Pharmacological doses of melatonin appear to be safe and generally well tolerated with no serious side effects.

For background information *see* BIOLOGICAL CLOCKS; BRAIN; ENDOCRINE MECHANISMS; EYE (VERTEBRATE); HORMONE; LIGHT; PHOTORECEPTION; PINEAL GLAND; REPRODUCTIVE BEHAVIOR in the McGraw-Hill Encyclopedia of Science & Technology.

David E. Blask

Bibliography. J. Arendt, Melatonin and human rhythms, *Chronobiol. Int.*, 23:21-37, 2006; D. E. Blask et al., Melatonin-depleted blood from premenopausal women exposed to light at night stimulates growth of human breast cancer xenografts in nude rats, *Cancer Res.*, 65:11174-11184, 2005; G. C. Brainard and J. P. Hanifin, Photons, clocks, and consciousness, *J. Biol. Rhythms*, 20:314-325, 2005; A. Lewy et al., Circadian uses of melatonin in humans, *Chronobiol. Int.*, 23:403-412, 2006; S. R. Pandi-Perumal et al., Melatonin—nature's most versatile biological signal?, *FEBS J.*, 273:2813-2838, 2006; R. J. Reiter (ed.), Melatonin, *Endocrine*, 27:87-212, 2005; R. J. Reiter (ed.), Melatonin: Medicinal chemistry and therapeutic potential, *Curr. Top. Med. Chem.*, 2:113-209, 2002.

Metabolic engineering

Metabolic engineering is the directed improvement of cellular properties or products by modifying specific biochemical reactions or introducing new ones using recombinant DNA technology. It is a subset of biochemical engineering, which has industrial and medical applications. Metabolic engineering is used by biotechnology, pharmaceutical, chemical, energy, and food companies to design or optimize microorganisms that overproduce particular chemicals. Products via metabolic engineering include drugs, enzymes, food, commodity chemicals, and fuels. Metabolic engineering has also been used to study human diseases.

General principles. Metabolic engineering evolved from genetic engineering as a systematic analysis of cellular enzymatic reactions. With the advent of recombinant DNA technology, biological products of chemical or medicinal value, which previously were isolated from plants or animals in low yield, could be produced in microorganisms at reduced cost. This was accomplished by transferring the DNA that encodes the enzymes for the desired reaction into the new host cell. While this great achievement made the commercial production of many compounds possible, scientists quickly realized that only in certain cases could high yields be realized. As a discipline, metabolic engineering looks beyond the enzyme that forms the final product to understand how the rest of the cell behaves as this reaction is added. By understanding the metabolic reaction network—the complex set of chemical reactions that take place in the cell—reactions can be identified that would increase the output of the desired product (see **illustration**). This systems approach is necessary, as the cell is very complex. Normally, a cell will regulate the reaction network to provide the components it needs to survive and grow, which often diverts reactants away from the desired reaction.

Methodology. Metabolic engineering broadly relies on a three-step approach to optimize the productivity for the desired biomolecule. Measurement, analysis, and genetic change are used together to build knowledge about the cellular system to make productivity improvements. Because metabolic engineers want to look at the cell as a whole, experimental techniques must simultaneously measure the levels of many biomolecules, in particular metabolites (the reactants and products in the metabolic reaction network). Aside from measuring metabolites, it is also important to calculate the many fluxes—how fast reactants are moving through the different enzymatic reactions. Metabolic flux analysis (MFA) is the experimental and theoretical framework that can measure many of these fluxes simultaneously. By determining the many fluxes through the cellular reaction network, the engineer can determine the optimal ways to change the metabolic network so that molecules can be diverted toward the desired reactions.

The most effective targets for modification may not be identifiable in the complex and large data sets produced in the measurement phase. If there are many steps in the same pathway, it may not be obvious which reaction(s) is limiting the throughput. As a result, metabolic and regulatory pathway analysis is often necessary to identify these less apparent modifications. Metabolic control analysis (MCA) is a theoretical framework that determines which parts of the network will have the largest effect on the productivity. A flux control coefficient (FCC) is a measure of the sensitivity of the flux to an enzyme's concentration. An enzyme with a large FCC is limiting the pathway and, if increased, would increase

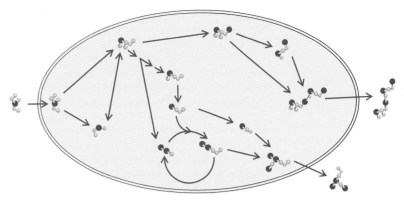

Representation of a metabolic reaction network, showing a cell which takes up sugars and other chemicals from the environment. Each arrow represents an enzymatic reaction that converts the reactant to the product. An actual cell has thousands of reactions that are critical for cellular function and making many potentially valuable products.

the throughput of the pathway. On the other hand, enzymes with a low FCC have no effect on the flux. MCA analyzes the fluxes after changing the enzyme level and determines which reactions are rigid, or unchangeable, and which are elastic. Enzymes that have a large FCC, or are rigid, become targets for modification.

Genetic changes are made to the cell guided by the previously described analysis. Typically, the changes will be either an increase or a decrease in the reaction rate of one or more steps in the metabolic network. Changing these reaction rates can be accomplished by using one of several methods, all of which rely on the central dogma of biology that states that DNA is transcribed into RNA which is translated into proteins, including the enzymes in the metabolic network. By changing the amount of DNA or RNA for a specific enzyme, one can potentially change the amount of the enzyme and the total reaction rate of that enzyme in the cell. To eliminate a reaction altogether, the DNA that encodes the enzyme can be deleted. Each gene also has a promoter that controls how many RNA copies are made from the DNA. The promoters can be altered to change the amount of RNA and thereby enzyme in the cell. Additional copies of the DNA can be added to the cell in the form of small circular pieces of DNA, called plasmids, to increase the amount of enzyme. Biologists have developed many techniques that allow metabolic engineers to easily change DNA in the cell. After the appropriate modifications are made, the cycle begins again by reanalyzing the reaction network, followed by further analysis and genetic changes. After several iterations, the yield of the desired product can be greatly increased.

Inverse metabolic engineering. Inverse metabolic engineering was first described by Jay Bailey in 1996. In contrast to classical metabolic engineering in which systematic changes are made to known genes which affect the metabolic pathway, inverse metabolic engineering entails measuring cellular properties of mutants in a genetic library (a collection of cells with different genetic changes in each cell) to discover cells with the most desirable properties (for example, accumulation of a desired product). In contrast to classical strain improvement which uses genetic libraries with cells containing random untraceable mutations, mutants identified using inverse metabolic engineering have traceable mutations which engineers can locate and endow upon other cells. The inverse metabolic engineering method allows engineers to determine particular genetic mutations that could not be discovered with a more directed technique. It can be challenging to use traditional metabolic engineering to identify all genes that affect the metabolism of a cell, since not only does the complex metabolic network itself need to be considered but also other regulatory mechanisms within the cell. As a result, inverse metabolic engineering can be a method for discovering new genes to target with traditional metabolic engineering.

Applications. Metabolic engineering has many applications, ranging from pharmaceuticals to food to chemicals. Traditionally, drugs have been manufactured either through organic synthesis or by extracting the chemicals from organisms, such as plants, that naturally produce them. However, both methods have issues with producing high yields. In chemical synthesis, only a certain percentage of the end product has the correct stereochemistry (a property of the arrangement of atoms in a molecule), so separation techniques must be used, adding cost to the product. When microorganisms are used to produce the drugs, reactions can often be engineered in such a way that they produce only the active chemical with the proper stereochemistry. An application is the use of *Rhodococcus* to produce stereochemically pure 1,2-indandiol, a precursor of the AIDS drug Crixivan. An example of a drug that is normally extracted from natural sources in low yields is the antimalarial drug artemisinin. When the sole source for this drug is the sweet wormwood plant, the drug is unaffordable for the world's poorest people who need it the most. An *Escherichia coli* strain has been engineered by adding enzymes from yeast and the sweet wormwood so that the bacteria can produce significant amounts of a chemical precursor to artemisinin, reducing cost for affordable mass distribution of the drug.

Metabolically engineering bacteria also allows the use of cheap feedstocks for the production of nutrients. For example, recombinant *E. coli* cells are used to produce lycopene (an antioxidant added to vitamins) instead of using tomatoes, which naturally produce lycopene but cost more to grow. Food crops have also been metabolically engineered to provide inexpensive nutrients. For example, Monsanto has genetically modified rice to produce a variety that is rich in iron and β-carotene, which is converted to vitamin A in the human body. The development of this special type of rice will help to reduce malnutrition in developing countries.

In the chemical industry, metabolic engineering provides a way to reduce the world's dependence on petroleum by using renewable feedstocks, such as corn, which have more stable prices than oil and reduce total carbon dioxide emissions. Yeast has been genetically engineered to convert ethanol from plant waste. This technology may allow the use of ethanol to become more widespread. Archer Daniels Midland is developing a process based on recombinant *E. coli* to produce poly(3-hydroxybutyrate), a biodegradable plastic that could replace polypropylene. The development of efficient processes for the production of biofuel from renewable resources is presently attracting a lot of attention. Metabolic engineering is a key technology for attaining these goals.

For background information *see* BIOCHEMICAL ENGINEERING; DEOXYRIBONUCLEIC ACID (DNA); ENZYME; GENE; GENETIC ENGINEERING; MOLECULAR BIOLOGY; PLASMID in the McGraw-Hill Encyclopedia of Science & Technology.

Keith E. Tyo; Benjamin L. Wang;
Gregory Stephanopoulos

Bibliography. J. E. Bailey, Toward a science of metabolic engineering, *Science*, 252:1668–1675, 1991; J. E. Bailey et al., Inverse metabolic engineering: A strategy for directed genetic engineering of useful phenotypes, *Biotech. Bioeng.*, 52:109–121, 1996; R. M. Raab, K. Tyo, and G. Stephanopoulos, Metabolic engineering, *Adv. Biochem. Eng. Biotech.*, 100:1–16, 2005; G. Stephanopoulos, A. A. Aristidou, and J. Nielsen, *Metabolic Engineering: Principles and Methodologies*, 1998; G. Stephanopoulos and J. J. Vallino, Network rigidity and metabolic engineering in metabolite overproduction, *Science*, 252:1675–1681, 1991.

Modern human origins

Genetic, paleontological, and archeological data indicate that modern humans evolved in Africa and then spread to other parts of the world. Genetic data, primarily from mitochondrial deoxyribonucleic acid (mtDNA) and Y chromosomes, firmly place the exodus from Africa between 65,000 and 50,000 years before present (BP). Archeology adds crucial evidence that humans reached northern Australia by 55,000–50,000 BP. The earliest human fossils from Australia come from Lake Mungo in the southeast and date to 45,000 BP. Y chromosomes and mtDNA indicate an early rapid movement from Africa to Australia, probably along the coastline of the Indian Ocean. Genetic, archeological, and fossil evidence shows that modern humans occupied the rest of Eurasia between 50,000 and 30,000 BP. In Africa, fossils indicate that modern humans evolved gradually from archaic ancestors often described as *Homo heidelbergensis, Homo helmei*, or more generally "archaic *Homo sapiens*." This process began by 400,000 BP and resulted in the emergence of anatomically modern people in Africa by 195,000–150,000 BP.

Anatomically modern humans first spread from Africa into Israel around 100,000 BP, but were

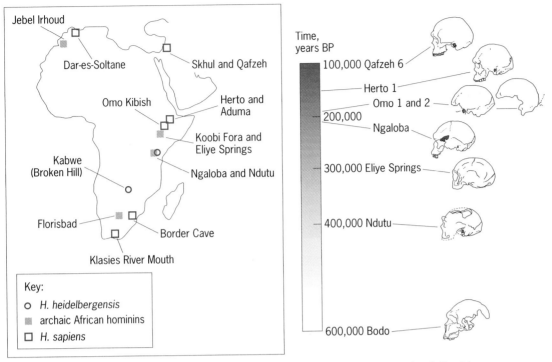

Human fossil record in Africa, documenting the gradual evolution of the anatomical features that distinguish modern humans.

locally replaced by Neanderthals by 60,000 BP. Why this earliest movement from Africa failed while the later movement succeeded so spectacularly is a matter of considerable debate. Though controversial, some genetic evidence suggests that the success of the later modern humans may have been due to the evolution of fully articulate speech capable of communicating abstract and symbolic thought. Archeological evidence may contradict this hypothesis, however, by suggesting that the capacity for art and symbolic behaviors, which serve as archeological proxies for language, were fully developed in African populations by at least 120,000 BP.

Fossils. The human fossil record in Africa documents the gradual evolution of the anatomical features that distinguish modern humans (see **illustration**). These include a protruding triangular chin; a high, nearly vertical forehead; reduced size of the browridges; a large endocranial capacity averaging 1350 cm³ (82 in.³); a flexed cranial base; a vertically short face tucked beneath the frontal lobes of the brain rather than protruding anteriorly; a pelvis of modern shape; and probably a reduction in body mass relative to the Middle Pleistocene *H. heidelbergensis*.

Harvard anthropologist Daniel Lieberman has proposed that many of the cranial changes that define anatomically modern morphology, including the high, domed cranium and small browridges and face, stem from a few key developmental changes that reduced the amount of growth that occurs in the anterior cranial base. Recent research has demonstrated

that modern humans and Neanderthals largely follow the same patterns of cranial growth, but do so from subtly different cranial forms present at birth. Thus, the changes in growth responsible for the evolution of modern cranial form probably affected development in utero. The flexed cranial base and small face of modern humans may have also been essential to produce the modern configuration of the vocal tract. Both features may also be linked, albeit controversially, to the evolution of language.

The evolution of modern cranial form begins in Africa with the emergence of *H. heidelbergensis* by at least 600,000 BP. *Homo heidelbergensis* differs from its predecessor and likely ancestor *H. erectus* in having an enlarged endocranial capacity of 1200 cm³ (73 in.³) versus an average of 1000 cm³ (61 in.³) in *H. erectus*. African specimens of *H. heidelbergensis*, including those from the sites of Bodo (Ethiopia), Kabwe (Zambia), and Ndutu (Tanzania), also possess a set of features in their temporal bones that give them a configuration similar to modern humans and different from *H. erectus*. These features include a high, arched course of the temporal suture, reduced massiveness of the ectotympanic plate, and an ossified styloid process.

Fossils morphologically intermediate between *H. heidelbergensis* and *H. sapiens* occur during this period between 300,000 and 150,000 BP in Africa. These include specimens from Eliye Springs and Koobi Fora in Kenya, Ngaloba and Eyasi in Tanzania, Florisbad in South Africa, Omo in Ethiopia, and Jebel Irhoud in Morocco. This "intermediate" group is sometimes placed in the species *H. helmei*. Brain size, which underwent an exponential increase in size over time in the Middle Pleistocene, reached modern size in specimens of this group. However, each of these specimens possesses a unique combination of modern and archaic features, and this poses three problems: whether these fossils formed a single, interbreeding population, where to draw the line between these specimens and *H. heidelbergensis*, and which population(s) was ancestral to modern humans.

The earliest anatomically modern human in Africa (and the world) is Omo 1, a specimen from southeastern Ethiopia, dated to 195,000 BP by a combination of argon-argon (Ar-Ar) dates on volcanic tuffs and correlation with sediments in the detritus cone of the Nile River in the Mediterranean Sea. The same age has been assigned to the appreciably more archaic Omo 2 cranium, raising questions about the population structure of the time. The next oldest specimens of modern humans are three crania from Herto in the Middle Awash of the Ethiopian Rift Valley dated to between 160,000 and 154,000 BP. They are anatomically modern, yet they are robust, with extremely long braincases and large browridges. The best-preserved cranium also has a vertically deep face. These features distinguish the cranium from recent humans and led the discoverers to place the specimens in a new subspecies, *H. sapiens idaltu*.

Later fossil finds from Africa, including those from Aduma (Ethiopia), Taramsa Hill (Egypt), Klasies River Mouth (South Africa), and Dar-es-Soltan (Morocco), are also anatomically modern, but many retain morphological details that are rare or absent in modern populations. A poor fossil record between 80,000 and 20,000 BP in Africa currently hinders investigations of morphological differentiation of modern African populations.

Genetics. Genetic data offer additional important information on the origin and spread of modern humans. MtDNA and Y chromosomes provide fairly clear pictures of the timing of the origin of modern humans in Africa and their later spreading out of Africa, but these pictures do not completely agree with each other and, importantly, clash with data derived from autosomal DNA (that is, the DNA in the cell except for the X chromosome, the Y chromosome, and the mtDNA). The woman whose mtDNA is the last common ancestor of all living humans' mtDNA lived in Africa 171,500 ± 50,000 BP. The last common ancestor of all human Y chromosomes occurred much more recently at 59,000 BP with a 95% confidence interval of 40,000–140,000 BP.

Studies of nuclear DNA have revealed substantial variation in human populations in Africa compared with the rest of the world. However, the data also document ancient, divergent sequences in non-African populations that are either not present or exceedingly rare in sub-Saharan Africans to establish that at least some interbreeding with archaic Eurasian populations occurred as modern humans spread from Africa.

One nuclear gene, *FOXP2*, is of particular interest because it may provide clues to the origin of language. Rare mutations in *FOXP2* reduce an individual's ability to form complex, grammatical sentences and also interfere with fine-scale control of facial movements. Although *FOXP2* is highly conserved in primates, its evolution has accelerated in humans. The human form of the allele appears to have originated in the last 120,000 years. Most human autosomal genes display greater variation, making it likely that the human form of *FOXP2* has been acted upon by a recent episode of natural selection.

Archeology. During the period in which modern humans first appeared in Africa and spread from the continent, African populations used a related group of flake-based stone technologies collectively grouped into the Middle Stone Age. The Middle Stone Age developed from the Acheulean (lower Paleolithic) between 300,000 and 200,000 BP and is marked by the disappearance of large bifaces (for example, hand axes) and the increasing use of prepared-core technologies (production of stone tools via earlier prepared stone cores). The fossil modern humans from Herto are associated with a late Acheulean industry, albeit one that has many small, flake-based tools made via prepared-core technology.

Controversy surrounds the issue of when fully "modern" behavior emerged. Many archeologists argue that evidence for modern behavior appears earliest in the African Middle Stone Age and may comprise the archeological correlates of the evolution of modern humans. The evidence for modern behavior includes increased planning; utilization of far-ranging, complex social networks supported by fully articulate language; the use of items of personal adornment and red ochre as symbolically encoded markers of group identity; improved ability to hunt large or dangerous game animals; and refinements in technology with the use of bone tools and the best available raw materials for chipped stone artifacts. Sally McBrearty and Alison Brooks argue that archeological traces of "modern" behavior appear gradually, beginning 500,000–300,000 BP. Other archeologists, like Richard Klein, favor a much later appearance of fully modern behavior capacities, perhaps fuelled by the spread of *FOXP2* and a modern capacity for language.

For background information *see* AFRICA; EARLY MODERN HUMANS; FOSSIL HUMANS; FOSSIL PRIMATES; MOLECULAR ANTHROPOLOGY; PALEOLITHIC; PHYSICAL ANTHROPOLOGY in the McGraw-Hill Encyclopedia of Science & Technology.   Osbjorn M. Pearson

Bibliography. R. G. Klein, *The Human Career: Human Biological and Cultural Origins*, 2d ed., University of Chicago Press, 1999; D. E. Lieberman et al., The evolution and development of cranial form in *Homo sapiens*, *Proc. Nat. Acad. Sci. USA*, 99:1134–1139, 2002; S. McBrearty and A. S. Brooks,

The revolution that wasn't: A new interpretation of the origin of modern human behavior, *J. Hum. Evol.*, 39:453–563, 2000; S. Pääbo, The mosaic that is our genome, *Nature*, 421:409–412, 2003; J. H. Relethford, *Genetics and the Search for Modern Human Origins*, Wiley-Liss, New York, 2001; T. D. White et al., Pleistocene *Homo sapiens* from Middle Awash, Ethiopia, *Nature*, 423:742–747, 2003; M. H. Wolpoff, Paleoanthropology, 2d ed., McGraw-Hill, Boston, 1999.

Monitoring bioterrorism and biowarfare agents

While all biothreat agent research in the United States has been of a defensive nature since the 1972 Geneva Convention, many foreign countries have continued to develop and stockpile weaponized biothreat agents. Numerous U.S. government agencies and laboratories as well as public and private research institutions have spent decades designing and developing technologies to safeguard the nation from biological attacks. This research has included developments for public health monitoring and intervention, as well as technology for direct detection of biothreat agents or their toxins. These biological agents include bacteria such as *Bacillus anthracis*, *Yersinia pestis*, *Francisella tularensis*, *Vibrio cholerae*, and *Escherichia coli* O157:H7; viruses such as Ebola and *Variola* (smallpox); and toxins such as ricin, botulinum, and Staphylococcal enterotoxin B to name a few.

As technology in general has improved over the last few decades, so has biothreat monitoring and detection. Improvements in computers, laser optics, nanotechnology, microfabrication, bioengineering, biochemistry, and genomics have had significant effects on our ability to design sensors to monitor not only our environment but also our food, medicine, and selves. People who work on the front lines of public defense (first responders, soldiers, and clinic/hospital staff) need instruments that are portable and user-friendly, that can reliably detect biothreat agents rapidly with high sensitivity and low false-positive rates, and that preferably can detect several agents simultaneously. No one instrument has yet been able to incorporate all the needed and desired capabilities, but systems under development today may achieve these goals.

Detection platforms and chemistries. Biothreat detection platforms, also called biosensors, currently use numerous different types of chemistries to detect and monitor agents of interest. These platforms can be grouped into a few main categories, namely those using nucleic acids, antibodies, or intrinsic biochemical or physical target characteristics for detection (see **table**). Irrespective of the type of platform, most sensor technologies today rely on some style of colored or fluorescent reporter signal to indicate the presence of the biothreat target. Typically, fluorophores (attachable fluorescent molecules) such as Cy3, Cy5, quantum dots, upconverting phosphors, or various bioluminescence compounds are incorporated into the platform chemistries and produce a fluorescent signal when excited by a laser of appropriate wavelength. This fluorescent signal is detected and interpreted by either a human operator or computer. Other chemistries incorporate different colorimetric dyes, and a positive signal is indicated by a color change in one of the sample analysis ports.

Nucleic acid–based detection. Nucleic acid–based sensors typically target the deoxyribonucleic acid (DNA) or ribonucleic acid (RNA)—particularly messenger RNA (mRNA) and ribosomal RNA (rRNA)—of biothreat agents by means of signal amplification using polymerase chain reactions (PCRs). PCR uses specific short DNA sequences (primers) designed to target only the threat agent of interest and, if present, produces a detectable fluorescent signal within a short amount of time. In addition to PCR, microarray or "lab-on-a-chip" platforms use single-stranded DNA to capture rRNA from targets and produce a fluorescent signal. PCR and microarray technologies usually require cumbersome equipment and advanced skills for proper functioning. However, there are several versions of handheld PCR and microarray platforms currently in development. These units are relatively small [less than 10 lb. (4.5 kg)] and can be used by

Examples of platform types used in biosensors for detection of biothreat agents

Platforms	In use*	In development†
Nucleic acid–based detection		
Polymerase chain reaction (PCR)	X	X
Nucleic acid sequence–based amplification (NASBA)	–	X
Microarrays	X	X
Immunological-based detection		
Biochip arrays	–	X
Cantilever and acoustic wave	–	X
Surface plasmon resonance	–	X
Evanescent wave	X	X
Enzyme-linked immunosorbent assay (ELISA)	X	
Microarrays	X	X
Smart Tickets	X	X
Physical or biochemical-based detection		
Mass spectrometry	–	X
Raman spectrometry	–	X
Bioluminescence	X	–
Electronic noses	–	X
Substrate utilization	X	–
Fatty acid methyl ester (FAME)	X	–

*Platforms are available in some format commercially.
†Platforms are not currently available commercially or are in beta-testing stages.

someone who is wearing heavy protective gloves. Major improvements in microfluidics and thermo-control devices have significantly enhanced the capabilities of these types of units for field applications. Although the nucleic acid–based platforms are sensitive, rapid, and very specific in their detection, they have two main drawbacks: the sample must be very clean (all inhibiting compounds removed), and the platforms are useful only for detecting bacterial or viral threats. Biothreats such as toxic chemicals or biological toxins are not detectable due to a lack of DNA in the samples.

Immunological (antibody)–based detection. Antibody-based platforms (also known as shape recognition technology) borrow from and utilize elements of the innate and adaptive mammalian immune system for pathogen detection. Antibodies are used as highly specific and highly sensitive affinity probes in a number of biosensors to detect bacteria, viruses, and toxins. As antibodies are typically developed via exposure to the antigen (target) of interest, recent advances in molecular biology have been boons to improving and speeding the development process without unnecessary exposure. Techniques known as biopanning allow researchers to explore large protein libraries for molecules that bind to specific antigens and, once found, can be mass-produced via recombinant techniques or chemically synthesized. Platforms that use antibodies for antigen recognition typically use them either as a capture molecule or as a reporter molecule, and sometimes together in a sandwich format. Capture antibodies are chemically bound or mounted on an inert surface and are allowed to freely associate with the sample being tested and to bind any target antigen present. After incubation, the sample is removed, a reporter antibody is added, and the sample is analyzed for the presence of the signal generated by reporter label if target antigens were present (numerous labels are available depending on the type of sensor used). The most portable format available is the lateral flow device or dipstick test, similar to those found in most pregnancy kits available to the public. The limitations of antibody assays are related to the efficiency of the antigen-antibody complex formation and detection of the bound complexes. In addition, these types of detection assays are usually limited to testing one antigen at a time. However, recent advances in multiplexing chemistries have resulted in some simultaneous assays being performed on automated systems. One advantage over that of nucleic acid methods is that live cells can often be recovered after the assay and stored for future study or confirmation.

Biochemical- or physical-based detection. Biochemical- or physical property–based platforms depend on intrinsic characteristics of the target itself. These platforms do not require the addition of biological components and are based on properties such as size, light refraction/absorbance, metabolic end products, or enzymatic activities. Physical parameters of biothreat targets can be determined and in-

struments calibrated to detect any samples which contain the same pattern recognition signature. Samples are collected, concentrated, and processed into the necessary format (cell suspensions, chemical lysis, vaporization, gas/vapor collection) and passed by a laser, which creates the fingerprint. For biochemical detection platforms, typically detection is determined by the production or inhibition of assay compounds. These are quite sensitive; however, they are susceptible to influences of unknown or unanticipated substances present in the samples. There has been a great deal of research done to optimize pattern generation and recognition for different formats; however, for any target to be "recognized" and detected, it must first be tested and entered into the system. This protocol leads to the possibility that an agent closely related, but not identical, to one used in calibrating the units would not be recognized. Additional limitations to these platforms are that the samples need to be highly concentrated (high level of target in sample volume), have a potential for lack of specificity in mixed or complex samples (multiple signatures in one sample), and require complex pattern recognition software for analysis.

Detection considerations. Some of the advantages and difficulties with each detection platform type have been mentioned; however, there are other considerations that need to be taken into account when designing these detection systems. Currently the main constraint for all platforms is the limit of detection (how few targets can be detected). If the targets are present in high-enough numbers, any one of the aforementioned platforms would work for detection. The problems arise in detecting small numbers of targets in large volumes of sample [for example, 100 cells in 100 liters (26.4 gallons) of water]. One dilemma facing researchers today is how sensitive are the platforms necessarily to be versus how sensitive is actually practical or possible. The question of foremost concern is what are the limits of detection needed to protect human health. Is it necessary or even advisable to try to detect the presence of 100 target cells, if it takes 100,000 to make someone ill?

Another limitation to any of the aforesaid technologies is the current inability to detect genetically altered organisms or toxins. For scientists to design and optimize these platforms, they must use available organisms, genomic data, and toxins. If an unscrupulous scientist were to uniquely create or modify a biothreat agent, it could very well be missed by the detection systems already in place. Many research groups are currently working on potential ideas for modifying the current technologies in ways that would allow an "unknown" or "modified" threat to be detected. Exactly how does one go about designing a detection system for something that is not available or does not exist yet? This is the new frontier in biothreat detection.

For background information *see* ANTHRAX; ANTIBODY; BIOSENSOR; FORENSIC BIOLOGY; NUCLEIC

ACID; SMALLPOX; TOXIN in the McGraw-Hill Encyclopedia of Science & Technology.   Joyce M. Simpson

Bibliography. R. M. Atlas, Bioterrorism: From threat to reality, *Annu. Rev. Microbiol.*, 56:167–185, 2002; R. J. Hawley and E. M. Eitzen, Jr., Biological weapons: A primer for microbiologists, *Annu. Rev. Microbiol.*, 55:235–253, 2001; S. S. Iqbal et al., A review of molecular recognition technologies for detection of biological threat agents, *Biosens. Bioelectr.*, 15:549–578, 2000; D. V. Lim et al., Current and developing technologies for monitoring agents of bioterrorism and biowarfare, *Clin. Microbiol. Rev.*, 18:583–607, 2005.

Multicomponent coupling

Multicomponent couplings (MCCs), also called multicomponent reactions (MCRs), are special types of synthetic organic reactions in which three or more starting materials react to generate a single product. The product forms either sequentially in a tandem or domino fashion, or all at once. All or most of the atoms of those starting materials are incorporated into the newly formed product. MCCs are highly convergent, as a new molecule is formed from several starting molecules in one operation. One-component and two-component reactions are divergent, and complex molecules have to be synthesized by many sequential reaction steps, each involving one or two components.

The concept of multicomponent coupling can be found in nature. Many building blocks of life are believed to have resulted from multicomponent couplings. For example, adenine, one of the major constituents of DNA and RNA, may have been formed from five molecules of isocyanic acid. The other nucleic acid bases may have been formed in similar reactions involving isocyanic acid and water.

Historical context. Although multicomponent couplings appear to be recent, such reactions can trace their origin to as early as 1850 with the discovery of the Strecker synthesis of α-cyanoamines. Many well-known MCCs are named reactions. Examples include the Hantzsch dihydropyridine synthesis (1882), Radziszewski imidazole synthesis (1882), Hantzsch pyrrole synthesis (1890), Biginelli reaction (1891), Mannich reaction (1912), Robinson tropinone synthesis (1917), Passerini reaction (1921), Bucherer-Bergs hydantoin synthesis (1934), Asinger reaction (1958), and Ugi reaction (1959) [see **illustration**]. It is noteworthy that both the Passerini and Ugi reactions feature isocyanide (R—C≡N) functionality in one of the reactants. The terminal carbon of this functionality serves a dual role as a nucleophile and an electrophile.

The result of an MCC is often crucially dependent on the nature of the solvent, catalyst, concentration of reactants, and proportion of reagents used. These variables render the optimization of reaction conditions more demanding, compared with sequential

Examples of multicomponent coupling reactions.

reaction schemes. In designing multicomponent coupling reactions, the nature of the individual components plays a key factor. New MCCs are found by building a large collection of compounds, known as a chemical library, from combinatorial chemistry or by combining existing MCCs. I. Ugi has also developed the concept of unifying multicomponent reactions. New MCCs can be created by combining two or more MCCs in one pot. The Ugi four-component reaction may be viewed as the combination of a Mannich-type three-component reaction using amine A, aldehyde or ketone B, and acid C, with a Passerini-type three-component reaction that uses aldehyde or ketone B, acid C, and isonitrile D. Ugi and coworkers reported in 1993 a seven-component coupling. This reaction (1) can be viewed as the combination of an Asinger and Ugi reaction.

Ugi-Domling seven-component

(1)

The recent demands in drug discovery for more diverse small molecules and their efficient synthesis have directed the attention of chemists to this largely unexplored area of chemistry. In the past few years, a variety of novel MCCs have been discovered and applied to the synthesis of biologically active molecules. Traditional drug discovery requires a robust assay of target activity and a collection of compounds for testing. The identification of candidate compounds usually requires high-throughput screening of collections of diverse small molecules or a group of structurally selected compounds with known or predicted activity against a target. Easily automated one-pot MCCs have become a powerful tool for producing organic compounds with diverse substitution patterns. Using MCCs with a small set of staring materials, very large libraries can be built in a short time.

Green chemistry. Currently, increasing environmental consciousness worldwide has led to the search for environmentally benign methods for chemical synthesis of organic compounds via green chemistry. In this connection, efforts have been made to reduce or eliminate hazardous substances during manufacture or use of chemical products. Therefore, the development of chemical reactions that are atom-efficient and environmentally friendly with enhanced selectivity are attractive for the synthetic chemists. The ideal synthesis should allow the construction of the desired product in as few steps as possible, in good overall yield, and by using environmentally acceptable reagents. The synthetic variables that have to be optimized are time, cost, overall yield, simplicity of performance, safety, and environmental impact. Most synthetic schemes still use a simple step-by-step approach to convert a starting material into a final product, in which intermediate products are isolated and purified for the next conversion step. Such traditional practice has the disadvantages of being time-consuming, with the likelihood of producing large amounts waste. MCCs convert three or more educts (starting materials) directly into their products by one-pot operations and therefore are more efficient than multistep syntheses because both the synthetic and work-up procedures need to be done only once. The other attractive features of MCCs are the formation of several new bonds in one pot, the attainment of higher atom economy and chemoselectivity, and the generation of a low level of by-products, compared to classical stepwise syntheses.

Synthesis of propargylamines. Chiral nitrogen-containing compounds are widely found in nature, and many of them show important biological properties. Therefore, the development of methods for synthesizing diastereomerically and enantiomerically pure amines is a major objective for organic chemists. Traditionally, in a manner parallel to the addition of carbanion to the carbonyl group of aldehyde and ketones, the Grignard-type addition of organometallic reagents to imines ($R_2C=NR$), or imine derivatives, is an important and efficient process to afford many useful nitrogen-containing compounds. The formation of propargylamines by the addition of an acetylenic moiety to a $C=N$ double bond is one such process. However, the reactive metal acetylenic moieties are often prepared by deprotonation of terminal alkynes by using a strong base (which is often an organometallic reagent) in a separate step. In addition, many organometallic reagents are usually not easy to handle because the reaction must be done in an anhydrous solvent under an inert atmosphere at low temperature. Such practice contradicts the principles of green chemistry on several counts, such as (1) the requirement of an organic solvent in each step which could potentially increase volatile organic emissions, (2) the need of the anhydrous conditions which requires excess drying agents that could lead to unwanted waste, (3) the presence of a strong base which necessitates the protection of function groups such as hydroxyl and carboxylic acid groups, and (4) the stoichiometric amount of metal associated with the strong base which would lead to a large amount of metal waste. In view of these drawbacks, an alternative reaction between imines or imine derivatives and a terminal alkyne via catalytic activation of the C-H end group without using organometallic reagent would provide an ideal solution. Even more desirable is to be able to use an aqueous medium to minimize the environmental impact. Toward this goal, C. J. Li and C. Wei recently developed a highly efficient, direct three-component coupling of aldehyde, alkyne, and amine to give propargylamines. The attractiveness of this methodology is twofold. First, the reaction can be done in water, organic solvents, ionic liquids, or under neat conditions (no solvent is used). Second, in the presence of a chiral catalyst prepared in situ chiral, propargylamines in high enantiomeric excess can be obtained.

The bimetallic system $CuBr/RuCl_3$ has been found to be an efficient catalyst for the three-component coupling of aldehydes, anilines, and terminal alkynes in aqueous media to give propargylamines [reaction (2)]. In those cases where the imines were easily

$$R^1 - CHO + Ar - NH_2 + R^2 \equiv \xrightarrow[\text{60-90°C, water}]{CuBr, RuCl_3}$$

(2)

hydrolyzed in water, the additions were found to be highly effective under neat condition.

Gold and silver have also been found to be good catalysts in the three-component coupling of aldehydes and dialkyl amines and terminal alkynes. It is noteworthy that no co-catalyst or activator is needed for the gold-catalyzed reaction. Less than 1 mol % of gold catalyst is sufficient to generate an excellent yield of the corresponding propargylamines. Dialkyl amines are good for the reaction, and aromatic aldehydes react more efficiently and nearly quantitative yields are obtained in most cases. The properties of the solvent significantly affect the reaction. In most cases, water is the best solvent, and the reactions are generally very clean with almost quantitative yields, while the use of organic solvents usually leads to low conversions and a considerable amount of by-products. With silver catalyst, the three-component couplings involving aliphatic aldehydes give both higher conversions and better yields than aromatic aldehydes. The reactions (3) proceed well in either

$$R^1 - CHO + R_2 \equiv + R^3_2NH \xrightarrow[\text{100°C, water}]{AuBr_3 \text{ or } AgI}$$

(3)

water or organic solvent. *See* GOLD-CATALYZED REACTIONS..

With the use of chiral tridentate bis(oxazoline)pyridine (structure **1**) with copper as the catalyst, enantioselective three-component coupling of aldehyde, alkyne, and amine can be accomplished. The reaction (4) can be executed in a remarkably simple manner by mixing an aldehyde, aniline, and a terminal alkyne with the catalyst in one pot. The yields and enantioselectivity are generally good.

$$R^1 - CHO + Ar - NH_2 + R^2 \equiv \xrightarrow[\text{toluene or water}]{Cu(OTf) / \text{ligand 1}}$$

(4)

(1)

Enantioselective three-component couplings to synthesize propargylamines have also been reported by P. Knochel, A. H. Hoveyda, E. M. Carreira, and others.

Outlook. With respect to productivity, yield, convergence, and ease in execution, MCRs continue to occupy an important position in organic synthesis, particularly in combinatorial chemistry. Therefore, the discovery and development of novel MCRs are currently receiving growing interest from various research groups in the chemical industries. Design of novel and useful MCRs remains a challenge for organic chemists.

For background information *see* ALDEHYDE; ALKYNE; AMINE; CARBOXYLIC ACID; CATALYSIS; COMBINATORIAL CHEMISTRY; COMBINATORIAL SYNTHESIS; ELECTROPHILIC AND NUCLEOPHILIC REAGENTS; ENANTIOMER; GRIGNARD REACTION; KETONE; NITRILE; ORGANIC SYNTHESIS; STEREOCHEMISTRY in the McGraw-Hill Encyclopedia of Science & Technology.

Chunmei Wei

Bibliography. H. Bienayme et al., Maximizing synthetic efficiency: Multi-component transformations lead the way, *Chem. Eur. J.*, 6:3321, 2000; I. Ugi, Recent progress in the chemistry of multicomponent reactions, *Pure Appl. Chem.*, 73:187, 2001; C. Wei et al., The discovery of A3 (aldehyde, alkyne and amine) coupling and asymmetric A3 coupling, *Synlett*, 1472, 2004; C. Wei et al., The first silver-catalyzed three-component coupling of aldehyde, alkyne, and amines, *Org. Lett.*, 5:4473, 2003; C. Wei et al., A highly efficient three-component coupling of aldehyde, alkyne, and amines via C-H activation catalyzed by gold in water, *J. Amer. Chem. Soc.*, 125:9584, 2003.

Multiple sclerosis

Multiple sclerosis (MS) is a chronic, degenerative, inflammatory disease of the central nervous system. The best early description of MS was that of Jean-Martin Charcot in 1868, who called the disease *sclerose en plaques*. He noted three cardinal symptoms of MS, known as Charcot's triad—ataxia (coordination problems), dysarthria (slurred speech), and tremor—and associated these symptoms with gray discolorations distributed throughout the brain and spinal cord. Today, MS is recognized as causing a variety of potentially disabling neurological symptoms, including diminished strength, coordination, or sensation; visual changes; mood and cognitive problems; and loss of bowel or bladder control.

Pathophysiology. MS appears to be an autoimmune disease in which the immune system attacks myelin in the central nervous system. Myelin is a fatty substance which is wrapped around axons, the nerve cables which connect neurons to each other, and enhances the conduction of nerve impulses. In MS, an inflammatory cascade is initiated which causes breakdown of the blood-brain barrier, a layer of endothelial cells which normally prevents large molecules or cells from entering the central nervous system. This allows activated T (thymus-derived) lymphocytes, other inflammatory cells, and signaling molecules known as cytokines to enter the central nervous system. Ultimately, by mechanisms that are not completely understood, these inflammatory cells or their products damage both the myelin wrapping and, to a variable extent, the axons themselves. Earlier in the disease course, remyelination leads to some clinical improvement. Unfortunately, remyelination is limited and generally not as effective as the original myelin. Over time the scars referred to as plaques accumulate, and patients are no longer able to recover well from attacks.

Common symptoms. The plaques in MS may occur in many locations in the white matter (myelin is yellowish-white) of the central nervous system. Typical locations include the spinal cord (causing numbness, tingling, weakness, and bowel and bladder disturbance), optic nerves (painful loss of vision), brainstem (vertigo, double vision), cerebellum (incoordination, slurred speech), and the areas next to the ventricles, deep in the brain. Thus, a wide variety of symptoms are possible. Later in the disease course, muscle spasms, emotional lability, mood disorders, short-term memory and other cognitive difficulties, and fatigue often become prominent symptoms.

Geographical distribution. In North America, northern Europe, and Australia, about one in 1000 people suffers from MS. Disease frequency is much lower in Asia, Africa, and the Middle East. This disparity is not well understood, but with notable exceptions there appears to be a geographical gradient worldwide whereby people closer to the Equator (warmer climates) have a significantly lower chance of developing MS than those farther away (colder climates).

This gradient has been attributed to several environmental factors, including sun exposure, vitamin D metabolism, diet, toxins, and viral illnesses, although none of these potential links has been conclusively established. Interestingly, these geographic tendencies seem to be most important in childhood. If a patient moves to a different climate after the age of 15, they retain the risk for the geographic area from which they moved. In addition, as with most autoimmune diseases, about twice as many women as men have MS.

Disease subtypes. Disease course is highly variable in MS, but most patients fall into one of four categories of disease progression. About 85% of patients, especially young women, initially have attacks (also called relapses or exacerbations) of symptoms lasting days to weeks and followed by complete or incomplete remission of symptoms—so-called relapsing-remitting MS (RRMS). These patients may be relatively stable until they have a new attack, which could happen weeks, months, or years later. After a variable number (3–20) of years, a significant proportion of these patients will transition into secondary progressive MS (SPMS), when they stop having new attacks but gradually decline neurologically. Much of the disability that accrues over time occurs during SPMS. Primary progressive MS (PPMS) is described in about 10–15% of patients with no clear attacks but slowly worsening progression of symptoms, often with significant leg weakness and bowel and bladder dysfunction, from the onset of disease. A small number of patients have a combination of both progressive and rare superimposed attacks, referred to as progressive relapsing MS (PRMS). Prognosis is variable, but MS patients generally do somewhat better overall if they are young at diagnosis and female, and if they have few attacks, mostly sensory problems, and fewer lesions seen on imaging studies. Less favorable prognosis is associated with those diagnosed at a later age, males, African-Americans, and those with many attacks or PPMS, motor problems, and more lesions seen on imaging studies. In general, MS does not limit lifespan but may result in significant disability over time.

Relapses. Patients may have new symptoms (a new attack) or often will have recurrence of old symptoms (a pseudoattack) with elevation of body temperature, infection, or stress. An example of a pseudoattack is a patient with previous optic neuritis transiently losing visual function (Uhthoff's phenomenon) with exercise and raised core body temperature. Also, when large populations of MS patients are studied, true attacks seem to be more prevalent during the warmer months. Most attacks, however, appear randomly and with no apparent cause. There is no evidence that vaccinations, such as influenza, varicella (chickenpox), tetanus, hepatitis B, or other commonly given vaccines, increase the risk of MS attacks. Interestingly, the third trimester of pregnancy is associated with a decrease in relapse rate, suggesting a protective hormonal effect.

Diagnosis. Although formal diagnostic criteria have changed over time, the diagnosis of MS has always required "dissemination of lesions in time and space" within the central nervous system, with objective abnormalities on the neurological examination, and no other cause identified to account for the symptoms. The role of laboratory testing for MS has evolved significantly over the last 50 years, first with the identification of abnormal production of immunoglobulins (referred to as oligoclonal bands) in the cerebrospinal fluid. Later, it was found that stimulation of the visual, auditory, or somatosensory systems was slowed in MS patients, and identification of slowing confirmed lesions as demyelinating in nature. In the last 25 years, neuroimaging, especially magnetic resonance imaging (MRI) of the brain and spinal cord, has not only enhanced our ability to diagnose MS but also altered our view of the nature of the illness. MRI has reinforced that many pathological lesions in the brain are unaccompanied by new clinical symptoms, that brain atrophy is common and important early in the disease, and that the number of lesions and brain atrophy correlate with progression of disability over time. Adopted in 2001 and updated in 2005, the newest diagnostic MS criteria (McDonald criteria) explicitly allow the use of laboratory tests, especially MRI, to assist in determining dissemination in space and time, and this allows for earlier diagnosis and treatment of MS.

Differential diagnosis. Many diseases may mimic MS, including systemic lupus erythematosus, Sjögren's syndrome, antiphospholipid antibody syndrome, neurosarcoidosis, Behçet's disease, Lyme disease, other central nervous system infections, and degenerative disorders such as amyotrophic lateral sclerosis (Lou Gehrig's disease). While all of these diseases can have a variety of neurological manifestations, many mimics of MS also have symptoms outside of the central nervous system, while MS symptoms are restricted to the central nervous system. The paraclinical data such as MRI and the cerebrospinal fluid profile also point to the diagnosis of MS, and laboratory studies in blood often eliminate other potential causes.

Treatment. A variety of approaches are used in treating MS. Chronic MS symptoms, including fatigue, incontinence, spasticity, and depression, are treated with drugs for those specific problems. Physical, occupational, and speech therapy are often helpful in maximizing activities of daily living, especially gait. Acute attacks of relapsing-remitting MS are treated with high-dose corticosteroids, generally intravenous methylprednisolone (IVMP) for 3-5 days, which will hasten the recovery of neurological function. If IVMP fails to shorten the duration of a severe, disabling attack, plasma exchange over a 2-week period is frequently helpful.

There are six therapies approved by the US Food and Drug Administration which are used to prevent new attacks and slow the onset of disability in relapsing forms of MS (see **table**). Four are injected by patients at home. Three of these drugs are interferons—interferon β-1a (Avonex and Rebif), and interferon β-1b (Betaseron). These medications are derived from human cytokines, naturally occurring proteins produced by white blood cells, which are induced in

Therapies approved by the FDA to prevent new attacks and slow the onset of disability in relapsing forms of multiple sclerosis*					
Drug	Mode of delivery	Frequency	Indication	Proven benefits	Adverse effects
Interferon β-1a (Avonex®)	Intramuscular injection	Once weekly	RRMS	Prevents new relapses by about 1/3; reduces new MRI lesions; slows progression of disability	Common: flulike symptoms, injection site reactions Rare: liver/bone marrow toxicity, depression
Interferon β-1a (Rebif®)	Subcutaneous injection	Three times weekly	RRMS	Prevents new relapses by about 1/3; reduces new MRI lesions; slows progression of disability	Common: flulike symptoms, injection site reactions Rare: liver/bone marrow toxicity, depression
Interferon β-1b (Betaseron®)	Subcutaneous injection	Every other day	RRMS	Prevents new relapses by about 1/3; reduces new MRI lesions; slows progression of disability	Common: flulike symptoms, injection site reactions Rare: liver/bone marrow toxicity, depression
Glatiramir acetate (Copaxone®)	Subcutaneous injection	Daily	RRMS	Prevents new relapses by about 1/3; reduces new MRI lesions	Common: injection site reactions Rare: postinjection reaction
Mitoxantrone (Novantrone®)	Intravenous infusion	Quarterly (every 3 months)	Severe RRMS or SPMS	Reduces rate of new relapses; reduces new MRI lesions; slows progression of disability	Common: bone marrow suppression, hair loss, nausea, fatigue Rare: cardiac toxicity, leukemia
Natalizumab (Tysabri®)	Intravenous infusion	Every 4 weeks	Severe or refractory RRMS	Prevents new relapses by 1/2 to 2/3; reduces new MRI lesions by over 80%; slows progression of disability	Rare: postinfusion anaphylactic reaction (about 1% of patients); PML (3 patients)

*RRMS = relapsing-remitting MS; SPMS = secondary progressive MS; MRI = magnetic resonance imaging; PML = progressive multifocal leukoencephalopathy.

the human body in response to infection and other biological factors. It is not entirely understood how these drugs work, but they have an immunomodulatory function and decrease the inflammation involved in the pathogenesis of MS. The fourth drug is glatiramir acetate, a mixture of polypeptides that is thought to work by shifting the T lymphocyte response away from attacking myelin. These four drugs are injected at home by the patient and are modestly effective at preventing new attacks (about a one-third reduction) and slowing the progression of disability in MS patients (only shown convincingly in the interferon products). They are distinguished from each other primarily by frequency of dosing and side effects, and there is some evidence that higher dose and frequency of administration of interferon may be slightly more efficacious than lower dose and less frequent administration. Mitoxantrone (Novantrone), a chemotherapy drug that also is helpful in secondary progressive MS and severe relapsing-remitting MS, is limited in its use due to risks of damaging heart muscle and inducing leukemia.

The sixth drug, natalizumab, is a monoclonal antibody which prevents trafficking of white blood cells across the blood-brain barrier, and is a novel agent in the treatment of MS. Approved for use in late 2004, it was withdrawn from the market in early 2005 because three patients taking the drug in clinical trials contracted progressive multifocal leukoencephalopathy, a serious brain infection typically seen in patients with compromised immune systems. The infection was fatal in two of the patients. The patients were also taking other medications that affect the immune system, so natalizumab has been reapproved only for use alone. Now reapproved for use in relapsing-remitting MS, it most likely will be restricted in practice to those patients who have failed or were intolerant of one or more standard MS medications and/or those patients with bad prognosis.

For background information *see* CENTRAL NERVOUS SYSTEM; IMMUNOGENETICS; INTERFERON; MEDICAL IMAGING; MULTIPLE SCLEROSIS; NERVOUS SYSTEM (VERTEBRATE); NERVOUS SYSTEM DISORDERS; NEUROIMMUNOLOGY in the McGraw-Hill Encyclopedia of Science & Technology.

John Corboy; Karen Rollins

Bibliography. F. D. Lublin and S. C. Reingold, Defining the clinical course of multiple sclerosis: Results of an international survey, National Multiple Sclerosis Society (USA) Advisory Committee on Clinical Trials of New Agents in Multiple Sclerosis, *Neurology*, 46(4):907–911, 1996; T. J. Murray, *Multiple Sclerosis: The History of a Disease*, Demos Medical Publishing, New York, 2005; J. H. Noseworthy et al., Multiple sclerosis, *N. Engl. J. Med.*, 343(13):938–952, 2000; G. Ogawa et al., Seasonal variation of multiple sclerosis exacerbations in Japan, *Neurol. Sci.*, 24(6):417–419, 2004; C. H. Polman et al., Diagnostic criteria for multiple sclerosis: 2005 revisions to the "McDonald Criteria," *Ann. Neurol.*, 58:840–846, 2005; C. H. Polman et al., *Multiple Sclerosis: The Guide to Treatment and Management*, 6th ed., Demos Vermande, New York, 2006.

Nanometrology

The continuing progression of miniaturization technology for electronic devices has brought us into the nano regime. In our daily lives, we are using nanodevices, which consist of artificial nanostructures less than 100 nanometers in size. To fabricate these nanodevices accurately, we have to measure at the nanometer scale various values, such as length and resistivity. Despite the importance of nanoscale measurement technology, or nanometrology, we have only limited solutions. New metrological methods are now urgently required for the further development and improvement of nanodevices. Recently, new tools for nanometrology have been developed.

The most fundamental tool for metrology is a scale for length. A scale with nano-order accuracy, or nanoscale, is necessary to determine absolute length in nanotechnology applications. A second important tool is the miniaturized four-point probe, which is used for measuring the sheet resistance of conductive materials.

Nanoscale. Various microscopic technologies, such as transmission electron microscopy, scanning electron microscopy, and scanning probe microscopy, are used to observe nanostructures. The length in the microscopic image is affected by the specifications of the microscope, such as its focal length, aberration, and sample drift. Thus, the length in the microscopic image must be adjusted using a standard scale in order to determine the correct value.

In the International System of Units (SI), the basic unit of length, a meter, is now defined as the length of the path traveled by light during a very short defined time. Since it is impossible to implement this method in nanometrology, we need a scale of length that can be used in the nanometer region. A lattice grating with 1-micrometer pitch is conventionally used as a scale for micrometrology. However, a scale with an even smaller pitch is required for nanometrology, because the image size is often smaller than 1 μm^2.

Figure 1a and b show micrographs of the cross section of a nanoscale made from single-crystal silicon (Si) with 100-nm and 50-nm pitch, respectively. Conventional fabrication methods for the grating, such as mechanical scratching using a ruling engine and lithographic patterning based on holographic interference, cannot be used to create nanostructures smaller than 100 nm. As a result, the line patterns were delineated with high accuracy by electron beam lithography. The lithographic patterns were transferred to a Si {110} substrate by means of anisotropic etching, which is a kind of wet etching. What is notable about this nanoscale is that the cross-sectional shape is a well-defined rectangle. As shown in Fig. 1c, the sidewall of the Si nanostructure is defined by the Si {111} surface using anisotropic etching with high selectivity. The deviation of the line width of this Si nanostructure is small, less than 1 nm in standard deviation. The nanostructure in the nanoscale is the straightest ruler that can be made artificially.

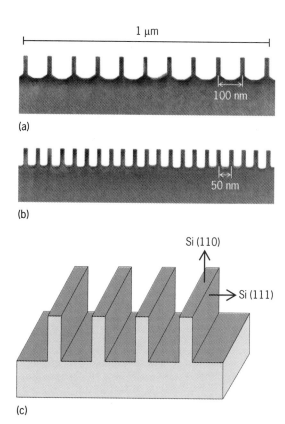

Fig. 1. A nanoscale for nanometrology. Micrographs of the cross section of a nanoscale with (*a*) 100-nm pitch and (*b*) 50-nm pitch. (*c*) Schematic of the nanoscale fabricated by anisotropic etching on Si{110} substrate.

Nanotool. Resistivity is the most fundamental measurement for understanding electrical transport phenomena in conductive materials. The four-point-probe method is widely used to measure resistivity, or sheet resistance, with high accuracy. In a four-point-probe measurement, four probes, aligned with the same pitch, contact a sample surface. A conven-

tional four-point probe is a macroprobe because the distance between the probes is 1 mm in almost all instruments. Since the minimum distance of recently developed probes is 100 μm, it is impossible to measure the microscopic properties of conductive materials. Downsizing of conventional probes is limited by the precision constraints of machining technology. A new type of miniaturized four-point-probe system to measure micro- and nano-order electrical properties is required.

There are two ways to realize a four-point probe for the micro- and nanometer regimes. One is a multiprobe system based on multiple scanning mechanisms. The other system integrates multiprobes on a single scanner. The former type of system requires another microscope to confirm the contact position of each probe. Furthermore, the total system is large and complex. On the other hand, a single scanner system requires only the modification of an SPM (scanning probe microscopy) probe. The modified probe, fabricated using nanolithography, can be called a nanotool.

A nanotool for nanometrology is a miniaturized measurement system integrated on an SPM microprobe. The conventional SPM microprobe has only one probe tip on a single cantilever. The width of the cantilever is typically 40 μm. While the cantilever width is smaller than that of a human hair, it is large enough to integrate functional devices fabricated using nanolithography. Recently, M. Nagase and coworkers (Nippon Telegraph and Telephone Corp.) developed various types of four-point nanoprobes integrated on SPM cantilevers.

Figure 2 shows the schematic and the scanning electron micrograph of a four-point nanoprobe integrated on an SPM cantilever using focused ion beam nanolithography. Four carbon probes with nanosprings (multi-nanoprobe) are integrated on a Si cantilever with aluminum (AL) electrodes. The diameter of the carbon probe is 110 nm, and the

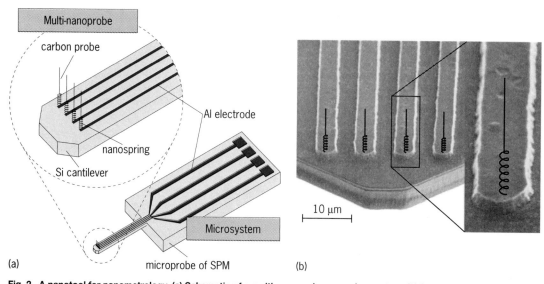

Fig. 2. A nanotool for nanometrology. (*a*) Schematic of a multi-nanoprobe on a microsystem. (*b*) Scanning electron micrograph of the nanotool. Nanoprobes with nanosprings made from diamondlike carbon are fabricated using focused ion beam nanolithography.

diameter of the nanospring is 380 nm. The carbon probe, made from diamond-like carbon (DLC), is electrically conductive and mechanically hard. These properties are preferable for the electrical probe.

Since their measured stiffness is almost the same as that of a standard steel spring, it is expected that these nanosprings will compensate for the height differences between the probes and the sample surface. Although the heights of the probes shown in Fig. 2b differ from each other in the range of 200 nm, it was confirmed that electrical contact was established between all four probes and the sample. This result shows that the nanosprings on the Si cantilever are actually acting as tiny mechanical devices.

The nanotools shown are the miniaturized versions of existing macroscale metrology tools, bridging macro- and nanotechnology.

For background information *see* CRYSTAL STRUCTURE; ELECTRICAL RESISTANCE; ELECTRICAL RESISTIVITY; ELECTRON MICROSCOPE; INTEGRATED CIRCUITS; NANOTECHNOLOGY; RESISTANCE MEASUREMENT; SCANNING TUNNELING MICROSCOPE; SURFACE PHYSICS in the McGraw-Hill Encyclopedia of Science & Technology.   Masao Nagase

Bibliography. M. Nagase et al., Carbon multiprobe on a Si cantilever for pseudo-metal-oxide-semiconductor field-effect-transistor, *Jap. J. Appl. Phys.*, 45:2009–2013, 2006; M. Nagase et al., Carbon multiprobes with nanosprings integrated on Si cantilever using focused-ion-beam technology, *Jap. J. Appl. Phys.*, 44:5409–5412, 2005; M. Nagase et al., Metrology of atomic-force microscopy for Si nanostructures, *Jap. J. Appl. Phys.*, 34:3382–3387, 1995; M. Nagase et al., Nano-four-point probes on microcantilever system fabricated by focused ion beam, *Jap. J. Appl. Phys.*, 42:4856–4860, 2003.

New carcharodontosaurid

The word "dinosaur" often conjures an image of a gigantic ferocious reptile. This image has been changing drastically in recent years, as progressively more species of dinosaurs have been found that are smaller than dogs. Many of the small meat-eaters are feathered, stark evidence of their close relationship to birds. Even though the average size of known dinosaurs has been getting smaller overall, larger ones continue to be found.

Discovery of Mapusaurus roseae. Three of the largest flesh-eating dinosaurs discovered in recent years belong to the family of theropod dinosaurs (which include all carnivorous dinosaurs) known as the Carcharodontosauridae. Claims have been made that *Carcharodontosaurus* (from northern Africa), and *Giganotosaurus* and *Tyrannotitan* (from Argentina) rival, and possibly even surpass, *Tyrannosaurus rex* in size. This is true in certain dimensions, such as skull length. Carcharodontosaurids have longer, narrower skulls than tyrannosaurids. However, in other dimensions, including brain size and the lengths of hindlimb bones, *Tyrannosaurus*

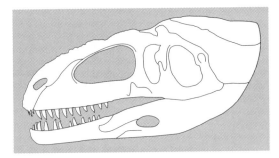

Reconstructed skull of *Mapusaurus roseae*.

rex has yet to be toppled from its long-standing reign as "king of the tyrant reptiles."

In 2006, the discovery of another carcharodontosaurid was announced in Argentina. *Mapusaurus roseae* came from a dinosaur excavation in northern Patagonia (dated to about 100 million years ago). Parts of nine skeletons of *M. roseae* (see **illustration**) were recovered from one of the largest dinosaur quarries in South America, between 1997 and 2000. Some of the bones collected were larger than the same bones in the closely related *Giganotosaurus*, suggesting that *Mapusaurus* may have been the largest of the theropods [probably exceeding 40 feet (12 m) in length]. This seems appropriate, because *Mapusaurus* came from the same rocks that produced *Argentinosaurus*, a plant-eating sauropod that is the heaviest dinosaur on record [an estimated 70 tons (63 metric tons)]. One would expect that it would take the largest hunters to tackle the largest prey; *Mapusaurus* versus *Argentinosaurus* would have been a titanic battle.

Paleontologists are fortunate when they find many specimens of the same species in the same place, particularly if they represent animals of different ages. This makes it possible to study individual variation, and to see what changes may have taken place in body proportions as the animal grew up. Like tyrannosaurids in the Northern Hemisphere, *Mapusaurus* was a relatively slender and lightly built animal when it was young. Its elongate hindlimbs suggest that it was able to run much faster than the adults.

Implications. Carnivorous dinosaurs are usually the rarest animals in any fauna because if there are too many meat-eaters, they eat all of the plant-eaters and starve to death. In most dinosaur localities, large theropods make up only 5–10% of the fauna. If we assume that there was only one carnivore (*Mapusaurus*) for every nine plant-eaters, one would expect to find on average nine skeletons of plant-eaters for every skeleton of a carnivorous dinosaur. To find two *Mapusaurus* skeletons together would be highly improbable, unless there were at least 18 skeletons of herbivores also present. The probability of finding two *Mapusaurus* skeletons together without any other animals is only 1/10 multiplied by 1/10, that is, 1 in 100. In other words, it could happen, but it is only likely to happen once in a hundred quarries where multiple animals were recovered. To find three *Mapusaurus* skeletons in the same quarry without any plant-eaters is even less likely (one in

1000). However, remember that we found nine *Mapusaurus* skeletons in the same quarry, with no other animals represented. The probability of this happening is only one chance in 1,000,000,000. Although possible, it seems highly unlikely that nine *Mapusaurus* died coincidentally in the same place at the same time.

Our working hypothesis is that the nine *Mapusaurus* specimens were living together at the time they died. Perhaps they were hunting together (instead of being solitary hunters) when they were killed catastrophically. Unless further evidence comes to light, we may never know what happened. However, similar accumulations of skeletons of the large tyrannosaurids *Albertosaurus*, *Daspletosaurus*, and *Tarbosaurus* have been found in Canada, Mongolia, and the United States. All of these sites are telling similar stories, which greatly increases our confidence in suggesting that the Cretaceous was a time when dinosaurs had complex behavior that led to the formation of family groups or packs. Perhaps this was something that happened only at certain times of the year. Or perhaps the packs were permanent. We may never know, but the investigation of this ancient "crime scene" has been informative and has supplied us with new insights into carcharodontosaurids.

For background information *see* CRETACEOUS; DINOSAUR; EXTINCTION (BIOLOGY); FOSSIL; MESOZOIC; PALEONTOLOGY; SAURISCHIA in the McGraw-Hill Encyclopedia of Science & Technology.

Philip J. Currie

Bibliography. R. A. Coria and P. J. Currie, A new carcharodontosaurid (Dinosauria: Theropoda) from the Upper Cretaceous of Argentina, *Geodiversitas*, 28:71–118, 2006; R. A. Coria and L. Salgado, A new giant carnivorous dinosaur from the Cretaceous of Patagonia, *Nature*, 377:224–226, 1995; P. J. Currie, Possible evidence of gregarious behavior in tyrannosaurids, *Gaia*, 15:271–277, 2000; P. J. Currie et al., An unusual multi-individual tyrannosaurid bonebed in the Two Medicine Formation (Late Cretaceous, Campanian) of Montana (USA), in K. Carpenter (ed.), *The Carnivorous Dinosaurs*, pp. 313–324, Indiana University Press, Bloomington, 2005; F. E. Novas et al., A large Cretaceous theropod from Patagonia, Argentina, and the evolution of carcharodontosaurids, *Naturwissenschaften*, 92:226–230, 2005; P. C. Sereno et al., Late Cretaceous dinosaurs from the Sahara, *Science*, 272:986–991, 1996.

Nobel prizes

The Nobel prizes for 2006 included the following awards for scientific disciplines.

Chemistry. The chemistry prize was awarded to Yves Chauvin of the Institut Français du Pétrole, Robert H. Grubbs of the California Institute of Technology, and Richard R. Schrock of the Massachusetts Institute of Technology for the development of the catalytic olefin metathesis method in organic synthesis.

Olefins, or alkenes, are hydrocarbons that contain one carbon-carbon double bond. A catalyst is a substance that speeds up a chemical reaction without being consumed. Olefin metathesis is a catalyzed reaction in which double-bonded atom groups are exchanged, resulting in the formation of two new olefins, as shown in the simplified reaction below,

$$RHC = CH_2 + RHC = CH_2 \underset{\longleftarrow}{\overset{catalyst}{\longrightarrow}} RHC = CHR + H_2C = CH_2$$

known as cross metathesis. In addition to cross metathesis, there are ring-opening, ring-closing, and ring-opening cross metathesis reactions, to name some. Today, metathesis reactions are used to produce polymers with novel properties as well as complex molecules for the pharmaceutical and biotechnology industries. Many of these molecules could be made by less efficient, multistep processes. However, the efficient use of raw materials by metathesis reactions represents not only a breakthrough in organic synthesis but also an important application of green chemistry.

Although catalyzed metathesis of polymers was known in the 1950s, the mechanism was not understood until 1971, when Chauvin explained that metal alkylidene (metal-carbon double bond) complexes act as catalysts for the multistep exchange of atom groups between two alkenes via metallocycloalkane (ring) intermediates, resulting in two new alkenes, plus the regenerated catalyst.

In 1990, Schrock reported the first efficient metathesis catalysts. These were molybdenum- and tungsten-alkylidene complexes, with the general formula

$$[M(= CHMe_2Ph)(= N—Ar)(OR_2)]$$

where R are sterically bulky groups. Schrock's catalysts are still the most active known, but they are unstable in the presence of oxygen or water.

In 1992, Grubbs reported a ruthenium-based metathesis catalyst that was stable in air, water, and alcohols. Its general formula was

$$[RuCl_2(PR_3)(= CH—CH(= CPh_2)]$$

where R are phenyl groups. Since then, Grubbs has improved upon his catalyst structure. These improved catalysts, known as Grubbs' catalysts, are commercially available and have become widely used in organic synthesis.

For background information *see* ALKENE; CATALYSIS; HETEROGENEOUS CATALYSIS; HOMOGENEOUS CATALYSIS; ORGANIC SYNTHESIS; REACTIVE INTERMEDIATES in the McGraw-Hill Encyclopedia of Science & Technology.

Physics. The physics prize was awarded to three scientists in the field of optics. Roy J. Glauber (Harvard University) was awarded half the prize "for his contributions to the quantum theory of optical coherence." The other half of the prize was shared by John L. Hall (JILA, and the National Institute of Standards and Technology, both in

Boulder, Colorado) and Theodor W. Hänsch (Max-Planck-Institut für Quantenoptik in Garching, Germany, and Ludwig-Maximilians-Universität in Munich) "for their contributions to the development of laser-based precision spectroscopy, including the optical frequency comb technique."

Glauber pioneered the application of quantum physics to optical phenomena. Light provides a prime example of the dual nature of quantum objects. Its wavelike properties are described by the classical electromagnetic theory of James Clerk Maxwell, but in 1905 Albert Einstein proposed that light also occurs in energy packets (quanta or photons). Although quantum electrodynamics (QED), a comprehensive theory encompassing electromagnetic radiation, was developed in the late 1940s and was central to the development of particle physics, for many years it was seldom used to treat visible light. However, in 1954 Robert Hanbury Brown and Richard Q. Twiss, while investigating an interferometric method to measure the angular sizes of stars, discovered a correlation between signals from separated light detectors, and understood this phenomenon (called bunching) to be a consequence of quantum theory.

In 1963, Glauber presented a comprehensive method, based on QED, of interpreting optical observations. It explained not only bunching but also an opposite effect called antibunching whereby, in some situations, pairs of photons can occur less frequently than in a random signal. Glauber's work formed the basis of the new field of quantum optics. It explains the fundamental difference between thermal light sources such as light bulbs and coherent sources such as lasers. Quantum optics has been applied in the creation of squeezed quantum states, quantum cryptography, quantum computation, recording of ultraweak signals in high-precision experiments, and fundamental tests of quantum theory.

Lasers and masers provide sources with extremely sharply defined frequencies that are used to probe the structure of atoms, molecules, and atomic nuclei. Full advantage of the laser's precision can be taken only when the laser output frequency is stabilized using techniques developed by Hall. He has used stabilized lasers to improve experimental tests of special relativity. Since 1972, Hänsch has pioneered the spectroscopy of the hydrogen atom, measuring its $1s$-$2s$ transition and the Rydberg constant with unprecedented precision. The work of Hall, Hänsch, and many others has led to rapid advances in optical laser spectroscopy such that its precision is now overtaking that of microwave atomic clocks at a level of 1 part in 10^{15}.

In the 1970s, Hall and many others were involved in determining the speed of light with unprecedented accuracy by multiplying the frequency of a stabilized laser source, measured by comparison with the cesium atomic clock standard, by its wavelength, measured in terms of the spectroscopic definition of the meter. In 1983, the meter was redefined as the distance traveled by light in a stated fraction of a second, thereby fixing the speed of light and

making frequency and wavelength measurements equivalent. However, measuring optical frequencies around 10^{15} Hz by relating them to the cesium standard near 10^{10} Hz involved long chains of highly stabilized lasers combined with microwave sources. Since such measurements could be performed by only a few specialized laboratories, simpler ways were needed to accurately measure optical frequencies.

Such a method is provided by the frequency comb. A mode-locked laser emits a train of femtosecond pulses. Their Fourier transform is a series of peaks (the "teeth" of the comb), equally separated in frequency by the pulse repetition rate f_R of the laser, and f_R can be easily measured by detecting the optical pulse train on a fast photodiode. Around 1999, Hänsch realized that such lasers could be used to measure an optical frequency directly with respect to the cesium clock by comparing it with the comb frequencies. He and his colleagues demonstrated that these frequencies were equally separated with extreme precision. A problem arose in determining the common offset of the comb frequencies from multiples of f_R. A simple solution was found by Hall, with techniques later refined by himself and Hänsch. They involve broadening the comb spectrum in a highly nonlinear optical fiber to cover at least an octave of optical frequencies, and beating a doubled frequency against the original frequency to measure the frequency offset.

The frequency comb technique should enable future optical frequency measurements with a precision approaching 1 part in 10^{18}, leading to a new optical standard of time, and to improved satellite navigation systems, spectroscopic comparisons of matter and antimatter, and tests of possible changes in the fundamental "constants" of nature.

For background information *see* ATOMIC CLOCK; ATOMIC STRUCTURE AND SPECTRA; COHERENCE; FREQUENCY MEASUREMENT; LASER; LASER SPECTROSCOPY; LIGHT; PHYSICAL MEASUREMENT; QUANTUM ELECTRODYNAMICS; QUANTUM MECHANICS; RYDBERG CONSTANT; SQUEEZED QUANTUM STATES in the McGraw-Hill Encyclopedia of Science & Technology.

Physiology or medicine. J. Robin Warren (Royal Perth Hospital, Perth, Australia) shared the Nobel Prize for Physiology or Medicine with Barry J. Marshal (Heliocobacter pylori Research Laboratory, Queen Elizabeth II Medical Centre, Nedlands, Perth, Australia) for their discovery of the bacterium *Helicobacter pylori* and its role as the primary cause of peptic ulcer disease and gastritis, or inflammation of the stomach. The discoveries led to a simple drug regime that permanently cures what had been a chronic and painful illness. Their work also initiated a paradigm shift in the study of other chronic inflammatory diseases, from a focus restricted to genetic and lifestyle causes, to investigations into possible microbial origins of chronic inflammation and more debilitating disease.

As a pathologist in 1979, Warren noticed that colonies of spiral-shaped bacteria appeared in about half of patients' biopsies from the lower part of the stomach, or antrum, that he examined. At the time, peptic ulcer disease was blamed on lifestyle factors, such as stress and spicy foods, which were believed to increase acid production and irritate the stomach. Warren made the critical observation, however, that the previously unknown bacteria he had found were always associated with inflamed areas of the mucosa, or stomach lining, suggesting a relationship between the two.

As a young clinician at the same hospital, Marshall became interested in Warren's findings. Together the researchers first confirmed that Warren's initial observations held true across a larger sample size of biopsies from 100 patients. Marshall then attempted to culture the bacterium for further study, succeeding in growing the new species, later named *Helicobacter pylori*, in 1982. The bacterium infects only humans, although related members of its genus infect animals such as cats, ferrets, cheetahs, and pig-tailed macaques.

Helicobacter pylori is a gram-negative bacillus that uses its spiral shape and several whiplike flagella to burrow into the protective mucus layer coating the gastric epithelium of its human host. It can secrete the surface enzyme urease, which neutralizes and protects it from the caustic action of stomach acid. Once in contact with the epithelium, the bacillus produces a molecule called adhesin to bind itself to the epithelial cells.

People vary genetically in their susceptibility to infection with the bacteria, and individual strains of *H. pylori* also possess genetic variation in their virulence. Some strains can manipulate host cells into initiating the cascade of responses causing inflammation, for example. The bacteria are also able to suppress or fend off the host's immune response by various means, such as producing a toxin that kills immune system T-cells. Such abilities allow the bacteria to produce the chronic infections characteristic of the species.

In subsequent studies—all using basic techniques such as endoscopy, silver staining, and bacterial culture—Marshall and Warren found that the bacteria were present in the vast majority of patients suffering from gastritis or ulcers of the stomach or duodenum (the first section of the small intestine, immediately posterior to the stomach.) Despite accumulating evidence, however, the scientific community was slow to accept their idea that bacteria could cause gastric disease. At one point, aiming to sway doubting colleagues, Marshall dosed himself with live *H. pylori*, developed gastritis, and promptly cured his infection with antibiotics and bismuth.

It is now widely accepted that 50% of all people carry *H. pylori*, with the value much higher in nonindustrialized nations. The first infection is most often transmitted from mother to child at around 2 years of age, with an increase in infection rate of about 1% per year in the developed world. Only 10–15% of those infected with the bacteria develop gastritis or ulcers of the duodenum or stomach, however. Peptic ulcers most often occur in the duodenum, and inflammation is often restricted to the antrum. A more widespread infection that inflames the large,

upper portion of the stomach, or corpus, however, can eventually lead to stomach cancer.

Antibiotics and acid-secretion inhibitors remain the best treatment for the infection today, and little doubt remains in the estimate that 80–90% of ulcers are caused by the bacillus. The discovery of the role of *H. pylori* in gastric disease and the additional research it spawned have now begun to illuminate causative relationships between chronic infection, inflammation, and cancer.

For background information *see* ANTIBIOTIC; GASTROINTESTINAL TRACT DISORDERS; HELICOBACTER; INFLAMMATION; MEDICAL BACTERIOLOGY; STAIN (MICROBIOLOGY); ULCER; VIRULENCE in the McGraw-Hill Encyclopedia of Science & Technology.

Ocean birth through rifting and rupture

The formation of a new ocean basin begins with the rupture of a more than 100-km-thick (60-mi) continental plate, but only after millions of years of heating and stretching. The deep, fault-bounded valleys above the zones of stretching and heating are called continental rift zones. Fortunately for Earth's inhabitants, the rate of geological processes is extremely slow, and the rupture occurs in episodes separated by hundreds of years. Volcanic and earthquake activity in these episodes affects only a sector of the long narrow rift zones, producing a regular along-axis rift segmentation that is maintained in subsequent episodes. In September–October 2005, a 60-km-long (40-mi) segment of the East African rift system in Ethiopia experienced an intense period of localized deformation. Over 162 moderate [body wave magnitude (mb) > 4.5 to 3.9] earthquakes and an explosive volcanic eruption occurred over a 3-week period. Subsequent field, remote sensing, and modeling studies showed that molten rock (magma) was intruded into the plate beneath this 60-km-long rift segment, with cracks and faults forming in the brittle rocks above the narrow zones of magma injection. Thus, we directly observed a rare event—the injection of an approximately 8-m-wide (26-ft) column of magma that will quickly freeze to form a new strip of ocean floor. Continued monitoring of the activity will provide vital information for seismic and volcanic hazard mitigation in East Africa.

Rifting and rupture processes. Continents are regions with an approximately 40-km-thick (25-mi), buoyant, quartz-rich crust that is weaker than the 8-km-thick (5-mi) basalt-rich crust of oceanic plates. Both oceanic and continental crust are underlain by an approximately 100-km-thick (60-mi) layer of olivine-rich mantle. Tectonic plates comprise both of these layers (**Fig. 1**). The plates move at velocities of a few centimeters per year (the rate of fingernail growth) relative to one another in response to forces exerted by the convecting mantle beneath them, interactions between them, and lateral density contrasts within their interiors.

When an extensional (pulling) force is applied to a continental plate, long narrow basins flanked by mountains more than 1000 m (3300 ft) in height form as the plate stretches and thins. These seismically and sometimes volcanically active zones are called continental rift basins (Fig. 1*a*). Mantle rocks rise up under the thinning plate, melting as they decompress. The molten rock, or magma, is less dense than the surrounding rock, and this buoyancy force assists the plate stretching. With few exceptions, the buoyant melt rises up through the plate, transferring heat and weakening the plate. Some of this magma reaches the surface via fissures or volcanoes, and some solidi-

fies in narrow vertical sheets (dikes) or subhorizontal sheets (sills) within the crust. This basaltic magma intrusion process increases the density of the thinning crust, and the basins subside. Tensional forces on the plates sometimes cause brittle faulting of the crust, a process usually marked by earthquakes. Plate stretching is achieved through the combined processes of movement on dipping faults and dike intrusion.

If the forces are maintained for several millions of years, the plate can thin to a fraction of its original thickness. Few continental rift zones actually progress to rupture and createl new oceanic lithosphere. The forces required for rupture are more than 100 megapascals, the equivalent of 500 jumbo-jet engines pulling on a square meter of the plate, and must be maintained for millions of years. With progressive episodes of rifting, the distribution of faulting and magmatism localizes to a narrow zone within the ever-expanding rift basin. The mantle beneath the thinning plate rises to shallower levels, decompresses, and melts. The creation of a new column of basaltic crust sandwiched between the stretched continental crust signals the onset of sea-floor spreading, which may occur above sea level during the first few million years. The denser oceanic crust eventually will subside below sea level as it cools and thickens at a rate proportional to the square root of the time since formation.

The geological record of the rifting and rupture process is preserved along the edges of many ocean basins, such as the eastern and western coasts of the Atlantic Ocean. The ancient sites of plate rupture along these passive continental margins are masked by more than 10 km (6 mi) of sedimentary rocks, and Earth scientists have only recently recognized that most margins experienced voluminous volcanic eruptions prior to plate rupture. An alternative approach to the study of continental breakup and the birth of an ocean basin is to study continental rift zones that are in the process of rupture.

Afar Depression. The East African rift system in the Afar Depression is one of few places worldwide where the process of continental rupture is occurring on land, affording an opportunity to directly observe and quantify the plate separation process. Large tracts of the depression lie at or below sea level, attesting to a long history of stretching and magma intrusion. Another unique feature of this rift system is its position above or near a large deep-seated zone of anomalously hot mantle rocks that both heat the overlying plates and provide a long-lived supply of melt. Within the Afar Depression, three rift systems intersect. The Red Sea rift separates stable Africa (Nubian plate) from the Arabian plate. The Gulf of Aden rift separates eastern Africa (Somalian plate) from the Arabian plate. And the Ethiopian rift system separates the Nubian and Somalian plates (**Fig. 2**, inset). Rupture has already occurred in the Asal rift, the westernmost arm of the Gulf of Aden rift, which experienced a volcanic eruption and several earthquakes in 1978.

Little was known about the rates and timing of events elsewhere in the Afar Depression, partly

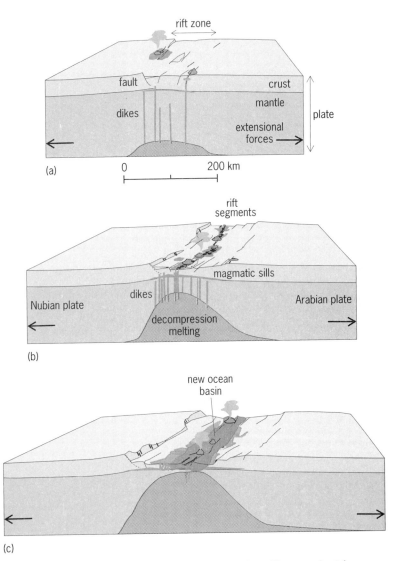

Fig. 1. Three-stage model for the rupture of continental plates. The approximately 150-km-thick (90-mi) continental plates comprise two layers: a weak quartz-rich crust and a stronger olivine-rich mantle. If the forces within, beneath, or at the edges of the plates are large and extensional, the layers will begin to stretch and thin. (*a*) The layered plate begins to thin through brittle and ductile deformation. The mantle rocks beneath the plate rise to replace the thinning lithosphere, and some decompression melting of the mantle may occur. A small volume may reach shallow crustal levels (volcanoes), and some may be accreted at the compositional boundary between the crust and mantle lithosphere. Deformation in the brittle crust occurs by slip along dipping surface, called normal faults. (*b*) With increasing time and strain, the lithosphere will continue to thin by faulting and ductile deformation. Molten rock rises through the heated and weakened plate through vertical cracks to the surface. Most melt does not reach the surface, but freezes in vertical cracks (dikes) that are perpendicular to the extension direction, or in thin sheets (sills) between layers. (*c*) Sea-floor spreading. The thinned and heated lithosphere is now too weak to support the plate-pulling stresses, and melt rises to the surface, producing columns of new oceanic crust. The stretched crust loses heat and subsides; it is now a passive continental margin.

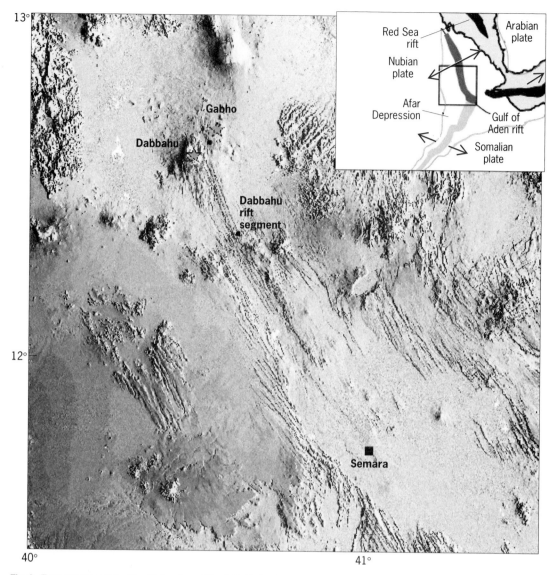

Fig. 2. Topographic relief of the 60-km-long (40-mi) Dabbahu rift segment within the Afar Depression. Inset shows directions of plate divergence between the stable African (Nubian), Arabian, and Somalian plates. (*After T. Wright et al., 2006*)

because of the harsh desert environment and limited access to it. Globally we have a short time sequence of observations, with written records of earthquakes and volcanic eruptions spanning just a few centuries in much of Africa, and instrumental recordings spanning about the past 100 years. Thus, with rifting cycles of centuries, some aspects of the rifting process remain undocumented.

2005 Dabbahu rift episode. The September-October volcano-tectonic crisis in the Afar Depression is the largest rifting sequence to have occurred on land, since the 10-year-long Krafla, Iceland, rifting episode, which occurred in a well-developed mid-ocean ridge. The spatial extent and amount of opening are comparable to the Krafla event (**Fig. 3**). Earthquakes centered on a discrete rift segment, the Dabbahu (Boina) segment, began on September 14, 2005, with a cluster of 98 damaging earthquakes occurring on September 24 and 25 (Fig. 2). On September 26, a volcanic vent opened between two strato-volcanoes

near the northern end of the Dabbahu rift segment (**Fig. 4**). Local pastoralists watched as some scarps (cliffs) moved up to a meter; field teams documented 5 m (16 ft) of offset on faults near the center of the Dabbahu segment (**Fig. 5**). By comparing satellite radar images acquired before and after the earthquake swarms, we have detected deformation along and across a 60-km-long (40-mi) segment of the rift. Although only a small volume of magma was erupted at the surface, the satellite and surface data showing up to 1-m-wide (3-ft) open fissures and faults cannot be explained by the energy released in the earthquakes alone. Buoyant magma injected into the plate provides rift opening with few earthquakes; dike injection can account for the energy deficit. The rifting episode was probably triggered by the injection of a 60-km-long and approximately 8-m-wide (20-ft) sheet of basaltic magma (dike), with faults slipping above the dikes. The basalts did not reach the surface in this episode.

Fig. 3. View south-southeast along one of many open fissures near the central part of the Dabbahu segment. Some cracks are open fissures, and some are faults with vertical displacements of up to 6 m (20 ft). These formed during the September-October 2005 rifting episode. (*Photo by C. Ebinger*)

Fig. 4. View north toward the Gabho volcano showing the ∼600-m-long (2000-ft) volcanic vent that opened in blasts beginning September 26, 2005. On the right (eastern) side of the image, open fissures and normal faults cut the ashes erupted from the vent. The more than 2-km-long (1.2-mi) fault at the right corner slipped over 0.5 m (1.6 ft) on September 26. (*Photo by E. Baker, Royal Holloway University of London*)

Fig. 5. View north-northwest from the central part of the eastern flank of the Dabbahu rift segment. The Dabbahu volcano is about 30 km (18 mi) from this site. The steep scarps were formed by many episodes of slip along dipping fault planes; some faults show more than 3 m (10 ft) of movement in the September-October episode. The faults displaced basaltic lavas (dark rocks) and small pockets of windblown ash and dust (white rocks). (*Photo by C. Ebinger*)

Outlook. Has the Dabbahu rift segment episode stopped? Preliminary results from Global Positioning System (GPS) measurements show continuing deformation in the Dabbahu segment. We draw insights from the 10-year Krafla, Iceland, sequence. A smaller dike intrusion in 1975 marked the beginning of the Krafla rifting episode; as in Ethiopia, the magma did not reach the surface. It was followed by 20 more dike events, many of which produced curtains of fire as they erupted. By analogy to Krafla, there may be more to come over the next few years. Ethiopian geoscientists and international colleagues are monitoring earthquake activity, ground deformation, vent emissions, and satellite images to understand the longer-term response of the plates to this intense applied stress, as well as to detect any resurgence in potentially explosive volcanic activity.

For background information *see* AFRICA; EARTH CRUST; EARTH INTERIOR; EARTHQUAKE; FAULT AND FAULT STRUCTURES; MAGMA; PLATE TECTONICS; REMOTE SENSING; VOLCANO in the McGraw-Hill Encyclopedia of Science & Technology.

Cynthia Ebinger; Gezahegn Yirgu; Tim Wright; Eric Calais; Elias Lewi

Bibliography. A. Abdallah et al., Relevance of Afar seismicity and volcanism to the mechanics of accreting plate boundaries, *Nature*, 282:17–23, 1979; A. Rubin and D. Pollard, Dike-induced faulting in rift zones in Iceland and Afar, *Geology*, 16:413– 417, 1988; F. Sigmundsson, *Iceland Geodynamics: Crustal Deformation and Divergent Plate Tectonics*, Springer-Praxis, 2006; T. Wright et al., Magma-maintained rift segmentation at continental rupture in the 2005 Afar dyking episode, *Nature*, 442:291– 294, 2006.

Optical character recognition

Optical character recognition (OCR) systems convert digital images of text into symbolic strings (for example, ASCII). Digital images of text are typically created by scanning paper documents containing printed or handwritten text, from video images of scenes with text, or by digital pens. The OCR-generated symbolic form of the paper documents allows multiple users to simultaneously access, search, edit, and transmit the contents of these pages. In fact, with the help of machine translation and speech synthesis systems, the information in paper documents can now be accessed by people across language boundaries and by those with visual impairment.

Anatomy of system. A typical OCR system analyzes a document image in a sequence of steps. During the preprocessing stage, the document is "straightened" by estimating the rotation angle of the document and then transforming the original document image by the estimated rotation angle. Next, a suite of image processing filters are applied to "clean" the image by removing pixel noise or filling "holes." At this stage, the multilevel gray-scale image is "thresholded" (based on whether a pixel exceeds a certain darkness) to create a binary black/white image if the following stages work on binary images; otherwise, the image is left as gray-scale.

After the document image has been preprocessed, gross physical entities, such as the number and location of the columns, paragraphs, lines, footers and headers of the document, are extracted. This is done by computing the distribution of the distances between blobs of black pixels, their linear alignment, and their size distribution. Next, the system extracts rough logical structure attributes, such as reading order (which zone should be read after which), section/subsection nesting, identifying caption/figure pairs, identifying tables, and list items. Given scanned images of a multipage document, the layout analysis module ideally should be able to create a hierarchical representation of all the entities or items in the document.

Once gray-scale or binary images of the individual lines of text in a document page are obtained, a recognition algorithm converts each text-line image into a sequence of symbols. The recognition algorithm typically uses statistical language models, dictionaries of words in the language the document is written in, and classifiers based on local image features to produce the symbolic text.

The above description is that of OCR systems during recognition mode. However, before one can use an OCR algorithm, the algorithm has to be trained. Training entails collecting a representative sample of document images and their corresponding manually generated symbolic text. This training dataset is then used to learn the parameters of the OCR system.

Applications of technology. Some OCR technologies, such as bar-code readers, form readers, signature verification, postal address recognition, and printed text recognition, are very mature. The newer OCR applications are fueled by the Internet, including book sales by searching inside the book, and making historical documents and books available.

Bar-code readers are ubiquitous in retail stores. Bar codes encode the identity of a product using a sequence of black and white stripes of different width. A bar-code sensor is swiped across the black-and-white pattern, and the recognition system extracts the pattern information from the varying light intensity across the bar code.

Form readers extract textual fields from a scanned form (such as tax forms). The system uses an "unfilled" form as a template, aligns the new scanned form and the template, and locates the filled-in text. The textual fields are then sent to a "recognizer" to convert the text into symbolic form.

The number of checks that need to be processed by banks daily is so huge that the task cannot be done by humans. OCR systems have been used by banks for a long time for automated signature verification and extracting the payee name and money amount. Check processing systems extract geometric features from an individual's signature and use it to train a model. The model is then used to verify that the signature read is similar to the signatures used in training. The name and currency amount are extracted by a process similar to that used by form readers.

All personal digital assistants (PDAs) and cell phones now use technology similar to "graffiti," where users can enter information by drawing symbols made up of simple strokes. The simple symbol recognition technology can be seen in most systems requiring handwritten data entry.

Another success story is that of postal address recognition. Most countries, including the United States, use automated printed postal address recognition to route physical mail. The OCR systems use prior knowledge of the possible street names within specific zip codes to improve the recognition accuracy.

Commercial general-purpose multilingual OCR systems, like OmniPage and ABBYY, allow users to scan magazine articles in a variety of languages and easily convert them to symbolic text. These are available in most Latin-based languages as well as Chinese, Korean, Japanese, and other non-Latin languages.

In pre-Web days, users searched for books using a catalog. This has changed, however, as Amazon has scanned and OCR'd millions of books, allowing potential buyers to search for books related to specific terms. The search-engine company Google has a similar service. In addition, Google is currently in the process of scanning books in the Library of Congress and making them available to the public over the Web in multiple languages.

CAPCHA is a new authentication technology that capitalizes on the robustness of the human visual system and the weakness of OCR systems. The idea is simple. An image of a text string is presented to a user trying to create a new account, and then the user is asked to type in the corresponding string. The image of the text string image is warped so that only humans can recognize it and not automated systems. Most online systems currently use this to authenticate new users and logins.

Functionality of systems. While speed and memory requirements are standard metrics by which computing systems are evaluated and compared, the possibility of incorrect recognition of text calls for another dimension of evaluation—accuracy. But since each OCR application is very different and uses very different context, each application typically needs to be evaluated differently.

To evaluate the accuracy of general printed-text OCR systems, a set of document image pages is collected, and the correct symbolic text corresponding to those documents is typed in by humans. Then the OCR system is run on the document images, and the output text is compared to the human-entered text. Various types of errors, such as deletion, insertion, and substitution, are computed and reported.

When comparing the technology across languages using the above methodology, one comes across a problem. Since the performance results are tied so intimately to the dataset used for evaluation, which document collection should be used in each of the languages? Ideally, for the evaluation results to be comparable across languages, the evaluation documents should be translations of documents in one of the languages. Creating such a corpus can be laborious, expensive, and error-prone. Recently, researchers have suggested using the Bible as a linguistic corpus for evaluating OCR systems across languages. In this study, it was found that OCR systems perform well in Latin-based languages and poorly in languages like Arabic and Chinese due to the presence of connected characters or large vocabularies (see **illustration**).

Experimental methodology similar to that outlined above has been used for assessing OCR systems for numeral recognition, address block recognition, forms processing, and noisy text retrieval.

Current research problems. While the problems of OCR have been worked on for many decades, these systems break down if the image quality is very low, as is the case with a low-resolution fax or multiple generations of a photocopied document. While humans are able to extract meaningful information from such documents, OCR systems perform abysmally.

The availability of new page layout tools, such as PageMaker and InDesign, has created yet another problem that current OCR systems do not handle well. These new page layout products allow users to create multilayered documents with text of different fonts and color in each layer. In such documents, separating the foreground from background and the context (line/word) for each character is difficult. Analyzing these documents is still challenging for OCR systems.

Since the output of OCR systems is noisy, any information system that uses the output of OCR systems should be able to handle noisy text. For example, information retrieval systems should be able to index various versions of the word *cloud*, such as *doud*, *cloucl*, *doucl*, which an OCR system might produce. Similarly, any computational linguistic module that extracts names of people or places from text should not break down when presented with noisy output from OCR systems. While information retrieval systems have been able to adapt to handle noisy text to some degree, linguistic systems are still very brittle. Interestingly, noisy text can be produced not just by OCR systems but also by human-generated text on the Web, which can have numerous spelling and grammatical errors.

Street signs, car license plates, billboards, store names, and product labels in a grocery store are examples of text fragments that we all see and read. While we can segment and read the text easily, OCR systems cannot. Solving this problem can help in automated navigation and foreign country travel, and help the visually impaired.

One old problem that still defies researchers is that of recognizing handwritten text. Personal letters and memos, historical documents, annotations, are all handwritten. There is a lot of variability in the handwritten text from one person to another. The OCR system recognition accuracy for handwritten documents is far below that of current recognition accuracy of printed text.

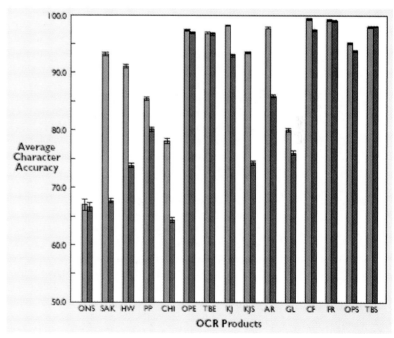

Language	OCR Product Name	Abbr. Product Name
Arabic	Onset2	ONS
	Sakhr3	SAK
Chinese	Hw99	HW
	PenPower2.0	PP
	CHIOCR	CHI
English	OmniPage8.0	OPE
	TextBridge98	TBE
Japanese	KanjiOCR2.0	KJ
	KanjiScan1.0	KJS
Korean	Armi4.0	AR
	Glnun97	GL
Russian	Cuneiform2000	CF
	FineReader4	FR
Spanish	OmniPage8.0	OPS
	TextBridge98	TBS

(a)

(b)

Multilingual OCR product evaluation on noise-free images and degraded images. (*a*) Product names, their abbreviated names, and the text language they recognize. (*b*) Plot showing the average character recognition accuracy for each product. The light bar represents accuracy for noise-free images, and the dark bar represents accuracy for degraded images. The values are in percentage. (*T. Kanungo et al., The Bible and optical character recognition, Communications of the ACM, vol. 48, no. 6, pp 124–130, 2005*)

Currently a user cannot use a formal language to specify or describe a specific layout structure/template to an OCR system. For example, if a class of documents has a similar physical look (number columns, location of footers and headers, how title and authors are printed), there is no way to communicate this information to the system, and thus there is no way of improving the accuracy of the layout-understanding module.

Outlook. Much of the progress in the database community is due to availability of open source systems like PostgreSQL and MySQL. Besides being robust database systems that are used worldwide, these systems can be used for research, experiments, and testing new algorithms. Open-source OCR systems are still nowhere close to commercial products like ABBYY and OmniPage. Availability of an open-source, retargetable, multilingual OCR system will allow researchers to push their ideas into use faster and simultaneously allow the end users to benefit from the research.

The Web and the Internet are going ahead with the notion of Web services and service-oriented architecture. The idea is that on the Web there will be service providers, service consumers, and others who will compose new services from basic services. To make OCR-as-a-service a possibility, standards are needed. Currently, each OCR system produces output in standard ASCII or PDF formats. However, the output of an OCR system can be quite complex with text regions, captions, and hyperlinks. Currently there are no standards for OCR output. Similarly, OCR systems have numerous parameters that need to be set to generate a specific output. To have a service where results can be reproduced, the OCR service provider will have to get these OCR parameters, in addition to the document images, before it can deliver the output. Thus, standards will have to be established to exchange the output between OCR service providers and consumers.

Scientific progress can happen only if we regularly monitor the state of the technology. This monitoring should be quantitative like that done by NIST/TREC for the information retrieval community. While there have been attempts by various organizations to run such regular tests, most of them have been short-lived.

For background information *see* ALGORITHM; CHARACTER RECOGNITION; COMPUTER VISION; IMAGE PROCESSING; INTERNET; PROGRAMMING LANGUAGES; WORLD WIDE WEB in the McGraw-Hill Encyclopedia of Science & Technology.

Tapas Kanungo

Bibliography. T. Kanungo et al., The Bible and optical character recognition, *Commun. ACM*, 48(6):124–130, 2005; T. Kanungo et al., A statistical, non-parametric methdology for document degradation model validation, *IEEE Trans. Pattern Anal. Machine Intell.*, 22:1209–1223, 2000; R. Kasturi and L. O'Gorman, *Document Image Analysis*, IEEE Computer Society Press, 1995; S. Mao and T. Kanungo, Stochastic language models for style-directed layout analysis of document images, *IEEE Trans. Image Process.*, 12:583–596, 2003.

Oxygen and evolution of complex life

Without having the theoretical framework establishing its atomic nature, adroit experimentalists such as Joseph Priestley, Daniel Rutherford, and Antoine Lavoisier had laid foundation to the great importance of dioxygen (O_2) [referred to herein simply as oxygen] by the end of the eighteenth century. Vital air—*l'air vital*, as frequently referenced in Lavoisier's notebooks—was known even then to be required for life as we observe it, as, with a few notable exceptions, all complex macroscopic life breathes oxygen. Although anaerobic fermentation may have been practiced since the dawn of civilization, it was not until Louis Pasteur's discovery of anaerobic microorganisms that the necessary role of oxygen was rigorously questioned. Insights from the geological record and, more recently, the reconstruction of the evolutionary history of genes and genomes reveal that anaerobes are not simply oddities of adaptation to unique environments such as the oxygen-free guts of ruminants, but are the modern vestiges of a biosphere that was once utterly anoxic (that is, without oxygen) and dominated by anaerobes.

The paramount importance of oxygen to modern organisms manifests from several of its key features. Atomic oxygen is the most abundant element on Earth, third overall in the universe, and is common to almost every class of biological molecules. The electronegativity of oxygen confers a degree of charge polarity upon otherwise nonpolar hydrocarbons, increasing their solubility in water and improving their ability to interact and react. O_2 is also notable for its strength as an oxidant (that is, an oxidizing agent)—its energetic potential for participating in redox reactions is surpassed only by a few other relatively common compounds such as chlorine (Cl_2) and fluorine (F_2). Of particular importance to life is the fact that oxygen is essentially an oxidized form of water. Whereas strong oxidants such as Cl_2 and F_2 are reduced to strong acids, oxygen is reduced to an innocuous form, water.

History of oxygen on earth. On human time scales, the Earth's atmosphere appears roughly balanced at 78.1% N_2, 20.9% O_2, 0.93% Ar, and 0.037% CO_2. Of course, this stasis is only short-term, as attested to by well-publicized anthropogenic increases in CO_2 and methane. Thus one might ask how the atmosphere has changed through geologic time and what effects these changes have had not only on climate and ultraviolet (UV) fluxes but on the organisms inhabiting ancient environments. Answering these questions requires peering back in time—that is, studying chemical and biological signatures that have been preserved in sediments, rocks, and ice sheets. For instance, various types of elements are transformed between reduced and oxidized states when oxygen is present, so their occurrence in minerals of known age can be used as a gauge of the amount of free oxygen in the atmosphere. An oxygenated atmosphere also results in distinct patterns of isotope fractionation—the enrichment of one stable isotope of an element over another in biogeochemical cycles—and also plays an important role in understanding when and how the age of oxygen began. Most importantly, because oxygen-induced changes were global in scale, multiple indicators, both isotopic and compositional, are actively being unearthed and unraveled.

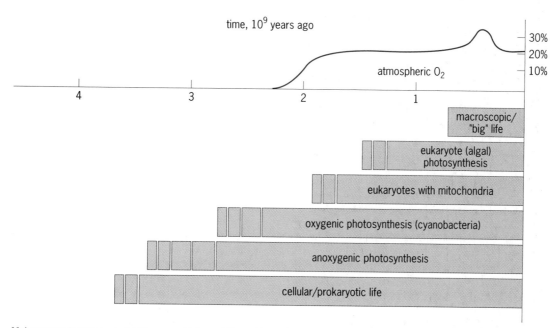

Major events in the nearly 4-billion-year history of life on Earth. Oxygen first became abundant about 2.4 to 2 billion years ago, although it did not immediately reach present atmospheric levels. Note that dating ancient evolutionary events has associated uncertainties, roughly indicated by broken bars, due to gaps or ambiguities in the geological record.

These multiple lines of evidence suggest that the transition from an anoxic to an oxic atmosphere began between 2.4 to 2 billion years ago (see **illustration**), a climactic time in the Earth's history as the oxygen buildup may have stimulated changes in the composition of the oceans as well as global glaciations. Remarkably, it is known that such massive amounts of oxygen could only have been produced on the early Earth biologically—specifically by the process of oxygenic (oxygen-producing) photosynthesis. Photosynthesis is a mechanism by which the energy of a photon can be used to drive chemical reactions, such as the generation of high-energy reduced compounds. Broadly, there are two types of photosynthetic organisms: oxygenic, whose photosynthetic reaction centers are energetic enough to strip electrons from water, converting it to O_2 in the process; and anoxygenic, whose reaction centers are not so energetic and so thereby rely on electron-donating molecules, such as hydrogen sulfide and reduced iron, that are not so difficult to oxidize as water. Only the former are capable of producing the oxygen that first contaminated the early atmosphere, and upon which aerobic life is dependent. Furthermore, it was not until oxygen first pervaded the atmosphere that a UV-screening ozone layer was formed, which had immediate consequences for adaptation to shallow marine and terrestrial environments.

At this point, evolutionary comparisons of photosynthetic organisms give some insight into this early era, supporting the hypothesis that anoxygenic photosynthesis was the evolutionary predecessor to oxygenic photosynthesis. In fact, the two processes have remarkable similarities in the reaction centers where light-to-chemical energy transduction takes place, as well as in the types and orientations of pigments within the proteins that make up their photosynthetic apparatuses. Furthermore, the only group of bacteria that are capable of oxygenic photosynthesis are the cyanobacteria (which, paralleling the endosymbiotic origin of mitochondria discussed below, also gave rise to chloroplasts and the ability of eukaryotic algae and plants to do photosynthesis). The geological evidence for oxygenic photosynthesis suggests that the ancestors of modern cyanobacteria had "invented" oxygenic photosynthesis by 2 billion years ago—and quite likely several hundred million years earlier, as the oxygen they produced would have been buffered by the excess of reductants in the early atmosphere and oceans.

Recent evidence suggests that, despite the atmosphere and shallow oceans becoming oxic not long after the aforementioned converging evidence at 2 billion years ago, the deep oceans may have remained anoxic for another 1.2 billion years. This is argued based on high sulfide concentrations inferred for the deep oceans during this period, which would have acted as a sink for O_2. The resulting stratified ocean is analogous to the modern Black Sea, wherein the shallow waters (less than 80 m or 262 ft) are well mixed and oxygenated, but the deep waters are anoxic and rich in sulfide—a bastion for anaerobic life and a preservative against the oxidative decay of sunken ships. Because oxygen greatly accelerates redox cycling of important elements, the oxidation of the deep oceans that is thought to have occurred about 800 million years ago may have released a vast reservoir of previously "locked up" elements such as phosphorus and sulfur, possibly contributing to the ensuing diversification of life leading up to the Cambrian explosion.

Oxygen's influence on biology. The best indicator of what life was like prior to the rise of oxygen comes from modern anaerobic microbial communities. Despite the pervasiveness of oxygen, oxygen-free environments are not difficult to find; the guts of many animals, including humans, harbor anoxic zones and even entire compartments, as in the case of ruminants such as cows. Fermentative processes in the food industry rely crucially on the absence of oxygen. In nature, finding anoxic zones is often as simple as digging down into sediments, where arrival is often greeted by the acrid aroma of sulfide. Each of these environments is a niche for obligate anaerobes, organisms who are quickly killed by even low concentrations of oxygen and whose ancestors have been evading O_2 for some 2 billion years. Most obligate anaerobes are prokaryotes, although a few groups of single-celled (protistan) and fungal eukaryotes have adapted to oxygen-free niches.

To a small extent, anaerobes can mitigate accidental exposure to oxygen and its especially damaging by-products using enzymes such as catalase, superoxide dismutase, and dioxygen reductase. However, these protective enzymes probably arose after oxygen became widespread, arguing that the initial appearance of oxygen in the Earth's atmosphere and oceans would have been nothing short of catastrophic for life 2 billion years ago. Organisms would have been killed outright, would have retreated into anoxic environments, or would have been under enormous evolutionary selection pressure to develop mechanisms by which to detoxify oxygen and its by-products and repair oxidative damage.

Aerobic organisms express high levels of dioxygen reductases, such as cytochrome oxidase, that rapidly and specifically reduce O_2 to water. This reaction is coupled through the electron transport chain to the oxidation of organic matter, resulting in an energy yield the likes of which were not possible prior to the availability of oxygen [up to a 20-fold increase in adenosine triphosphate (ATP) generation over anaerobic fermentation]. Lynn Margulis has championed the hypothesis that the development of eukaryotic cells depended crucially on this energetic leap forward. Not having any intrinsic ability to cope with oxygen, ancestral anaerobic eukaryotes developed symbiotic relationships with aerobic bacteria, which would ultimately become inseparable intracellular oxygen-respiring structures now known as mitochondria. This energetic switch from low-yield fermentative reactions to high-energy aerobic respiration has been argued as a requirement for life to grow large and complex, although aerobes still retain the ability to do fermentative reactions when oxygen is limiting (such as during strenuous exercise, leading to lactic acid formation).

Moreover, oxygen had a remarkable influence on aerobes at the biochemical level. O_2 is explicitly required in the biosynthesis of many molecules on which humans and other complex organisms are dependent, including vitamin A, cholesterol and its steroid derivatives, and vitamin C (although the ability to synthesize vitamin C has been lost in primates and guinea pigs, rendering them dependent on fruits and vegetables). The importance of oxygen both to the energetics and to the biochemistry of life on Earth underscores its use as a proxy for detecting planets capable of supporting large complex life forms. NASA has identified it as a key biosignature, and ozone—an easier-to-detect analog of O_2—is a primary target of its upcoming *Terrestrial Planet Finder* orbiters.

As alluded to in the discussion of elemental markers for the rise of oxygen, the bioavailability of many important compounds and metals has been altered as the Earth has become more oxidized. Several trace metals of importance to life have undergone sharp changes in solubility. Molybdenum, copper, zinc, and vanadium are more soluble in their oxidized than in their reduced forms. Many so-called metalloenzymes, dependent on these metals for their catalytic function, probably evolved only after the metal became widely available due to favorable solubility. Conversely, iron is much less soluble in its oxidized form. While its necessity in many enzymes, such as heme-containing and electron transfer proteins, was established early in the evolution of life, organisms in many modern environments, such as the oceans, are iron-limited. Indeed, one of the ways in which oceanic photosynthetic organisms have circumvented this iron limitation has been through the evolution of redox transfer proteins that utilize more readily available copper than iron.

Outlook. Oxygen has played a pivotal role in the development of life on this planet and may plausibly be key in the evolution of complex life elsewhere in the universe. The oxygen in the Earth's atmosphere results almost entirely from a single biological process—oxidation of water by oxygenic photosynthesis—which, to date, can only be poorly replicated by human engineering. Examination of the geological record suggests that the Earth's atmosphere is quite dynamic and has been influenced and profoundly altered by the very life it helps support. Although atmospheric oxygen levels have been closely coupled to the carbon cycle, the possible effects of recent increases in anthropogenic inputs to atmospheric CO_2 are very poorly understood. For example, 300 million years ago, atmospheric oxygen may have reached concentrations 50% or more above present levels, resulting in supersized biota but no doubt with dramatic consequences on intracellular oxidative damage and the proliferation of wildfires on an unprecedented scale. Alternatively, the minimal growth requirements and adaptability of cyanobacteria and algae have attracted attention as oxygen factories that could be used in long-term space colonization or possibly even in terra-forming planets (that is, making planets habitable for humans), provided there is a steady supply of light, carbon dioxide, and water. Although technically out of reach at present, these far-reaching ideas suggest that the role of oxygen and oxygen-producing organisms in transforming environments may just be getting off the ground.

For background information *see* ANIMAL EVOLUTION; CYANOBACTERIA; MITOCHONDRIA; ORGANIC

EVOLUTION; OXYGEN; OXYGEN TOXICITY; PHOTO-SYNTHESIS; PLANT EVOLUTION in the McGraw-Hill Encyclopedia of Science & Technology.

Jason Raymond

Bibliography. R. E. Blankenship, *Molecular Mechanisms of Photosynthesis*, Blackwell Science, 2002; N. Lane, *Oxygen: The Molecule That Made the World*, Oxford University Press, 2004; L. Margulis, *Symbiotic Planet: A New Look at Evolution*, Basic Books, 2000; D. T. Sawyer, *Oxygen Chemistry*, Oxford University Press, 1991; R. J. P. Williams and J. J. R. Frausto Da Silva, *The Natural Selection of the Chemical Elements*, Oxford University Press, 1996.

P-bodies

P-bodies, or processing bodies, are discrete cytoplasmic foci (chief centers of a morbid process) composed of messenger ribonucleic acid–protein complexes containing a subset of proteins involved in mRNA decay. Concentrated foci of mRNA decay enzymes were initially reported in mammalian cells and subsequently were further expanded in the yeast *Saccharomyces cerevisiae*. Recent studies indicate that P-bodies serve a dual functional role. In addition to mRNA decay, P-bodies are sites of mRNA storage for translationally silenced mRNAs, including noncoding microRNA-mediated translational silencing in mammals.

Gene expression is initiated in the cell nucleus to produce preliminary RNA transcripts that undergo extensive processing events to generate the mature mRNAs. Processed mRNAs exit the nucleus to serve as substrates for the translation machinery to synthesize proteins in the cytoplasm. An important step in the regulation of gene expression is the nonrandom and systematic degradation of the mRNA. An mRNA strand harbors elements to protect its termini from nonspecific degradation. It contains an unusually linked guanosine residue at the 5′ end (or beginning) referred to as the 5′ cap, and adenosine residues at the opposite end (3′ end) referred to as the polyadenylated [poly(A)] tail. Degradation of an mRNA generally proceeds through two distinct exoribonucleolytic pathways that require the removal of the poly(A) tail (deadenylation) as a prerequisite step. In the 3′ decay pathway, degradation of the mRNA continues from the 3′ end until occurs a residual cap structure, which is subsequently hydrolyzed by a scavenger decapping enzyme, DcpS. In the 5′ decay pathway, the deadenylated mRNA is decapped by the Dcp2 decapping enzyme exposing the 5′ end of the mRNA to the exoribonuclease Xrn1 that degrades the body of the decapped mRNA in the 5′ to 3′ direction.

In addition to the exonucleolytic pathways, mRNA decay can initiate through an internal cleavage by sequence-specific endoribonucleases or by a microRNA-guided endonuclease. This initial cleavage step generates RNA substrates for the exonucleolytic decay pathways. MicroRNAs are short

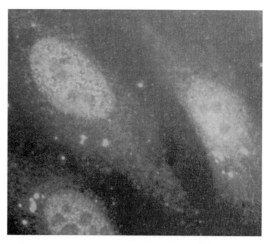

P-bodies are discrete cytoplasmic foci. The P-body component Dcp2 was immunostained to localize P-bodies in the cytoplasm of human cancer cells.

noncoding RNAs that generally hybridize to their target mRNA with partial complementary, and either promote translational silencing or direct cleavage of the mRNA by the RNA-induced silencing complex (RISC). It is becoming evident that utilization of the endonucleolytic pathway for the regulation of mRNA stability is more prevalent than previously appreciated. Interestingly, many of the factors involved in the 5′ to 3′ decay pathway, including Dcp2, Xrn1, and components of the RISC, are contained in P-bodies. In contrast, enzymes involved in the 3′ decay pathway are not apparent in P-bodies, indicating that mRNA decay is compartmentalized.

Components. P-bodies are concentrated foci present in the cytoplasm of eukaryotic cells (see **illustration**). They consist of mainly cytoplasmic proteins assembled on mRNAs destined for decay or translational silencing. They were initially shown to contain Dcp2, Xrn1, and a protein that facilitates decapping in yeast, Dcp1. This list has subsequently significantly expanded (see **table**). P-bodies lack a membrane and are dependent on a subset of proteins, including GW182 and the eIF4E transporter eIF4-T or the decapping stimulator Ge-1/Hedls, as well as mRNAs for their structural integrity. However, the protein components are dynamic and in constant flux, moving in and out of P-bodies, and consequently the size and number of P-bodies are also variable. In yeast, there are fewer and smaller P-bodies during logarithmic (exponential) growth (the period during which cells divide at a constant rate), while they are relatively larger and in greater numbers during stationary growth (the period following logarithmic growth when the number of viable organisms remains constant for a time). Therefore, they vary with cellular growth and mRNA translation state.

Functional role in mRNA decay. The presence of mRNA decapping enzymes and the exoribonuclease Xrn1 in P-bodies is indicative that these cytoplasmic structures are active sites of mRNA degradation. Following the completion of translation, an mRNA can be shunted to P-bodies for eventual decay. A block of

Proteins known to be present in mammalian P-bodies			
5′ to 3′ decay	RNA-induced silencing	Translation	Others
Dcp1a	Ago1	eIF4E	GW182
Dcp1b	Ago2	eIF4E-T	RAP55
Dcp2	Ago3		TTP
Edc3	Ago4		BRF1
Ge-1/Hedls			CPEB1
Lsm1-7			FAST
rck/p54			MOV10
Ccr4			TNRC6B
Xrn1			APOBEC3F
			APOBEC3G
			Upf1
			SMG5
			SMG7

translation elongation to trap mRNAs on translating polysomes (a polysome is a complex of ribosomes bound together by a single mRNA molecule) inhibits their entry into P-bodies and leads to their rapid disappearance in both human and yeast cells. Similarly, a block in the initial step of mRNA decay, deadenylation, can also lead to their disappearance. Conversely, inhibiting the subsequent decapping and 5′ to 3′ exonuclease activities results in an accumulation of mRNA decay intermediates and consequently an increase in the size and number of P-bodies. Collectively, compelling evidence indicates that mRNAs that have completed their translational life cycle are substrates for mRNA decay in P-bodies. Therefore a functional link exists between mRNA translation, P-bodies, and mRNA decay.

Functional role in translational silencing/mRNA storage. It has become evident recently that not all mRNAs within P-bodies are destined for mRNA decay as was initially thought. Several reports have now demonstrated that P-bodies can also serve as sites for the accumulation of translationally silenced mRNAs. In mammals, microRNAs contained within the RISC can direct transcript-specific translational silencing. A functional link between P-bodies and microRNA-mediated silencing was revealed by the presence of key RISC proteins, Ago1 and Ago2, as well as both the microRNAs and corresponding target mRNAs within P-bodies. Furthermore, cellular manipulations that disrupted P-body assembly (downregulation of GW182) also abrogated microRNA-mediated translation silencing. Additional evidence to support a storage role of translationally silenced mRNAs in P-bodies was provided in yeast. Upon nutrient deprivation, mRNA translation is inhibited, and the mRNA is transported to P-bodies. Remarkably, upon improved conditions conducive to cell growth, the silenced mRNA exits the cytoplasmic foci and returns to polysomes to resume translation. Therefore, P-bodies can serve both mRNA silencing and decay functions. What distinguishes the two fates for the mRNA is unknown and remains an active area of research.

The transient translational repression of mRNAs is reminiscent of stress granules. Stress granules are cellular substructures that form to sequester mRNAs and prevent their translation during a stress response. These structures only form under various cellular stress conditions, and they dissociate and release the mRNA to the translation apparatus upon return of the cell to normal conditions. Stress granules are larger and stationary aggregates, while P-bodies are smaller and mobile. These structures have many proteins in common. Moreover, under oxidative stress conditions in human cells, P-bodies can transiently interact with stress granules. These interactions enable the exchange of mRNA and protein components between the two compartments and perhaps coordinate the fate of mRNAs following an extracellular stimulus.

Potential role in human disease. A link between P-bodies and human disease first emerged from the discovery that patients suffering from a motor and sensory polyneuropathy produced autoantibodies to a protein termed GW182. Identification and characterization of the GW182 protein revealed that it was cytoplasmic and concentrated in distinct foci initially termed GW-bodies that were subsequently shown to be identical to P-bodies. Moreover, a small percentage of patients with primary biliary cirrhosis develop antibodies directed against other P-body components, Ge-1/Hedls and RAP55. The presence of autoantibodies directed against P-body proteins in different human disorders further underscores the significance of this cytoplasmic structure. A correlation between P-bodies and cancer is also emerging, albeit circumstantial at present. The expression level of a P-body component, rck/p54, is elevated in certain tumors. Furthermore, the requirement of P-bodies for microRNA function and the significant role of microRNAs in both tumorigenesis and tumor suppression demonstrates a central role for P-bodies in cancer. In some cancers, microRNAs involved in the downregulation of oncogenes are underexpressed, while in other instances microRNAs that repress tumor suppressor genes are more abundant. Furthermore, the degradation of the tumor necrosis factor-α mRNA is dependent on the coordinate function of a P-body component, tristetraprolin (TTP), and microRNA16 (miR16) that serve to facilitate degradation of the mRNA. An in-depth understanding of the signaling pathways and molecular mechanisms governing P-body formation will provide novel insights into the intricate control of eukaryotic gene expression.

For background information *see* CELL (BIOLOGY); GENE; GENE ACTION; GENETIC CODE; NUCLEOPROTEIN; RIBONUCLEIC ACID (RNA); RIBOSOMES in the McGraw-Hill Encyclopedia of Science & Technology.

Megerditch Kiledjian

Bibliography. P. Anderson and N. Kedersha, RNA granules, *J. Cell Biol.*, 172:803–808, 2006; M. Brengues, D. Teixeira, and R. Parker, Movement of eukaryotic mRNAs between polysomes and cytoplasmic processing bodies, *Science*, 310:486–489, 2005; J. Coller and R. Parker, Eukaryotic mRNA decapping, *Annu. Rev. Biochem.*, 73:861–890, 2004; N. Cougot, S. Babajko, and B. Seraphin, Cytoplasmic foci are sites of mRNA decay in human cells, *J. Cell Biol.*,

165:31–40, 2004; J. Liu et al., A role for the P-body component GW182 in microRNA function, *Nat. Cell Biol.*, 7:1261–1266, 2005.

Paper recycling

Interest in paper recycling accelerated when the *U. S. Federal Acquisition, Recycling, and Waste Prevention Executive Order* (Executive Order 12873) was issued on October 20, 1993. This directed all federal agencies to purchase papers containing recycled fibers. Other state and local agencies and many organizations, including large corporations, followed suit. The paper industry responded by investing millions of dollars in deinking plants and recycling upgrades. The American Forest and Paper Association's recycling goals of 40% by 1995 and 50% by 2000 were achieved ahead of their respective schedules. A new goal is to increase the recovery rate to 55% by 2012. A higher recovery rate is important as export demand increases, particularly in countries such as China and India. Paper-recycling science and technology has advanced sufficiently that the paper industry is able to consume more recovered papers and produce high-quality recycled paper and board products, including corrugated boxes, cardboard boxes, newsprint, tissue, towels, napkins, and printing papers.

Recovered paper grades. There are many choices of making paper or board products from recovered papers. The fiber source for manufacturing paper products may be derived from trees (virgin fibers) or recovered papers. Many paper grades contain a mixture of virgin and recycled fibers. Examples of fiber sources for manufacturing paper and paperboard include:

Wood
Softwoods (coniferous trees)
  Pine
  Spruce
  Fir
  Hemlock
Hardwoods (deciduous trees)
  Oak
  Maple
  Aspen

Recovered papers
OCC (old corrugated containers)
ONP (old newspapers)
CPO (computer printouts)
OMG (old magazines)
OWP (office papers)
Commerical/industrial/converter material
  Boxboard cuttings
  Envelope cuttings
  Printer's trimmings

The term "wastepaper" should not be used as it is not really a waste but a raw material for making useful products. The preferred terms are recovered paper, scrap papers, secondary fibers, and recycled papers.

The Institute of Scrap Recycling Industries (ISRI) publishes a list of 51 recycled-paper grades, as well as the level of unacceptable paper contaminants (outthrows) and nonpaper contaminants (prohibitive materials) in each grade. In a bale of old newspaper, for example, brown bags and board are considered outthrows, while plastics are classified as prohibitive materials.

Process technology. Processing recovered papers involves many steps such as pulping, deflaking, screening, cleaning, deinking, flotation, washing, dispersion, kneading, bleaching, water clarification, and rejects handling.

Pulping and deflaking. Once the recovered paper is accepted, it goes to the pulper where, through proper operating conditions and accessory equipment, it is defiberized without significant disintegration of contaminants. Important parameters include stock consistency, temperature, low versus high pulping intensity, and the configuration of the pulper.

Most modern pulpers are equipped with auxiliary equipment to remove contaminants without breaking them into small pieces. The auxiliary equipment includes a ragger to remove wire and string, a junker for large contaminants, and a secondary pulper. A stream is bled off at the secondary pulper for mild defiberizing. High-density contaminants accumulate in a chamber with a double-valve arrangement, while stock is sent back to the pulper or is rotary-screened.

Most pulpers have a high-speed rotor for defibering. A pulper that is gaining in popularity, particularly for newsprint deinking, is the drum pulper. Drum pulpers are continuous pulpers in which the papers are lifted and dropped on a hard surface several times to achieve defibering. Because of the absence of a high-speed rotor, the drum pulper does not have any cutting action, and correspondingly the disintegration of contaminants is minimal. As a result, many contaminants, such as plastics and bookbindings, remain virtually intact and are rejected by the associated rotary screen.

Screening and cleaning. Coarse screens with holes and fine screens with slots are used to remove contaminants, based primarily on their size. Holes are generally 1.38 mm (0.055 in.) or larger, although screens with somewhat smaller hole size have been used. Fine screens have slots with widths ranging from 0.10 to 0.60 mm (0.004 to 0.024 in.). Generally, as the slot size decreases, contaminant removal increases but so does fiber loss.

Most pressure screens operate with mass reject ratios of 15–30%. To minimize fiber loss, second and third screening stages must be used. How these stages are arranged is very important to the contaminant removal efficiency of the system. In a conventional cascade system, recirculation of contaminants between stages is quite common, and this can be detrimental to overall system efficiency. The use of a forward-flow arrangement is recommended whenever possible to avoid this problem.

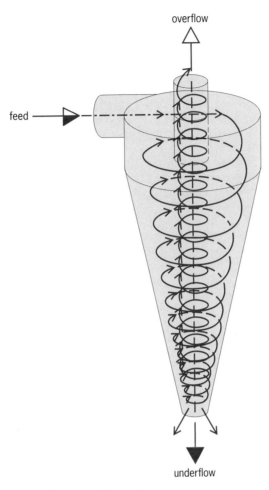

Hydrocyclone separation system.

In addition to the screening operation, cleaners or hydrocyclones are used to remove contaminants based on their density differences (see **illustration**). A slurry containing fiber, water, and contaminant is injected into cyclone at high velocity. The slurry swirls around in the cyclone, accelerating as it passes through the conical section. Under the action of centrifugal force, high-density particles migrate toward the cyclone wall, while low-density particles move toward the core in the center of the cyclone.

As the slurry traverses along the conical section, a portion of it is swept along with the overflow core, while the remainder continues the journey toward the underflow. Thus, a proportionally higher concentration of high-density particles will be in the underflow than in the overflow, and vice versa; that is, a proportionally higher concentration of low-density particles will be in the overflow than in the underflow.

Cleaners are classified as high-, medium-, or low-density, with their use dependent on the density and size of the contaminants they are removing. A high-density cleaner is used to remove nuts, bolts, paper clips, and staples. These high-density or forward cleaners are usually located immediately after the pulpers. For medium-density contaminants, smaller-diameter cleaners are used. As the hydrocyclone di-ameter decreases, efficiency in removing small-size contaminants increases. For practical and economic reasons, the 75-mm-diameter (3-in.) cyclone is the smallest cleaner used in the paper industry.

Reverse hydrocyclones and throughflow cleaners are used to remove low-density contaminants. A disadvantage of reverse hydrocyclones is that 55% of the flow is in the reject stream, and therefore secondary and tertiary stages are needed to recover the usable fiber. This problem does not occur in throughflow cleaners. The accepted and rejected materials come out at the same end in throughflow cleaners. The reject stream is only 10% by volume and 2% by mass. However, the contaminant removal efficiency of reverse hydrocyclones is usually higher than that of the throughflow cleaners. Another problem with throughflow cleaners is that they are somewhat prone to plugging due to the narrow gap at the exit. Rotating body cleaners are effective for the removal of low-density contaminants. The reject ratio of these cleaners is so low that there is no need for a second or third stage.

Other types of cleaners are available. In the core bleed cleaner, both high- and low-density contaminants are removed in a single unit operation. High-density contaminants are removed through underflow, while low-density contaminants are concentrated in the center and removed through the core tube. Accepted materials are removed through the annular space between the core tube and the overflow tube.

Deinking, flotation, and washing. Deinking, a process of removing inks from recycled pulp, involves three steps: detachment of ink from fibers, removal of detached ink from pulp, and clarification of the water to dispose of the removed ink and contaminants.

Mechanical, chemical, and thermal forces are used to detach the ink from the fibers. This is usually carried out in pulpers where strong agitation provides shear force. Steam or hot water and deinking chemicals are added to help dislodge ink from the fibers. Some of the chemicals used in deinking are listed in the **table** together with their dosages and function.

The detached ink is removed from the stock by screening, cleaning, flotation, and washing. In flotation deinking, air bubbles are used to collect and remove ink particles from the slurry. Surface chemistry plays an important role in flotation deinking, as in the case of the flotation process used in the mineral processing industry. Washing deinking is somewhat similar to the washing cycle of a laundry machine. In this case, fine ink particles and dirt wash out with the water, leaving behind relatively ink-free fibers.

The removal efficiency of a given process will depend on the ink particle size, shape, and density. Based on particle size alone, large particles (greater than 150 micrometers) will be efficiently removed by screens, intermediate particles (approximately 25 to 150 μm) will be removed by cleaners and flotation, and small particles (less than 25 μm) will be removed by washers. However, there are many exceptions. For example, flat disk-shape toner particles

Deinking chemicals, their dosages and functions

Chemical	Dosage, kg/ton	Function
Alkali	10–20	Helps in fiber swelling and ink release
Silicate	15–25	Acts as a dispersant for released ink and serves as a buffering agent
Surfactant	2–15	Emulsifies or forms micelles and helps in the detachment of ink
Dispersant	2–10	Keeps the detached ink in suspension
Peroxide	5–20	Prevents yellowing of ground wood fibers; brightens fibers
Collector	2–10	Assists in collecting ink particles on air bubbles in flotation deinking
Displector	2–15	Combines functions of dispersants and collectors

or soft gel-like ink particles can pass through screen openings. Flexographic ink disperses into fine particles under alkaline conditions and can be trapped in fiber crevices and lumens, making them difficult to remove by washing or any other known process.

Dispersion and kneading. The objective in dispersion is to break up contaminants and inks further so they will be invisible to the naked eye in the final product. Important parameters to consider in dispersion are consistency, temperature, and pressure. Consistencies of 25-30% are used, while temperatures range from 70 to 80°C (160 to 180°F) at atmospheric pressure. In some instances, higher temperatures are used.

The purpose of dispersion in processing old corrugated container stock is to break up waxes and hot melts. A somewhat more common application of dispersion and kneading is in the deinking of office papers. Some of the inks, such as photocopy and laser-printed papers, are relatively difficult to detach because they are strongly bonded to the paper. Mechanical, thermal, and chemical forces imparted to the pulp during dispersion and kneading help detach these types of inks. However, when ink particles are fragmented into smaller particles and dispersed, brightness will decrease. In this instance, dispersion should be followed by an ink removal step to improve brightness.

Bleaching. The intent of bleaching is to attack the color bodies and lignin in the pulp to improve brightness and color of the pulp. Usually, oxidative bleaching agents such as hydrogen peroxide, oxygen, and ozone, and reductive bleaching agents such as sodium hydrosulfite and formamidine-sulfinic acid are used for this purpose.

Water clarification. Water clarification and reuse is important in the recycling mills. Water reuse conserves freshwater as well as energy. Proper use of coagulating and flocculating chemicals is important for achieving high efficiency in water clarification. Most recycling mills use dissolved air flotation (DAF) for internal water clarification and reuse.

Primary clarification of an effluent from a paper mill usually involves sedimentation or DAF. The objective is to reduce the concentration of suspended solids. The suspended solids concentration is reduced from 1000–3000 ppm in the influent to less than 300 ppm after clarification. Dissolved organic material in the effluent (measured as biochemical oxygen demand, or BOD) is generally reduced by secondary or biological treatment.

Rejects handling. The amount of residue generated by a paper or board mill can be as high as 40% and depends on many factors such as the size of the mill, raw material used, and type of product manufactured. Mill residue has a reasonable amount of heating value, and many mills are currently burning the residue to recover energy. Fluidized-bed incineration is gaining in popularity for this reason. Innovative and creative use of the residue and ash from incineration can be environmentally as well as economically beneficial to a mill.

Stickies. Adhesives, waxes, and binders used in coatings and inks are commonly known as stickies. Stickies continue to be a problem for the recycling industry. New developments in terms of methods for the quantification of stickies, additives to pacify stickies, and modification of adhesives to make them recycling-friendly are progressing at a rapid pace. The U.S. Postal Service and USDA Forest Products Laboratory have contributed significantly to testing and identifying environmentally benign adhesives.

Effect of recycling on paper properties. Many wonder how many times can one recycle paper. This is not a major concern in countries like the United States where there is sufficient infusion of virgin fibers in the system. It is also not a major concern in recycling newspapers, as theses fibers contain sufficient amount of original plant lignin to protect them from degradation during papermaking and recycling. Fibers used in corrugated containers, as well as printing and writing grades, degrade on recycling. This is of particular concern in countries where the recovery rate exceeds 80%. Some of the approaches available for restoring the properties of recycled fibers include refining, fractionation, and the use of bonding agents such as starch.

For background information *see* INK; PAPER; PRINTING; RECYCLING TECHNOLOGY; WOOD PRODUCTS in the McGraw-Hill Encyclopedia of Science & Technology.   Mahendra Doshi

Bibliography. M. R. Doshi (ed.), *Recent Developments in Paper Recycling: Stickies*, Doshi & Assoc., Inc., 2002; M. R. Doshi and J. M. Dyer (eds.), *Paper Recycling Challenge*, vol. I: *Stickies*, 1997, vol. II: *Deinking & Bleaching*, 1997, vol. III: *Process Technology*, 1998, vol. IV: *Process Control & Mensuration*, Doshi & Assoc., Inc., 1999; L. Gottsching and H. Pakarinen (eds.), *Recycled Fiber and Deinking*, Tappi, 2000; R. W. J. McKinney, *Technology of Paper Recycling*, Chapman and Hall, Great Britain, 1995; *Proceedings of TAPPI Recycling Symposium*, March 5–8, 2000, TAPPI Press, Washington, DC, 2000; R. A. Spangenberg (ed.), *Secondary Fiber Processing*, TAPPI Press, Atlanta, 1993.

Perfluorooctanoic acid and environmental risks

Perfluorooctanoic acid [PFOA; $CF_3(CF_2)_6COOH$] is a member of the perfluoroalkyl acids (PFAA) family of chemicals, which consist of a carbon backbone typically four to fourteen carbons in length and a charged functional moiety (primarily carboxylate, sulfonate, or phosphonate). Many chemical intermediates (such as alcohols and amides) can be derived for commercial uses, but these intermediates ultimately break down to PFOA or its sister compound, perfluorooctane sulfontate (PFOS). PFAA are distinguishable from another class of perfluorocarbons, the perfluoroalkanes, primarily used clinically for oxygenation and respiratory ventilation. PFAA are relatively contemporary chemicals, in use only in the past 50 years, and until recently have been considered as biologically inactive.

Carbon-fluorine bonds are among the strongest in organic chemistry, and fully fluorinated hydrocarbons (perfluorocarbons) are stable in air at high temperature, nonflammable, and not readily degraded by strong acids, alkalis, oxidizing agents, or photolysis. The stability of these chemicals renders them practically nonbiodegradable and persistent in the environment. The fluorine moiety of PFAA provides extremely low surface tension and contributes to their unique hydrophobic and oleophobic (a dislike of both water and oil) nature. The physical properties of PFAA thus render these chemicals ideal surfactants. Although all PFAA share some surfactant properties, the eight-carbon chemicals PFOS and PFOA are most effective.

These products are found in over 200 industrial and consumer applications, ranging from water-, soil-, and stain-resistant coatings for clothing fabrics, leather, upholstery, and carpets, to oil-resistant coatings for paper products approved for food contact, electroplating, electronic etching-bath surfactants, photographic emulsifiers, aviation hydraulic fluids, fire-fighting foams, paints, adhesives, waxes, and polishes. PFOA is used as an emulsifier in the production of polytetrafluoroethylene, fluoropolymers, and fluoroelastomers.

Historically, the production and use of PFOS (3500 metric tons in 2000) dwarfed those of PFOA (estimated at 500 metric tons). However, because the major manufacturer of PFOS, 3M, phased out production in 2002, the global production of this chemical dropped precipitously to 175 metric tons by 2003. In contrast, the global production of PFOA escalated to 1200 metric tons per year by 2004 and has presumably become the most common PFAA in commerce. In 2006, the U.S. Environmental Protection Agency initiated the PFOA Stewardship Program, in which the eight major companies in the industry have committed to reduce facility emissions and product contents of PFOA and related chemicals on a global basis by 95% no later than 2010, and to work toward eliminating emissions and product content of these chemicals by 2015.

Naturally occurring fluorinated organic compounds are rare. PFOA, PFOS, and other PFAA are synthetic chemicals, primarily produced by two methods. The Simons electrochemical fluorination process produces fluorinated molecules of various carbon chain lengths and a mixture of linear, branched, and cyclic isomers. An alternative production method involves telomerization of tetrafluoroethylene units, which always yields straight-chain telomer alcohols [$F(CF_2CF_2)_nCH_2CH_2OH$] that can be converted into final products for commercial application. A telomer is a low-molecular-weight polymer in which the terminal group differs from the repeating unit. Thus, PFAA found in the environment are composed of a family of target compounds as well as by-products of various chain lengths and isomers.

Public concerns. Public concerns about the environmental risks of PFOS, PFOA, and other PFAA began to mount in the late 1990s, primarily because of five factors: (1) environmental persistence of these chemicals, (2) their ubiquitous distribution, even to remote regions of the world, (3) detection of PFOS, PFOA, and possibly other PFAA in humans, (4) presence of these compounds in a variety of wildlife populations, and (5) toxicity findings of PFOS and PFOA from laboratory animal models.

Environment. PFOA and PFOS have been detected in surface waters worldwide. These include the Tennessee River downstream from a fluorochemical manufacturing plant, drinking-water sources near a production plant in West Virginia, the Great Lakes, rainwater from an urban center in Canada, coastal waters in south China and Korea, and samples from a number of cities in Japan. Typically, the PFOA levels were in the parts per trillion range, although concentrations found in West Virginia tended to be higher (about 3 parts per billion). PFOA, PFOS, and a slew of PFAA intermediates were also found in San Francisco Bay area sediments and sludge.

Although PFOA and PFOS have low volatility and vapor pressure, particulates containing PFOA can be detected in ambient air, ranging from $0.07-0.9\ \mu g/m^3$ among different Japanese cities to $0.12-0.9\ \mu g/m^3$ in a sampling area near a fluoropolymer manufacturing facility in the United States. PFOA, PFOS, and perfluorohexane sulfonate (PFHxS) have been detected in indoor dust samples from Canadian homes, averaging about 100, 450, and 400 ppb (ng/g), respectively.

Our understanding of the sources of PFOA and PFOS in the environment is only in its infancy. Because of the popular uses of PFOA and PFOS in consumer and industrial products, it may not be so surprising to find these pollutants in urban areas. On the other hand, the detection of these perfluorochemicals in remote regions of the world is quite unexpected. Recently, two theories have emerged for the fate and transport of these chemicals around the world. The first is long-range transport by oceanic currents which is supported by the finding of a number of PFAA in the surface water of the Atlantic and Pacific oceans, South China Sea, Sulu Sea, and the Labrador Sea, in parts per quadrillion (pg/L) levels, with PFOA being the major PFAA detected, followed by PFOS. Concentrations of PFAA decrease by 2–4 orders of magnitude from coastal waters to offshore; additionally, traces of PFOA and PFOS were detected in deep-sea water. The second theory is atmospheric transport and transformation of precursor chemicals. While the volatility of PFOA and PFOS is nominal, the volatility of their precursors and derivatives is high at normal temperature and pressure. According to one model, perfluorinated telomer alcohols (11–165 pg/m^3) and sulfonamides (22–400 pg/m^3) have been found in the North American troposphere, and their estimated atmospheric lifetime of 10–20 days is sufficient to account for the widespread hemispheric distribution. These precursors of PFOA and PFOS can be oxidized by hydroxyl radicals in the atmosphere, thereby providing an explanation for their presence in remote locales.

Wildlife. J. P. Giesy and K. Kannan were among the first to report the global presence of PFAA in wildlife. PFOS and, to a lesser extent, PFOA, have been detected in most of the species examined. These include marine mammals, fishes, birds, freshwater mammals, turtles, frogs, and field rodents. Follow-up reports confirmed distribution of PFOS and PFOA in wildlife from various parts of the world, such as Japan, the Pacific Ocean, Baltic Sea, Dutch Sea, Mediterranean Sea, the coasts of Columbia and Greenland, the Canadian Artic, Antarctica, the Great Lakes, and United States rivers and coasts. These perfluorinated chemicals were found primarily in the liver and blood of the animals (in ppb ranges), although their presence in seabird eggs has been reported. Levels of PFOS, PFOA, and other PFAA are exceptionally high in polar bears and arctic foxes (ppm levels in the liver). Trophic transfer of PFAA in wildlife is not well understood, although PFOS and its precursors have been detected in the food web in fresh and marine waters. The major source appeared to be sediments rather than water, and bioaccumulation/biomagnification was evident in the top trophic-level organisms. An apparent lack of bioaccumulation of PFOA at the top of the food web was noted.

Humans. The presence of organic fluoride in humans was first reported over 30 years ago. Advances in analytical techniques in the past few years has enabled the identification of individual PFAA in various matrices at sub-ppb levels. Equipped with improved detection, G. W. Olsen and coworkers reported serum concentrations of 0.94–1.32 ppm for PFOS and 0.90–1.78 ppm for PFOA in production workers at 3M fluorochemical manufacturing plants. Based on continuous monitoring of retired fluorochemical workers, the mean half-life eliminations for PFOS and PFOA were estimated at 5.4 and 3.8 years, respectively.

A broad survey of individual blood samples from Red Cross donors and other cohorts indicated mean serum concentrations of PFOS, PFOA, and PFHxS of 30–40, 4–5, and 2–5 ppb, respectively. A recent study compared human blood between 1974 and 1989 from a community health study, and reported an increase of PFOS from 29.5 ppb to 34.7 ppb, of PFOA from 2.3 ppb to 5.6 ppb, and of PFHxS from 1.6 ppb to 2.4 ppb. When results from 1989 were compared to the samples collected from the Red Cross in 2001, no significant difference was noted, suggesting that the PFAA levels in the human blood did not increase significantly after 1989. A. M. Calafat and coworkers reported similar values of PFOS (31.1 ppb), PFOA (11.6 ppb), and PFHxS (2 ppb) from limited pooled serum samples in United States residents collected from 1990 through 2002. Samples collected by the Centers for Disease Control and Prevention (CDC) from a larger cohort of the 2001–2002 National Health and Nutrition Examination Survey (NHANES) indicated a small gender difference (males > females) and differential distribution among ethnic groups of these PFAA in the United States. A full report on 11 perfluorochemicals from the NHANES will be provided by CDC in 2007.

Presence of PFOS, PFOA, and other PFAA in human blood has also been reported in Japan, China, Columbia, Brazil, Belgium, Italy, Poland, India, Malaysia, and Korea. Serum values of PFAA in these countries were comparable to those observed in the United States, with PFOS levels typically higher than those for PFOA and PFHxS. A recent study from Japan indicated a gender difference in PFOS and PFOA, along with a dramatic increase of PFOS and PFOA, by factors of 3 and 14, respectively, between 1978 and 2003.

Unlike other persistent organic pollutants, PFAA levels in children were found to be comparable to those in adults. The presence of PFOS and PFOA in umbilical cord blood and breast milk, suggesting a maternal source, may contribute to the body burdens of infants and children.

Potential health effects. The toxicology of PFOA has been recently reviewed. Several epidemiologic and medical surveillance studies were done on workers employed at various manufacturing sites in the United States. Most of the studies focused primarily on males. To date, these studies have not shown an association between exposure to PFOA and health outcomes.

In studies with laboratory rodents, liver toxicity, immunotoxicity, developmental toxicity, and liver, pancreas, and testicular tumors have been observed after exposure to PFOA. In rodents and nonhuman primates, mortality occurs at higher doses in newborns as well as adults. The toxicity profiles of PFOA and PFOS in animal models share a number of similarities, suggesting that some of these features may be common to all PFAA. Mechanisms underlying the adverse effects of PFOA are poorly defined at present, although several possibilities such as the involvement of peroxisome proliferator-activated receptors, cell-cell communication, mitochondrial bioenergetics, and hormonal disruption have been proposed.

Environmental risks. The Environmental Protection Agency has been assessing the potential human health effects of PFOA for several years, and recently released a draft risk assessment for review by its Scientific Advisory Board. However, there are significant uncertainties with the existing information, and further research is needed to achieve a full understanding of the potential risks to environmental sources of PFOA.

[The information in this document has been funded by the U.S. Environmental Protection Agency.

It has been subjected to review by the National Health and Environmental Effects Research Laboratory and approved for publication. Approval does not signify that the contents reflect the views of the Agency, nor does mention of trade names or commercial products constitute endorsement or recommendation for use.]

For background information *see* ENVIRONMENTAL TOXICOLOGY; EPIDEMIOLOGY; FLUORINE; FLUOROCARBON; FOOD WEB; HALOGENATED HYDROCARBON; MUTAGENS AND CARCINOGENS; POLYFLUOROOLEFIN RESINS; TOXICOLOGY; TROPHIC ECOLOGY in the McGraw-Hill Encyclopedia of Science & Technology.
Christopher Lau; Jennifer Seed

Bibliography. A. M. Calafat et al., Perfluorinated chemicals in selected residents of the American continent, *Chemosphere*, 63:490-496, 2006; A. M. Calafat et al., Perfluorochemicals in pooled serum samples from United States residents in 2001 and 2002, *Environ. Sci. Technol.*, 40:2128-2134, 2006; J. P. Giesy and K. Kannan, Global distribution of perfluorooctane sulfonate in wildlife, *Environ. Sci. Technol.*, 35:1339-1342, 2001; J. P. Giesy and K. Kannan, Perfluorochemical surfactants in the environment, *Environ. Sci. Technol.*, 36:146A-152A, 2002; K. Harada et al., The influence of time, sex and geographic factors in levels of perfluorooctane sulfonate and perfluorooctanoate in human serum over the last 25 years, *J. Occup. Health*, 46:141-147, 2004; G. L. Kennedy et al., The toxicology of perfluorooctanoate, *Crit. Rev. Toxicol.*, 34:351-384, 2004; H.-J. Lehmler, Synthesis of environmentally relevant fluorinated surfactants—A review, *Chemosphere*, 58:1471-1496, 2005; G. W. Olsen et al., Human donor liver and serum concentrations of perfluorooctanesulfonate and other perfluorochemicals, *Environ. Sci. Technol.*, 37:888-891, 2003; G. W. Olsen et al., Historical comparison of perfluorooctanesulfonate, perfluorooctanoate, and other fluorochemicals in human blood, *Environ. Health Perspect.*, 113:539-545, 2005; K. Prevedouros et al., Sources, fate and transport of perfluorocarboxylates, *Environ. Sci. Technol.*, 40:32-44, 2003; D. R. Taves, Evidence that there are two forms of fluoride in human serum, *Nature*, 16:1050-1051, 1968; T. J. Wallington et al., Formation of $C_7F_{15}COOH$ (PFOA) and other perfluorocarboxylic acids during the atmospheric oxidation of 8:2 fluorotelomer alcohol, *Environ. Sci. Technol.*, 40:924-930, 2006.

Phylogeny of bryophytes

The emergence of land plants approximately 480 million years ago was a monumental event in the history of life. It had major impact on the Earth's environment, for example, changing atmospheric carbon dioxide and oxygen levels, lowering surface temperature, producing soil, and increasing mineral nutrient release from land crust into oceans. Through interacting coevolution with animals (consumers) and fungi (decomposers), the plants as primary producers initiated development of the entire modern terrestrial ecosystems and thus fundamentally changed the course of evolution of life. To understand how all these events happened, plant evolutionary biologists have long been interested in reconstructing the evolutionary history, or phylogeny, of early land plants. Most of these plants fall into the category of bryophytes, which include three groups, liverworts, mosses, and hornworts (all characterized by the lack of true roots, stems, and leaves). Over the last few years, the massive infusion of molecular biology techniques (for example, automated DNA sequencing and bioinformatic tools) into systematics has significantly improved our understanding of relationships among early land plants. At present, it is clear that (1) all land plants share a common origin, (2) they evolved from green algae that resemble today's charophytes (aquatic plants that are found on the lake shore), and (3) bryophytes preceded vascular plants during early evolution of land plants. Since bryophytes represent the transitional group between algae and vascular plants, their phylogeny becomes especially relevant in our understanding of the origin and early evolution of land plants. In particular, two questions have been explicitly pursued by plant systematists over the last few years: (1) do liverworts or hornworts represent the earliest land plants, and (2) are mosses or hornworts sister groups to vascular plants?

Analysis. For the first question, earlier studies of morphology identified liverworts as the sister to all other land plants. Later evidence from distribution of mitochondrial introns and auxin conjugation patterns throughout land plants confirmed the result. For the second question, cladistic analyses of morphological data had suggested mosses to be sister to vascular plants. However, several recent studies analyzing sequences of a few genes or entire chloroplast genomes from a relatively small number of taxa have shown either bryophytes to be monophyletic (having evolved from a single interbreeding population) or hornworts to be sister to all other land plants, with the question of which lineage is sister to vascular plants left largely unanswered. Further, investigation of sperm development in major plant lineages seems to support the hornwort-basal hypothesis. One major problem of all these studies is that they examined only a small number of taxa. In reconstructing a phylogeny of land plants, which originated so long ago and include well over a half-million species, use of about two dozen species in these analyses raises a question of sampling error in the final results. To address this concern, one recent study sampled nearly 200 species of green algae and land plants, with particular heavy representation of liverworts, mosses, and ferns (the three species-rich lineages of early land plants), and used six genes from chloroplast, mitochondrial, and nuclear genomes. It also analyzed two other data sets, one of 28 mitochondrial intron positions in 16 algae and land plants, and the other of three dozen chloroplast genomes. Analyses of all three data sets strongly supported bryophytes as a paraphyletic group (containing some of the descendants from a common ancestor) to vascular plants and identified liverworts as the sister to all other land plants. Analyses of the first and last data sets provided moderate to strong support to hornworts being the sister to vascular plants (see **illustration**). This study also confirmed monophylies of all three bryophyte lineages, some of which had been questioned in earlier single-gene analyses. Although a possibility of sampling error still exists in this study, it is much less likely to occur here than in previous ones, as it used three very different data sets with different taxon sampling schemes and character sources.

Liverworts. The liverwort position identified in this most recent study is supported by several lines of independent evidence: (1) Morphologically, liverworts lack stomata (pores facilitating transpiration of water in plants to regulate body temperature and to assist water and nutrient flow), while all other land plants have them. (2) Several mitochondrial group II introns (pieces of DNA that are present in genes but removed before messenger ribonucleic acids are translated into polypeptides) have been found to be lacking in liverworts as in green and other algae but present in all other land plants. (3) Liverworts have a primitive way of utilizing auxin (a plant growth hormone), breaking it down after using it, whereas all other land plants evolved a more sophisticated way of using it and storing it for reuse. (4) Analyses of chloroplast genome sequences, when some critical taxa such as lycophytes are sampled to represent vascular plants properly, also recovered the liverworts-basal topology.

Hornworts. The position of hornworts as the sister to vascular plants has been suggested only in several recent molecular phylogenetic studies, but the sparse taxon sampling in all these studies left some doubt as to whether the result has been affected by taxon sampling, which often has dramatic effect on phylogenetic results. The study with dense taxon sampling of early land plants mentioned above removed this doubt by sampling all three bryophyte lineages and vascular plants densely, and obtaining the result from two of the three data sets. Furthermore, there are several morphological characters, particularly those related to sporophyte (spore-bearing plant) development, that seem to support this new position of hornworts. For example, hornworts have persistently chlorophyllous and nutritionally largely independent sporophytes, and surface cells of the hornwort sporophyte foot also show rhizoid-like behavior (rhizoids are rootlike structures that help to hold a plant to a substrate). It is common knowledge that development of roots to absorb water and nutrients from the ground is a necessary step to establish a free-living sporophyte. In fact, biennial, nearly free-living sporophytes, with the gametophytic tissues around the base of the sporophyte discolored and more or less collapsed, have been found in the wild for a hornwort species, *Anthoceros fusiformis*, in southern California. The excised sporophytes even survived independent of the gametophyte on sterile soil for 3 months. In contrast, the sporophytes of liverworts and mosses are strictly matrotrophic (taking

Phylogenetic relationships among three bryophyte lineages. Shown here are representatives of liverworts (*Marchantia macropora*), mosses (*Acrophyllum quadrifarium*), and hornworts (*Anthoceros agrestis*), along with charophytes (*Chara vulgaris*), which represent the closest algal relative of land plants, and vascular plants (*Cyathea smithii*), which evolved from a bryophytic ancestor.

nutrition from the mother) on the gametophytes, and no report of free-living sporophytes has ever been made of either group. The nutritionally largely independent sporophytes of hornworts figured prominently in the discussion on the origin of alternation of generations (the phenomenon in which two alternating forms with different amounts of DNA make up a complete life cycle, and one form, known as a sporophyte, produces spores, which grow into individuals of the other form, called gametophytes) in land plants in the first half of the twentieth century, but had been left out in the cladistic analysis–oriented studies of early land plant phylogeny in the 1980s and 1990s. The reason is perhaps that characters such as photosynthetic capability and longevity of sporophytes are quantitative and polymorphic, and thus difficult to code in cladistic studies. It is unfortunate that such important information has been overlooked in pursuing an understanding of early land plant evolution. The new prospect of taking a whole-organism biology approach in the postphylogeny era to study evolution of plants will hopefully bring more characters like these into the light and subject them to further investigation using new molecular techniques, eventually leading to a deeper and more complete understanding of the origin and early evolution of land plants.

Mosses. The position of mosses has been changed from being sister to vascular plants as suggested by morphological cladistic studies to being an intermediate lineage between liverworts and hornworts. This change calls for a reevaluation of some characters that were purported to be homologous between mosses and vascular plants, for example, hydrom and leptom (the water- and food-conducting tissue of mosses, respectively) in mosses and xylem and phloem (the water- and food-conducting tissue of higher plants, respectively) in vascular plants. In fact, the claimed homology between sporophytes of mosses and vascular plants was already called into question before the recent molecular studies.

Phylogeny hypothesis. The phylogeny of early land plants reconstructed by recent molecular studies sheds significant new light on our understanding of alternation of generations in land plants, which has been elaborated by the antithetic hypothesis. According to this hypothesis, the diploid sporophyte generation was interpolated into the life cycle of charophytes through a delay in meiosis after fertilization. The sporophyte generation expanded as bryophytes evolved, accompanied by structural elaboration and progressive sterilization of potentially sporogenesis tissues (which may become columella, that is, tissue having a sterile axial body), and ultimately became a dominant generation in the life cycle of vascular plants. Hornworts were envisioned as the transitional bryophytes to vascular plants by some advocates of this hypothesis. While the charophytic ancestry of land plants was recognized several decades ago, the hornwort position in this evolutionary scenario has never been seriously considered since its initial proposal. Identification of hornworts as the sister group to vascular plants now highlights the importance of development of nutritionally largely independent sporophytes in facilitating the transition from gametophyte to sporophyte as the dominant generation in life cycle of early land plants. This phylogenetic hypothesis provides a timely guide for evolutionary developmental studies to investigate several aspects that are essential for establishment of a free-living sporophyte: persistence of photosynthesis in sporophytes, longevity of meristems (formative plant tissues composed of undifferentiated cells capable of dividing) in sporophytes, and root initiation and development on the sporophyte foot. The evidence from these evolutionary developmental studies can further test the robustness of the phylogenetic hypothesis postulated by the recent molecular studies. Ultimately, our understanding of early land plant evolution is likely to be significantly enhanced through multidisciplinary research on the obscure yet highly interesting bryophytes.

For background information *see* BRYIDAE; BRYOPHYTA; BRYOPSIDA; PLANT EVOLUTION; PLANT KINGDOM; PLANT TRANSPORT OF SOLUTES; PRIMARY VASCULAR SYSTEM (PLANT) in the McGraw-Hill Encyclopedia of Science & Technology.

Bin Wang; Yin-Long Qiu

Bibliography. P. G. Gensel and D. Edwards (eds.), *Plants Invade the Land*, Columbia University Press, New York, 2001; P. Kenrick and P. R. Crane, *The Origin and Early Diversification of Land Plants: A Cladistic Study*, Smithsonian Institution Press, Washington, DC, 1997; Y.-L. Qiu et al., The deepest divergences in land plants inferred from phylogenomic evidence, *Proc. Nat. Acad. Sci. USA*, 103:15511–15516, 2006; Y.-L. Qiu et al., The gain of three mitochondrial introns identifies liverworts as the earliest land plants, *Nature*, 394:671–674, 1998; K. S. Renzaglia et al., Vegetative and reproductive innovations of early land plants: Implications for a unified phylogeny, *Phil. Trans. Roy. Soc. London, Ser. B Biol. Sci.*, 355:769–793, 2000.

Phytochrome

Phytochrome is a photoreceptor—a pigment used by plants and other photosynthetic organisms to detect light. After perception of light, phytochromes act to regulate light-dependent growth and development. Phytochromes have been identified in prokaryotes, fungi, algae, and plants.

Photoperception. Light is one of the most important signals for photosynthetic organisms. Plants, in particular, are sessile organisms that cannot simply move to more favorable conditions during adverse periods, and thus monitor many aspects of their light environment, including light quality (wavelength or color), quantity (amount of light), duration (photoperiod or hours of exposure to light), and direction. Plants integrate all of this information about the light environment to determine the best pattern of cellular metabolism, growth, and development for the environment in which they are growing. Plants possess a number of photoreceptors that allow them to detect light throughout the visible and near-ultraviolet (near-UV) spectra. Phytochromes and other plant photoreceptors regulate growth and development throughout multiple stages of the life cycle of plants from seed germination to flowering and seed production by mature plants. Phytochromes detect light primarily in the red and far-red regions of the visible spectrum.

Biosynthesis. Phytochromes consist of two components, the apoprotein (the protein of a conjugated protein exclusive of its prosthetic group needed for functionality; about 1100 amino acids in length) and its covalently attached light-absorbing chromophore, a plastid-derived linear tetrapyrrole or bilin. The apoprotein contains two conserved domains—an N-terminal photosensory domain and a C-terminal histidine-kinase-related domain. The chromophore is covalently attached to a conserved cysteine in the photosensory domain of the apoprotein. Functional phytochromes depend upon two independent subcellular biosynthetic pathways (**Fig. 1**). The apoprotein of phytochrome is encoded by a small family of nuclear genes. Five phytochrome genes, *PHYA–PHYE*, exist in the widely studied model plant *Arabidopsis thaliana*. After transport of the messenger ribonucleic acid (mRNA) to the cytosol, phytochrome apoproteins are synthesized on cytosolic ribosomes. A single bilin chromophore attaches to all apoproteins. This chromophore, phytochromobilin (PΦB), is synthesized entirely in the plastid compartment of higher plants and is translocated to the cytoplasm, where it binds and is covalently attached to apoproteins (Fig. 1). Holophytochrome, or photoactive phytochrome, is synthesized in the ground state as the red-light-absorbing (Pr) form and functions as a dimer. Recent findings indicate that, in addition to homodimerization of the different phytochrome isoforms, PHYB–PHYE also heterodimerize, thereby increasing the diversity of this class of photoreceptor.

Photochromicity. Phytochromes are photochromic biliproteins that exist in two stable, spectrally distinct forms—the red-light-absorbing Pr form and the far-red-light-absorbing Pfr form. These two forms are interconvertible by the absorption of red and far-red light by the covalently attached chromophore. Photoconversion occurs due to rotation around a double bond between the C and D rings of the bilin moiety in response to light absorption. This conversion results in slight color changes in the protein. The Pr form absorbs maximally at 660–670 nm, while the Pfr form absorbs maximally at about 730 nm. Absorbance of red light by the ground-state Pr form converts it to the Pfr form of phytochrome, which is thought to be the predominant biologically active form. In the absence of light, the Pfr form can revert to the Pr form through a process called dark reversion. This process is thought to be important for the regulation of phytochrome signaling by limiting the accumulation of photoactive phytochrome.

Phytochrome responses in plants. One of the earliest noted phytochrome responses was the red/far-red-reversible regulation of seed germination. In the Pr form, phytochrome inhibits seed germination. Upon absorption of red light, phytochrome converts to the Pfr form that stimulates or induces germination. This is a reversible response in which the last color of light to which the seed is exposed determines the signaling state of phytochrome and whether or not seed germination occurs. Phytochromes are also involved in hypocotyl and stem elongation, leaf development, and greening.

Phytochromes also regulate responses to day length or photoperiod. Some plants flower under short days, while others flower under long days. A third class of plants are called day-neutral and flower independently of the length of the day or night period. Using night-break experiments, this photoperiodic response has been largely shown to be controlled by phytochromes. If the night period is interrupted by a red light pulse, phytochromes reset the timing mechanism and flowering is disrupted. However, if the red light exposure is followed by far-red light, phytochromes return to the inactive state and flowering continues as normal.

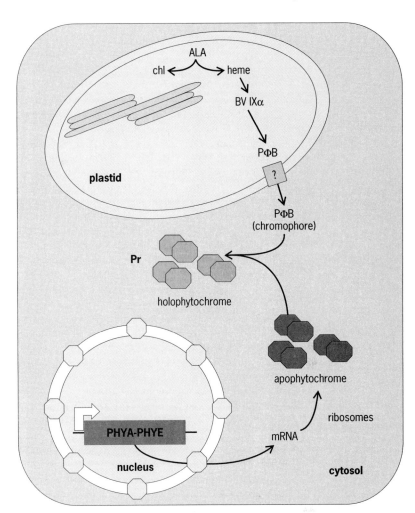

Fig. 1. Model of phytochrome biosynthesis. Apophytochromes are encoded by a small nuclear gene family (*PHYA–PHYE*). The phytochrome chromophore, phytochromobilin (PΦB), is synthesized in plant plastids and transported to the cytosol via an unknown mechanism. Holophytochrome is formed in the cytosol by the binding and subsequent covalent attachment of PΦB to apophytochrome.

Another noted and highly important phytochrome-regulated response in plants is shade avoidance. The shade avoidance response allows plants to detect when they are growing in the shade of a neighbor and to adjust their growth accordingly to maximize exposure to light. Direct sunlight is rich in red light, while light reflected from neighbors is enriched for far-red light due to the preferential absorption of red light for photosynthesis by the neighbor that is growing in full sunlight. Absorbance of far-red-enriched light by the phytochromes of shaded plants results in stem elongation that allows the shaded plant to propel itself out of the shade of its neighbor.

Other organisms. Phytochromes exist in organisms as simple as unicellular prokaryotes, in addition to their presence in eukaryotic green algae such as *Mesotaenium caldariorum* and higher plants. The first prokaryotic phytochrome-like gene was identified by David Kehoe and Arthur Grossman in the cyanobacterium *Fremyella diplosiphon* as a regulator of a light-dependent adaptational process known as complementary chromatic adaptation. Following

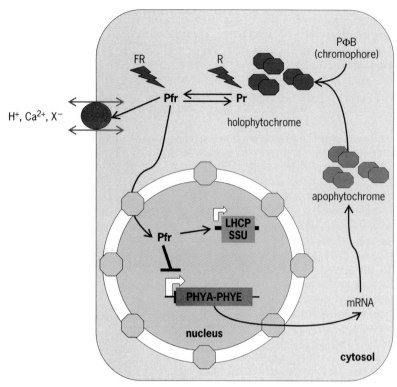

Fig. 2. Phytochrome molecular functions in plant cells. Phytochromes are synthesized in the red-light-absorbing ground state (Pr). Upon absorption of red light, phytochromes photoconvert to the far-red-light-absorbing form (Pfr). Phytochromes have multiple biochemical outputs, including effecting changes at the membrane that result in fluxes of calcium ions (Ca^{2+}) or protons (H^+) and translocation of Pfr into the nucleus, where it effects gene expression through interactions with transcription factors such as PIF3.

this report, the release of several sequenced prokaryotic genomes led to the identification of phytochromes in a wide range of cyanobacteria, including *Synechocystis* sp. PCC 6803 and *Anabaena* sp. PCC7120, as well as in other prokaryotes, including *Deinococcus radiodurans, Agrobacterium tumefaciens, Rhodopseudomonas palustris,* and *Pseudomonas aeruginosa.* Apart from the original photoreceptor identified in *F. diplosiphon,* many of these receptors have not been linked definitively to biological phenomena in the organisms in which they are found.

Where definitive studies have been conducted, analyses of these different classes of phytochromes show that most of these molecules adopt two spectrally distinct forms and possess kinase activity. Major distinctions observed for these molecules are the site of chromophore attachment and the identity of the chromophore molecule that attaches to the apoprotein. Higher-plant, green-algal, and many cyanobacterial phytochromes use phytochromobilin (higher plants) or phycocyanobilin (all others characterized to date) as the chromophore that attaches to a conserved cysteine in the recognized chromophore-binding domain of the photosensory portion of apophytochromes. Bacterial phytochromes, conversely, have been shown to use biliverdin as the chromophore that attaches to a conserved N-terminal cysteine residue. The use of the N-terminal residue as the biliverdin chromophore at-

tachment site in bacteriophytochromes was recently confirmed in the crystal structure obtained for the chromophore-binding domain of the *D. radiodurans* bacteriophytochrome.

Molecular mechanisms. Absorption of light by phytochrome leads to a conformational change that initiates a signal transduction cascade that results in changes in cellular metabolism, growth, and development. Phytochromes appear to have distinct classes of biochemical outputs. The first consists of changes at the membrane level resulting in fluxes of protons, calcium ions, or other second messengers (**Fig. 2**). How these ion fluxes and second messengers are directly related to phytochrome-dependent responses has yet to be definitively determined.

An additional phytochrome output in response to light is autophosphorylation. Light-dependent phosphorylation is directed by the histidine-kinase-related domain of phytochromes. Prokaryotic phytochromes have been shown to possess histidine kinase activity, while higher-plant phytochromes are serine/threonine kinases. Recent results also show that phosphatases act to remove the phosphate that is added. Together, the activity of such phosphatases and the intrinsic kinase activity of phytochromes are expected to result in phosphorylation being used as a posttranslational modification that regulates phytochrome signal output. The phosphorylation state of phytochromes presumably affects their ability to interact with downstream signaling components, including transcription factors and proteins such as PKS1 (phytochrome kinase substrate 1). PKS1 is a cytoplasmic protein that has been shown to physically interact with phytochromes, resulting in phosphorylation of PKS1 and negative regulation of phytochrome activity.

The most recently identified and most definitively characterized phytochrome output at the molecular level occurs via the light-induced movement of phytochrome into the nucleus. Phytochromes, particularly PhyA and PhyB, translocate into the nucleus in the Pfr form. In the nucleus these molecules form speckles that are necessary for physiological function. PhyC, PhyD, and PhyE appear to be found in the nucleus constitutively; however, the abundance of nuclear speckles is influenced by light. Once in the nucleus, Pfr interacts with a number of different components, including transcription factors such as PIF3 (phytochrome interacting factor 3), to effect changes in gene expression, which in turn leads to downstream phytochrome-dependent physiological changes. The observation that phytochromes interact with transcription factors and form speckles in the nucleus led to a compelling hypothesis that these nuclear speckles represent active transcriptional complexes that regulate various aspects of light-dependent growth and development.

Phytochrome movement into the nucleus is associated with its cell autonomous activities. Subcellular localization assays and the intracellular injection of signaling intermediates such as cyclic guanosine monophosphate, (cGMP), calmodulin, and calcium support cytosolic, cell autonomous functions of phytochromes. In addition, phytochromes mediate intercellular or cell nonautonomous responses. Such intercellular roles for phytochromes were largely discovered through the early use of microbeam irradiation to photoactivate phytochromes in discrete plant tissues. Such experiments are designed to induce a phytochrome response in a localized region of the plant and to assay for a particular phytochrome-mediated response at discrete sites in the plant. The most commonly recognized cell nonautonomous phytochrome response is the photoperiodic induction of flowering. Such floral induction experiments established that the photoperiod-dependent induction of flowering in photoperiod-sensitive plant species involves the perception of light by leaf-localized phytochromes followed by transport of an inductive substance from leaves to the apical meristem of plants. Furthermore, leaf-localized phytochromes have been shown to participate in shade-avoidance responses that result in changes in cell elongation in stems. Such studies support the existence of tissue-specific pools of phytochromes, which regulate discrete aspects of light-dependent growth and development through intercellular coordination of growth and development. The mechanisms of such intercellular phytochrome signaling are still under investigation.

Practical applications. Phytochrome photoactivity has largely been studied by the isolation of apoprotein, chromophore biosynthetic, or downstream signaling mutants. Phytochrome mutants display defects in photomorphogenesis under white, red, and/or far-red light. While many of these aberrant responses are detrimental to plant fitness, others could be potentially useful, such as the ability to override neighbor detection. Most crops are grown under relatively high-density conditions. Many plants grown under these conditions allocate a great deal of energy to detecting neighbors and initiating shade-avoidance responses. Genetic and transgenic studies have been performed that alter phytochrome signaling in order to develop crops that exhibit increased performance under high-density growth conditions. A greater understanding of the signaling pathways regulated by phytochromes will enhance our ability to generate high yielding crops.

For background information *see* LIGHT; PHOTOMORPHOGENESIS; PHOTOPERIODISM; PHOTORECEPTION; PHYTOCHROME; PLANT GROWTH; SIGNAL TRANSDUCTION in the McGraw-Hill Encyclopedia of Science & Technology. 　　Beronda L. Montgomery

Bibliography. A. Nagatani, Light-regulated nuclear localization of phytochromes, *Curr. Opin. Plant Biol.*, 7:708–711, 2004; P. H. Quail, Phytochrome photosensory signalling networks, *Nat. Rev. Mol. Cell. Biol.*, 3:85–93, 2002; N. C. Rockwell, Y. S. Su, and J. C. Lagarias, Phytochrome structure and signaling mechanisms, *Annu. Rev. Plant. Biol.*, 57:837–858, 2006; L. C. Sage, *Pigment of the Imagination: A History of Phytochrome Research*, Academic Press, New York, 1992; I. Schepens, P. Duek, and C. Fankhauser, Phytochrome-mediated light signalling in *Arabidopsis*, *Curr. Opin. Plant Biol.*, 7:564–569, 2004; J. R. Wagner et al., A light-sensing knot revealed by the structure of the chromophore-binding domain of phytochrome, *Nature*, 438:325–331, 2005.

Plant hormone receptors

Plant hormone receptors are proteins to which hormones bind, triggering responses in the target cells which express the receptors.

There are basically five plant hormone classes—abscisic acid, auxin, cytokinin, gibberellin, and ethylene—all of which are simple organic molecules that have been chemically defined for many decades. More recently, additional hormones such as brassenosteroids and jasmonic acid have been identified, and there are likely to be many more given the complexity of plant secondary metabolism (metabolism of secondary plant products that are not necessary for survival, but that may be useful to the plant, such as pigments, scents, flavors, etc.). Despite years of effort and detailed knowledge of the chemical structure and physiological and developmental effects of plant hormones, dissecting hormone signal transduction pathways has lagged, and plant hormone receptors have been particularly elusive. Recent advances in this area have identified typical transmembrane receptor kinases for ethylene, cytokinin, and brassenosteroids, similar to those known for many well-characterized signaling pathways. However, in 2005 unusual soluble nuclear receptors for auxin and gibberellin were discovered. Because of their unconventional properties, it is perhaps not surprising that these receptors have been difficult to identify.

Receptor-mediated signal transduction. The receptors for gibberellin and auxin have been identified recently as a hormone sensitive lipase (HSL)–like protein called gibberellin insensitive dwarf1 (GID1) from rice, and a small family of F-box proteins (an F-box is a protein motif of ~50 amino acids functioning as a site of protein-protein interaction) typified by transport inhibitor response1 (TIR1) from *Arabidopsis* (a small flowering plant of the mustard family often used to study plant biology), respectively.

In both these pathways, targeted protein degradation is of central importance. Both gibberellin and auxin signal transductions are mediated by promoting the degradation of proteins to activate hormone response. The addition of a polyubiquitin chain to a protein targets it for degradation by the 26S proteasome. This requires three different enzymes: ubiquitin activating enzyme (E1), ubiquitin conjugating enzyme (E2), and ubiquitin protein ligase (E3). It is the E3 that is responsible for specifically selecting the target protein. One class of E3 ubiquitin protein ligase is the SCF (SKP1, cullin, and F-box containing) complex type protein. The F-box component of the SCF confers target specificity. Both the gibberellin and auxin receptors are involved in regulating hormone-dependent interactions between specific SCFs and their target proteins.

Identification of TIR1 as a receptor for auxin.

Auxin plays an essential role in many plant processes, including pattern formation in embryogenesis, tropisms (orientation movements of a sessile organism in response to stimuli), organ development, apical dominance (inhibition of lateral bud growth by the apical bud of a shoot), and lateral root initiation. Auxin was chemically defined in the 1930s. Since then, many different techniques have been used in the hunt for its receptor(s). Biochemical approaches to isolate proteins with high affinity for auxin uncovered a range of different proteins, among which auxin binding protein1 (ABP1) has been particularly well characterized. However, there is no evidence for any involvement of ABP1 in auxin-regulated gene expression, and its role is apparently limited to membrane-level responses.

The hunt for auxin signaling components by genetic screens has flourished since the adoption of *Arabidopsis* as a model system, and this approach has proved to be highly successful. A wide range of mutants with altered responses to auxin have been identified, resulting in the characterization of many components of the auxin signaling pathway. This includes the auxin receptor TIR1, which was originally identified in mutant screens for *Arabidopsis* plants with altered auxin-related responses. Although the molecular identity of TIR1 as an F-box protein has been known since 1998, it has taken several more years of research to link TIR1 with auxin perception. This is perhaps a result of its being the first example of a protein with dual receptor, F-box protein function.

Role of TIR1 in auxin signaling. Central to auxin signaling are the two families of proteins: the auxin response factor (ARF) family of DNA-binding transcription factors and the Aux/IAA (indoleacetic acid) family of transcriptional corepressors.

Q-rich (glutamine-rich) ARFs promote the expression of auxin responsive genes by binding to specific auxin response elements in their promoter regions. Dimerization of ARFs with Aux/IAAs blocks ARF function, resulting in transcriptional repression. In the presence of auxin, the Aux/IAA proteins are degraded, releasing the repression (**Fig. 1**).

TIR1 is the F-box protein in an SCF-type ubiquitin-protein ligase complex that targets the Aux/IAA proteins for degradation in an auxin-dependent manner. Such signal-triggered degradation mechanisms are common throughout eukaryotes, but in all previous cases the signal triggers modification of the target protein, promoting its interaction with the relevant SCF. In contrast, auxin acts by binding directly to TIR1, promoting the interaction between TIR1 and Aux/IAA proteins. As a result, Aux/IAA proteins are ubiquitinated, targeting them for degradation. This releases ARFs from repression, allowing them to promote expression of auxin-inducible genes.

Auxin signaling F-box proteins (AFBs). *Arabidopsis* plants lacking the TIR1 protein have only mild defects in growth and development. This suggests that there are other receptors for auxin that can compensate for TIR1. There are over 700 F-box proteins in *Arabidopsis*. At least three other proteins, auxin signaling F-box proteins (AFBs), have been identified that are closely related to TIR1 and interact with Aux/IAAs in an auxin-dependent manner. While mutant plants lacking single AFB proteins have mild phenotypes, plants lacking all four of these proteins (the three AFBs and the TIR1) have severely affected development, leading to embryo lethality. This indicates their overlapping functions in auxin response.

Identification of GID1 as a gibberellin receptor. First identified in the 1930s, gibberellins are tetracyclic diterpenoids involved in a diverse range of processes from seed germination to stem elongation. The gibberellin receptor gibberellin insensitive dwarf1 (GID1) was identified through forward genetic approaches in rice. GID1 is a soluble protein found in the nucleus. It shares homology to the hormone sensitive lipase (HSL) family; however, GID1 lacks the amino acids essential for enzyme activity. GID1 has a high affinity for gibberellin. Loss-of-function

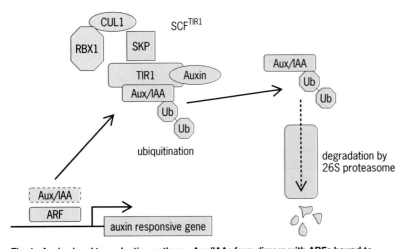

Fig. 1. Auxin signal transduction pathway. Aux/IAAs form dimers with ARFs bound to promoters of auxin regulated genes, causing transcriptional repression. Repression is released when auxin binds to TIR1, promoting the interaction with Aux/IAAs, which are ubiquitinated and subsequently degraded.

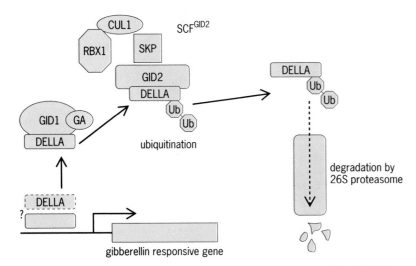

Fig. 2. Gibberellin signal transduction pathway. DELLA proteins repress gibberellin (GA) response in an as yet undefined mechanism. Gibberellin binding to GID1 promotes interaction of GID1 with DELLA proteins. The DELLA proteins are then ubiquitinated by the SCF^{GID2} and subsequently degraded, releasing repression.

mutations in GID1 confer dwarfism and gibberellin insensitivity, while overexpression of GID1 leads to long spindly plants characteristic of a gibberellin overdose.

Role of GID1 in gibberellin signaling. In a similar mechanism to that described for auxin, the gibberellin receptor is involved in the gibberellin-dependent degradation of proteins that function in the gibberellin signaling pathway. GID1 is involved in promoting the interaction of specific target proteins to an SCF-type E3 protein ubiquitin ligase in an as yet undefined mechanism.

The targets for gibberellin-dependent GID1-mediated degradation are proteins of the DELLA (named after a conserved N-terminal stretch of amino acids) family, which act as repressors of the gibberellin response. In the presence of gibberellin, DELLA proteins are targeted for degradation through their interaction with an SCF complex containing the F-box proteins of the family defined by GID2 (SCF^{GID2}) in rice and SLY (SLEEPY) in *Arabidopsis*. Gibberellin binds to GID1 in the nucleus to promote GID1 interaction with the DELLA proteins. This interaction somehow promotes the interaction of DELLA proteins with SCF^{GID2}, resulting in the ubiquitination and degradation of the DELLA repressor proteins by the 26S proteasome (**Fig. 2**). In the absence of DELLA proteins, repression is lifted, triggering gibberellin responses.

Soluble hormone receptors. These recent advances have illustrated the diversity of plant hormone receptors ranging from canonical transmembrane receptor kinases to soluble nuclear proteins in families not previously implicated in signal perception. As more hormones and their receptors are characterized, it will be interesting to see the full range of options used by plants to perceive their local environment. Intriguingly, the *Arabidopsis* RNA-binding protein FCA (flowering time control protein in *Arabidopsis*), which has previously been shown to regulate the transition of plants to flowering, has recently been shown to bind abscisic acid, and hence may act as a receptor in the flowering response.

For background information *see* ABSCISIC ACID; APICAL DOMINANCE; AUXIN; CYTOKININS; ETHYLENE; GIBBERELLIN; PLANT GROWTH; PLANT HORMONES in the McGraw-Hill Encyclopedia of Science & Technology.   Lynne Armitage; Ottoline Leyser

Bibliography. G. Badescu and R. Napier, Receptors for auxin: Will it all end in TIRs?, *Trends Plant Sci.*, 11:217–223, 2006; A. Bishopp, A. P. Mahonen, and Y. Helariutta, Signs of change: Hormone receptors that regulate plant development, *Development*, 133:1857–1869, 2006; D. Bonetta and P. McCourt, A receptor for gibberellin, *Nature*, 437:627–628, 2005; L. Hartweck and N. Olszewski, Rice GIBBERELLIN INSENSITIVE DWARF1 is a gibberellin receptor the illuminates and raises questions about GA signaling, *Plant Cell*, 18:278–282, 2006; G. Parry and M. Estelle, Auxin receptors: A new role for F-box proteins, *Curr. Opin. Cell Biol.*, 18:152–156, 2006; A. Woodward and B. Bartel, A receptor for auxin, *Plant Cell*, 17:2425–2429, 2005.

Pluto

Formerly considered the outermost planet, tiny Pluto is now recognized to be one of the largest and closest members of a disk of icy planetesimals that surrounds the solar system beyond the orbit of Neptune. For this reason, the International Astronomical Union recommended reclassifying Pluto in a new category, "dwarf planet," in 2006.

Pluto was discovered on February 18, 1930, by C. W. Tombaugh at the Lowell Observatory, Flagstaff, Arizona, on photographic plates taken with a special astronomical camera as part of a systematic search. The presence of a planet beyond Neptune's orbit had been predicted independently by P. Lowell and W. H. Pickering (among others) on the basis of analyses similar to those that had led U. J. Leverrier and J. C. Adams to the prediction of Neptune. It was thought that there were perturbations in the motion of Uranus that Neptune alone could not explain. Pluto was found surprisingly near its predicted position, since it is now known that the mass of Pluto is far too small to have caused the (evidently spurious) perturbations.

Charon, Pluto's largest satellite, was discovered in 1978 by J. W. Christy and R. Harrington at the nearby Flagstaff station of the U.S. Naval Observatory. H. Weaver and a team of astronomers using the *Hubble Space Telescope* discovered two additional satellites, Hydra and Nix, in 2005 (**Fig. 1**).

Orbit. Pluto's orbit (**Fig. 2**) has a semimajor axis (mean distance to the Sun) of 5.9×10^9 km (3.7×10^9 mi), an eccentricity of 0.25, and an inclination of $17.1°$. The inclination and eccentricity are larger than those of any of the planets. At a mean orbital velocity of 2.95 mi/s (4.7 km/s), it takes Pluto 248.0 years to make one revolution around the Sun. The large orbital eccentricity means that at perihelion Pluto is closer to the Sun than Neptune, but the orbits of the two bodies do not intersect.

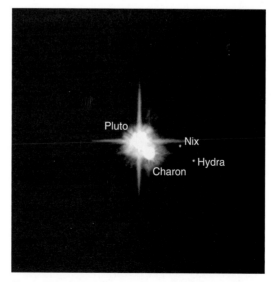

Fig. 1. Pluto and its three known moons imaged by the *Hubble Space Telescope* on February 15, 2006. (*NASA, ESA, H. Weaver (JHUAPL), A. Stern (SwRI), HST Pluto Companion Search Team*)

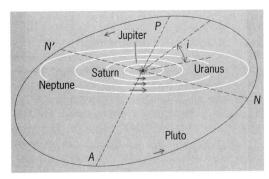

Fig. 2. Orbit of Pluto, a perspective view to show the inclination *i* and eccentricity of the orbit. *A*, aphelion; *P*, perihelion; *NN'*, line of nodes. (*After L. Rudaux and G. de Vaucouleurs, Astronomie, Larousse, 1948*)

Telescopic appearance. Pluto is visible only through fairly large telescopes, since its visual magnitude at mean opposition is 15.1, thousands of times fainter than our visual threshold. Periodic variations in its brightness demonstrate that the surface of Pluto is covered with bright and dark markings and indicate a period of rotation of 6.4 days. This is longer than the rotational period of any of the planets except Mercury and Venus.

Size and mass. The radius of Pluto is 1153 km (716 mi), making Pluto significantly smaller than the Earth's Moon (whose radius is 1738 km or 1080 mi), while Charon's radius is 604 km (375 mi). Charon is thus about half the size of Pluto itself, making this the most closely matched pair in the solar system. The density of Pluto is 2.0 g/cm^{-3}, while that of Charon is 1.65 g/cm^{-3}. These densities suggest compositions of ice and rock within the range exhibited by Saturn's regular satellites. Nix and Hydra are too small to permit determinations of size and mass.

Surface and atmosphere. The near-infrared spectrum of Pluto reveals absorption features of solid nitrogen (N_2), carbon monoxide (CO), and methane (CH_4) ice. From these absorptions, it is possible to conclude that nitrogen is the dominant ice on Pluto's surface (some 20–100 times more abundant than the other ices), which means it must also be the major constituent of the planet's tenuous atmosphere. The nitrogen absorption in the spectrum indicates a surface temperature near 38 K ($-391°$F), consistent with an atmospheric surface pressure roughly 10^{-5} times that on Earth. Such a pressure was deduced directly from observations of a stellar occultation in 1988, which argues against the possible presence of a thick haze in Pluto's lower atmosphere. (If such a haze exists, it could affect the radius measurement and hence the deduction of the planet's mean density.) Its size, surface temperature, composition, and thin nitrogen atmosphere make Pluto very similar to Triton, Neptune's largest satellite. However, there are important differences such as the absence of frozen carbon dioxide (CO_2) on Pluto.

Seasons on Pluto are modulated by the large eccentricity of the orbit. Pluto reached perihelion in 1989, which brought it closer to the Sun than Neptune. This means that during the 1988 occultation Pluto was observed at the warmest time in its seasonal cycle, when the atmospheric pressure might be expected to be at its maximum value.

However, observations of another stellar occultation by Pluto in 2002 revealed that the atmosphere was actually twice as massive as it had been in 1988. This result had been anticipated by some scientists, who had predicted that the changing orientation of Pluto as it moved around its orbit would expose polar deposits of frozen atmospheric gases to the Sun, if such deposits existed. The atmospheric pressure would then increase as the polar deposits sublimed. The same seasonal change in atmospheric pressure is observed on Mars, where the dominant gas is carbon dioxide (CO_2), which condenses out at the winter polar cap. On Pluto, this seasonal effect will soon be overwhelmed by the decrease in temperature as the planet moves farther from the Sun in its eccentric orbit, and the atmospheric pressure will gradually diminish as nitrogen again freezes out on Pluto's surface. Thus, to observe these processes it would be advantageous to explore the planet with a spacecraft as soon as possible, and such a mission is currently underway.

Satellites. Pluto has three small satellites—Charon, Nix, and Hydra—with Charon far larger than the other two (see **table**; **Fig. 3**). Demonstrating another striking difference with Pluto, Charon apparently has little if any frozen methane on its surface, which instead exhibits the spectral signature of water ice. The orbit of Charon is unique in the solar system in that the satellite's period of revolution is identical to the rotational period of the planet. Thus an inhabitant of Pluto who lived on the appropriate hemisphere would see Charon hanging motionless in the sky. From the opposite hemisphere the satellite would be invisible. (On Earth, artificial satellites are put into such geostationary orbits for communications and television broadcasting.)

Satellites of Pluto						
Name	Orbit radius, km (mi)	Period, days	Eccentricity	Inclination*	Radius, km (mi)	Magnitude at mean opposition
Charon	19,571 (12,161)	6.4	~0	~0°	604 (375)	17.3
Nix	48,675 (30,245)	24.9	~0	~0°	~45 (28)?[†]	24.6
Hydra	64,780 (40,252)	38.2	~0	~0°	~45 (28)?[†]	24.4

*With respect to Pluto's equatorial plane.
[†]Estimated from assumed reflectivity.

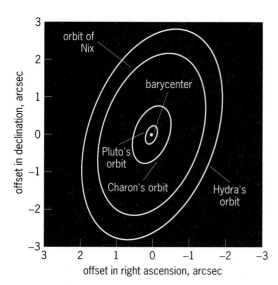

Fig. 3. Orbits of Pluto's satellites. *(After H. A. Weaver et al., Discovery of two new satellites of Pluto, Nature, 439:943–945, 2006)*

Pluto and the Kuiper Belt. In 1992, the first of what is proving to be a huge number of icy planetesimals (200 km or 125 mi in diameter or less) was discovered orbiting the Sun at distances comparable to or greater than the orbit of Pluto. These are members of the Kuiper Belt, a disk-shaped distribution of material left over from the formation of the solar system. These objects appear to be representative of the solid components required to initiate the growth of giant planets. Held in the cosmic deep freeze of trans-Neptunian space, they must be among the most primitive, unprocessed objects still circulating in the solar system. Collisional fragments occasionally make their way into the inner solar system, where they become the short-period comets. In the course of this journey, they lose some of their most volatile constituents but still represent a unique resource for studying the early days of the solar system. The European Space Agency (ESA) mission Rosetta is scheduled to carry out a detailed investigation of a comet in 2012.

The existence of these small bodies (Kuiper Belt objects) finally provides a proper context for Pluto itself. Because of its small size and the high inclination and eccentricity of its orbit, Pluto has been an "ugly duckling" among the planets. As one of the largest members of the Kuiper Belt, however, it becomes a majestic "swan," swimming serenely at the boundary between the planets and this disk of debris left over from the formation of the giant planets.

By the middle of 2006, astronomers had discovered six Kuiper Belt objects with diameters of approximately 1000 km (625 mi), comparable in size to Charon, and at least one, Eris, that may be slightly larger than Pluto. They have also found that several of the Kuiper Belt objects, including Eris, have satellites. Of the larger bodies, Sedna and Eris are particularly interesting. The diameter of Sedna is 1000–1500 km (625–940 mi), still poorly known as it depends on an assumed reflectivity. (The more reflec-

tive the surface, the smaller the object needs to be to produce the same brightness.) Sedna's orbit has a very high eccentricity of 0.85 and an inclination of 12°, but it is the perihelion distance of 76 astronomical units that is most remarkable: Sedna's closest approach to the Sun is more than twice as distant as that of Pluto and is the farthest of any known Kuiper Belt object. This orbit is well beyond the gravitational influence of Neptune, and raises the questions of how Sedna got there and whether there are still more large bodies in such distant orbits. Eris is indeed larger, but its orbit lies somewhat closer to the Sun. Current models of solar system formation do not allow enough material at these enormous distances to form objects the size of these two Kuiper Belt objects, so they must have been scattered from closer orbits. Perhaps other large icy bodies suffered the same fate, and ongoing searches will reveal them.

In 2003, the National Aeronautics and Space Administration (NASA) authorized the *New Horizons* Pluto-Kuiper Belt mission. A single spacecraft was launched on January 19, 2006, and is scheduled to fly past Pluto and Charon in 2015, making measurements and recording images. The spacecraft will then venture forth into the Kuiper Belt and attempt to fly past one or more of these small distant worlds.

For background information *see* COMET; KUIPER BELT; NEPTUNE; PLANET; PLANETARY PHYSICS; PLUTO; SATELLITE (ASTRONOMY); SOLAR SYSTEM in the McGraw-Hill Encyclopedia of Science & Technology.
Tobias C. Owen

Bibliography. J. K. Beatty, C. C. Petersen, and A. Chaikin (eds.), *The New Solar System*, 4th ed., Cambridge University Press, 1999; W. G. Hoyt, *Planets X and Pluto*, University of Arizona Press, 1980; D. Morrison and T. Owen, *The Planetary System*, 3d ed., Addison Wesley, 2003; S. A. Stern and J. Mitton, *Pluto and Charon: Ice Worlds on the Ragged Edge of the Solar System*, Wiley, 1998; S. A. Stern and D. J. Tholen (eds.), *Pluto and Charon*, University of Arizona Press, 1997.

Precast and prestressed concrete

In the field of structural concrete construction, two basic concepts are generally applied in practice: conventionally reinforced and prestressed concrete.

In conventionally reinforced concrete, ordinary un-prestressed reinforcing steel is integrated with the concrete matrix of a structural element such as a slab, beam, column, or wall for the purpose of providing resistance to concrete strains and cracking. The reinforcing steel acts only when the concrete is strained under tension or compression as it is fully bonded to the concrete.

In prestressed concrete, predetermined forces are imposed on the concrete structural member by special high-tensile steels prior to the members functioning as a load-supporting element. These preemptive prestressing forces are calculated and positioned to mitigate the anticipated stresses that subsequently occur during the service loading of the member, thus

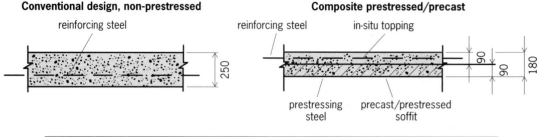

Fig. 1. Flat slabs (one-way span). Design for live load = 4 kPa (83.6 lb/ft²). Clear span = 8 m (26 ft).

Material	Conventional design, non-prestressed	Prestressed/precast	Material savings
Concrete	0.25 m³/m²	0.18 m³/m²	28%
Reinforcing steel	18.30 kg/m²	6.20 kg/m²	45%
Prestressing steel	—	3.85 kg/m²	45%

reducing the net final stresses developed and allowing a significant reduction in the required size of the concrete elements.

Both conventionally reinforced and prestressed concrete elements can be fabricated as precast units, which can then be integrated to form a structural framing system.

Construction methods. Over the past several decades, precast and prestressed concrete construction methods have produced considerable savings in materials, labor, and construction time.

The reduction in required material quantities due to prestressing produces lighter-weight building structures. Since gravitational and seismic loading forces are proportional to structure weight, a lighter-weight building will enjoy additional reductions in required structural sizes for all vertical and lateral force-resisting elements such as columns, beams, slabs, moment frames, shear walls, and foundation elements. *See* DESIGNING FOR AND MITIGATING EARTHQUAKES.

For high-rise buildings, a reduction in concrete sizes will also result in shallower structural concrete deck framing depths, which in turn will reduce the floor-to-floor height requirements and the overall height of the building structure. This will in turn reduce overturning resistance requirements under wind and seismic loading, while simultaneously lowering the cost of vertical elevator and stair shafts, mechanical and electrical risers, exterior façade cladding, and other vertical elements of construction.

Prestressing usually is performed with special high-tensile steels that provide maximum strength with a minimum quantity of steel, resulting in significant savings in steel tonnage and cost. Generally, the prestressing steel strength is approximately 4.5 times stronger than ordinary reinforcing steel, but costs only twice as much per unit weight as ordinary reinforcing steel.

Prestressing of beams and slabs is most economically applied in long-line precasting mass-production operations. Because of the inherent efficiencies of mass-production techniques, large reductions in labor, formwork, and hardware costs can be realized.

The precast prestressed structural and nonstructural concrete units are usually in production prior to and during site and foundation preparation work. This enhances the speed of construction as upper-floor structural elements can be in production while site and foundation works are in progress.

By its production logistics, precast units usually have had substantial curing periods and significant time for initial shrinkage of the concrete material to occur prior to installation. Since the major structural and architectural components are inherently precured during precasting, the finished structure, composed of integrated precast structural units with in-situ concrete, will be subjected to less shrinkage forces and consequential cracking damage in the building frame.

Benefits. Precast structural units are usually designed to minimize temporary formwork and shoring. As a result, additional savings in construction time and costs can be realized.

The overall savings in construction materials will have a significant impact on the environment. To manufacture one ton of cement, the power required involves the emission of approximately one ton of carbon dioxide (CO_2) into the atmosphere, thus compounding the problem of global warming. Savings in construction materials would then mean a reduction in the total consumption of fossil fuels, for which the supply is diminishing rapidly. Cement and steel constitute the majority of the construction materials used in precast and prestressed concrete construction. Among principal construction materials, cement production is the third most polluting process, while steel production ranks higher than cement in terms of fuel consumption and CO_2 emission.

The actual material savings can be analyzed by considering two basic structural elements: a typical floor slab and a typical floor beam sample. In the case of a conventional in-situ floor slab design, considerable savings in materials can be realized when a composite prestressed and precast slab system is used. For instance, when prestressed/precast construction is applied to a slab system spanning 8 m (26 ft) and designed to support a live load of 4 kPa (83.6 lb/ft²), a savings of 28% in concrete and 45% in steel materials can be realized (**Fig. 1**). The environmental and natural-resource impact of this concept is very significant, especially since many millions of square meters of slab construction are in progress on the planet each day.

In the case of conventionally designed reinforced concrete beams, significant savings in materials can also be realized by precasting and prestressing techniques. For example, a precast and prestressed beam spaced at 4 m (13 ft) on centers and spanning 12 m (39 ft) and designed to support a load of 4 kPa (83.6 lb/ft²) can result in a saving of 60.8% in concrete and 66% in steel quantities over a conventionally designed un-prestressed beam cast in-situ supporting a similar span and loading (**Fig. 2**).

Floating ocean platforms. Through innovation with precast and prestressed concrete, some latest trends are now focused toward the development and construction of floating ocean platforms used to extract minerals, energy, and other natural resources. Precast and prestressed concrete platforms have been constructed to support phosphate processing plants, floating liquefied propane gas (LPG) processing and storage facilities, and oil exploration platforms that are transported afloat and grounded for drilling (**Fig. 3**).

For ocean platforms, the size and weight of prestressed and precast concrete construction will provide the greatest dynamic stability due to its large inertial advantage. Long-term durability of concrete construction in an ocean environment has been proven by actual service of existing prestressed precast concrete platforms over the last several decades.

To exploit the greatest source of energy in the world, floating platforms for the support of ocean thermal energy conversion (OTEC) plants are also being planned with precast and prestressed concrete for the main hull structure. These plants will convert the difference in temperature between the warm sea surface water and the cold deep ocean water [approximately 1000 m (3300 ft) depth] into electrical energy through a system of heat-transfer units and turbines. This warm sea surface water derives its energy from the Sun and is concentrated along the equatorial belt between the Tropic of Cancer and Tropic of Capricorn girdling the Earth. Along this equatorial belt, the energy received daily from the

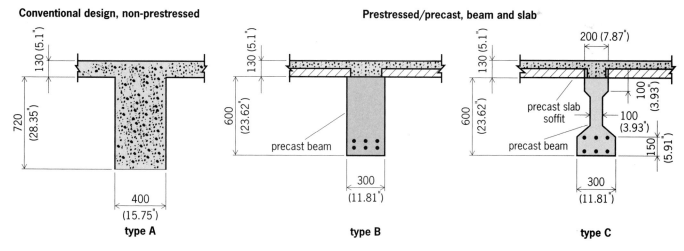

Material	Conventional design, non-prestressed Type A	Prestressed/precast Type B	Material savings from type A to type B	Prestressed/precast Type C	Material savings from type B to type C	Material savings from type A to type C
Concrete	0.288 m³/m	0.18 m³/m	37.5%	0.113 m³/m	37.2%	60.8%
Reinforcing steel	42.0 kg/m	6.20 kg/m	66%	6.0 kg/m	—	66%
Prestressing steel	—	8.47 kg/m	66%	8.47 kg/m	—	66%

Fig. 2. Beams. Span = 12 m (39 ft). Clear spacing = 4 m (13 ft) center to center. Live load = 4 kPa (83.6 lb/ft²). Other dimensions are in millimeters (inches).

Fig. 3. Concrete Island Drilling System (CIDS) platform constructed with precast and prestressed concrete elements designed to withstand the severe ice-floe pressures of the Arctic. The CIDS platform replaces the building of gravel islands, which cost approximately $100 million to construct, whether oil was found or not. In contrast, the Concrete Island Drilling System cost only $75 million to construct and has been reused for the past 20 years to explore numerous locations along the north Alaskan coastline in the Beaufort Sea without showing any deterioration in the hull. In addition, it leaves no footprint on the seabed. The CIDS is presently operating along the east coast of Russia.

Sun is equivalent to approximately 10,000 times the total amount of energy consumed worldwide in a day. The cold water circulating about the deep ocean comes from the polar regions. This clean thermal energy source is literally inexhaustible, and prestressed precast concrete can play a major role toward its exploitation.

The energy derived from OTEC operations is developed in a closed-cycle system with no emissions or environmental pollution. This pure form of making energy can also be used to manufacture liquid hydrogen, a fuel highly regarded for its purity and considered as a possible replacement for fossil fuels. The hydrogen fuel cell concept will produce electric power with no pollution, as the by-product of hydrogen-oxygen combustion is water. Hydrogen fuel cells can be used to power automobiles, buses, trucks, trains, power plants, ships, spacecraft, and many other vehicles and equipment. In theory, approximately two-thirds of the fossil fuels now used could be replaced by hydrogen-powered fuel cells.

Outlook. The economic and environmental benefits of precast and prestressed concrete can be significant if a coordinated engineering educational program in this regard is implemented. This educational program is critical and should extend worldwide with the cooperation of international bodies.

The greatest impetus to such a program would be its economic benefit. When an economic advantage is available, it becomes very attractive to users who stand to benefit financially from cost and time savings. To maximize profits, owners and developers of a building project prefer to save as much cost and time as possible. The owner or developer would also likely prefer engaging architects and engineers who can provide these amenities of construction

material—cost and time savings—assuming esthetic and functional requirements are equally met.

Architects and engineers worldwide may then be attracted to learning the tools of precast and prestressed concrete construction in order to meet the needs of owners and developers and enjoy added demand for their services. This could enhance greater use of methods that save construction material and time while actually preserving the environment and limited supply of natural resources.

The entire construction and associated manufacturing industry, including its practicing professionals as well as governing bodies and academia, should explore known and yet-to-be-discovered methods of protecting the environment and natural resources by developing new technology. This would also include the development of renewable energy resources, new materials, as well as the recycling of existing materials in order to survive the anticipated material shortages and environmental problems of the future.

For background information *see* ARCHITECTURAL ENGINEERING; BUILDINGS; CONCRETE; CONCRETE BEAM; CONSTRUCTION ENGINEERING; FLOOR CONSTRUCTION; PRECAST CONCRETE; PRESTRESSED CONCRETE; REINFORCED CONCRETE; STRESS AND STRAIN in the McGraw-Hill Encyclopedia of Science & Technology.   Alfred A. Yee

Bibliography. ACI Committee 357, *State-of-the-Art Report on Barge-Like Concrete Structures*, ACI 357.2R-88, American Concrete Institute, 1988; *PCI Design Handbook: Precast and Prestressed Concrete*, 5th ed., Precast/Prestressed Concrete Institute, 1999; A. A. Yee, Design considerations for precast prestressed concrete building structures in seismic areas, *PCI J.*, 36(3):40-55, 1991; A. A. Yee, Honeycomb units for barges and floating platforms, *Struc. Eng. Prac.*, 1(1):89-93, 1982; A. A. Yee, Prestressed concrete for buildings, *PCI J.*, 21(5):112-157, 1976; A. A. Yee, Social and environmental benefits of precast concrete technology, *PCI J.*, 46(3):14-19, 2001.

Precipitation scavenging

Precipitation scavenging is an efficient mechanism of cleaning the atmosphere. It includes all the processes that take up pollution in the liquid drops or ice crystals of clouds and deposit it onto the ground in rain or snow.

The Earth's atmosphere is composed mainly of nitrogen and oxygen (**Table 1**). In addition to some noble gases, it contains trace gases in varying quantities (**Table 2**).

The atmosphere also contains small particles ranging from nanometers up to a few micrometers in diameter. Their total concentration ranges from 100 cm^{-3} in clean air, up to several 10,000 s cm^{-3} in polluted air. In heavily polluted areas, mass concentration values as high as 2000 μg/m^3 have been reported, while in clean air the mass concentrations are generally a factor of 100 smaller. Aerosol particles smaller than 1 μm are generally formed by condensation processes from the gas phase and by

TABLE 1. Chemical composition of the Earth's atmosphere

Gas	Molar fraction (dry air), [mol mol^{-1}]
Nitrogen (N$_2$)	0.78
Oxygen (O$_2$)	0.21
Argon (Ar)	0.0093
Carbon dioxide (CO$_2$)	365×10^{-6}
Neon (Ne)	18×10^{-6}
Helium (He)	5.2×10^{-6}
Methane (CH$_4$)	1.7×10^{-6}
Krypton (Kr)	1.1×10^{-6}
Water vapor (H$_2$O)	Variable

TABLE 2. Typical concentration range of atmospheric trace gases

Component	Min molar fraction, 10^{-9} (clean air)	Max molar fraction, 10^{-9} (polluted air)
SO$_2$	1	200
CO	120	10,000
NO	0.01	750
NO$_2$	0.1	250
O$_3$	20	500

aggregation. The gases can be of natural or anthropogenic origin. However, giant aerosol particles (larger than 1 μm) have a mechanical origin (such as sea spray, windblown dust, volcanoes, plant debris, and diesel engines), which is mostly natural.

Even though natural and anthropogenic activities constantly add trace gases and particles to the atmosphere, their concentrations remain relatively stable in the ranges given above. In part, this is due to dry deposition—particles and gases settle and deposit on the surfaces, which are mainly located at the ground. Another very important cleaning mechanism of the atmosphere is linked to clouds. They quite efficiently take up (scavenge) gases and particles into the liquid phase, and when precipitating they deposit the pollutant mass on the ground. In the following, we will review the mechanisms that incorporate pollutants into the cloud hydrometeors and discuss their efficiency. But first, the cloud hydrometeors and the dominating processes will be briefly reviewed.

Clouds. Clouds form when air ascends vertically and, due to the expansion cooling of the air, the water vapor condenses and forms cloud droplets. These droplets grow further by condensation, and then collide and coalesce with each other until they become sufficiently heavy to fall against the updraft velocity that is suspending them. Depending on the height of cloud base and the temperature conditions, they might reach the ground as rain. If the cloud temperature is sufficiently below freezing, ice crystals develop. Ice crystals also grow by water vapor deposition, and collision with other droplets and ice crystals. If they become sufficiently heavy, they will reach the ground in solid or liquid form, as a function of the temperature below the clouds. During their entire lifetime, cloud hydrometeors (droplets or ice particles) take up pollution in particulate and

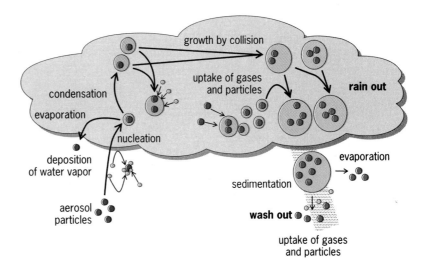

Fig. 1. Microphysical and scavenging processes in liquid clouds.

gaseous form and deposit it on the ground together with precipitation. **Figure 1** depicts these processes for liquid clouds.

Nucleation scavenging of drops. The atmosphere does not allow the formation of drops by agglomerating water vapor molecules alone. To form a tiny drop with a 10^{-3}-μm radius, assembled of 800 molecules, would require a relative humidity of 200%. Consequently, in the atmosphere, droplets form on existing nuclei. Aerosol particles present in the air mass provide these necessary nuclei.

Depending on their size, chemical composition, and the ambient relative humidity, aerosol particles take up a certain amount of water. So even in the absence of a cloud, the atmosphere is full of tiny droplets, representing swollen aerosol particles, which can be visible in the atmospheric as haze.

At a given size, particles carrying hygroscopic material can take up more water. When a given aerosol particle has passed a critical size (around 1 μm), it is called activated and serves as a cloud condensation nucleus. Activated particles do not require any further increase in relative humidity to grow and are considered cloud droplets. This includes essentially all dry aerosol particles larger than 0.02 μm (shaded area in **Fig. 2**) in a given cloud volume. Because these large aerosol particles contain the main particle mass

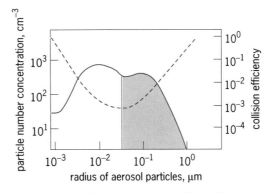

Fig. 2. Aerosol particle number-size distribution (the shaded part is activated and forms drops) and the qualitative behavior of the collision efficiency of aerosol particles with drops.

(about 85–90%), a huge part of the scavenging of particulate pollutants proceeds this way.

Impaction scavenging of aerosol particles. Inside the cloud, unactivated aerosol particles exist between the nucleated drops (the nonshaded part in Fig. 2). These particles can collide with the hydrometeors and become incorporated in the cloud particles. But since most of the particle mass was already scavenged by nucleation, this process does not contribute a significant pollutant mass in precipitation scavenging. However, when the hydrometeors fall toward the Earth's surface, they meet unperturbed aerosol particles (solid line in Fig. 2). Here, the collision with aerosol particles can contribute 20–40% to the aerosol particle loading of the rain on the ground, depending on the height of cloud base. The qualitative dependency of the collision curve of drops as a function of the aerosol size is displayed in Fig. 2 (broken line). Note that the efficiency of the collision process is high for very small particles and very large particles. For the small particles, the Brownian motion of the particles is the dominant process, with increased collision as a function of decreasing particle size. For the larger sizes, the inertial impaction dominates, which increases with increasing drop sizes. In the middle is the Greenfield gap, where the collision efficiency is at a minimum, meaning that only a few particles of theses sizes are captured.

Scavenging of gases. In addition to the particles, numerous trace gases are present in the atmosphere (Table 2). Gases are taken up by drops because they are more or less soluble. The maximum amount of a gas that can be taken up into water is a function of the Henry's law coefficient H, which gives the ratio of the concentration of the gas in water (mol L^{-1}) with respect to the concentration in air (atm). When the gas is very soluble (for example, HNO_3, where $H = 2.1 \times 10^5$ mol L^{-1} atm^{-1} at 298K), the gas is very efficiently scavenged. When the gas is almost insoluble (for example O_3, where $H = 9.4 \times 10^{-3}$ mol L^{-1} atm^{-1} at 298K), the gas mostly remains in the air. Henry's law describes an equilibrium between the concentration in the air and the liquid. Once in the liquid phase, most gases are destroyed by chemical reactions, and thus an equilibrium is never achieved. Consequently, more and more gas can be taken up into the cloud drops. Only the drop lifetime (maximum of 30 min) will limit the gas scavenging.

Summary of cloud processes. Atmospheric pollution is made of aerosol particles and trace gases. Both are taken up into cloud hydrometeors during the lifetime of a cloud. The processes acting inside the clouds are called "in-cloud scavenging" or "wash out." The dominating processes are the nucleation of drops on existing aerosol particles. This transfers almost 90% of the particulate pollution mass inside the cloud volume in the liquid phase. In addition, the liquid drops take up gases according to their solubility and the ongoing chemical destruction in the liquid phase. Once the temperature decreases significantly below 0°C (32°F), the drops freeze. Once frozen, the uptake of particles and gases is significantly reduced, as the solid crystals have difficulty retaining the colliding aerosol particles and dissolving gases.

Through collision, the hydrometeors can become sufficiently large and heavy to fall against the prevailing updrafts toward the ground. During the time of fall, the big drops continue to take up particles and gases. This is called "below cloud scavenging" or "rain out."

Globally, it is estimated that below-cloud scavenging contributes around one-third of the pollution mass in the rain, while the remaining two-thirds is taken up inside the cloud. Roughly the same ratio seems to exist in ground precipitation between the pollution mass that was contributed by the particulate pollution mass (two-thirds) and the gaseous pollution mass (one-third). However, these numbers just represent orders of magnitude, and will be different for the different chemical components that the particles and gases are made of. Also, these numbers will be influenced by the presence of the ice phase, the size of the cloud, and the intensity of precipitation. All these scavenging processes contribute to the cleaning process of the atmosphere and a deposition of pollution onto the ground.

Since only half of the liquid water of the cloud precipitates and the rest evaporates, pollution in the cloud hydrometeors is released back into the air. This pollution mass is also significantly altered with respect to the initial air—the particles are generally larger and have more chemical species attached to them due to nonvolatilized gases. Also, the trace-gas concentration is altered. Thus, precipitation scavenging also influences air pollution and atmospheric chemistry.

For background information *see* AEROSOL; AIR POLLUTION; ATMOSPHERE; ATMOSPHERIC CHEMISTRY; CLOUD; CLOUD PHYSICS; PRECIPITATION (METEOROLOGY) in the McGraw-Hill Encyclopedia of Science & Technology.   Andrea I. Flossmann

Bibliography. H. R. Pruppacher and J. D. Klett, *Microphysics of Clouds and Precipitation*, 2d rev. ed., Kluwer Academic, 1997; J. H. Seinfeld, *Atmospheric Chemistry and Physics of Air Pollution*, Wiley, 1986; P. Warneck, *Chemistry of the Natural Atmosphere*, vol. 41, Academic Press, 1988.

Printable semiconductors for flexible electronics

Large-area flexible electronic systems, recently referred to as macroelectronics, have attracted a great deal of attention in the last decade because of their potential applications in various areas, such as paper-like roll-up displays, conformable x-ray imagers, and steerable fold-up radio-frequency antennas, where conventional integrated circuits on wafer supports do not provide the necessary form factors. For these and other systems, thin substrates made of electrically insulating polymers represent an ideal choice, as demonstrated in various prototype systems that use thin-film-type semiconductor materials, ranging from small-molecule and polymeric organics to amorphous silicon. The low degree of crystallinity associated with these materials and the unfavorable nature of charge transport in them (for example, hopping mode as opposed to bandlike) results in relatively low performance and in some cases uncertain reliability and reproducibility, both of which limit their applications. Recently, a completely different approach has been explored, whereby high-quality single-crystalline inorganic semiconductors are directly printed onto a plastic substrates, followed by depositing and patterning other materials at relatively low temperatures (less than 250°C) to construct flexible transistors and circuits, with performances approaching that of similarly scaled wafer-based systems.

Fabrication of semiconductor micro/nanostructures. Single-crystalline micro/nanostructures of semiconductors can be fabricated from conventional wafers of these materials through "top-down" approaches in which lithographic patterning and etching create the desired structures from the near-surface regions of the wafers. Semiconductor structures generated in this manner inherit the high quality of the original wafers in terms of well-controlled doping type, carrier concentration, dimensions, and crystallinity, which in turn enables excellent electrical properties in devices using them. Two etching strategies have been used to produce these kinds of structures (**Fig. 1**). The most straightforward approach involves

isotropic etching of sacrificial layers of multiple-layered wafers to release thin semiconductor structures (Fig. 1a). Typical examples include Si ribbons generated from silicon-on-insulator (SOI) wafers and GaAs ribbons from GaAs wafers with epitaxial layers of AlAs and GaAs (top surfaces), where lithographic steps are followed by removal of SiO_2 and AlAs, respectively, to release thin semiconductor ribbons. Figure 1b shows a scanning electron microscope image of GaAs ribbons with thicknesses of 270 nm, widths of 5 μm, and lengths of up to several centimeters, clearly showing the mechanical flexibility of these ribbons.

In another approach, similar lithographic steps, followed by anisotropic etching of bulk wafers along certain crystalline planes, generates ribbons/wires from standard wafers. Figure 1c depicts a process for generating Si ribbons from a (111) Si wafer. The first step defines shallow trenches with side walls terminated by (110) planes by lithography and reactive ion etching. Coating the top surfaces of the plateaus and parts of the sidewalls of the trenches with thin resists of SiO_2, Si_3N_4 or Au, followed by etching in hot potassium hydroxide (KOH) aqueous solution generates thin ribbons of Si. Figure 1d presents a scanning electron microscope image of ribbons fabricated by this process. In a related but simpler process, anisotropic etching that yields reverse mesas (that is, structures with newly formed

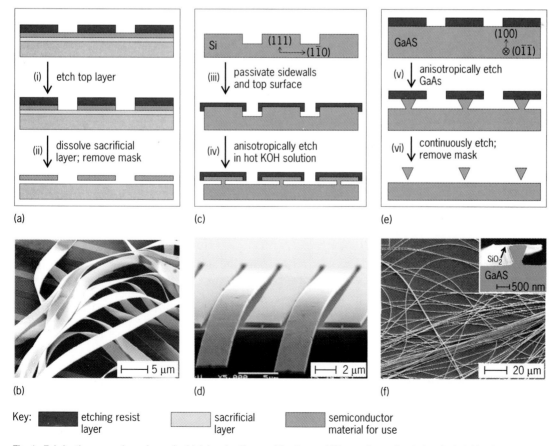

(a)   (c)   (e)

(b)   (d)   (f)

Key: ▮ etching resist layer   ▯ sacrificial layer   ▯ semiconductor material for use

Fig. 1. Fabrication procedures (a, c, e) which involve the combined use of lithography and wet chemical etching to generate micro/nanostructures of single-crystalline inorganic semiconductors. Scanning electron microscope images of structures fabricated using these approaches: (b) GaAs ribbons, (d) Si ribbons, and (f) GaAs wires with triangular cross sections.

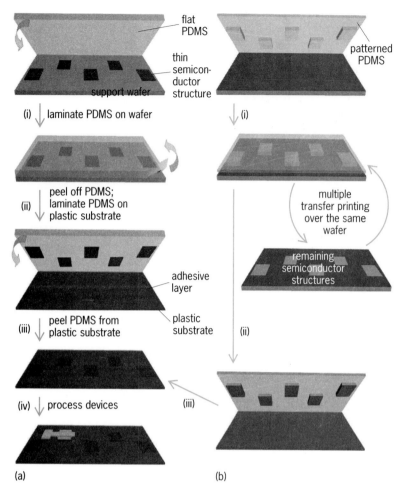

flat PDMS

thin semiconductor structure

patterned PDMS

support wafer

(i) laminate PDMS on wafer

(i)

(ii) peel off PDMS; laminate PDMS on plastic substrate

multiple transfer printing over the same wafer

remaining semiconductor structures

adhesive layer

plastic substrate

(iii) peel PDMS from plastic substrate

(ii)

(iv) process devices

(iii)

(a)   (b)

Fig. 2. Steps for transfer printing micro/nanostructures generated with the procedures of Fig. 1 onto plastic substrates using polydimethylsiloxane (PDMS) stamps with (a) flat and (b) patterned surfaces. The continuous semiconductor layers provide a simple means to indicate the arrays of wires/ribbons.

Transfer printing of micro/nanostructures. Micro/nanostructures fabricated via the procedures of Fig. 1 can retain the positional and orientational order defined by the lithography by using designs that leave these elements anchored to the wafer at their ends. Ordered wires/ribbons produced in this manner can be transfer-printed onto substrates, such as plastic sheets, using elastomeric [polydimethylsiloxane (PDMS)] stamps as transfer elements. **Figure 2** shows the major steps of the printing process. First, laminating a flat PDMS slab on the surface of a wafer with patterned patches of wires/ribbons generates a conformal contact. Generalized adhesion forces (or strong chemical bonds, with appropriately designed surface chemistries) bond the semiconductor micro/nanostructures to the surface of the PDMS (step i). Peeling the PDMS stamp away from the wafer transfers all of the micro/nanostructures to the stamp (step ii). Placing this "inked" stamp against a plastic substrate coated with a thin layer of adhesive (such as epoxy resins or photocurable polymers), activating the adhesive (curing the polymer), and then peeling back the stamp completes the printing (step iii). In a related approach, control of peel rate enables transfer without adhesives. The printed arrays of semiconductor micro/nanostructures can be processed into thin-film transistors (TFTs) through traditional photolithography and deposition of other materials that are compatible with the plastic substrate. If high-temperature processing, such as annealing of ohmic contacts for GaAs, doping of Si, and growth of thermal oxides on Si, is required, the related steps can be carried out on the wafer before transfer. By repetitive printing in a step and repeat fashion, it is possible to process devices over areas on the plastic substrate that are much larger than the size of the wafer (Fig. 2b). A patterned PDMS stamp can pick up ribbons/wires at selected areas by spatially registering the posts; the remaining micro/nanostructures on the wafer can be printed in subsequent steps.

Examples and electrical characterization of flexible TFTs. The printed arrays of wires/ribbons can serve as active materials for high-performance electronic devices. **Figure 3a** shows an optical image of a 25-μm-thick polyimide sheet, covered with an array of metal oxide semiconductor field-effect transistors (MOSFETs) fabricated with 290-nm-thick Si ribbons with contact doped regions. (Note their flexibility.) The adhesive layer used is polyamic acid, a liquid that can be converted into electronic grade polyimide by baking. The dielectric material is SiO_2 formed by low-temperature (about 250°C) plasma-enhanced chemical vapor deposition (PECVD). Source, drain, and gate electrodes use Cr/Au (5/100 nm) deposited by electron-beam evaporation. Figure 3b and c shows the geometry of the devices. These transistors exhibit electrical behavior similar to that of typical MOSFETs fabricated on wafers. Figure 3d shows the dependence of current flow from source to drain (I_{DS}) of a transistor with a channel length (L_c) of 9 μm, channel overlap distance (L_o) of 5.5 μm, and channel width (W_c) of 200 μm on the drain voltage

sidewalls that have an acute angle relative to the original surface of wafer) forms free-standing wires with triangular cross sections. This approach has been applied to wafers of III–V compounds (for example, GaAs), with top (100) surfaces and zinc blende face-centered cubic lattices. Figure 1e shows the case of GaAs wires formed by anisotropic etching of a GaAs wafer with H_3PO_4 (85 wt %)/13 H_2O_2 (30 wt %)/12 H_2O ($v/v/v$). In the fabrication, mask lines of suitable materials (such as photoresist and SiO_2) are patterned along the $(0\bar{1}\bar{1})$ crystallographic direction on the surface of a (100) GaAs wafer. Figure 1f presents a typical scanning electron microscope image of GaAs wires with width of about 400 nm formed using 50-nm-thick SiO_2 as the etching mask. The curved configurations of some of the wires indicate their excellent mechanical flexibility. Careful studies show that roughness (about tens of nanometers) on the etched surfaces originates from slight edge roughness and angular misorientation of the resist lines, together with intrinsic roughness associated with the etching. In many cases, the active devices rely only on the flat top surfaces of the wires, such that performance is not adversely affected by the roughness of the sidewalls.

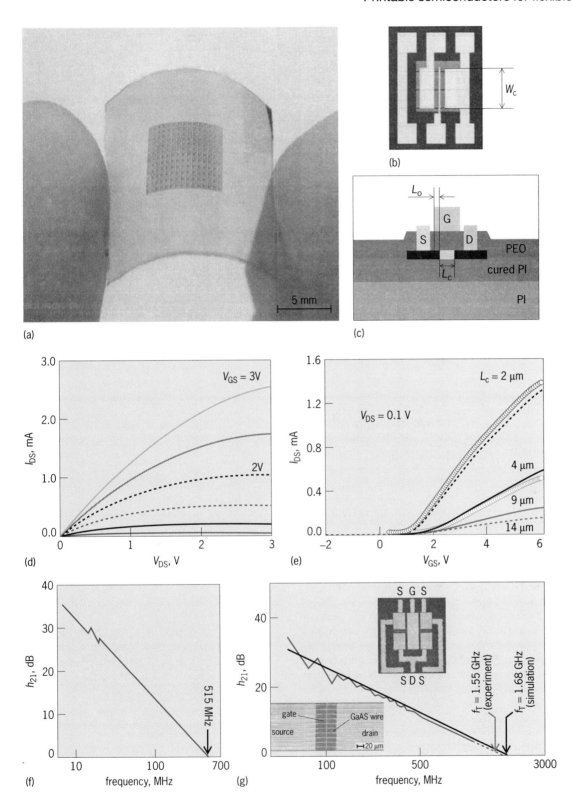

Fig. 3. Images and electrical characteristics of transistors fabricated with (*a–f*) Si ribbons on polyimide substrates and (*g*) GaAs wires on polyester substrates. PEO represents the SiO$_2$ layer grown through plasma-enhanced chemical vapor deposition.

(V_{DS}) at different gate voltages (V_{GS}). Figure 3*e* shows the transfer curves from devices with different channel lengths. The linear regime mobilities extracted from this type of device are as high as about 500 cm^2/V · s, which are higher than those possible with other approaches. Figure 3*f* shows the result of

microwave measurement over a transistor with $L_c = 2$ μm, $L_o = 1.5$ μm, and $W_c = 200$ μm. The unity current gain frequency (f_T) is about 500 MHz for this device. This frequency can be increased by reducing L_c and/or L_o, but this approach is not attractive for many macroelectronic systems because it requires high-resolution lithography, which can be difficult to achieve in a cost-effective manner on large-area plastic substrates. By contrast, metal-semiconductor field-effect transistor (MESFET) devices made of GaAs wires can offer high speeds even with coarse patterning resolution and limited capacity for overlay registration, the latter of which is of particular advantage for plastic substrates since they often do not show good dimensional stability over large areas. MESFETs (as images shown in the insets of Fig. 3g) with L_c of 50 μm, W_c of 150 μm, and gate length of 2 μm fabricated with GaAs wires (width of about 2 μm) integrated with ohmic stripes, exhibit f_T in gigahertz regime, that is, 1.55 GHz (Fig. 3g). These MESFETs use flexible poly(ethylene terephthalate) [polyester] substrates with thin polyurethane adhesive layers. This level of high-frequency operation indicates a potentially promising pathway to large active antennas operating at ultrahigh frequencies.

Mechanical characterization of TFTs. Good bendability is an important characteristic of many of the envisioned applications. The flexibility of transistors fabricated with single-crystalline micro/nanostructures was systematically evaluated by squeezing the plastic substrates with specially designed mechanical stages to generate concave (compressive strains on top device surface) and convex surfaces (tensile strains) [**Fig. 4a**]. Figure 4b shows the variation of linear regime device mobility in a silicon transistor, normalized by the value in the unbent state μ_{0eff} with bending radius and surface strain. The results indicate only small changes in device performance in this range of strains. Repetitive bending for many cycles (fatigue studies) confirm the robustness and durability of these devices (Fig. 4c).

Integration of flexible TFTs into circuits. High-performance individual transistors of the type described above can be integrated into circuits with desired functionalities for particular applications. For example, integrating five Si-ribbon-based n-MOSFETs with channel widths of 200 μm (serving as drivers) and five devices with channel widths of 30 μm (serving as loads) forms a five-stage ring oscillator. L_c and L_o of all transistors are 4 and 2 μm, respectively. **Figure 5a** and b show an optical image of a typical and its equivalent circuit. Figure 5c presents the measured waveform for the ring oscillator shown in Fig. 5a at supply voltage $V_{dd} = 4$ V. The circuit exhibits a maximal oscillation frequency of 8.2 MHz, corresponding to a stage delay of 12 ns. The oscillation speed of circuits can be further increased by reducing the contact overlaps and channel lengths.

Outlook. Flexible transistors that show high performance on plastic substrates can be achieved using printed single-crystalline inorganic micro/nanostructures of various semiconductors, such as Si, GaAs, and GaN, generated from high-quality single-crystal bulk wafers. Electrical and mechanical measurements indicate that the devices provide a promising class of building block for flexible macroelectronic applications. Preliminary results

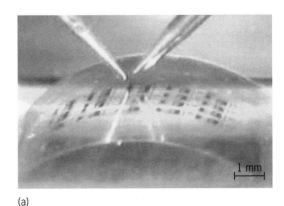

(a)

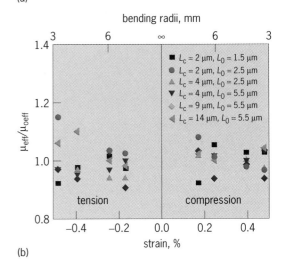

(b)

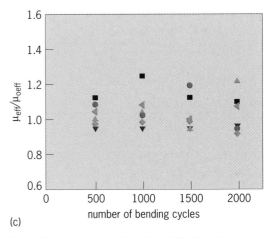

(c)

Fig. 4. Mechanical characterization of Si-ribbon-based transistors on polyimide substrates.

imply that individual TFTs of this kind can be integrated into complex circuits with appropriate functionalities for particular applications. Further improvement in device design and processing steps to increase performance, particularly in terms of operation speed, as well as improved device yields represent current directions for research in this field.

[Acknowledgment: The work was partially supported by the Defense Advanced Projects Agency under Contract No. F8650-04-C-7101 and by the U.S.

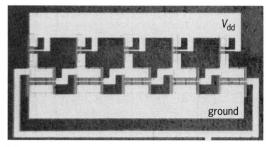

(a)

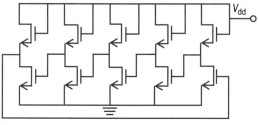

(b)

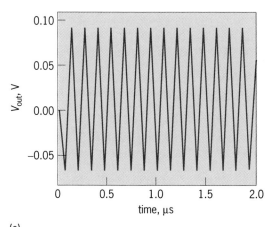

(c)

Fig. 5. Five-stage ring oscillator fabricated with Si-ribbon-based transistors. (*a*) Optical image, (*b*) equivalent circuit, and (*c*) output waveform.

Department of Energy under Grant No. DEFG02-91-ER45439. Argonne National Laboratory's work (for Y. Sun) was supported by the U.S. Department of Energy, Office of Science, Office of Basic Energy Sciences, under contract DE-AC02-06CH11357.]

For background information *see* CIRCUIT (ELECTRONICS); EPITAXIAL STRUCTURES; MICROWAVE SOLID-STATE DEVICES; POLYMER; SEMICONDUCTOR; TRANSISTOR; VAPOR DEPOSITION in the McGraw-Hill Encyclopedia of Science & Technology.

Yugang Sun; Jong-Hyun Ahn; John A. Rogers

Bibliography. J.-H. Ahn et al., High-speed mechanically flexible single-crystal silicon thin-film transistors on plastic substrates, *IEEE Electr. Device Lett.*, 27(6):460–462, 2006; K. J. Lee et al., Large-area, selective transfer of microstructured silicon: A printing-based approach to high-performance thin-film transistors supported on flexible substrates, *Adv. Mater.*, 17:2332–2336, 2005; S. Mack et al., Mechanically flexible thin-film transistors that use ul- trathin ribbons of silicon derived from bulk wafers, *Appl. Phys. Lett.*, 88:213101, 2006; E. Menard et al., A printable form of silicon for high performance thin film transistors on plastic substrates, *Appl. Phys. Lett.*, 84(26):5398–5400, 2004; R. H. Reuss et al., Macroelectronics: Perspectives on technology and applications, *Proc. IEEE*, 97(7):1239–1256, 2005; R. F. Service, Inorganic electronics begin to flex their muscle, *Science*, 312:1593–1594, 2006; Y. Sun and J. A. Rogers, Fabricating semiconductor nano/microwires and transfer printing ordered arrays of them onto plastic substrates, *Nano Lett.*, 4(10):1953–1959, 2004; Y. Sun et al., Gigahertz operation in flexible transistors on plastic substrates, *Appl. Phys. Lett.*, 88:183509, 2006.

Reducing human error in medicine

Human error in medicine contributes to adverse events that threaten patient safety. Certain errors tend to recur across adverse events in health care, reflecting unmet perceptual, cognitive, and behavioral needs as users interact with systems to accomplish patient care goals. For example, slips such as overdosing patients result from inputting the wrong number into infusion pumps or misreading medication orders because critical information is not highlighted or similar symbols are confused on the interface or in the order.

Human factors researchers have made important contributions to patient safety by raising awareness of the role of preventable medical errors in adverse events, and developing and evaluating approaches to mitigating these errors. This is due in part to analytical tools (for example, task analysis) that identify user needs and the extent to which they are met by health care systems.

Framework for analyzing error. The 1999 Institute of Medicine report, *To Err Is Human: Building a Safer Health System*, estimated that more than 98,000 deaths per year are attributable to medical error. As in other high-risk professions, such as aviation, it is misleading to equate human error with human fault. For example, outcomes labeled adverse events in medicine may be viewed as "bad" only in hindsight. Given the uncertain nature of patients' responses to treatment, it is likely that many adverse events occur in spite of appropriate diagnosis and treatment.

A definition of human error remains elusive. According to one definition, it is substandard human performance that should have been recognized by the practitioner as substandard at the time it occurred.

Progress in reducing errors depends on frameworks that specify methods for measuring and analyzing error and guiding design and training interventions. D. G. Morrow and coworkers have combined the error taxonomy of psychologist James Reason with an information-processing stage model. Events are detected depending on the extent that they are attended to, and then interpreted (for example, understanding and integrating multiple symptoms in diagnosis). Such operations require knowledge stored in long-term memory and involve processes in working memory. Perception and diagnosis typically lead to action choice and execution. Errors occur when perceptual, cognitive, and action needs are not supported by the environment. Errors can be categorized (1) as mistakes, such as when an incorrect intention to act is chosen; (2) as slips, such as when the correct action is intended but an unintended action is executed; and (3) as lapses, such as when an intended action that should be done is not. Mistakes occur because practitioners have inappropriate knowledge or rules of action. Slips often result from poor design that encourages improper and unintended action. This taxonomy is useful to the extent that it guides mitigating strategies. Knowledge-based mistakes can be addressed by better displays, training, and possibly automated decision aids. Slips can be addressed by better interface design, achieving, for example, compatible mappings from perception to action. It is also clear that error in medicine, as in other complex settings, must be addressed by focusing on the systems within which practitioners work, because errors are fostered by high levels of patient load, poorly designed environments, and other latent factors in systems. Human factors research aims to reduce errors and adverse events by addressing both individual and system components of medical settings.

Medical devices. Medical devices are responsible for many errors that lead to adverse events, although the exact proportion is difficult to estimate because of inconsistency in how adverse events are reported. Manufacturers have incorporated more complex software into devices to increase their diagnostic and therapy value, but accompanying interfaces are not always designed with user capabilities in mind. This is clearly illustrated by research on drug infusion pumps.

Drug infusion pumps allow controlled delivery of drug therapy by practitioner or patient. Recent advances allow programming of the delivery in precise intervals and quantities, improving efficiency of drug delivery to patients but also creating new opportunities for error. For example, inadequate interface design can tax users' cognitive abilities during setup and therapy management. Poor feedback during drug parameter entry can lead to programming drug errors. Inconsistent meaning of messages can lead to confusion, and complex programming sequences related to entering medication concentration and rate can overload operators' working memory—all revealed in careful investigations. New interface designs addressing these concerns have reduced programming time and errors, compared to existing interfaces. Similar human factors problems have been identified with other devices such as patient monitors, glucose monitors, and electronic defibulators.

Medication errors. Adverse drug events (ADEs)—failure of the correct medication, dose, or modality to reach a patient—may be the most adverse common event. At least 10% of hospital patients in the United States experience ADEs, resulting in many injuries or deaths per year. About one-third are preventable because they are associated with errors that occur when practitioners prescribe medications, transcribe orders, or dispense and administer medications to patients. Perceptual, cognitive, and behavioral processes are involved in each of these stages. For example, physicians diagnose patients and develop treatment plans that include medications, which involve perceiving and interpreting cues (for example, symptoms) in terms of knowledge about illness and treatment. Implementing plans involves actions such as writing medication orders. Transcribing orders and administering medications depend on comprehension and mental computation. These processes may fail when limited attentional resources create needs that are not met by the health care environment.

Ordering problems account for almost half of medication errors. Ordering wrong doses can involve failure to retrieve or integrate medication or patient information. Wrong medications are also ordered—a slip error made more likely by the increasing number of medications, many with similar names. More common than inaccurate orders are poor prescription practices, sometimes resulting in incomplete orders with sloppy writing and nonstandard abbreviations that increase the likelihood of transcription errors.

Ordering errors can be mitigated by supporting physicians' perceptual, cognitive, and behavioral needs. For example, computerized physician order entry (CPOE) systems, integrated with clinical decision support systems (CDSSs), reduce reliance on vulnerable cognitive processes such as retrieval and integration of drug and patient knowledge, and reduce poor prescribing practices. However, even as they reduce some errors, these systems can create new ones. Ordering errors are also addressed by lower-tech strategies such as preprinted prescription forms that support effective prescribing.

Medication administering as well as ordering errors are addressed by human factors research and interventions. Medication errors are reduced by helping nurses, physicians, pharmacists, and patients collaborate (such as standard procedures for checking and resolving orders and including pharmacists on physician rounds). More generally, there is growing recognition that poor collaboration contributes to many medical errors, reflecting the need for communication procedures, collaborative training programs, and other approaches to improve surgical teamwork, hand-offs during shift changes, and other points of collaborative vulnerability.

Decision support. Errors related to physicians' diagnosis and treatment decisions also contribute to adverse events. While solutions involving clinical practice guidelines have been proposed, computer-based decision supports are increasingly used. Such supports may assist with diagnosis, such as a smart alert that infers not only something wrong with a patient but also the nature of the problem. These supports may explicitly or implicitly recommend certain forms of action or treatment.

Despite documented success of computer-assisted decision supports, the overall pattern of success has been less than satisfactory. For example, a meta-analysis of clinical decision support systems evaluation studies found little evidence for improved patient outcomes. While more studies demonstrated improved practitioner performance, success rates were higher when the study was conducted by system developers than by independent investigators, suggesting possible bias. Clinical decision support systems were most effective when providing unrequested clinical reminders rather than depending on the practitioner's initiative. Clinical decision support systems may not be adopted. Designers who do not fully understand the nature of the practitioner's task environment develop systems that do not readily integrate into the practitioner's workflow. And within domains as uncertain as medicine, even well-designed diagnostic tools may be imperfect in their guidance, leading to distrust of the system.

Outlook. Progress is needed in addressing medical error. Challenges include a gap in the precision of models that identify needs of individual users, and models that identify needs of teams interacting with complex health care settings. Frameworks are needed that encompass collaboration among practitioners in order to analyze error in the context of routine work practices, which will guide interventions that target error at multiple levels.

For background information *see* DRUG DELIVERY SYSTEMS; HUMAN-COMPUTER INTERACTION; HUMAN-FACTORS ENGINEERING; MEDICAL CONTROL SYSTEMS; MEDICINE; PERCEPTION in the McGraw-Hill Encyclopedia of Science & Technology.

Daniel G. Morrow; Christopher D. Wickens

Bibliography. M. R. Cohen (ed.), *Medication Errors: Causes, Prevention, and Risk Management*, Jones & Bartlett, 1999; R. I. Cook and D. D. Woods, Operating at the sharp end: The complexity of human error, in M. S. Bogner (ed.), *Human Error in Medicine*, pp. 255–310, Erlbaum, 1994; A. X. Garg et al., Effects of computerized clinical decision support systems on practitioner performance and patient outcomes, *JAMA*, 293:1223–1238, 2005; A. Gawande and D. Bates, The use of information technology in improving medical performance, Part I: Information systems for medical transactions, *Medscape General Medicine*, vol. 2, pp. 1–6, February 2000; L. T. Kohn, J. M. Corrigan, and M. S. Donaldson, *To Err Is Human: Building a Safer Health System*, National Academy Press, Washington, DC, 1999; D. G. Morrow, R. North, and C. D. Wickens, Reducing and mitigating human error in medicine, in R. S. Nickerson (ed.), *Reviews of Human Factors and Ergonomics*, vol. 1, pp. 254–296, Human Factors and Ergonomics Society, 2006; J. Reason, *Human Error*, Cambridge University Press, 1990.

Remote sensing of fish

With the worldwide decline of fish in the oceans, it is important to be able to remotely image and monitor fish populations, as well as accurately estimate their abundances. A better understanding of the behavior and dynamics of fish populations, including their response to environmental and anthropogenic pressures, is essential for effective management of marine fisheries. Recently, an acoustic imaging technique was developed, called ocean acoustic waveguide remote sensing (OAWRS), for instantaneously imaging and rapidly locating fish and other biomass in the ocean over wide areas spanning the continental shelf. This new technology will influence the way in which fish and other biological organisms in the ocean are surveyed and studied. It will provide a more global approach for studying fish and marine ecosystems, analogous to an optical or radar satellite system for sensing the Earth's surface.

In the ocean, sound can travel much greater distances with far less absorption than light and any other form of radiation. Under water, limited visibility makes sound the primary means for remote sensing and imaging. Marine mammals, dolphins, and whales naturally use sound to echolocate food, to navigate their surroundings, and to communicate. Devices and instruments that use sound to probe the ocean environment are called sonars. Since the 1930s, sonars have been used to detect and study the abundances and behavior of fish and other biological organisms in the ocean.

All fish-finding sonars consist of one or more projectors that generate a short burst of sound. The sound travels through the water column until it is reflected and scattered off fish (fish echo) and then received by one or more underwater microphones called hydrophones. A projector that can also receive echo signals is called a transducer. Based on the travel time of the received echoes, the distance of the fish from the sonar system can be determined. A sonar system's range of detection depends on the frequency of sound it transmits. Lower sound frequencies undergo less absorption and can travel longer ranges and detect more distant objects than higher sound frequencies. Most fish have air-filled swim bladders, which are responsible for reflecting over 90% of the acoustic energy incident on them. Some fish, such as the Atlantic mackerel, have no swim bladder but still scatter sound with their bone and flesh, producing weaker echoes.

Conventional fish-finding sonars. Until recently, the standard technique for surveying fish used a combination of capture-trawl methods, along with acoustic surveys with a high-frequency echo sounder. An echo sounder is a device with a narrow acoustic beam, like a searchlight, that is directed vertically downward. It is towed by a ship along line transects typically several to tens of kilometers apart in a back-and-forth pattern to produce images of the ocean in depth and range along the ship's track (**Fig. 1**). Echo sounders have been used commercially for finding fish since the 1950s, and they operate at high frequencies, above 20 kHz and up to 600 kHz. The survey rates of echo sounders are small, typically 0.2 km²/h (0.08 mi²/h). This rate is determined by the resolution footprint, or insonified area, of the echo sounder at any instant, which is highly localized to within several tens of meters of the tow ship

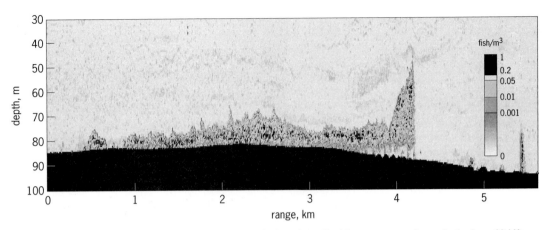

Fig. 1. Echogram of the water column as a function of depth along the path of the survey vessel, acquired using a 38-kHz echo sounder on the United States eastern continental shelf on May 14, 2003. It took the survey vessel roughly 30 min to acquire the data shown. Echos from fish in water depths of 50-85 m (160-280 ft) are clearly visible. The range resolution of the system is 6 m (20 ft), and the depth resolution is 0.5 m (1.6 ft). (*Courtesy of Redwood W. Nero*)

and by the tow speed of roughly 5 m/s (16 ft/s). As a result, echo sounders tend to undersample the environment in both time and space, leaving highly aliased and ambiguous records of fish activity and abundances. Since the 1970s, side-scan sonars have been used to image fish. The side-scan sonar uses two echo sounders to project the sound beam sideways from the ship, providing a horizontal aerial coverage that is roughly 10 times larger than a conventional echo sounder along the ship track. In the late 1980s, multibeam sonars were developed with multiple sound beams pointing in many different directions to provide wider angular coverage of the water column surrounding the sonar. This enabled the system to image both pelagic and demersal fish, which are found close to the sea surface or sea bottom, respectively. The multibeam sonar extends the two-dimensional imaging of the echo sounder to allow

limited three-dimensional imaging of the morphology of a fish school. The conventional fish-finding sonars (CFFS), described so far, rely on direct waterborne propagation paths from the sonar to an object. As a result, the maximum detection ranges for these systems are limited to within roughly a kilometer of the survey vessel. For surveying large areas of the ocean, the biggest challenge in using CFFS, due to their limited aerial coverage, is to figure out exactly where the fish are located. Since conventional sonar systems are not capable of distinguishing fish species from their echoes, capture-trawl sampling methods are necessary for classifying the fish groups.

Ocean acoustic waveguide remote sensing. In recent years, ocean acoustic waveguide remote sensing (OAWRS) has been developed and applied to study fish populations on the New Jersey continental shelf (**Fig. 2a**). For the first time, it revealed

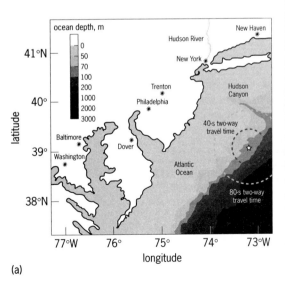

(a)

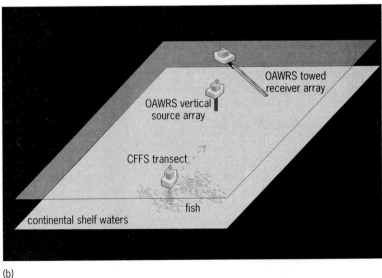

(b)

Fig. 2. OAWRS 2003 survey of the New Jersey continental shelf. (*a*) Location of survey. The star shows the geographic location of the OAWRS source array for the image in Fig. 3. (*b*) Deployment of the system, with a vertical source array transmitting low-frequency acoustic pulses in an omnidirectional horizontal azimuth. Scattered returns from fish and other features in the environment are continuously received by a horizontally towed array of hydrophones. Also shown is a conventional high-frequency echo sounder towed by a ship for concurrent line-transect surveys of fish in the water column directly underneath the vessel.

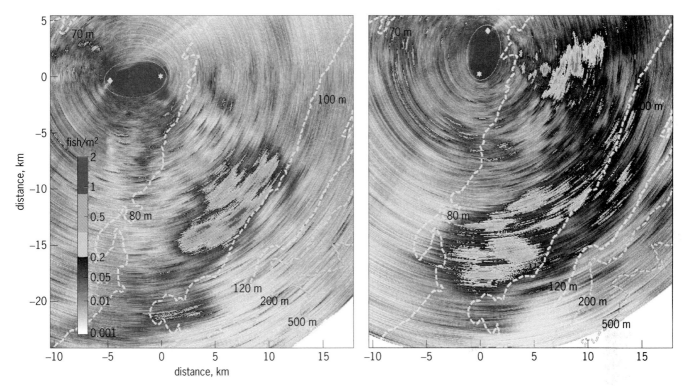

Fig. 3. Two instantaneous OAWRS images of massive fish shoals on the United States eastern continental shelf on (a) May 14, 2003, 09:32 EDT and (b) May 15, 2003, 08:38 EDT. The location of the source array is shown as a star for geographic reference in Fig. 2a and here at the origin. The location of the receiving array is shown as a diamond. The source transmitted 1-s linear frequency modulated (LFM) pulses from 390 to 440 Hz. Several tens of millions of fish were instantaneously imaged. The receiving array has a 2.6° azimuthal resolution at the array broadside, the direction normal to the array axis (along the bearing of fish in a), and the range resolution is 30 m (100 ft), after averaging. Local fish population densities were obtained from the scattered field returns by compensating for two-way transmission loss in the range-dependent continental-shelf waveguide, the spatially varying resolution footprint of the OAWRS system, expected fish target strength, and source power. (*After N. C. Makris et al., 2006*)

the instantaneous horizontal structural characteristics of a very large fish shoal (school) spanning several tens to a hundred square kilometers in area, and the evolution of the shoal over time. The technique uses the ocean as a waveguide to channel sound over very long ranges, imaging the ocean at about a million times greater rate than conventional fish-finding methods. OAWRS uses a vertical source array to transmit a short broadband pulse at low frequencies (several hundred hertz up to several kilohertz) that spreads out omnidirectionally in a horizontal azimuth. The sound waves reflect from the sea surface and bottom to form standing waves in depth that are called waveguide modes, analogous to the modes of a vibrating guitar string. These modes propagate outward in range, suffering only cylindrical spreading loss rather than the spherical loss of the CFFS. Return echoes from fish and other environmental features are continuously received on a horizontal array of hydrophones that can be either towed from a ship (Fig. 2b) or deployed as a fixed structure off the sea floor. The echoes are charted in horizontal range from the receiving array using match filtering over the two-way travel time of the signal and in bearing, or horizontal azimuth, by plane-wave beamforming. Match filtering is a pulse compression technique that matches the return echoes with the original transmitted pulse over the signal bandwidth, which greatly improves the range resolution of the

system. Beamforming is a method that utilizes delay in the arrival time of a signal, measured by the hydrophones on an array, to determine the bearing of the signal. The resulting image is an instantaneous snapshot of the ocean environment over the two-way travel time of the echoes. The instantaneous aerial coverage of OAWRS is roughly circular with typical diameters of 60 and 120 km (37 and 74 mi), corresponding to the maximum two-way travel times of 40 and 80 s, respectively, for a sound speed of 1500 m/s (4900 ft/s) [Fig. 2a]. In the OAWRS system, sound is propagated to ranges about ten to a thousand times the waveguide depth, to image fish. It readily provides fish localizations horizontally, but it can be challenging to determine the depth location of fish in the water column with this system.

OAWRS was used during an experiment on May 14, 2003. As a result, a typical snapshot of the continental shelf environment on the east coast of the United States is shown in **Fig. 3**. The image reveals the instantaneous horizontal morphology of a very large fish shoal, representing probably the largest mass of animals ever imaged instantaneously. Several population centers interconnected by fish bridges through which the fish migrate can be seen in the image. OAWRS measurements of fish distribution were found to be in good correspondence with measurements made concurrently along line-transect surveys of the shoal with a conventional echo sounder.

The OAWRS imagery also revealed that fish in large shoals and smaller fish groups are structurally similar, with spatial distributions that follow a fractal or power-law spectral process. This enables accurate statistical predictions to be made of the instantaneous spatial distribution of fish populations over wide areas. Movies were made of the ocean environment by concatenating OAWRS imagery at a 50-s update rate, which provided information of the temporal evolution, behavior, and dynamics of such a massive shoal, again for the first time. Clustering, fragmentation, and dispersal of fish groups were found to occur over time scales of 5–10 min. This showed that OAWRS can provide an image of a fish population that is unaliased in both time and space, and that the temporal and spatial variations of large fish groups and shoals are too rapid to be tracked using CFFS methods. Fish density waves were observed in the movies, with speeds of up to 10 m/s (33 ft/s) due to a sequence of localized compaction and expansion events within the dense population centers. These density waves were found to propagate over the coherence width of the population centers of roughly 1–3 km (0.6–2 mi) before bouncing off the boundaries of this region. The waves may be used by fish to sense the spatial extent and maintain the coherence of the locally dense subgroup. They may also be used for communication, for instance, in response to predation. In Fig. 3, the fish species imaged by the OAWRS system are likely to be Atlantic herring, scup, hake, and black sea bass, based on annual trawl surveys in the area, along with visual observations made during the experiment.

Fish classification and abundance estimation. One of the challenges in fish research is to remotely identify fish species, characterize their sizes, and make estimates of their abundances in a given region. In CFFS systems, the echo returned from an individual fish is highly dependent upon the aspect of the fish imaged, relative to the sonar beam. This is because at high operating frequencies most fish have bodies and swim bladders that are large compared to the acoustic wavelength, making them highly directional scatterers. There is also shadowing of the acoustic waves from one fish to the next and multiple scattering in dense fish groups that must be accounted for with these systems. This, combined with the fact that the survey rates for CFFS are small, makes estimation of fish abundances challenging, since abundances need to be extrapolated to the vast areas where data are not available from these systems.

One advantage of the OAWRS system is that it uses low acoustic frequencies for imaging, and fish are compact and small compared to the acoustic wavelength. As a result, the scattering from fish becomes omnidirectional for both the swim bladder and body, independent of the fish aspect. Shadowing of the incident acoustic waves is also negligible at low frequencies. These advantages and the fact that entire fish shoals can be instantaneously imaged over very wide areas with OAWRS without temporal aliasing allow estimates of fish abundances to be made more readily.

All current sonar technologies, both CFFS and OAWRS, require the fish imaged to be identified and statistically characterized in order to make reliable estimates of abundances from the acoustic imagery. Apart from capture-trawl sampling of fish, fish groups may be identified using remote acoustic sensing methods. Most swim-bladder-bearing fish have acoustic resonance frequencies for scattering at less than 5 kHz for fish several tens of centimeters long. To characterize the fish, multifrequency systems can be used to measure the resonance frequency of the fish group. Many factors, such as bathymetric migration and feeding, may affect the resonance within the fish group, which must be taken into account. Other methods use the spatial distribution, intensity, oceanographic and environmental correlation, along with dynamics of the fish group echo for species identification and classification.

Future fish surveys. Fish-finding sonars have evolved over the last 70 years from the initial simple high-frequency echo sounders that survey the water column locally in depth, to the recently developed and more sophisticated low-frequency OAWRS system that can instantaneously image environments covering thousands of square miles. It is now possible to explore the ocean with various technologies of different capabilities to remotely map and study fish and marine ecosystems.

We envisage that in the near future the low-frequency OAWRS system will be deployed in various regions of the ocean as a wide-area system to remotely locate and monitor fish and other biomasses from long ranges. Higher-resolution local investigation of the depth distribution and classification of these dense biomass regions may be provided by high-frequency echo sounders and multibeam acoustic systems, along with underwater video cameras. These devices may be dynamically deployed from mobile platforms, ships, and submersibles or mounted off the sea bottom. The unaliased spatial and temporal mapping of the ocean with OAWRS, combined and integrated with data from high-frequency sonars and video imagery, can provide the knowledge base needed for more accurate and reliable studies of fish behavior, dynamics, and abundances.

For background information *see* CONTINENTAL MARGIN; ECHO SOUNDER; HYDROPHONE; MARINE BIOLOGICAL SAMPLING; MARINE ECOLOGY; MARINE FISHERIES; MATCHED-FIELD PROCESSING; POPULATION ECOLOGY; REMOTE SENSING; SONAR; UNDERWATER SOUND; WAVEGUIDE in the McGraw-Hill Encyclopedia of Science & Technology.

Purnima Ratilal; Deanelle T. Symonds

Bibliography. D. R. Gunderson,; *Surveys of Fisheries Resources*, Wiley, New York, 1993; D. N. MacLennan and E. J. Simmonds, *Fisheries Acoustics*, Chapman & Hall, London, 1992; N. C. Makris et al., Fish population and behaviour revealed by instantaneous continental shelf-scale imaging, *Science*, 311:660–663, 2006; P. Ratilal et al., Long range acoustic imaging of the continental chelf environment: The acoustic clutter reconnaisance experiment

2001, *J. Acous. Soc. Amer.* 117:1977–1998, 2005; M. Wilson, Acoustic waveguide sonar finds enormous fish shoals, *Phys. Today*, 20–22, April 2006.

Scientific drilling in hotspot volcanoes

The term "hotspot" describes a small region of intense volcanism that is located far from a plate boundary. The essential features of a hotspot are embodied in the Hawaiian island chain. The rate of eruption of basalt lava in and near the island of Hawaii is higher than in any other comparable-size area on the Earth, and yet the island is located in the middle of the Pacific plate, thousands of miles from the nearest plate boundary. The standard explanation for the Hawaiian volcanism is that a plume of exceptionally hot solid rock material is rising within the mantle under Hawaii. The plume is envisioned as a roughly cylindrical region, 200–300 km (120–190 mi) wide, extending from about 100 km (60 mi) below the Earth's surface all the way down to the core-mantle boundary 2900 km (1800 mi) below the surface. For Hawaii, the plume has been nearly stationary for the last 60 million years, whereas the Pacific plate has been moving at 8–10 cm/year (3–4 in./year) to the northwest over the plume. This situation has resulted in a chain of volcanoes more than 5000 km (3000 mi) long that extends from the island of Hawaii to the northwestern-most part of the Pacific. Many volcanoes on other planets in the solar system, such as Mars and Venus, are believed to form by processes similar to those that produce hotspot volcanoes on the Earth.

Significance of hotspot volcanoes. Hotspot volcanoes have special significance for understanding the workings of the deep Earth. Other types of volcanoes are produced by plate tectonic processes, including sea-floor spreading at mid-ocean ridges, subduction at plate margins (for example, the Andes and the Cascades), and continental extension (for example the Great Basin of the western United States). The magma erupted from the non-hotspot volcanoes comes from the uppermost mantle [down to about 200 km (120 mi) below the surface], so the chemistry and mineralogy of the lavas tell us about the composition of the upper mantle. The idea of using volcanoes to study the Earth's mantle began in the 1960s and has developed into a major activity of geochemists and petrologists. The attraction of hotspot volcanoes from this perspective is that if, as widely thought, the mantle plumes that produce them originate in the lower mantle, perhaps from near the core-mantle boundary, then hotspot volcanoes can tell us about the deepest levels of the mantle. Information of this sort may be the only way we can test models for the composition and history of the deep mantle.

Drilling project. A major project to study hotspot volcanoes is the Hawaii Scientific Drilling Project (HSDP). The objective of this project is to drill and sample 4 km (2.5 mi) down into one of the youngest Hawaiian volcanoes, Mauna Kea. The core brought up by drilling provides samples of lava erupted from the volcano over the last 600,000 years or more, during which the volcano has been drifting over the top of the Hawaiian mantle plume (**Fig. 1**). The drillcore samples give a nearly uninterrupted record of how the composition of the erupted magma changed in response to the movement of the volcano over the plume. Study of the samples allows the investigators to produce a picture of the internal structure of the plume, and this information can be used to test models for the composition and history of the deepest mantle.

In 1993, the HSDP drilled a pilot hole to a depth of 1039 m (3049 ft) on the eastern flank of the Mauna Kea volcano, adjacent to the shoreline on the east side of Hilo (**Fig. 2**). In 1999, a second hole was drilled to 3109 m (10,200 ft) just south of the Hilo airport and about 2 km (1.2 mi) south of the pilot hole. The second hole was deepened to 3300 m (11,000 ft) in 2004 and 2005. The target depth for drilling is 4000 m (13,000 ft). Almost 100% core recovery was achieved in both the pilot hole and the deep hole. The drill sites are located far from the volcano summit to minimize the likelihood of drilling

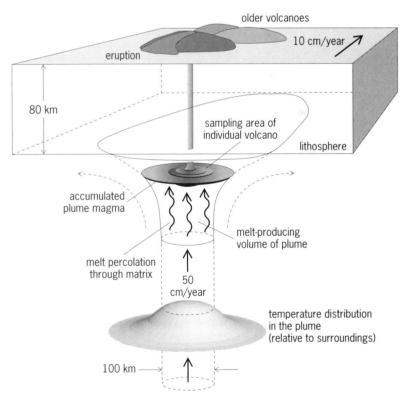

Fig. 1. Model for magma production by the Hawaiian mantle plume rising underneath the lithosphere of the Pacific plate. The hot cylinder of solid rock comes from deep in the mantle, probably from the core-mantle boundary. The temperature is highest in the middle of the plume and decreases outward. The middle of the plume is rising at a velocity of about 50 cm/year (20 in./year). The upward velocity decreases to zero as the plume approaches the base of the lithosphere, so the plume spreads out just below the lithosphere. The motion of the plate smears the plume out in the direction of the plate motion. At a depth of about 150 km (90 mi), the rising plume begins gradually to melt, and the liquid that is produced percolates upward through partly liquid rock as far as it can go, probably up to 90 km (56 mi) depth. Accumulated magma is then periodically funneled into a conduit through which it can rise to the surface and erupt. The magma fed to an individual volcano comes from only a fraction of the area of the melting region, so as the volcano drifts over the plume, the erupted lava presents a systematic sampling of a swath over the plume.

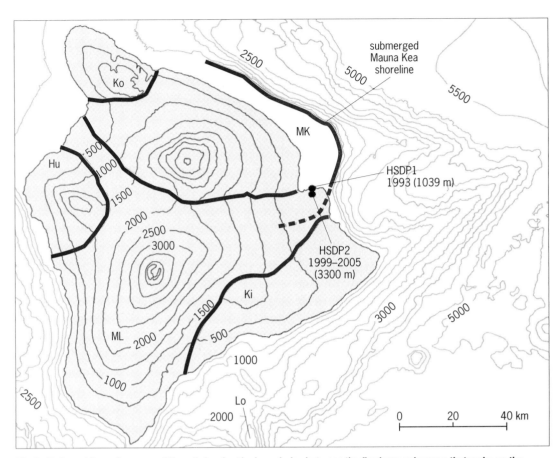

Fig. 2. Bathymetric contour map of Hawaii showing the boundaries between the five large volcanoes that make up the island, and the locations of the two drilling sites of the HSDP. Both drill sites are on lava from Mauna Loa volcano (ML). The thickness of the Mauna Loa lavas at the drill sites is 260–280 m (850–920 ft), and below that depth lavas from Mauna Kea (MK) are present all the way to the old ocean floor 6.5 km (4 mi) down. The land area of the Mauna Kea volcano once was much greater than at present. The old shoreline (color) has subsided below the sea surface, and the southern side of the volcano has been covered with lavas from younger Mauna Loa. Older volcanoes, Kohala (Ko) and Hualalai (Hu), extend under Mauna Kea. Consequently a drill hole on the northwestern side of Mauna Kea would not remain in the Mauna Kea volcano but would cross into lavas from one of the older volcanoes. Kilauea volcano (Ki) is currently the most active Hawaiian volcano, and Loihi (Lo), the youngest volcano, has not yet grown tall enough to break the sea surface. Loihi, Kilauea, and Mauna Loa are highly active volcanoes. Mauna Kea and Hualalai still have sporadic small eruptions. Kohala, the volcano farthest to the northwest, is considered extinct.

into igneous intrusive rocks, to avoid high temperatures in hole, and for logistical reasons. A team of 30 investigators from the United States, Canada, and Europe has been studying the rock samples and making measurements in the well.

Findings and observations. The HSDP studies have resulted in several new insights about hotspot volcanoes. One important finding is that the Mauna Kea lavas are about twice as old as inferred from study of the volcano surface. The Hawaiian volcanoes grow more slowly and are active for a longer period of time than previously thought. The new data support a fairly simple model for how the mantle plume supplies magma to the volcanoes. This model (Fig. 1) can be used to simulate the growth of the island of Hawaii surprisingly accurately (**Fig. 3**). Another surprise is that the underground temperature beneath Hilo is low. At a depth of 1000 m (3300 ft), the temperature is only about 12°C (54°F). Apparently, cold seawater is circulating through the flanks of the volcano fast enough to cool the rocks. The well has also intersected a number of pressurized aquifers—some fresh and some salty—all the way down to 3000 m

(9800 ft). The presence of these aquifers implies that the hydrology of Hawaii is more complex and interesting than had previously been recognized. There are also traces of microbiota in the rocks down to 2600 m (8500 ft).

One of the most significant observations with regard to the deep mantle is that the isotopic geochemistry of Mauna Kea lavas changes systematically with depth in the hole (and with age of the lava). In particular, the helium isotope ratios (^{3}He/^{4}He) and the ^{208}Pb/^{204}Pb ratios increase with increasing age. The observations confirm many aspects of the model in Fig. 1, but it was not expected that the large changes in ^{3}He and ^{208}Pb would be accompanied by only slight changes in Nd, Sr, and Hf isotope ratios, and in ^{206}Pb/^{204}Pb and ^{207}Pb/^{204}Pb. Helium isotope ratios are also more highly variable than expected among lavas of roughly the same age. This information provides clues about the structure of the base of the Earth's mantle. One hypothesis is that the He isotope signal comes from a thin layer, perhaps only 10 km (6 mi) thick, at the base of the mantle, and that the anomalous helium may be leaking into this

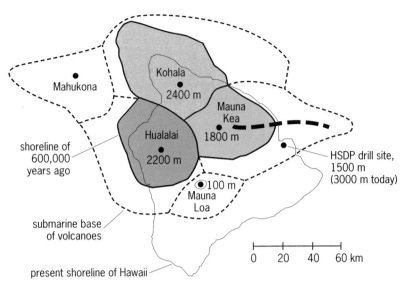

Fig. 3. Reconstruction of the island of Hawaii 600,000 years ago based on data from the HSDP drill cores, and a model for the ages and growth rates of Hawaiian volcanoes. The lavas in the HSDP drill core are about 600,000 years old at a depth of 3000 m (9800 ft) below sea level. Because the site has been sinking (or subsiding) at the rate of 2.5 m (8 ft) every thousand years, the lavas currently at a depth of 3000 m in the hole must have been at a depth of 1500 m (4900 ft) when they formed. Using the typical angle of the modern submarine slopes of the volcanoes, it is inferred that the drill site was located about 10 km (6 mi) offshore at 600,000 years ago. This inference locates the shoreline of the Mauna Kea volcano at that time. The sizes of the other volcanoes are estimated using a model to calculate how fast the volcanoes were growing at different times during the past 600,000 years, subtracting that growth from their current size, and accounting for subsidence. The circles locate the summits of the volcanoes; the Mahukona volcano was not above sea level. Kilauea and Loihi volcanoes had not yet begun to form at 600,000 years ago. Mauna Loa, now almost 4000 m (13,000 ft) high, had recently breached the sea surface.

layer from the Earth's core. Another hypothesis is that the He signal comes from a stagnant layer at the bottom of the mantle that has been in existence for billions of years. Because the isotopic signals of the other elements differ from those of He, they must come from a thicker layer at the base of the mantle and originate in a different way from the helium signal.

Scientific drilling in other Hawaiian volcanoes and at other hotspots could be valuable for further testing hypotheses about the origins of mantle plumes and the chemical and isotopic properties of the deep mantle.

For background information *see* EARTH INTERIOR; GEOCHEMISTRY; HOT SPOTS (GEOLOGY); LAVA; MAGMA; PLATE TECTONICS; VOLCANO in the McGraw-Hill Encyclopedia of Science & Technology.

Donald J. DePaolo; Edward M. Stolper;
Donald M. Thomas

Bibliography. G. F. Davies, *Dynamic Earth: Plates, Plumes and Mantle Convection*, Cambridge University Press, 1999; F. Press et al., *Understanding Earth*, W. H. Freeman, 4th ed., 2003.

Sedimentology of tsunami deposits

On December 26, 2004, an earthquake estimated at between magnitude 9.0 and 9.3 resulted from rupture of over 1000 km (620 mi) of the Andaman Subduction Zone. The rupture also generated a tsunami that caused damage throughout the Indian Ocean, killing some 250,000 people and causing billions of dollars in damage. The tsunami was the most devastating in more than 40 year and one of the largest such events on record.

Though disastrous, the tsunami presented an unparalleled opportunity for geoscientists to study a large tsunami. One aspect was to study the sediments left in the passage of the waves. Although one previous transoceanic tsunami (1960 Chile) was among the first to be studied by sedimentologists, the 2004 South Asia tsunami represents the first time that sedimentary deposits have been studied systematically both near to and far from the tsunami's source.

Such studies are vital to understanding the hazard posed by low-frequency high-risk events, like tsunamis, because sediments left in the wake of tsunamis are often the only prehistoric record that a coastline has been struck. In locations with a relatively short written history (such as the west coast of the United States and Canada), tsunami deposits may be the only way of assessing the magnitude and frequency of large tsunamis. It is vital, then, to understand how to recognize the deposits of tsunamis and to use them in an attempt to determine the size of the waves that transported them.

One location where tsunami deposits have been used to estimate the frequency and magnitude of ancient earthquakes is Hokkaido, the northernmost island of Japan. F. Nanayama and coworkers reported finding 17 sheets of sand in marsh sediments dating back 7000 years. Because these sheets contain marine shells and resemble the deposits of modern tsunamis, they concluded that these sand sheets are tsunami deposits. Because the sheets extend more than 3 km (2 mi) farther inland than any historical tsunami, the researchers suggest that these sheets are evidence for earthquakes larger than any seen historically along the coast of Hokkaido (where historical large earthquakes have ranged from magnitude 7.8 to 8.2).

Until the 2004 South Asia tsunami, Hokkaido deposits, like many others found around the world, were identified based on comparisons to smaller modern tsunamis, including tsunamis in 1992 (Indonesia and Nicaragua), 1996 (Indonesia and Peru), 1998 (Papua New Guinea), and 1999 (Vanuatu), and on modern field studies of the 1960 Chile tsunami. This marks the first time that geoscientists have the deposits of a large tsunami coupled with data on the flow characteristics of that tsunami, and only the second time they have access to the known deposits of a tsunami capable of crossing ocean basins.

2004 South Asia tsunami deposition. The 2004 South Asia tsunami was a global phenomenon, entering all oceans of the world. Damage, however, was restricted to the Indian Ocean, where the tsunami struck Indonesia within about 15 minutes of the earthquake, Sri Lanka and Thailand within about 2.5 hours, and India about 3 hours after the earthquake. The waves reached areas more than 35 m (100 ft) above sea level in Sumatra, Indonesia, and flooded coastal lowlands to a depth of more than 20 m (66 ft). In Sri Lanka and Thailand, more than 700 km (4400 mi) from the earthquake epicenter, waves reached more than 6 m (20 ft) above sea level and flooded coasts with more than 2.5 m (8 ft) of water.

The tsunami deposited sediment virtually everywhere reached by water, ranging in size from individual coral boulders removed from offshore reefs to mud picked up from fields and deposited farther inland. Along most coastlines, the deposits consist of sheets of sand transported inland from beaches (**Fig. 1**). The sheets are usually 10–30 cm (4–12 in.) thick along the coast of Sumatra and are commonly laminated and become finer upward, although local deposition exceeds 80 cm (32 in.) and some deposits become coarser upward. Similar sand sheets have also been reported in Sri Lanka and Thailand.

Deposits along the Sumatran coast consist of poorly sorted coarse sand that becomes progressively finer landward and upward (**Fig. 2**). In some places, the sand sheet contains complex internal stratigraphy, including structures (most commonly finding upward pulses within the sand sheet) that may record the passage of up to three individual waves. Within each pulse, the sediments are commonly laminated—this sedimentary structure is caused when sediment rains down from the water column, rather than rolling and hopping along the bottom (in contrast, deposits of coastal storms commonly contain sedimentary structures consistent with sand rolling along the bottom). Although the deposit is thin or absent near the shoreline (where the soil has been eroded to bedrock), it maintains an average thickness of 10–15 cm (4–6 in.) starting about 50 m (160 ft) from the shore and extending nearly to the limit reached by the tsunami, suggesting that

Fig. 1. Sediment deposit from the 2004 South Asia tsunami in Sumatra. Note lamination of the deposit, suggesting that sand grains fell from the water column rather than rolling along the bottom, and the changes in grain size upward, suggesting multiple pulses of sedimentation were associated with individual waves.

the tsunami was initially erosive and became more depositional inland. The sand tends to selectively fill in low spots, most dramatically in preexisting stream channels.

Most of the sand in the tsunami deposit is indistinguishable from material found on the beach shortly after the tsunami. However, the tsunami deposit also contains grain sizes and grain compositions not found on the present beach. These grains become more prominent landward and upward in the deposit, suggesting that some other source of sand was progressively exploited and intermixed with the beach sand by the tsunami. The deposit also contains fresh shells of subtidal marine organisms, suggesting that at least part of the deposit came from below the low-tide line.

Hokkaido deposition. The Kuril Subduction Zone lies along the southeastern coast of Hokkaido. Earthquakes are common along this subduction zone, and many of them (most recently a magnitude-8.0 earthquake in 2003) are tsunamigenic. The 2003 tsunami was no higher than 4 m (13 ft), and did not run inland more than about 1 km (0.6 mi). More importantly, little sediment was deposited by this wave.

In contrast, coastal marshes along the southeastern coast of Hokkaido contain 17 sand sheets in 7000 years of deposition at one location (F. Nanayama and coworkers, 2003), and 8 sand sheets in 3000 years of deposition at a location 200 km (120 mi) southwest of Nanayama's study site (S. Woodward and coworkers, 2005) [Fig. 2]. These sand sheets contain marine

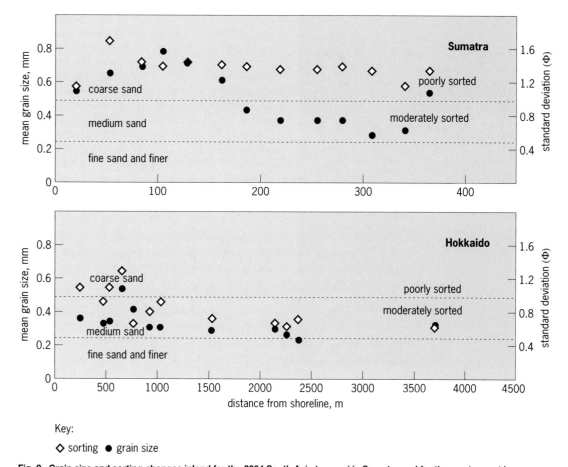

Fig. 2. Grain size and sorting changes inland for the 2004 South Asia tsunami in Sumatra and for the most recent large tsunami in Hokkaido (Fig. 3, layer A). The grain size changes are similar for both tsunamis, with the coarsest grain sizes occurring in about the same location relative to the total inundation length. Although probably a larger tsunami, the South Asia tsunami did not reach as far inland as did the Hokkaido tsunami because the shoreline was much steeper where the Sumatra data were taken.

fossils, become finer landward and upward (**Fig. 3**), selectively infill low spots, and contain the same type of lamination found in the 2004 South Asia tsunami deposits, suggesting that they too were deposited by tsunami.

The Hokkaido deposits are also consistent with the South Asia deposits in that they contain grain sizes and compositions unavailable on the modern beach. Although the beach may have been different when these deposits were created, it is also possible that these were picked up from offshore and deposited inland, as were the South Asia deposits. The uppermost of these sand sheets (Fig. 3, layer A) contains blocks of overturned peat, suggesting that the tsunami initially eroded sediment near shore (as did the South Asia tsunami) and deposited some of those materials inland. More importantly, the uppermost layer penetrates more than 4 km (2.5 mi) inland, and overtopped a coastal cliff more than 10 m (33 ft) high, suggesting that it results from a far larger tsunami than has been recorded historically along that coast.

Many of Woodward's sand sheets are correlative to Nanayama's uppermost sheets. Nanayama has two sand sheets between volcanic ash layers deposited about A.D. 970 and in 1663; Woodward has three, but one is primarily composed of rounded pumice gravel, and may therefore be relatively local. Simi-

larly, Nanayama has four layers between A.D. 970 and a volcanic ash layer deposited about 2700 years ago, whereas Woodward has three. All of these layers can be found some 3 km (2 mi) from the modern shoreline in Nanayama's study area, and 2.5 km (1.5 mi) from shore in Woodward's study area, suggesting that there may have been five tsunamis in the last 2700 years capable of inundating kilometers inland over 200 km (120 mi) of coastline. No historical earthquake has generated such a tsunami in Hokkaido, leaving open the possibility that the Kuril Subduction Zone can produce large earthquakes (greater than magnitude 8.2) that can create large damaging tsunamis.

Outlook. Great tsunamis, like the 2004 South Asia tsunami, carry the potential for enormous loss of life and property. However, their infrequency makes it difficult to understand how often a given area might be exposed to such a hazard, or if an area might be subject to the hazard at all. In these instances, sand sheets like those left along the Hokkaido coast may be the only evidence of the passage of great tsunamis. Interpreting coastal sediments, including rejecting sand sheets not made by tsunamis and attempting to place limits on the size of tsunamis that did leave deposits, requires knowledge gained from modern great tsunamis like the one that struck South

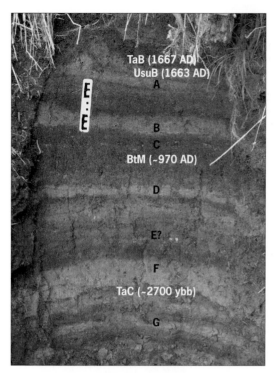

Fig. 3. Sediment deposits from Hokkaido tsunamis. Black letters represent sand sheets, white letters represent volcanic ash layers with known ages. Here, seven tsunami deposits are present within about 3000 years of marsh deposits. As with the 2004 South Asia deposits, the sediment fines both landward and upward, and contains marine organisms. Scale bar is 15 cm long.

Asia in 2004. Data on not only what these sediments look like but also what they reveal about the way sediment is moved by tsunamis will enable more detailed reconstruction of ancient tsunamis, which in turn will make clear the hazard faced along the world's coastlines.

For background information *see* EARTHQUAKE; MARINE SEDIMENTS; NEARSHORE PROCESSES; PALEOCEANOGRAPHY; SEDIMENTOLOGY; SUBDUCTION ZONES; TSUNAMI in the McGraw-Hill Encyclopedia of Science & Technology.   Andrew Lathrop Moore

Bibliography. K. Goto et al., Preliminary results of field survey on the Indian Ocean tsunami deposits in Thailand and Sri Lanka, *GSA Abstr. Prog.*, 37(7):74, 2005; A. L. Moore et al., Sedimentary deposits of the 26 December 2004 tsunami on the northwest coast of Aceh, Indonesia, *Earth, Planets, and Space*, 58(2):253–258, 2006; F. Nanayama et al., Unusually large earthquakes inferred from tsunami deposits along the Kuril trench, *Nature*, 424:660–663, 2003; Y. Tanioka et al., Tsunami run-up heights of the 2003 Tokachi-oki earthquake, *Earth, Planets, and Space*, 56(3):359–365, 2004; S. Woodward et al., Landward fining in late Holocene tsunami deposits from southeastern Hokkaido, *GSA Abstr. Prog.*, 37(7):74, 2005.

Sex differences in the brain

Efforts to understand the biological substrates of sex differences in cognition require examination of brain anatomy (structure) and physiology (function). Technological and methodological advances have increasingly enabled the examination in humans of the neurobiology of behavior across the lifespan. Starting in the early 1980s, a genre of safe methods for obtaining reliable measures of brain structure and function have become available. While most applications of these methods have been in people with different brain disorders, several sufficiently large-scale efforts have included healthy people and have examined sex differences in brain anatomy and physiology. Fewer studies have related such measures to cognitive performance; nonetheless, there is considerable convergence of replicated findings to support at least some hypotheses worthy of further refinement. This article will briefly describe the main findings from neuroimaging applications in which sex differences were established. These findings substantiate some hypotheses on neural substrates for sex differences in cognition.

Cranial tissue composition. One can begin the search for anatomic differences by measuring the tissue in the brain, which is composed of gray matter (GM; the somatodendritic tissue of neurons) and white matter (WM; the axonal compartment of myelinated connecting fibers). The brain is surrounded by cerebrospinal fluid (CSF), the slow circulating fluid that surrounds both cortical and spinal neuronal tissue. Initial studies that suggested sex differences in the composition of human neural tissue used a noninvasive procedure that can measure the proportion of tissue with fast blood flow (presumably gray matter). These studies showed rather substantial sex differences in percent gray matter in the cortical surface, with women having higher values. The main current method for studying brain anatomy is magnetic resonance imaging (MRI). Structural MRI studies use a variety of methods for segmentation of tissue into gray matter (both cortical and deep), white matter, and cerebrospinal fluid (**Fig. 1**). Replicating the earlier findings, higher % gray matter was found in women, but with MRI it was also possible to establish that men had higher % white matter and % cerebrospinal fluid (**Fig. 2**, top panels).

Sex differences in hemispheric asymmetries were also documented, with greater asymmetries in % gray matter and % cerebrospinal fluid in men compared to women (Fig. 2, bottom panels). % gray matter was higher in the left for men, white matter was symmetric, but % cerebrospinal fluid was higher on the right. No asymmetries were significant in women, and the difference in laterality gradients between men and women was significant. The hemispheric effects were quite small in absolute terms and did not overshadow the main sex differences in raw volumes. Thus, while men had higher % gray matter in the left relative to the right hemisphere whereas women have symmetric GM, women still had higher % gray matter than men in either hemisphere.

Few studies have examined sex differences in the correlation between cognitive performance and the volume of intracranial compartments. Anatomic findings may provide neural substrates for sex differences in cognition if volume correlates with performance

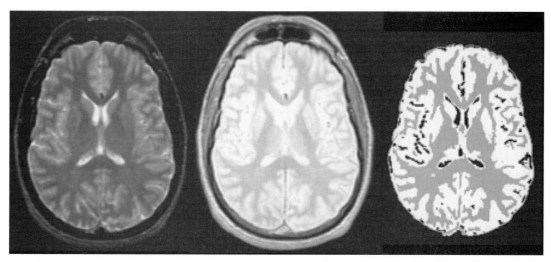

Fig. 1. Illustration of the MRI segmentation process showing an acquired T2-weighted image (left), a proton density image (middle), and the segmented image in which gray matter is depicted in white, white matter in light gray, and cerebrospinal fluid in black (right). (*From R. C. Gur et al., Sex differences in brain gray and white matter in healthy young adults: Correlations with cognitive performance, J. Neurosci., 19:4065–4072, 1999*)

on cognitive tasks. We examined whether our sample showed the reported sex difference of better verbal relative to spatial performance in women compared to men. Men and women did not differ in the global performance score. However, as expected, the verbal superiority index (verbal minus spatial) was positive in women and negative in men, and the two groups differed. Further supporting the functional significance of the neuroanatomic findings, cognitive performance correlated with intracranial volumes for the whole sample and similarly for men and women.

Corpus callosum. In contrast to globally lower white matter volume in women, there is some evidence that the largest white matter structure in the brain, the corpus callosum, is more bulbous in women. Cognitive and functional imaging studies have suggested a greater degree of hemispheric lateralization in males compared to females, while females displayed increased bilateral hemispheric activity for a variety of cognitive tasks. These studies seem to suggest enhanced interhemispheric communication in females and have motivated investigation into sexual dimorphism of the corpus callosum. Most investigators have examined the shape and size of the midsagittal section of the callosum as a surrogate for the structure's overall shape. To date, no consensus has been reached on the presence of such gender-based differences in the callosum. A possible reason for this controversy is the lack of standards in callosal analysis. Template deformation morphometry (TDM) is a relatively new method that avoids many of the pitfalls associated with more traditional measures of the callosum. By registering each subject to a template callosum, template deformation morphometry avoids the issue of normalizing callosal measurements to some arbitrary index of overall brain size. Template deformation morphometry demonstrated that the splenium of the callosum was larger in females than males, while a relatively larger genu of the callosum was found in men (**Fig. 3**).

The findings of the larger female splenium is consistent with the arrangement of the corpus callosum, where the front connects frontal brain regions, and the back connects posterior visual cortex regions. The neuropsychological literature indicates enhanced bihemispheric representation of verbal tasks in females, and the splenium would be involved in interhemispheric transfer of language processing.

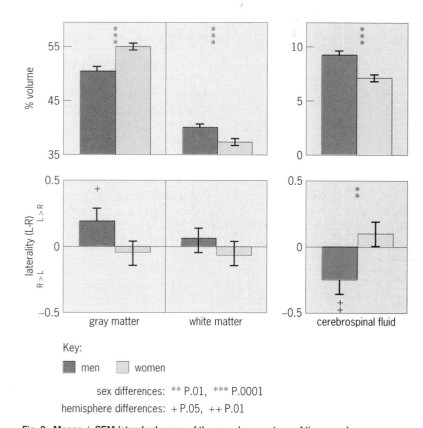

Fig. 2. Means ± SEM (standard errors of the mean) percentage of tissue and cerebrospinal fluid averaged bilaterally (top) and examined as a laterality index (bottom) in men and women. (*From R. C. Gur et al., Sex differences in brain gray and white matter in healthy young adults: Correlations with cognitive performance, J. Neurosci., 19:4065–4072, 1999*)

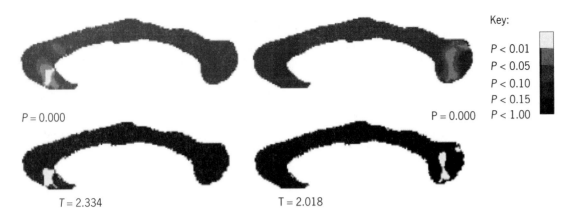

Fig. 3. Size comparison of male and female callosa. At top, raw pointwise *P* values, and at bottom corrected clusters with *P* values. Left side shows areas of larger male size, and right side shows areas of larger female size. (*From A. Dubb et al., Characterization of sexual dimorphism in the human corpus callosum, Neuroimage, 20:512–519, 2003*)

The finding of the larger male genu may relate to enhanced motor coordination in men, a finding also supported by the literature.

Longitudinal studies and studies of young infants may help elucidate causal direction of the relation between brain volume and cognition. J. N. Giedd et al. demonstrated sex differences in the pattern of gray and white matter development. The curves were similar but tended to peak at different ages. The peaks tended to be earlier (for example, in terms of peak gray matter volume for frontal cortex) for girls than boys, except for the temporal cortex, where girls peaked at a slightly older age. A notable pattern was that occipital gray matter had not yet peaked for males by age 22, but by about 13 for girls. White matter increased for both sexes from 4 to 22 years, but at a higher rate for males than females. The increased brain development period for males, especially with respect to white matter and occipital gray matter, is intriguing because these correlate with spatial performance in adults. The extended developmental period also makes male brain development more condition-dependent—good health is needed for a longer period of time to achieve full potential.

To summarize the anatomic studies, sex differences are evident across the age range, and there are sex differences in age effects on brain compartmental volumes. In general, women have higher percentage of tissue devoted to neuronal cell bodies and their immediate dendritic connections, while men have higher volume of connecting white matter tissue, with the exception of the splenium of the corpus callosum, which is more bulbous in women than men. Furthermore, male brains show greater volumetric asymmetries than female brains. The higher white matter volume seems associated with better spatial performance in men, while lower asymmetry seems associated with better language processing in women.

Brain physiology. The feasibility of studying neural substrates of behavior is enhanced by functional neuroimaging methods for measuring regional brain activity. Activation patterns are linked to performance on cognitive tasks requiring cognitive operations such as verbal, spatial, attention, memory, and facial processing. Performance measures are obtained during the physiological studies, permitting linkage of brain activity with task execution.

Cerebral blood flow. Sex differences have not been examined as extensively with functional as with structural imaging. Consistently across studies, women have higher rates of resting cerebral blood flow (CBF) than men. Sex differences in activation patterns are less consistent. Greater bilateral activation for language tasks was reported in females. For spatial tasks, better performance of males when solving the harder problems was associated with more focal activation of right visual association areas. In contrast, women recruited additional regions bilaterally for the harder spatial task. This finding was recently replicated and extended to mental rotation and numeric calculation tasks by K. Kucian et al., who likewise reported more distributed and bilateral recruitment of regions in women than men with increased task complexity. Similarly, G. Grön et al. reported that men and women used different brain regions in a three-dimensional virtual maze. Women used more parietal and prefrontal regions (the latter suggesting it was an effortful task), whereas the men relied more on the hippocampus, suggesting an automatic encoding of geometric-navigation cues.

Cerebral glucose metabolism. In contrast to higher global cerebral blood flow in women, resting cerebral metabolic rates for glucose (CMRglu) are equal in men and women. Sex differences are evident in the regional distribution of metabolic activity, with men showing higher glucose metabolism in all motor basal ganglia regions and cerebellum, as well as all subcallosal limbic regions, while women showed higher metabolism in the cingulate gyrus, a limbic region closer to language areas. As with MRI, women show more symmetric glucose utilization than men.

Neurotransmitter function. Another set of physiologic parameters that can be measured with functional neuroimaging is neurotransmitter function. Depending on the specific neurotransmitter, greater abundance or receptor availability can facilitate or inhibit brain function. Few studies included sufficiently large samples to examine sex differences. Of these, K. H. Adams et al. reported no sex differences in serotonin (5-HT) binding. However, sex differences were found in dopamine function. The dopamine transporter is the primary indicator of dopaminergic tone, and L. H. Mozley et al. investigated the relationship between cognition and dopamine transporter availability in healthy men and women. Women had higher dopamine availability in the caudate nucleus, and they also performed better on verbal learning tasks. Furthermore, dopamine transporter availability was correlated with learning performance within groups.

Behavioral effects of sex differences in the brain. The state of knowledge on the neurobiology of sex differences is far from enabling strong statements. Especially lacking are large-scale studies in healthy people where behavioral data are rigorously measured and related to brain anatomy and physiology. Nonetheless, several tentative hypotheses can be proposed.

A hypothesis based on the neuroanatomic data is that male brains are optimized for enhanced connectivity *within* hemispheres presumably in the anterior-posterior or dorsal-ventral directions, as afforded by overall higher white matter volumes, while female brains are optimized for communication *between* the hemispheres, especially in language processing and posterior brain regions, as indicated by the larger callosal splenia. Evolutionarily, this may have conferred an advantage to males in actions requiring rapid transition from perception (posterior) to action (anterior), incorporating limbic (ventral) input. For females, better interhemispheric communication confers advantage in language and the ability to better integrate verbal-analytical (left hemispheric) with spatial-holistic (right hemispheric) modes of information processing. This hypothesis can be tested directly with novel MRI methods using diffusion tensor imaging (DTI).

Biologically, females have higher cerebral blood flow and the same metabolic rates as men. This affords them "luxury perfusion" relative to metabolic demands, and may better equip them for sustained mental activity. The excess cerebral blood flow relative to CMRglu may also confer longevity of tissue, which may relate to the higher longevity of women. Activation studies support the notion that women perform better on tasks requiring bilateral activation, such as language processing, while men excel in tasks requiring focal activation of visual association cortex.

For background information *see* BRAIN; COGNITION; DEVELOPMENTAL PSYCHOLOGY; LEARNING MECHANISMS; MEDICAL IMAGING; NEUROBIOLOGY; SEX-INFLUENCED INHERITANCE; SEXUAL DIMORPHISM in the McGraw-Hill Encyclopedia of Science & Technology.　　　　　Ruben C. Gur; Raquel E. Gur

Bibliography. K. H. Adams et al., A database of [(18)F]-altanserin binding to 5-HT(2A) receptors in normal volunteers: Normative data and relationship to physiological and demographic variables, *Neuroimage*, 21:1105–1113, 2004; C. E. Coffey et al., Sex differences in brain aging: A quantitative magnetic resonance imaging study, *Arch. Neurol.*, 55:169–179, 1998; J. N. Giedd et al., Brain development during childhood and adolescence: A longitudinal MRI study, *Nat. Neurosci.*, 2:861–863, 1999; G. Grön, Brain activation during human navigation: Gender-different neural networks as substrate of performance, *Nat. Neurosci.*, 3:404–408, 2000; R. C. Gur et al., An fMRI study of sex differences in regional activation to a verbal and a spatial task, *Brain Lang.*, 74:157–170, 2000; R. C. Gur et al., Sex differences in brain gray and white matter in healthy young adults: Correlations with cognitive performance, *J. Neurosci.*, 19:4065–4072, 1999; A. Kastrup, Gender differences in cerebral blood flow and oxygenation response during focal physiologic neural activity, *J. Cereb. Blood Flow Metab.*, 19:1066–1071, 1999; K. Kucian et al., Gender differences in brain activation patterns during mental rotation and number related cognitive tasks, *Psychol. Sci.*, 47:112–131, 2005; L. H. Mozley et al., Striatal dopamine transporters and cognitive functioning in healthy men and women, *Amer. J. Psychiat.*, 158:1492–1499, 2001; B. A. Shaywitz et al., Sex differences in the functional organization of the brain for language, *Nature*, 373:607–609, 1995.

Simian virus 40

Simian virus 40 (SV40) was discovered in 1960 as a contaminant of poliovirus vaccines. Soon thereafter, it was shown to possess cancer-causing (oncogenic) activity. By the time the virus was recognized and removed from the vaccines, millions of people worldwide had been exposed to live SV40. The possible consequences of those exposures continue to be the topic of lively debate. There is strong evidence that SV40 is causing human infections today, including those in individuals too young to have been exposed to the early contaminated vaccines, but the potential role of the virus as an agent of human disease remains a focus of investigation. Other infectious agents are known to cause human cancer; the identification of another member of that select group would represent a significant advance in our quest to understand human carcinogenesis and its etiologies.

History. The early poliovaccines were prepared in cell cultures established from kidneys of rhesus monkeys. Asian macaques, especially rhesus monkeys, are the natural hosts of SV40, and their kidneys often are infected. The presence of SV40 in the kidney cell cultures used for vaccine production escaped detection because it failed to produce recognizable cell changes or cytopathic effects. The inactivation procedure used to produce the killed (Salk) poliovaccine failed to completely destroy SV40 infectivity, leaving residual live virus in many lots of the Salk vaccine. The live attenuated (Sabin) oral poliovaccine used in field trials before licensure was contaminated with higher levels of SV40 because there was no inactivation step in its production. All poliovaccines in use after 1963 were believed to be free from SV40 contamination. However, a report published in 2005

Fig. 1. Genetic map of polyomavirus SV40. The circular SV40 DNA genome is represented with the unique EcoRI site shown at map unit 100/0. Nucleotide numbers based on reference strain SV40-776 begin and end at the origin (Ori) of viral DNA replication (0/5243). The open reading frames that encode viral proteins are indicated. Arrowheads point in the direction of transcription; the beginning and end of each open reading frame are indicated by nucleotide numbers. Large T-antigen (T-ag), the essential replication protein, as well as the viral oncoprotein, is coded by two noncontiguous segments on the genome. t-ag = small t-antigen. (*From G. F. Brooks, J. S. Butel, and S. A. Morse, Medical Microbiology, 21st ed., Appleton & Lange, 1998*)

showed that post-1963 oral poliovaccines prepared in Russia and used perhaps as late as 1978 contained infectious SV40. It is not known how many people exposed to SV40-contaminated vaccines actually became infected with SV40, but as millions of people were exposed to contaminated vaccines, it is likely that a considerable number of human infections occurred. There may be other sources of human exposure to SV40 besides the vaccines, but none has been proven with the exception of isolated cases of contacts with monkeys.

Properties. SV40 is a small nonenveloped virus with a double-stranded DNA genome about 5000 base pairs (bp) in size (**Fig. 1**). It is classified as a polyomavirus; many members of this virus group establish persistent infections and possess oncogenic properties. SV40 encodes replication proteins, called T-antigens, that are necessary for viral replication and are able to stimulate resting cells to enter the cell cycle. Cellular S-phase components are essential in order for SV40 viral DNA replication to occur. SV40 large T-antigen accomplishes cell entry into S-phase by binding and functionally inactivating cellular tumor suppressor proteins, including p53 and several retinoblastoma (a malignant tumor of the sensory layer of the retina) family members (**Fig. 2**). The large T-antigen is a potent transforming protein (on-

coprotein) because of its ability to dysregulate cellular growth control mechanisms.

SV40 has a stable genome not prone to rapid mutational change. However, different viral strains exist, based on variations in the sequence encoding the C-terminal portion of the large T-antigen protein. Virus isolates may also differ in the structure of the regulatory region of the viral genome. Viruses with complex duplications/rearrangements in the regulatory region tend to replicate better in tissue culture cells than variants with simple regulatory regions, but the effects of these regulatory region differences on virus replication in vivo in an intact host are unknown. The existence of SV40 genetic variation raises the theoretical possibility that viral strains might differ in biological properties that affect the development of human disease. There is a well-known precedent with the human papillomaviruses; strains of papillomavirus differ dramatically in their ability to cause human cervical cancer.

SV40 does not appear to cause disease in healthy monkeys, but central nervous system pathology may develop in immunosuppressed animals. Syrian golden hamsters are susceptible to tumor induction by SV40. The present author's laboratory has found that strains of SV40 differ markedly in their tumor-inducing capacity. One SV40 strain that is highly oncogenic in hamsters, SVCPC, has been detected in several human tumors and was identified as a contaminant in the Russian oral poliovaccine. These findings suggest that the risk for disease development by a host may depend, in part, on which viral strain causes an infection. The laboratory is currently investigating which viral genomic regions and gene functions are important in determining SV40 tumorigenic potential.

It was discovered recently that SV40 encodes microRNAs (small ribonucleic acid molecules thought to regulate the expression of other genes) that accumulate late in infection and target viral early messenger RNAs for cleavage, resulting in reduced expression of viral T-antigens. Such wild-type virus-infected cells appeared to be less sensitive to killing by cytotoxic T-cells, relative to cells infected with an SV40 mutant unable to make microRNAs. Regulation of viral gene expression and escape of infected cells from immune destruction presumably would prolong the life of infected cells, and may be important in the pathogenesis of natural infections in vivo. Perhaps regulatory microRNAs contribute to the maintenance of persistent infections by polyomaviruses that hosts are unable to eradicate. The role, if any, that microRNAs may play in SV40 tumorigenesis remains to be elucidated.

SV40 infections and human disease. Several lines of evidence indicate that SV40 infections are occurring today in individuals not exposed to the contaminated vaccines. SV40 has been detected in healthy subjects, as well as in patients with malignant and nonmalignant diseases. Young children have been observed to excrete SV40 in urine and in stool, SV40 sequences have been detected in blood cells from both healthy adults and organ transplant recipients,

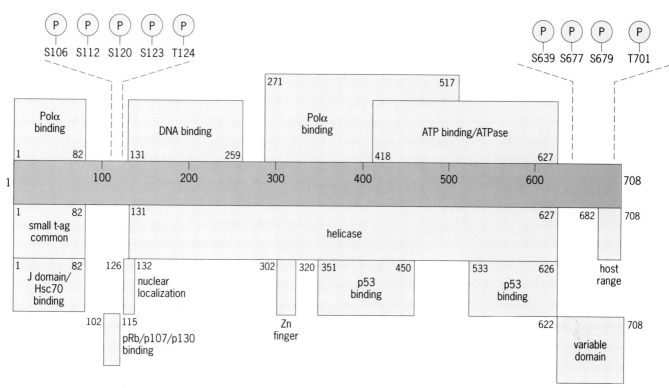

Fig. 2. Functional domains of SV40 large T-antigen. The numbers given are the amino acid residues; the numbering system for SV40-776 is used. Several regions of interest: "small t-ag common" is the region of T-antigen encoded in the first exon (the amino acid sequence in this region is common to both T-antigen and small t-antigen); "pRb/p107/p130 binding" is the region required for binding of the retinoblastoma (Rb) tumor suppressor protein and the Rb-related proteins p107 and p130; "p53 binding" are the regions required for binding the p53 tumor suppressor protein; "variable domain" is the region containing amino acid differences among viral strains. The circles containing P indicate sites of phosphorylation. (*From A. R. Stewart et al., Virology, 221:355–361, 1996*)

and SV40 DNA has been found in biopsies from adults with kidney disease and from allografts of pediatric kidney transplant recipients.

A major focus of investigation has been on SV40 links to human cancer. SV40-positive tumors have been identified in studies from a number of countries, from the United States to Costa Rica, Japan, and Italy, although the percentage of tumors that contained SV40 DNA varied widely. Other studies have yielded negative findings, including those from Finland, Turkey, and the United Kingdom (Scotland). This inconsistency among studies is the reason that scientific consensus has not yet been reached on the role of SV40 in human cancer. Such variability in detection of SV40 may reflect a number of factors, including the prevalence of SV40 infections in the populations being studied, technical differences in laboratory methodology, the types of samples analyzed, the presence of other factors that cause the same types of cancer, and viral strain differences which could affect disease causation. If there are significant geographical differences in the frequency of SV40 infections in humans, that phenomenon could account for the observed variability among studies, and might be explained by past usage of contaminated vaccines.

The types of human cancer reproducibly found associated with SV40 are mesotheliomas (tumors of the mesothelium lining the lungs, abdomen, or heart), brain tumors, osteosarcomas (bone tumors),

and lymphomas. These are the same types of tumors that are induced in the SV40 hamster tumor model. However, these types of tumors also arise in humans due to cancer-inducing insults other than SV40 infection. More is currently known about the molecular pathogenesis of mesotheliomas than the other tumors associated with SV40.

A question central to cancer causation is whether the virus is functionally important in the tumors in which it is detected. One indication of a functional effect by SV40 in human cancer comes from methylation studies; methylation is a mechanism of gene silencing in human tumors. In both mesotheliomas and lymphomas, aberrant methylation was observed in the promoter regions of certain tumor suppressor genes in SV40-positive tumors as compared to tumors negative for SV40 sequences. Another line of evidence comes from immunohistochemistry studies. The expression of large T-antigen in some SV40 DNA-positive lymphomas was detected, but not in SV40 DNA-negative samples (**Fig. 3**). The T-antigen staining reactions obtained in human tumors were not as homogeneous or intense as those with SV40-induced hamster tumors, and not all human tumor cells appeared to be positive, indicating that T-antigen does not accumulate as abundantly in the human cells as in the hamster tumor cells. Although simple detection of the protein does not prove functionality, from the decades of studies characterizing the potent nature of T-antigen, it would be

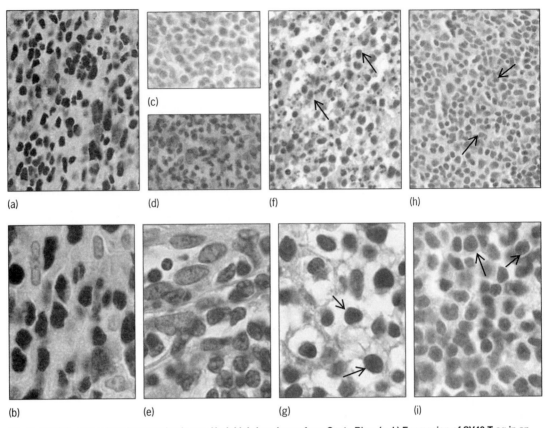

Fig. 3. SV40 T-antigen (T-ag) expression in non-Hodgkin's lymphoma from Costa Rica. (*a, b*) Expression of SV40 T-ag in an SV40-induced hamster tumor, stained with PAb101. (*c*) SV40 DNA-negative lymph node stained with PAb101. (*d, e*) SV40 DNA-negative diffuse large B-cell lymphoma from a 29-year-old male reacted with PAb101. (*f, g*) Expression of SV40 T-ag in an SV40 DNA-positive diffuse large B-cell lymphoma from a 45-year-old male, stained with PAb101. (*h, i*) Expression of SV40 T-ag in malignant B-cells in an SV40 DNA-positive diffuse large B-cell lymphoma from a 67-year-old female, stained with PAb101. Original magnification for *a, c, d, f, h,* 40×; *b, e, g, i,* 100×. (*From A. Meneses et al., Haematologica, 90:1635–1642, 2005*)

surprising if the viral oncoprotein were expressed in human lymphoid cells but had no effect on cellular properties. However, it is possible that T-antigen might become dispensable in advanced tumors as cellular mutations accumulate due to genomic instability and tumor progression occurs.

The prevalence of SV40 infections in different human populations remains unknown. Serologic assays have estimated a prevalence between 2 and 20%, but the actual prevalence is hard to gauge. The human immune response to SV40 is uncharacterized, and assays are complicated by the existence of cross-reactive epitopes (regions on an antigen's surface that trigger a corresponding antibody response) between SV40 and the widely prevalent human polyomaviruses BKV and JCV. SV40 neutralizing antibody titers in humans are generally low and tend to wane over time. Molecular-based assays could be a superior alternative to serological assays to detect SV40 infections, but the most informative specimens for molecular testing are not known because the natural history of human infections by SV40 is not well understood.

Future directions. It is a matter of public health importance to identify any causative role that SV40 may have in human disease, especially in those cancers that have been associated with SV40 markers. Important issues that need to be investigated include the prevalence of SV40 infections in different populations, the mode of transmission of virus from person to person, the pathogenesis of infections in susceptible hosts, and viral effects on cancer cells. New and better assays need to be developed to detect the presence of SV40 and the expression of viral genes in human specimens. Consensus agreement regarding a role for SV40 in the development of selected human cancers will lead to new diagnostic, therapeutic, and prevention strategies.

For background information *see* ADENO-SV40 HYBRID VIRUS; ADENOVIRIDAE; ANIMAL VIRUS; CANCER (MEDICINE); GENETIC ENGINEERING; ONCOLOGY; TUMOR VIRUSES; VIRUS, DEFECTIVE in the McGraw-Hill Encyclopedia of Science & Technology.

Janet S. Butel

Bibliography. D. Ahuja, M. T. Sáenz-Robles, and J. M. Pipas, SV40 large T antigen targets multiple cellular pathways to elicit cellular transformation, *Oncogene*, 24:7729–7745, 2005; M. K. Axthelm et al., Meningoencephalitis and demyelination are pathologic manifestations of primary polyomavirus infection in immunosuppressed rhesus monkeys, *J. Neuropathol. Exp. Neurol.*, 63:750–758, 2004; D. Bookchin and J. Schumacher, *The Virus and the Vaccine*, St. Martin's Press, New York, 2004; J. S. Butel, Viral carcinogenesis: Revelation of molecular mechanisms and etiology of human disease,

Carcinogenesis, 21:405–426, 2000; J. S. Butel and J. A. Lednicky, Cell and molecular biology of simian virus 40: Implications for human infections and disease, *J. Nat. Cancer Inst.*, 91:119–134, 1999; R. Cutrone et al., Some oral poliovirus vaccines were contaminated with infectious SV40 after 1961, *Cancer Res.*, 65:10273–10279, 2005; Z. H. Forsman et al., Phylogenetic analysis of polyomavirus simian virus 40 from monkeys and humans reveals genetic variation, *J. Virol.*, 78:9306–9316, 2004; A. Meneses et al., Lymphoproliferative disorders in Costa Rica and simian virus 40, *Haematologica*, 90:1635–1642, 2005; K. Stratton, D. A. Alamario, and M. C. McCormick, *Immunization Safety Review: SV40 Contamination of Polio Vaccine and Cancer*, National Academies Press, Washington, DC, 2003; C. S. Sullivan et al., SV40-encoded microRNAs regulate viral gene expression and reduce susceptibility to cytotoxic T cells, *Nature*, 435:682–686, 2005; J. A. Vanchiere et al., Frequent detection of polyomaviruses in stool samples from hospitalized children, *J. Infect. Dis.*, 192:658–664, 2005; R. A. Vilchez et al., Differential ability of two simian virus 40 strains to induce malignancies in weanling hamsters, *Virology*, 330:168–177, 2004.

Single silicate crystal paleomagnetism

Understanding the nature and origin of the geodynamo remains a central focus in geophysics. Significant progress has been made in defining the chronology of geomagnetic reversals, using the record frozen in oceanic crust when it forms and the history preserved in sedimentary strata. Another important aspect of the geomagnetic field is its long-term strength. There is general agreement that the strength of the field, when averaged over intervals of at least several millennia, has varied only by a factor of approximately 3 over the last 2.5 billion years. When compared to other physical properties of the Earth, the past field strength is relatively well known.

Scientists who study the ancient magnetic field (paleomagnetists) are constantly seeking better resolution. This improved view is necessary to understand a wide range of issues, including the nature of the recent decay of the dipole magnetic field, the potential relationship between field strength and the frequency of geomagnetic reversals, and the magnetic field (and magnetic shielding) of the young Earth.

Progress on understanding these topics has been slow because the demands of paleointensity experiments are severe. Magnetic particles must be very small (generally 50 nanometers to a micrometer), and consist of one or at most a few magnetic domains. The magnetization must be an original thermoremanent magnetization, locked-in when magnetic minerals cool through their Curie points [for example, 580°C (1080°F) for magnetite]; the rock must not carry a chemical magnetization acquired at a later time; and the magnetic particles, or the matrix in which they are contained, must not alter during the measurement of paleointensity—a nontrivial experimental demand.

Paleointensity experiments. A typical Thellier paleointensity experiment involves paired heating steps in which a sample is heated in the absence and presence of a known "laboratory" field. A comparison of how a sample loses and acquires a magnetization allows a calculation of the past field strength. These laboratory heating steps can easily induce changes in natural samples.

Bulk samples of lavas are typically used for paleointensity analyses, and experiments have demonstrated that some fresh modern flows do faithfully preserve a recording of geomagnetic field strength. However, as we proceed to older lava flows to examine the history of the geodynamo, the chances for alteration in nature and in the laboratory increase. Weathering can transform the rock as clays form. During paleointensity experiments, new magnetic minerals can form from these clays. The composition of the magnetic minerals can also change by weathering, transforming the original thermoremanent magnetization into a chemical remanent magnetization. These processes can create a bias toward low field strength values when whole rocks are used in paleointensity experiments.

In most lava, one can identify components that might be potentially good and poor magnetic recorders. Among the best recorders are single silicate crystals, especially plagioclase in lavas (**Fig. 1**). Such crystals do not have an intrinsic magnetization of interest, but they tend to contain minute magnetic mineral inclusions, which are ideal recorders. The measurement of feldspar crystals 1 mm in size and smaller is possible with the latest generation of high-resolution dc superconducting quantum interference device (SQUID) magnetometers.

This method has been tested using a modern lava flow of the Kilauea, Hawaii, volcano—the obtained field strength values agreed with the actual values known from magnetic observatory

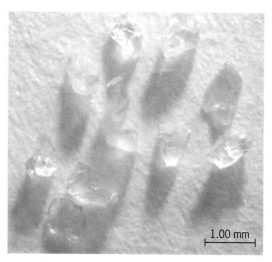

Fig. 1. Typical silicate crystals used in paleointensity experiments. (*Photo courtesy R. D. Cottrell*)

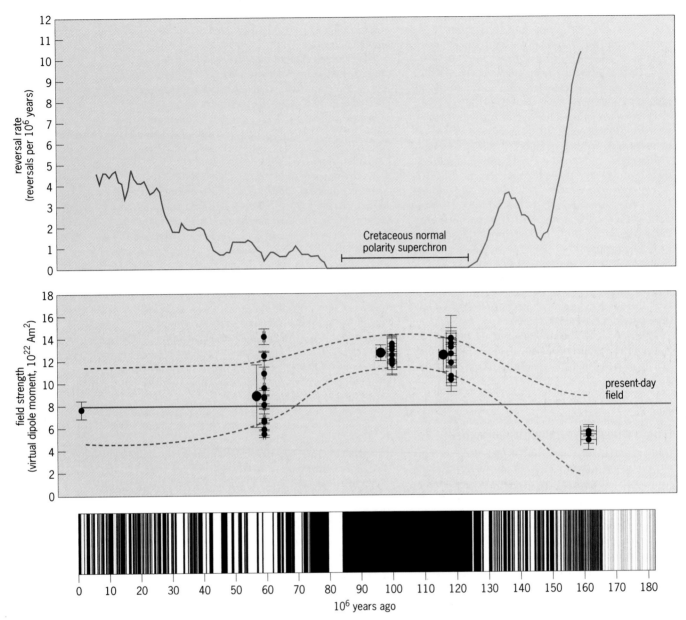

Fig. 2. Paleointensity based on single plagioclase measurements versus the chronology of geomagnetic reversal (black = normal polarity; white = reversed polarity) and an estimate of reversal rate. Small symbols show mean value derived from multiple experiments using separate crystals, whereas large symbols represent averages of the lava unit results. (*Modified from J. A. Tarduno et al., 2001, 2006*)

measurements. The technique has also been used to explore trends in geomagnetic field strength over the past 160 million years, especially with respect to the frequency of geomagnetic reversals.

Field reversals. The results of over 400 paleointensity experiments using plagioclase from continental flood basalt lavas in India and the Arctic, and lavas obtained from ocean drilling expeditions provide a synoptic view, suggesting an inverse relationship between field strength and reversal frequency (**Fig. 2**). Such a relationship was suggested in early theoretical work by Allan Cox, and is consistent with the results of some numerical simulations, including the Glatzmaier-Roberts dynamo. These analyses further support the idea that mantle convection controls the efficiency of the geodynamo on time scales

of tens-to-hundreds of millions of years.

The paleointensity data available to date hint at a mean field value higher than that suggested by other natural recorders, including lavas and submarine basaltic glass. The determination of this mean value is important for several reasons, such as a backdrop on which we can evaluate the rapid decrease in the dipole field strength that has been observed over the last 150 years. That is, the resolution of the mean field strength includes the related debate of whether the modern dipole field strength is falling from a short-term anomalously high value toward the mean, or from a value closer to the mean to a low level which might indicate the onset of field reversal. While limited in number, data from plagioclase support the latter view.

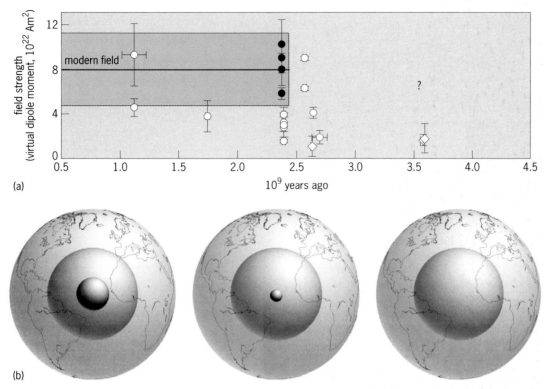

(a)

(b)

Fig. 3. Field values of rocks formed near the Proterozoic-Archean boundary. (*a*) Field strength based on plagioclase separated from dikes approximately 2.5 billion years old (mean values shown as solid symbols); other estimates based on whole-rock studies. Results shown at approximately 3.5 billion years ago (diamonds) are from rocks that carry a chemical remanent magnetization. The modern field intensity and its variation are shown by the line and shaded interval, respectively. (*b*) Scenario for growth of the solid inner core. (*Modified from A. V. Smirnov et al., 2003*)

Inner-core growth. Another aspect where single silicate crystal studies will play an important role is in the definition of the magnetic field of the early Earth and the initiation of growth of the solid inner core, which is thought to play a role in powering the dynamo and perhaps governing its geometry. Many, if not most, rocks greater than a few billion years old have experienced low-grade metamorphic conditions. While mild in geologic terms, this reheating can impart secondary magnetizations and instigate the formation of secondary chemical remanent magnetizations. Studies of single silicate crystals have already yielded promising results, including field values similar to those of the modern field in rocks formed near the Proterozoic-Archean boundary, 2.5 billion years ago (**Fig. 3**). When combined with directional results, these data suggest that a geodynamo similar to that of today's field was operating, consistent with commencement of inner-core growth. Defining the field for even older time intervals represents the frontier area for analyses of single silicate crystals.

For background information *see* BASALT; EARTH INTERIOR; FELDSPAR; GEODYNAMO; GEOMAGNETIC VARIATIONS; GEOMAGNETISM; LAVA; MAGNETIC REVERSALS; PALEOMAGNETISM; PROTEROZOIC; ROCK MAGNETISM; SQUID in the McGraw-Hill Encyclopedia of Science & Technology.   John A. Tarduno

Bibliography. G. A. Glatzmaier et al., The role of the Earth's mantle in controlling the frequency of geomagnetic reversals, *Nature*, 401:885–890, 1999;

G. Hulot et al., Small scale structure of the geodynamo inferred from Oersted and Magsat satellite data, *Nature*, 416:620–623, 2002; A. V. Smirnov, J. A. Tarduno, and B. N. Pisakin, Paleointensity of the early geodynamo (2.45 Ga) as recorded in Karelia: A single crystal approach, *Geology*, 31:415–418, 2003; J. A. Tarduno, R. D. Cottrell, and A. V. Smirnov, High geomagnetic field intensity during the mid-Cretaceous from Thellier analyses of single plagioclase crystals, *Science*, 291:1779–1783, 2001; J. A. Tarduno, R. D. Cottrell and A. V. Smirnov, The paleomagnetism of single silicate crystals: Recording geomagnetic field strength during mixed polarity intervals, superchrons and inner core growth, *Rev. Geophys.*, 41:RG1002, 2006.

Slow-spreading mid-ocean ridges

Over 60% of the Earth's surface is made of oceanic crust that formed by the spreading of the tectonic plates at mid-ocean ridges (MORs). Slow- and ultraslow-spreading mid-ocean ridges (spreading rates of <55 mm/yr) make up about 55% of the global ridge system; that is, roughly 30% of the Earth's current surface area has been created at slow-spreading ridges. In the last 10 years, exploration of slow- and ultraslow-spreading ridges has shown they are different from the better-understood fast-spreading ridges. Recent research suggests that ultraslow-spreading ridges are a separate type of mid-ocean ridge.

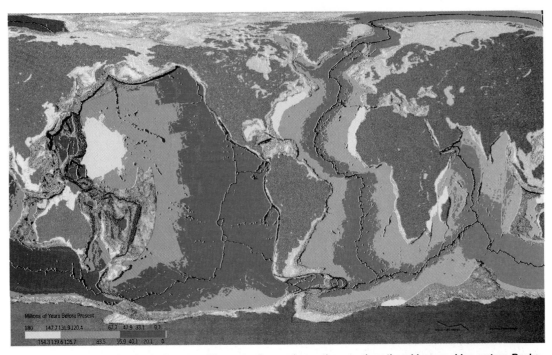

Fig. 1. Composite map showing bathymetry of the ocean floor and accretion age along the mid-ocean ridge system. Rocks making up the sea floor are younger at the center of the ridge and increase in age with distance from its center. (*From World Data Center for Marine Geology & Geophysics, Age of the Ocean Floor, Geology and Geophysics Report MGG-12, National Geophysical Data Center, Boulder, 1996, http://www.ngdc.noaa.gov/mgg/fliers/96mgg04.html*)

The Earth's mid-ocean ridge system is a 50,000–60,000-km-long (31,000–37,000-mi) mountain chain that encircles the globe between the continents, much like the seam on a baseball. This ridge system marks the divergent boundaries of the roughly dozen tectonic or lithospheric plates that form the surface of the Earth (**Fig. 1**). Mid-ocean ridges are the most active volcanic mountain chain on the Earth, creating oceanic crust and lithosphere at a rate of >3 km³/y. Rates of spreading at these divergent plate boundaries range from about 16 cm/yr at the East Pacific Rise, to <1 cm/yr at the Gakkel Ridge in the Arctic. The Mid-Atlantic Ridge (MAR) is perhaps the best-known mid-ocean ridge. This submerged mountain chain extends from the Arctic Ocean to beyond the southern tip of Africa, and is spreading at a rate that varies from about 3.6 to 1.6 cm/yr, and averages 2.5 cm/yr, or 25 km (15.5 mi) in a million years. The 1800-km-long (1100-mi) Gakkel Ridge represents the most northerly extension of the Mid-Atlantic Ridge. It is spreading at 1.5 to <1 cm/yr, and is considered an ultraslow-spreading ridge.

History. In 1912, Alfred Wegener proposed the theory of continental drift, based in part on the similarity in form of the coastlines of South America, North America, and Africa. This theory was dismissed due to a lack of knowledge about processes and mechanisms that might allow the continents to move across the oceans. In the 1950s, the submerged (average depth 2500 m or 8125 ft) mid-ocean ridge system was discovered through extensive shipboard surveys of the Atlantic sea floor between Europe and North America. This discovery, led by Marie Tharp

and Bruce Heezen, resulted in recognition of an enormous 16,000-km (10,000-mi) mountain chain running along the middle of the Atlantic Ocean. Tharp and Heezen also noticed that the axis of the Mid-Atlantic Ridge was marked by a 1.5-km-deep (1-mi) rift valley, flanked by normal faults. They began mapping the bathymetry (relief) of the ocean floor in 1947 at the Lamont Geological Laboratory, New York. For the first 18 years of their collaboration, Heezen collected shipboard data and Tharp drew the maps from that data (by tradition, women were not allowed onboard ships at that time). Additionally, Tharp used shipboard data from the Woods Hole Oceanographic Institution's research vessel, *Atlantis*, and seismographic data from submarine earthquakes. Their work represented the first systematic and comprehensive attempt to map the entire sea floor, and was published as the first physiographic map of the North Atlantic in 1957.

Ridge types. Mid-ocean ridges are classified based on their on their spreading rate, because observable features, including ridge morphology and crustal thicknesses, vary with plate spreading rates. Four main types of mid-ocean ridges are recognized: fast (>75 mm/yr), intermediate (~55–75 mm/yr), slow-spreading (20–55 mm/yr), and ultraslow-spreading (<20 mm/yr). Fast-spreading ridges (such as the northern and southern East Pacific Rise, with spreading rates of 15 cm/yr) are characterized by a narrow axial rise or ridge, smooth topography, and low relief (tens to hundreds of meters) [**Fig. 2a**]. In contrast, slow- and ultraslow-spreading MORs (such as the Mid-Atlantic, Southwest Indian, and Gakkel ridges) are characterized by a broad rift or median valley

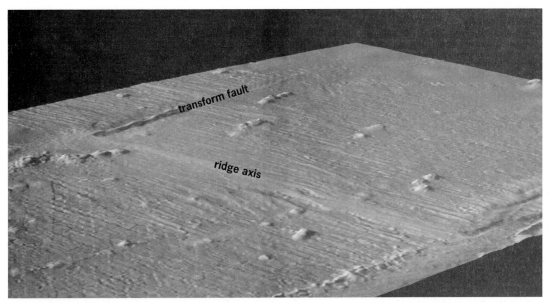

(a)

(b)

Fig. 2. Oblique view of fast- and slow-spreading mid-ocean ridges, showing differences in morphology along the ridge (each with 2× vertical exaggeration). (*a*) Fast-spreading East Pacific Rise at 19°S, viewed toward the north. (*b*) Slow-spreading Mid-Atlantic Ridge at 30°N and the Atlantis transform view toward the northeast. Images made with GeoMapApp software with multibeam sonar data. (*W. Haxby 2006, GeoMapApp; Marine Geosciences Data Management System, http://www.GeoMapApp.org/*)

that is ~1–3 km (0.6–2 mi) deep and up to 10–20 km (6–12 mi) in wide, with rough topography created by faulting (Fig. 2*b*). Ocean crust created by magmatism at the well-studied fast-spreading ridges is fairly uniform in thickness [~7 km thick (4 mi)]. In contrast, crustal thickness variations along slow- and ultraslow-spreading ridges ranges from nearly 0 to 6 km (3.6 mi). Even along a single segment of the Mid-Atlantic Ridge, thickness can vary from near zero at the ridge-transform intersection to >5 km (3 mi) near the ridge segment center. Another important difference is that mantle peridotite can be locally exposed at the sea floor in the rift valley of a slow-spreading ridge, as well as commonly at the inside-corner highs at ridge-transform intersections.

Neither fast- nor slow-spreading MORs form a straight/linear ridge system for significant distances. Instead, they are offset up to 1000 km (600 mi) by transform faults, which allow the two plates to move laterally past one another (Fig. 2). Most mid-ocean ridges are divided into hundreds of segments by transform faults. Along the Mid-Atlantic Ridge, transform faults and their off-axis traces (fracture zones) occur at an average interval of 55 km (34 mi), dividing the Mid-Atlantic Ridge into about 300 distinct segments.

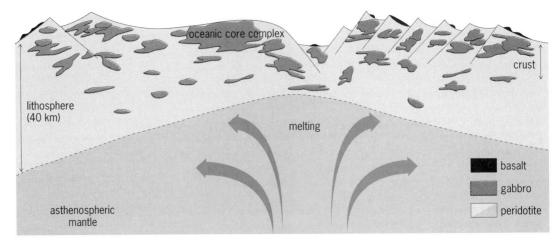

Fig. 3. Cross section of the crust and lithosphere at an ultraslow-spreading mid-ocean ridge.

As with fast-spreading mid-ocean ridges, researchers working on slow- and ultraslow-spreading ridges have recognized spectacular hydrothermal vent systems, including the Rainbow and TAG hydrothermal fields, and the serpentinite-hosted Lost City vent field near the Mid-Atlantic Ridge at 30°N. These hydrothermal vents transport heat and cations into the ocean and provide a substrate for chemosynthetic biological communities. Additional evidence for fossil hydrothermal vents (sulfide chimneys) along the ultraslow-spreading Gakkel Ridge was found via dredging in 2001. *See* LOST CITY HYDROTHERMAL FIELD.

Differences in slow-spreading and fast-spreading ridges. The reason why slow-spreading ridges are different from fast-spreading ridges is likely due to the increased importance of downward conductive cooling, below the Earth's surface. At fast-spreading rates, mantle/asthenospheric upwelling outpaces conductive cooling from above, but as the spreading rate decreases, so does the mantle upwelling rate. The effect of conductive cooling from the Earth's surface downward therefore becomes more important. This leads to the presence of a relatively thick (possibly up to 20 km or 12 mi) lithosphere beneath the axis of a slow-spreading ridge, compared to the very thin (<5 km or 3 mi) lithosphere beneath fast-spreading ridges. An additional effect of cooling arises if the ridge is adjacent to a large-offset transform. The combination of both these sources of cooling leads to along-segment variation in crustal thickness and bathymetry. Consequently, the center of a segment may show characteristics of a fast-spreading ridge, while the ends of a segment may show the characteristics of an ultraslow-spreading ridge.

One predictable effect of this thick lithosphere is that it inhibits the ability of magmas to traverse its entire thickness to very shallow crustal depths and form a continuously fed long-lived magma chamber. Instead, gabbroic magmas probably form relatively small plutonic bodies, including dikes or sill-like intrusions, throughout the entire thickness of the lithosphere (**Fig. 3**), increasing in volume toward the surface. Recent seismological investigations of ocean

crust to the west of the Mid-Atlantic Ridge suggest that gabbro plutons can be intruded at depths up to 60 km (37 mi) below the sea floor. Consequently, the style of crustal accretion at slow-spreading ridges differs from that at fast-spreading ridges. This difference is also visible in the style of volcanism seen at the sea floor. At slow-spreading ridges, the magma supply may be insufficient to generate large fissure eruptions, which are known to occur at fast-spreading ridges. Instead, large numbers of small volcanoes, or seamounts, form along the ridge axis. D. K. Smith and J. R. Cann showed that these are discontinuous and consist of small coalesced volcanic cones that range 50–600 m (160–2000 ft) in height, with an average height of ~60 m (200 ft). The melting history at slow- and ultraslow-spreading MORs can be addressed in part by examining the composition of the mantle peridotite recovered from these settings. The common presence of depleted mantle harzburgite along numerous segments of these ridges shows that the mantle has undergone a similar amount of melting as mantle rocks dredged from faster-spreading ridges, but the difference is that some of the melt at a slow-spreading ridge is trapped in the lithosphere.

The second predictable effect of a thick oceanic lithosphere is that more of the motion of the diverging plates at the axis of the MOR will be accommodated by extensional faulting rather than by magmatic intrusion. At a fast-spreading ridge, the potential gap formed when the plates spread apart is filled by continuous intrusion of magma at the ridge axis. At a slow-spreading ridge, magma does not easily reach the uppermost levels of the lithosphere; plate spreading is partly accommodated by extensional faulting, leading to the exposure of mantle peridotite at the sea floor and the much rougher bathymetry seen at slow-spreading ridges. A spectacular example of this is the generation of oceanic core complexes, major domes of oceanic lithosphere bounded by >10-km-long (6-mi) normal faults (Fig. 3). One of the best-known examples of these features is the Atlantis Massif at 30°N on the Mid-Atlantic Ridge (Fig. 2b). Although core complexes are rare at slow- and ultraslow-spreading ridges (accounting for ~10% of the sea floor), they have not been recognized at fast-spreading ridges. At ultraslow-spreading ridges, plate spreading may be almost completely taken up by extensional faulting, with vast amounts of mantle peridotite exposed on the sea floor. In these environments, the recognition of a petrological base to the oceanic crust becomes difficult because the uppermost part of the Earth is primarily composed of mantle peridotite, with some thin basaltic lavas and gabbro intrusions extending to at least 20 km (12 mi) depth.

For background information *see* ASTHENOSPHERE; EARTH CRUST; EARTH INTERIOR; FAULT AND FAULT STRUCTURES; GABBRO; HOT SPOTS (GEOLOGY); HYDROTHERMAL VENT; LITHOSPHERE; MID-OCEANIC RIDGE; PERIDOTITE; PLATE TECTONICS; PLUTON; SEISMOLOGY; TRANSFORM FAULT; VOLCANO in the McGraw-Hill Encyclopedia of Science & Technology.

Barbara E. John; Michael J. Cheadle

Bibliography. M. Cannat, Emplacement of mantle rocks in the seafloor at mid-ocean ridges, *J. Geophys. Res.*, 98:4163–4172, 1993; H. J. B. Dick, J. Lin, and H. Schouten, An ultraslow-spreading class of ocean ridge, *Nature*, 426:405–412, 2003; B. C. Heezen, The rift in the ocean floor, *Sci. Amer.*, 203:99–106, 1960; D. Lizarralde et al., Spreading-rate dependence of melt extraction at mid-ocean ridges from mantle refraction data, *Nature*, 432:744–747, 2004; D. K. Smith and J. R. Cann, The role of seamount volcanism in crustal construction along the mid-Atlantic Ridge at 24°–30°N: *J. Geophys. Res.*, 97:1645–1658, 1992; R. S. White, D. P. McKenzie, and K. O' Nions, Oceanic crustal thickness from seismic measurements and rare earth element inversions, *J. Geophys. Res.*, 97:19683–19715, 1992.

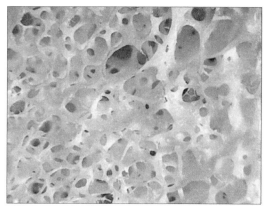
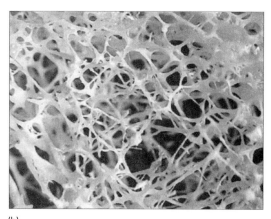

Fig. 1. Bone structure of (a) 54-year-old female and (b) 74-year-old female, showing spongy bone from the hip, with the degeneration of both the structure and density.

Smart restraint system (automotive)

Over the last decade, the quest to improve vehicle safety has intensified dramatically and is now used as a sales feature. Nevertheless, in the design of restraints, commercial pressures focus on meeting legal requirements for vehicle approval, even though such requirements use dummies that do not represent the range of car occupant shapes, sizes, and driving positions. A person with poorer skeletal characteristics may not be able to withstand the current fixed levels of restraint forces without sustaining injuries. Conversely, a person with better skeletal characteristics may be capable of withstanding greater levels of restraint forces.

As the criteria for assessing vehicle crashworthiness have changed from vehicle deformations and decelerations to include occupant-related parameters (such as body accelerations, forces, and deflections), recognition of the implications of human anatomical diversity has been slow. This is illustrated by the fact that while there are child and adult anthropometric devices (dummies) available for use in vehicle testing, for vehicle certification the test requirements are biased toward a 50th percentile adult male driver

representation. Consequently, it is easy to perceive that the safety systems in motor vehicles have been developed, tested, and approved for optimum use by a narrow band of the population whose physical characteristics are not representative of all people.

As more sensors are installed in motor vehicles, the ability to determine information about the driver, such as an indication of his or her mass, the position of the seat, and the position of the driver on the

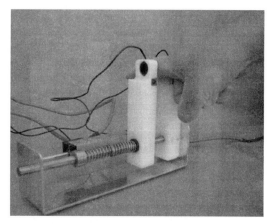

Fig. 2. Prototype BOSCOS ultrasound device.

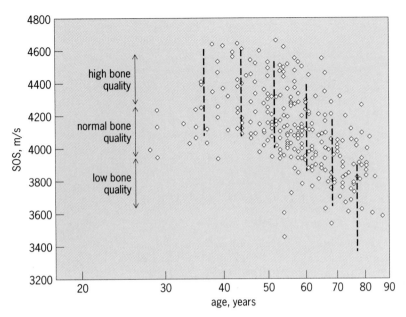

Fig. 3. Speed of sound (SOS) measurement values versus age for 295 volunteers.

seat, is much greater. However, even those parameters that can now be quantified give only limited information that can be used to extend the narrow optimum occupant protection band to a greater proportion of drivers. To extend this band, more data are needed about the individual occupants if they are to be better protected. The type of information that is needed concerns the physical injury tolerance limits of each individual so that restraint systems can be "tuned" by onboard processing to deliver the optimum protection for a specific crash/impact event. This means that maximum levels of protection can be delivered for each vehicle occupant, improving not merely survival but the likelihood of minimal injury.

Skeletal properties. Existing biomechanical data relating to human bone shows that with old age there are statistically significant reductions in load-carrying capability when compared with youth. H. Yamada showed that bones were able to resist only 78% of the mechanical forces applied to them by the age of 70–79, in comparison to their peak at 20–29. Gender was also shown to be important, with men generally

(but not exclusively) having greater load-carrying capability than women.

This reduction in biomechanical competence is supported by data from cadaver crash tests, which show that increasing age leads to greater probability of injury in the thorax and abdomen.

This reduction in mechanical properties is due to a multitude of factors, leading to a reduction of the overall density and structural competence, combined with changes in the biochemical makeup of the bone (**Fig. 1**).

One of the better parameters for assessing bone status is density. This parameter is used clinically for diagnosing low bone density and osteoporosis. A database of values from the population at large is needed with which to compare values from individuals to establish their relative bone status.

Bone-scanning equipment used in hospitals is large and heavy as well as impractical in a vehicle. A lightweight and practical alternative is necessary. In addition, since an in-vehicle system would be used by vehicle occupants on a daily basis, it is impractical to use equipment that exposes the user to ionizing radiation, as found in x-ray equipment. A suitable alternative is quantitative ultrasound. This is best known for its use in assessing fetal development during pregnancy, and according to popular belief, it is relatively risk-free.

Ease of use. To ensure that vehicle occupants use a scanning system, it must cause minimal inconvenience. For this reason, the scan needs to be done on a readily accessible bone site that is generally free from both clothing and jewelry. The finger's proximal phalanx bones are used in clinical tests as a means of assessing a patient's bone status and predicting fracture risk.

Prototype equipment. A prototype system designed to measure the proximal phalanx of the index finger has been developed (**Fig. 2**). The system, known as BOSCOS (bone scanning for occupant safety), works by positioning two ultrasound transducers on either side of the finger, with an ultrasound pulse transmitted through the finger between the transducers. The system takes a measurement of the separation of the transducers and the time taken for the ultrasound pulse to travel this distance. From this information, the speed of sound can be calculated. The speed of the ultrasound pulse is affected by the quality of the bone it passes through, with good-quality bone enabling the pulse to travel faster.

The system then compares the newly measured speed to a reference database, allowing for a quantitative evaluation of the subject's bone status in comparison to an expected normal. When the result indicated that the subject's measured bone status is below normal, the subject is deemed to have low bone quality and is therefore at higher risk of sustaining a fracture.

Demonstration of capability. An example of a reference database that could be used for comparing individual readings is shown in **Fig. 3**, where the speed of sound measured for the finger bone is shown with respect to age. The range of values shows how a

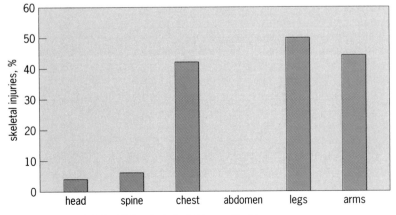

Fig. 4. Location of skeletal injury for belted drivers with airbag.

scanning system could be used to classify the population into at least three subgroups.

Application. Real-world in-depth crash injury data has been used to identify the types of crashes and the occupants who would benefit most from the BOSCOS system, as well as to determine the injury/cost reduction benefits of such a system. Real-world data were collected by the UK Co-Operative Crash Injury Study (CCIS), which samples accidents based on vehicle age and injury outcome. Accidents that occurred between 1995 and 2005 were examined.

The Abbreviated Injury Scale (AIS) is an anatomical scoring system where injuries are ranked on a scale of 1 to 6, with 1 being minor, 5 severe, and 6 a nonsurvivable injury. MAIS is the maximum AIS value for any body region. Around 57% of belted front-seat occupants who sustained an MAIS 2+ injury were involved in frontal impacts. Therefore, these impacts were seen as those where most occupants could benefit from an in-car scanning system. Application of the system to other impacts, particularly side impacts, would be a basis for future work.

Since the basis of a scanning system is to adapt the restraint system according to the skeletal strength of the occupant, it follows that skeletal injuries are those most likely to be reduced. Obviously, a reduction in skeletal injury resulting from "softer" restraints is also likely to be accompanied by a reduction in the occurrence of soft tissue injuries.

In the case of front impacts, skeletal injuries for drivers were concentrated in the chest, legs, and arm regions of the body (**Fig. 4**). The body regions of concern in this context are the chest and extremities. Since injuries to the chest are likely to pose a higher threat to life than those to the extremities, AIS 2+ chest injuries provided the focus for initial developments.

The relevant cases in the database indicated that 71% of all serious (AIS 2+) chest injuries were fractures to the ribs or sternum. In crashes where the crash severity was known, as determined by an equivalent test speed (ETS) calculation, 75% of injuries occurred at speeds lower than 56 km/h (35 mi/h), the current basis for European legislative testing. Since 96% of these cases below 56 km/h sustained little or no dashboard intrusion, less than 4 cm (1.6 in.), it is clear that there is the potential for an adaptive restraint system to provide significant benefit to chest injury risk.

Given that human bone strength decreases with age, it is expected that the benefits of a scanning system will be of greater magnitude to the weaker and elderly. With the aging populations in many countries, the societal benefit as a whole will increase as greater numbers of older drivers and passengers become exposed to the increased risk of injury attributable to a decrease in bone strength. The ability of a scanning system to measure bone strength means that sufferers of conditions such as osteoporosis will be detected and the restraint system will be tailored to them as much as is practicable.

Outlook. With an in-car scanning system that is able to assess an occupant's skeletal condition, it is feasible that more optimally adjusted restraint-system settings could be achieved and tailored for individual biomechanical limits. Significant development work is needed to bring such a system to fruition, but the potential benefits are significant, and the work will be one more step in the move toward smart restraint systems.

For background information *see* AGING; AUTOMOBILE; BIOMECHANICS; BONE; MEDICAL IMAGING; MICROSENSOR; OSTEOPOROSIS; SKELETAL SYSTEM; SKELETAL SYSTEM DISORDERS; ULTRASONICS in the McGraw-Hill Encyclopedia of Science & Technology.   Roger Hardy

Bibliography. A. Ekman et al., Dual x-ray absorptiometry of hip, heel ultrasound, and densitometry of fingers can discriminate male patients with hip fractures from control subjects, *J. Clin. Densitometry*, 5(1):79–85, 2002; J. Lenard, R. J. Frampton, and P. Thomas, The influence of European airbags on crash injury outcomes, *Proc. 16th ESV Conf.*, pp. 972–982, Windsor, 1998; G. M. Mackay et al., *The Methodology of In-depth Studies of Car Crashes in Britain*, SAE Tech. Pap. 850556, Society of Automotive Engineers, 1985; R. Mele et al., Three-year longitudinal study with quantitative ultrasound at the hand phalanx in a female population, *Osteoporosis Int.*, 7:550–557, 1997; G. Schmidt et al., in S. H. Backaitis (ed.), *Biomechanics of Impact Injury and Injury Tolerances of the Thorax-Shoulder Complex*, vol. PT-45, pp. 371–399, Society of Automotive Engineers, 1994; H. Yamada, in F. G. Evan (ed.), *Strength of Biological Materials*, pp. 255–271, Williams and Wilkins, 1970.

Smart skin

Sensitive skin is defined by one of its pioneers, Vladimir Lumelsky, as a large-area, flexible array of sensors with data processing capabilities. The microsensors are integrated to allow sensing several stimuli, mimicking a real skin, including but not limited to temperature, touch, and flow. Other requirements are flexibility and ability to stretch and wrinkle without any degradation in its sensing performance. Additional capabilities can be incorporated, such as biochemical sensing and decision making, toward the realization of smart skin. With the incorporation of organic and silicon thin-film transistors fabricated on flexible plastic substrates, a complete smart skin is achieved. The main difference between sensitive skin and smart skin is the additional capabilities of the latter to communicate the sensor output to a different site and to receive instructions from a central location, with perhaps local decision-making and control ability.

Applications and impact. The future for microsensors on flexible substrates looks very promising in both the civil and military fields. The fusion of integrated electronics and micro-electro-mechanical systems (MEMS) has achieved the goal of systems-on-chip, in which a single package contains sensors, actuators, and associated electronic circuitry on one chip. Electronics and systems on flexible substrates are gaining popularity and importance. Flexible substrates allow additional capabilities, thus enabling foldable electronics, as well as a complete foldable system-on-chip made of sensors, electronics, and actuators. This has the potential to lead to systems in fabrics and conformal systems for nonplanar and nonrigid surfaces with applications in the fields of defense, medicine, and industrial monitoring and testing.

The following represent some of the applications for sensitive skin.

Wearable sensors. There can be sensitive gloves for space and industrial applications where the temperature, touch, and flow are remotely sensed and translated inside the glove for the hands to feel. This would enable the person wearing the gloves to feel objects and grip them without bodily harm.

Wearable physiological monitoring systems. These could be integrated into a wristwatch or wrist strap, or woven into clothing for diabetics, soldiers, astronauts, or workers in harsh environments.

Minimally invasive surgery. Surgical procedures are performed on organs using very small incisions and surgical instruments. If these surgical instruments are equipped with sensitive skin, the surgeon can gain the ability to "feel" the organs, similar to the case of conventional surgeries.

Entertainment. There can be an entrainment suit for full immersion into virtual reality games.

Artificial skin. Currently, robotic instrumentation has to function in a structured environment, where everything is limited, to prevent damage to the robot or the environment by the robot. This is necessitated by the lack of distributed and complete sensing capability of robots. Another option is to have the robot semiattended, which defeats the purpose of having a robotic helper. If a flexible sensing skin is developed, robotic instrumentation can be made to function autonomously on the factory floor or home, or can be used for reconnaissance in warfare, security, and defense applications.

Sensitive prosthetic devices. These could be devices such as an artificial leg with a sensitive skin. The sense of temperature, pressure, and flow can be felt by another part of the body, or perhaps someday the output signals can be transmitted to the brain for direct sensing. This would provide a sense of feel to prosthetic devices that would help avoid damage and injury.

Fingerprint sensors for biometrics. These would be used to measure the characteristic of a finger or hand, provided by the unique thermal signature of the blood vessels and grooves.

Conformal sensor arrays. These sensor arrays can be attached to nonplanar surfaces, such as aircraft wings, to provide distributed multisensory data acquisition as a step toward a smart wing for micro- and macroaircraft.

Thermal sensors on flexible substrates. A microbolometer is a thermal sensor whose resistance changes with temperature such that monitoring of this resistance can be used to detect temperature or infrared radiation. Work on microbolometers on rigid substrates has been published and is now evolving into development on flexible substrates. The first devices on flexible substrate were made on prefabricated sheets of polyimide. This was problematic in that the polyimide substrate could not be kept planar during the fabrication process. The next generation of microbolometers was made by using a silicon wafer as the carrier for prefabricated polyimide sheets. This produces a flatter substrate, but bubble formation was experienced during processing, which introduced sporadic nonlinearities. The latest microbolometers on flexible substrates have incorporated spin-cast polyimide films on the silicon-wafer carriers. The liquid polyimide can be coated on silicon wafers using a conventional spin-casting procedure and cured to yield a very smooth working surface of desired thickness. The cured polyimide is thermally stable in the region of operation, is resistant to all the process chemicals, and has no adhesion problems for layers deposited on it. Moreover, it is fully compatible with the post complementary metal oxide semiconductor (Post-CMOS) process used in common integrated circuit (IC) fabrication. The polyimide has a glass transition temperature (T_g) of about 400°C (750°F), which allows a wider operational temperature range.

Arrays of thermal sensors on polyimides have been successfully fabricated and tested (**Figs. 1** and **2**). Addition of complementary sensory functions, such as tactile and flow, are in the works.

Packaging. Although thermal sensors on flexible substrates have successfully been developed and reported in literature, a stumbling block in realizing truly flexible sensors is the packaging. Like most MEMS sensors, micromachined microbolometers work best in a vacuum. Previously, vacuum operation meant rigid packaging—a limitation that defeats the purpose of flexible substrates. Device-level vacuum packaging has been proposed and has been investigated for making fully flexible systems on a chip. The primary idea involves building an optically transparent microcavity around an infrared sensing microbolometer and sealing it in a vacuum.

For flexible electronics, the package must be flexible and able to withstand handling, while

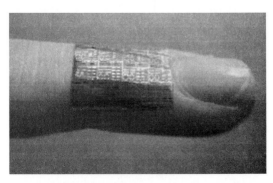

Fig. 1. Array of thermal sensors on a flexible polyimide substrate applied to a finger.

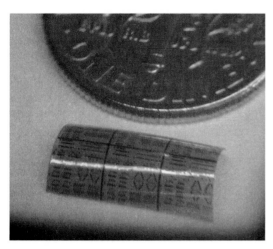

Fig. 2. Rolled-up smart skin, shown with a dime.

maintaining its integrity. For flexible electronics, it is beneficial for the package to be deposited during device fabrication. Rigid-substrate packaging may also see an advantage to depositing the package during device fabrication, especially in the case of vacuum packaging because of its high cost.

Currently, the packaging cost is about 50–90% of the total cost of a finished MEMS product. About half of the MEMS market is sensors. One of the major problems of sensor packaging is the need for the interaction of the sensor with the physical and chemical environment. This requires that the sensor either be immersed in the environment (as in the case of biochemical sensors) or be exposed through an encapsulant by means of a noninvasive medium such as an appropriate lens for radiation sensors. The minimum requirements that need to be met by sensor packaging are (1) to expose the sensor to the ambient environment "transparently," (2) to protect the sensitive electronics from the environment, and (3) to provide hermetic or vacuum sealing as needed. Recent defense and medical applications of sensors have added several other requirements

to sensor packaging, most of which are not met by conventional packaging techniques. The new generation of sensor systems should allow unobtrusive detection, requiring the package to be compact and lightweight. In addition, physical flexibility and compliance of the sensors and sensor packages holding these sensors would enable adhesion of the detectors on curved or flexible surfaces of common objects, such as furniture, cars, clothing, and building structures, that might have curved surfaces. The packaging should not cause any deterioration of the sensing ability. Moreover, low cost is essential for mass production, and disposable sensor systems are desired, if possible.

A novel design employing device-level vacuum encapsulation has been successfully developed and implemented for flexible infrared (IR) microbolometers (**Fig. 3**). The process used for device fabrication was fully CMOS-compatible. The packaging technique developed is scalable from a single device to an array of devices. Device-level packaged sensors with a vacuum cavity were fabricated in a process using two layers of sacrificial polyimide. As a preliminary attempt, silicon nitride (Si_3N_4) was chosen as the encapsulation material due its good structural and inert chemical properties and compatibility with the microbolometer fabrication process. However, the transmission properties of Si_3N_4 in the IR region are not optimum for its use as a window material. Therefore, the material above the detector was changed to aluminum oxide (Al_2O_3), which is transparent to IR radiation. The simplicity and design of the process could be used with many other materials for integrating different types of sensors on flexible substrates for smart skin applications.

For background information *see* BOLOMETER; ELECTRONIC PACKAGING; INTEGRATED CIRCUITS; MICRO-ELECTRO-MECHANICAL SYSTEMS (MEMS); MICROSENSOR in the McGraw-Hill Encyclopedia of Science & Technology.   Zeynep Celik-Butler; Donald P. Butler

Bibliography. S. A. Dayeh, D. P. Butler, and Z. Çelik-Butler, Micromachined infrared bolometers on flexible polyimide substrates, *Sensors and Actuators A*, 118:49–56, 2005; V.-J. Lumelsky, M. S. Shur, and S. Wagner, Sensitive skin, *IEEE Sensors J.*, 1:41, 2001; A. Mahmood, D. P. Butler, and Z. Çelik-Butler, Flexible microbolometers promise smart fabrics with imbedded sensors, *Laser Focus World*, pp. 99–103, April 2004; A. Yaradanakul, D. P. Butler, and Z. Çelik-Butler, Uncooled infrared microbolometers on a flexible substrate, *IEEE Trans. Electr. Dev.*, 49:930–933, 2002; A. Yildiz, Z. Çelik-Butler, and D. P. Butler, Microbolometers on a flexible substrate for infrared detection, *IEEE Sensors J.*, 4:112–117, 2004.

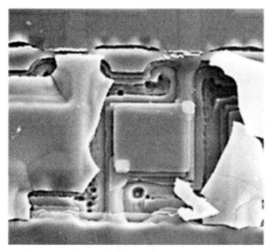

Fig. 3. Scanning electron micrograph of a self-packaged 60 × 60 μm^2 infrared detector. The encapsulation was intentionally broken with a probe tip.

Snail-eating caterpillar

The class Insecta (phylum Arthropoda) holds the largest share of the planet's biodiversity. Even conservative estimates suggest that approximately 80% of all species on Earth are insects. However, species diversity within the insects is not evenly distributed.

The success of the insects is largely due to their four most diverse orders: Coleoptera (beetles), Hymenoptera (wasps and bees), Diptera (flies), and Lepidoptera (butterflies and moths). The plethora of species of Coleoptera, Hymenoptera, and Diptera exhibit a wide range of life history strategies with plenty of predatory and plant feeding species making up their diversity. The Lepidoptera, however, are a stark contrast; they owe their diversity almost exclusively to plant-feeding species. Only 0.13% of all butterfly and moth caterpillars kill live prey. While it is remarkable that the Lepidoptera have achieved their tremendous diversity almost solely through herbivory, this imbalance begs the question: why are there so few predatory caterpillars? This question has yet to be answered, but every discovery of an additional species of extremely rare carnivorous caterpillar gives a little more insight into the workings and constraints of evolution.

Hawaii as a laboratory for evolution. The Hawaiian Islands are the most isolated landmass of their size on Earth. Perhaps it is this isolation that has sponsored evolutionary experiments which have not occurred on mainland areas. For example, Hawaii is home to a whole radiation of ambush predator *Eupethecia* moth caterpillars in the inchworm family Geometridae. The family occurs throughout the world—even the genus *Eupethecia* is widespread beyond Hawaii—but only in Hawaii have *Eupethecia* caterpillars evolved to eat other insects, which they surprise with a backward grab at astonishing speed.

Recently, our team discovered a group of unrelated predatory caterpillars lurking in Hawaiian rainforests, larvae that are perhaps even more unusual than the predatory *Eupethecia* inchworms. The small (8–10 mm or 0.3–0.4 in. in length at maturity) caterpillars of this new moth species belong to *Hyposmocoma*, a very large genus with over 350 identified species—all endemic to Hawaii. While being a predatory caterpillar automatically makes these animals an evolutionary anomaly, their choice of prey and method of capture have never been reported before in any lepidopteran.

Snail eating and web spinning. The caterpillars of *Hyposmocoma molluscivora* occur only in the wet forests on the east part of the island of Maui, on the slopes of the Haleakala Volcano. As is typical for *Hyposmocoma* (and many species in the family Cosmopterigidae, to which the genus belongs), *H. molluscivora* caterpillars spin a case of silk around their soft bodies, leaving only their head and legs protruding. The silk cases will usually have bits of lichen and dead leaf incorporated onto the surface, apparently to camouflage the larva while it hides in its case. Because the case covers most of the caterpillar's body, the larvae must drag themselves slowly along the leaves and stems of the trees in their wet forest habitat, rather like hermit crabs in speed and clumsiness. Fortunately, *H. molluscivora* is stalking native tree snails in the genera *Tornatellides* and *Auriculella*, and these snails are frequently found resting affixed to leaves. The caterpillars will not attempt to attack the snails while the snails are active. When

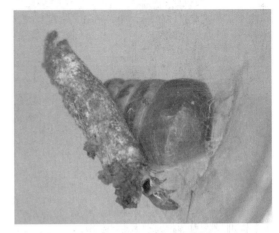

Fig. 1. *Hyposmocoma molluscivora* larva with prey.

the *H. molluscivora* caterpillar finds a resting snail, it very gently probes around the shell, possibly to confirm that the resident is still alive inside. Then the caterpillar gets to work frenetically spinning silk from glands just below its mouth, back and forth between the snail shell and the leaf, tying the snail to the leaf on which it is resting (**Fig. 1**). Gradually the caterpillar moves around toward the front of the shell and even climbs up onto it, spinning a net of silk that binds the snail. The snail may not even know the caterpillar has attacked until it is too late. We have seen snails try to move once the silk net has been spun, but they are unable to push their shells far up enough from the leaf to get their foot and eye stalks out; they give up and retreat into the false safety of their shells.

Once the caterpillar is satisfied that the snail is immobilized, it lumbers around to the front of the shell and orients its silk case so that it is facing the opening of the snail's shell. By pulling itself up into the snail's shell, the caterpillar prevents most parts of its body from being exposed, leaving only its protective silk case protruding from the snail's shell. With its silk case wedged into the front of the snail shell, the caterpillar stretches its elastic body out from its silk case and begins to feed on the snail (**Fig. 2**). As the snail retreats further into its shell, the caterpillar is able to safely pursue it, protected now by the snail's

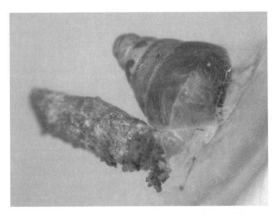

Fig. 2. *Hyposmocoma molluscivora* larva eating snail in shell.

softwood pallets, boxes, and crates exported from the United States, Canada, Japan, and China. There are fears that it, and other wood-borne pests, could spread around the world by means of international shipments and devastate forests.

Regulation on wood packaging for export. Unmanufactured wood articles exported to other countries pose a significant hazard of introducing plant pests, including pathogens, detrimental to agriculture and to natural, cultivated, and urban forest resources. The introduction of pests associated with solid wood packaging material (SWPM) is a worldwide problem. Because SWPM is very often reused, recycled, or re-manufactured, the true origin of any piece of SWPM is difficult to determine and thus its phytosanitary status cannot be ascertained. This often precludes national plant protection organizations from conducting useful specific risk analyses focused on the pests associated with SWPM of a particular type or place of origin and from imposing particular mitigation measures based on the results of such analysis.

For this reason, there is a need to develop globally accepted measures that may be applied to SWPM by all countries to practically eliminate the risk for most quarantine pests and significantly reduce the risk from other pests that may be associated with the SWPM. Phytosanitary standards are the province of the International Plant Protection Convention (IPPC). In March 2002, the Interim Commission on Phytosanitary Measures of the IPPC endorsed a standard prescribing uniform regulatory control of wood packaging materials moving in international commerce. This document, *ISPM No. 15, Guidelines for Regulating Wood Packaging Material in International Trade*, recognizes the inherent pest risks associated with the international movement of untreated wood packaging materials. Although this standard does not obligate countries to establish regulatory controls, the guideline represents a mechanism by which any country may establish regulatory controls in a manner that is internationally harmonized.

Regulated wood packaging material. In the IPPC Guidelines the term wood packaging material (WPM) is defined as "wood or wood products (excluding paper products) used in supporting, protecting or carrying a commodity (includes dunnage)." This definition is broader than the Animal and Plant Health Inspection Service (APHIS) term of solid wood packaging material. WPM includes manufactured wood such as plywood, veneer, and fiberboard, as well as loose wood materials such as shavings and excelsior (fine curled wood shavings used especially for packing fragile items). The IPPC Guidelines then distinguish between types of WPM that should be regulated because they present a risk (for example, raw wood pallets and dunnage) and types that should not be regulated because they present little risk (for example, manufactured wood and shavings).

Since it would be better to use a different term that applied only to the types of wooden materials used in packaging that should be regulated, APHIS added a definition of regulated wood packaging material. The new definition of regulated WPM closely resembles the current definition of SWPM: "Wood packaging materials other than manufactured wood materials, loose wood packaging materials, and wood pieces less than 6 mm (0.24 in.) thick in any dimension, that are used or that are for use with cargo to prevent damage, including, but not limited to, dunnage, crating, pallets, packaging blocks, drums, cases, and skids."

This definition of regulated WPM differs from the existing definition of SWPM in that it explicitly excludes manufactured wood materials, such as fiberboard, plywood, whisky and wine barrels, and veneer. The APHIS has never regulated such materials, but the definition of SWPM did not make that clear. The definition of regulated WPM also excludes pieces of wood that are less than 6 mm (0.24 in.) in any dimension. This exclusion will exempt from regulation many types of small boxes used to ship fruit or other articles.

Approved measures. Any treatment, process, or a combination of these that is significantly effective against most pests should be considered effective in mitigating pest risks associated with wood packaging material used in distribution. The choice of a measure for wood packaging material is based on consideration of the range of pests that may be affected, the efficacy of the measure, and the technical and/or commercial feasibility.

At present two measures have been approved: heat treatment and methyl bromide fumigation. Wood packaging material subjected to these approved measures should display the specified mark shown in the **illustration**.

Heat treatment (HT). Wood packaging material should be heated in accordance with a specific time-temperature schedule that achieves a minimum wood core temperature of 56°C (133°F) for a minimum of 30 minutes. Kiln-drying (KD), chemical pressure impregnation (CPI), or other treatments may be considered HT treatments to the extent that these meet the HT specifications. For example, CPI may meet the HT specifications through the use of steam, hot water, or dry heat. Heat treatment is indicated by the mark HT (replacing YY in the illustration).

Methyl bromide (MB) fumigation. The wood packaging material should be fumigated with methyl bromide. The treatment is indicated by the mark MB (replacing YY in the illustration). The minimum standard for methyl bromide fumigation treatment for wood packaging material is shown in the **table**.

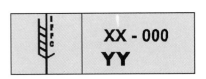

Marking for treatment of wood with approved measures.

Minimum standard for methyl bromide fumigation					
Temperature, °C and °F	Initial dose, g/m³ and lb/1000 ft³	Minimum required concentration, g/m³ and lb/1000 ft³			
		After 0.5 h	After 2 h	After 4 h	After 16 h
21/70 or above	48/3.0	36/2.25	24/1.50	17/1.06	14/0.875
16/61 or above	56/3.5	42/2.63	28/1.75	20/1.25	17/1.060
11/52 or above	64/4.0	48/3.00	32/2.00	22/1.38	19/1.190

The minimum temperature should not be less than 10°C (50°F), and the minimum exposure time should be 16 h.

Most significant pests targeted by HT and MB. Members of the following pest groups associated with wood packaging materials are practically eliminated by heat treatment and methyl bromide fumigation in accordance with the specifications listed above

Pest group
Insects
Anobiidae
Bostrichidae
Buprestidae
Cerambycidae
Curculionidae
Isoptera
Lyctidae (with some exceptions for HT)
Oedemeridae
Scolytidae
Siricidae
Nematode
Bursaphelenchus
xylophilus

Environmentally preferable alternative measures. The environmentally preferable alternative would be to prohibit importation of wood packaging material, which would virtually eliminate all associated pest risks, as well as the need for quarantine treatments. However, at present it would be technically and economically infeasible for many exporters, especially in developing countries.

The environment can be harmed by using methyl bromide, in which case recovery of the ozone layer may be delayed, or by not using methyl bromide, in which case agriculture and forested ecosystems, among other aspects of environmental quality, could be devastated unless other equally or more effective alternatives were strictly enforced (that is, heat treatment or use of substitute packaging materials). A considerable amount of research and development on methyl bromide alternatives has been conducted within the U.S. Department of Agriculture (USDA) and continues today. Although fumigation with methyl bromide has been accepted, it is expected to be abandoned in favor of alternative treatments in all cases in the future. Other treatments are being considered, including chemical pressure impregnation and irradiation.

Alternative measures considered for approval. Treatments that are being considered and may be approved when appropriate data become available include but are not limited to the following.

Fumigation
Phosphine
Sulfuryl fluoride
Carbonyl sulfide
Chemical pressure impregnation
High-pressure/vacuum process
Double vacuum process
Hot and cold open-tank process
Sap displacement method
Irradiation
Gamma radiation
X-rays
Microwaves
Infrared rays
Electron beam
Controlled atmosphere

The exporter must understand the country of destination's import regulation on wood packaging material. Up-to-date information for American shippers can be found at http://www.aphis.usda.gov/ppq/wpm/. Inspectors should look for the accredited mark, and shipments found not to comply can be delayed, refused entry, destroyed, or treated there.

For background information *see* FOREST PEST CONTROL; INVASION ECOLOGY; WOOD ANATOMY; WOOD ENGINEERING DESIGN; WOOD PROCESSING; WOOD PRODUCTS; WOOD PROPERTIES in the McGraw-Hill Encyclopedia of Science & Technology. Jongkoo Han

Bibliography. Canadian Food Inspection Agency, D-98-08, *Entry Requirements for Wood Packaging Materials Produced in All Areas Other Than the Continental United States*, Ottawa, 2005; FAO, *ISPM No. 15, Guidelines for Regulating Wood Packaging Material in International Trade*, Secretariat of the International Plant Protection Convention, Rome, 2002; Federal Registar, vol. 69, no. 179, *Importation of Wood Packaging Material*, pp. 55719–55733, Sept. 16, 2004; J. F. Hanlon, R. J. Kelsey, and H. R. Forcinia, *Handbook of Packaging Engineering*, 3d ed., Technomic Publishing, Basel, 1985; D. Twede and S. Selke, *Cartons, Crates and Corrugated Board: Handbook of Paper and Wood Packaging Technology*, DEStech Publications, Lancaster, PA, 2005.

Sonic drilling

Sonic drilling research began in the late 1940s with the goals of speeding up oil well drilling operations by adding vibration to the rotary motion of the drill pipe, and enhancing well development or rejuvenation.

Shell Oil Company continued the research and developed a downhole device that used a series of eccentric rotating weights driven by drilling fluid to generate vibration. Also developed were a high-horsepower vibrator for pile driving and a smaller oscillator for driving small piles, installing casing under roadways, and drilling seismic shot holes.

The next milestone was a Russian effort which, in a 1957 meeting of the Moscow Drilling Institute, brought notable attention to sonic technology with reports of 3 to 20 times the penetration rates as compared to conventional methods.

Hawker Sidley, a British manufacturer with operations in Canada, successfully manufactured the first modern sonic drill, producing 10 drills and 15 sonic heads. Their efforts were terminated, as there was not clear-cut commercial demand for the sonic drills.

In 1985, Northstar Drilling found that the capabilities of the sonic technology were a positive match for ground-water and soil cleanups. Using one of the Canadian-manufactured sonic drills, Northstar used sonic drilling technology for the first time in the United States at the Becker Country landfill in Minnesota. This was the beginning of sonic drilling used for environmental purposes.

Principle. Sonic drilling uses a high-frequency low-amplitude mechanical vibration, which is generated by two eccentric counterrotating weights in the sonic drill head. This intense vibration is cast down through the drill string until the drill string actually goes into resonance, similar to a tuning fork. The vibration is isolated from the drill rig itself by an air spring. Once the drill string is in resonance, it minimizes friction on the drill string and then places energy at the face of the drill bit. The boring is advanced via fracturing, shearing, and displacement of the formation. Rotation and down pressure are added to complement the vibration in the drilling process, as is necessary (**Fig. 1**)

Process. Sonic drilling uses a dual-tube system, consisting of an inner-string drill rod and sample barrel attached and a larger-diameter outer casing. The inner string of tooling is normally first advanced into the formation. Once the inner string is drilled down, the outer casing is drilled over the top to the inner string, and the inner string with the sample is extracted from the borehole, with the outer casing remaining in place to ensure that the integrity of the borehole is maintained so no formation caving occurs (**Fig. 2**). In this method, the borehole is advanced by repeating this sequence.

Benefits. There are a number of benefits in using sonic drilling technology.

Speed. Sonic drilling is twice as fast as conventional drilling, when sampling. This differential multiple, in speed of penetration and sampling rates, is even greater in troublesome formations such as heavy tills, hard layers, running sands, karst, and heaving conditions. In these instances, sonic is commonly four to five times faster than hollow stem augers.

Sample quality. Sonic drilling provides an overburden sample that is very high quality and clearly displays the lithology of the formation. It is also a complete sample, without the gaps that accompany conventional split-spoon and Shelby-tube sampling. The sonic sample is recognized as satisfying geotechnical measures, such as visual classification (ASTM D 2488), engineering classification (ASTM D 2487), moisture content (ASTM D 2216), mechanical analyses (ASTM D 422), and sieve analysis and Atterberg limits (ASTM D 4318). However, the sonic sample is considered a slightly disturbed sample and is not

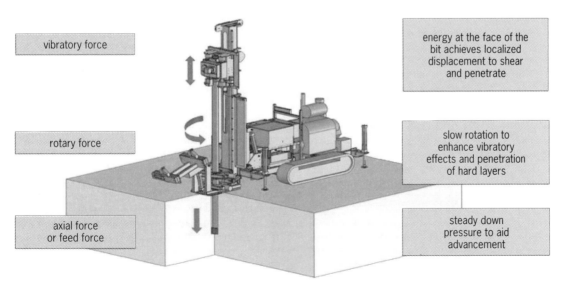

vibratory force

rotary force

axial force
or feed force

energy at the face of the bit achieves localized displacement to shear and penetrate

slow rotation to enhance vibratory effects and penetration of hard layers

steady down pressure to aid advancement

Fig. 1. Sonic drilling principle and operation.

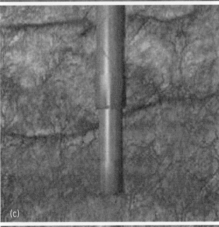

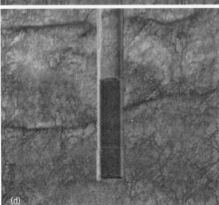

Fig. 2. Process of sonic drilling. (*a*) Sonic corebarrel advancing through overburden. (*b*) Sonic drilling through varied formations. (*c*) Telescoping tooling isolates contaminated aquifers and cases off a troublesome section of the borehole. (*d*) Retrieval of the sonically drilled sample.

adequate for direct shear, compression, or triaxial tests.

Minimal investigative derived waste (IDW). Compared to hollow-stem auger drilling, sonic drilling reduces IDW by 70–80%. To give an example of the significance of this reduction, if a hollow-stem auger drill were used to drill twenty 2-in.(5-cm) monitor wells to a depth of 50 ft (15 m), it would generate one 55-gallon drum of cuttings for every 15 ft (4.6 m) of hole drilled. Assuming a disposal cost of $500 per drum, using sonic drilling technology to drill twenty wells involved with "Class A" hazardous waste would reduce the cost by approximately $25,000.

Drills without fluids or with limited fluids. Sonic drilling has the advantage of being able to drill most formations without using fluids or using very limited fluids just to advance the casing. This is significant because for environmental applications there is less risk of a bias in sampling soils or dilution in sampling fluids with sonic technology.

Penetrating ability. Sonic drills can penetrate all overburden formations and drill into most bedrock, as well as clays, silts, sands, cobbles, hard layers, and bedrock float. This makes it a very versatile method of drilling, which is not subject to refusal and the resulting redrills common to hollow-stem augers. It is ideal for landfills because it can drill through wood, plastic, rubber, brick, cement, and other materials.

Depth capacity. Hollow-stem auger drills can be limited in depth capability to 100 ft (30 m) or less in difficult formations, whereas sonic drilling has the capability to drill to several hundred feet in depth. This is extremely meaningful when drilling for chlorinated solvents (high-density fluids), which can be deep in the water table. To delineate the extent of the contamination in soils or water, the ability to access deep levels with sampling tools is necessary. Sonic drilling is fully capable of providing soil and water samples from these depths. Sonic drilling depth capacity is also an advantage for remedial projects where a direct push drill cannot reach targeted areas.

Ability to isolate sections of the borehole. The dual-tube system used in sonic drilling prevents cross contamination and formation mixing. When drilling an exploratory boring through multiple aquifers, the outer casing can be vibrated down past an aquifer, while a bentonite grout mixture is pumped under pressure into the annular space on the outside of the casing. This process completely seals off the aquifers in the upper borehole, allowing for uncontaminated soil and/or water sampling at greater depths. The ability to telescope tooling downhole allows the boring to advance efficiently. Since troublesome sections of the boring can be cased off, this significantly increases production in tough formations.

Reduced well development time. Using sonic drilling technology, well development time is significantly less, especially in clay-laden formations where conventional drilling techniques tend to smear the formation or clog it with bentonite. The clean shearing action of sonic drilling minimizes smearing and reduces well development by as much as 50%.

Miniaturization of sonic technology. Sonic drilling initially used large drill rigs mounted on 45-ft-long (14-m) trucks. Over the past 5 years, sonic technology has been designed to be much more compact. Sonic drills can now be mounted on all-terrain vehicle platforms (14×7 ft or 2×4 m) and are capable of reaching difficult sites. Limited-access sonic drills are also available for use in tight quarters such as inside buildings. Other drilling platforms, such as barges, are compatible with sonic drills.

Case histories. The following are applications of sonic drilling.

Charnock MtBE-Chevron-Bechtel alliance. This is a classic example of sonic drilling used for a ground-water contamination investigation and cleanup. Leaking gasoline storage tanks and pipelines were determined to have contaminated the Charnock well fields near Los Angeles with the fuel additive methyl tertiary butyl ether (MtBE). This affected approximately 200,000 consumers and promoted great concern, since nearly all of California's drinking water comes from either ground water or surface water. Sonic drilling technology was used at this project because of its ability to recover a relatively undisturbed, continuous sample of overburden, which was key to understanding the intricate geology. This understanding was the basis of controlling and remediating the contamination. The ability to seal off upper aquifers and continue boring into deeper aquifers without cross contamination made sonic drilling an effective tool in delineating the extent of the contamination.

Crews advanced 10-, 9-, and 8-in. (25-, 23-, and 20-cm) sonic casing telescopically to seal off the individual aquifers. Because of these unique abilities to sample delicate geology and isolate contaminated areas, sonic drilling technology soon became the preferred method of the California Regional Water Board and the Environmental Protection Agency (EPA) for this project. Sonic drilling also generated minimal IDW, which substantially reduced the disposal costs of the drilling waste.

Pennsylvania Department of Transportation. A major bridge was failing because unacceptable settlement had occurred on an overpass replacement bridge near Allentown. The overpass was situated on geology containing karstic solution features. Micropiles were installed to stabilize this situation prior to construction; however, bridge settlement and sinkholes were causing major concern. Sonic drilling was used to parallel the existing micropiles and assist in defining the karstic rock profile. Sonic investigative boring was completed to over 525 ft (160 m) in depth. The boring and sampling were then used to evaluate remedial options. Sonic drilling was able to provide an accurate profile of the rock and voids defining the lithology below the bridge. Sonic drilling used casings to seal off the unstable and highflow karstic voids, soils, and rock seams. These sonic borings are naturally very straight and thus were ideal for installing instrumentation such as inclinometer pipe which measures any movement in the formation. When in proximity to a vulnerable structure, such as a bridge, sonic eliminates the use of air and water circulation, which is an advantage as it eliminates the possibility of erosion and any further damage to the bridge foundation.

Outlook. Sonic drilling has substantially grown in the environmental, construction, and mineral exploration markets over the past 20 years and presently is a fast-growing technology, with the number of sonic drills in operation increasing by tenfold in this period. The growth of sonic technology is based on its tangible advantages of being able to drill and sample all overburden formations faster, providing a better sample with less IDW. It is likely that its popularity within the environmental engineering community will continue to develop as enhancements and innovations are added. The environmental and civil engineer are both likely to use sonic drilling with increased frequency as the technology and benefits are better understood.

For background information *see* BENTONITE; CIVIL ENGINEERING; DRILLING, GEOTECHNICAL; ENGINEERING GEOLOGY; ENVIRONMENTAL ENGINEERING; KARST TOPOGRAPHY; ROCK MECHANICS; SOIL MECHANICS in the McGraw-Hill Encyclopedia of Science & Technology.   George Burnhart

Bibliography. J. P. Davis, Sonic drilling offers quality control and nondestructive advantages to geotechnical and construction drilling on sensitive infrastructure sites, *31st Annual Conference on Deep Foundations*, Washington, DC, 2006; T. Oothoudt, Tailing & mine waste sonic drilling applications, *Proceedings of the 6th International Conference on Tailings and Mine Waste '99, Fort Collins, CO*, 1999.

Space flight

Space flight in 2005 made a number of significant strides in both human and robotic activities, moving ahead in its increasingly dominant theme: progress toward human exploration ventures outside the Earth's boundaries. But based on the number of launches to orbit and the number of launched satellite payloads, the utilization of space in 2005, which had reached its lowest level in 2004 since 1961, remained on that level without showing signs of reversing this trend. For the second consecutive year, the number of space launches attempted worldwide totaled 55 (including three failed). However, in terms of commercial satellite sales, with 19 geostationary-orbit commercial communications satellites (comsats) ordered worldwide, 2005 brought a significant improvement over the 12 satellites ordered in 2004.

As the United States space budget managed to stay at a relatively stable level, a milestone was the return of the space shuttle to flight. But Russian launch services continued to dominate human flights. Commercial flights increased above the previous years level, and international space activities extended their trends of reduced public spending and modest launch services. A total of 52 successful launches worldwide (2004: 53; 2003: 60; 2002: 61; 2001: 57) carried 72 payloads (2004: 73). The three failed launches (up from two in 2004) were all Russian: a *Molniya-M* (Soyuz derivative), a *Volna* submarine-launched modified ballistic missile, and a *Rokot*.

In 2005, the National Aeronautics and Space Administration (NASA) completed a successful year of milestones and discoveries as the agency began to implement the Vision for Space Exploration, mandated by President Bush on January 14, 2004, America's long-term plan for returning astronauts to the Moon to prepare for voyages to Mars and other destinations in the solar system. The year 2005 included returning the space shuttle to flight, the announcement of plans for America's next-generation spacecraft, and numerous scientific milestones. The first shuttle mission to the *International Space Station* (*ISS*) since the Columbia loss in 2003 included breathtaking maneuvers, spacewalks, and tests of new procedures and safety equipment. The flight was successful, but engineers remained concerned about the shedding of some insulating foam material off the external tank and called for more work before the next shuttle mission, which slipped to mid-2006. Launched in January 2005, the *Deep Impact* spacecraft traveled approximately 429 million kilometers (268 million miles) to meet and collide with comet Tempel 1, while on Mars the twin rovers *Spirit* and *Opportunity* completed a full Martian year of exploration and discovery, which included evidence that water once flowed across the Martian surface and may still be there today. A new Mars mission, the *Mars Reconnaissance Orbiter* (*MRO*), was launched to rendezvous with the Red Planet on March 10, 2006; the *Cassini/Huygens* spacecraft made history at Saturn, landing the *Huygens* probe successfully on the moon Titan and touring the other Saturn moons for breathtaking photography and measurements; and the *Hubble Space Telescope* continued exploration and discovery.

The commercial space market leveled out in 2005 and reversed itself after its 2004 decline, begun in 2003 (after a surprising recovery in 2002 from the dramatic slump of the previous years). Of the 52 successful launches worldwide, about 23 (44%) were commercial launches (carrying 36 commercial payloads), compared to 19 (36%) in 2004 (2003: 20 [33%]). In the civil science satellite area, worldwide launches totaled six, down three from the preceding year.

Russia's space program, despite chronic shortage of state funding, showed continued dependable participation in the build-up of the *ISS*. This partnership had become particularly important after the shuttle stand-down caused by the loss of *Columbia* on February 1, 2003. Europe's space activities in 2005 rose over the previous year's total of three missions, with five flights of the Ariane 5 heavy-lift launch vehicle, which brought the number of successes of this vehicle to 25.

The People's Republic of China in 2005 successfully launched its second crewed spaceflight, bringing the total of crewed flights in this year to four, carrying 15 humans (2004: 6). This brought the total number of people launched into space since 1958 (counting repeaters) to 989, including 100 women, or 443 individuals (38 women), in a total of 249 missions. Some significant space events in 2005 are listed

---

TABLE 1. Some significant space events in 2005

Designation	Date	Country	Event
Deep Impact	January 12	United States	Launch on a Delta 2 to Comet Tempel 1 where the probe's Impactor part slammed into the comet on July 4 for analysis, observed by the Flyby part and Earth telescopes
Progress M-52/17P	February 28	Russia	Crewless logistics cargo/resupply mission to the *International Space Station* (*ISS*), on a Soyuz-U rocket
Soyuz TMA-6/ISS-10S	April 14	Russia	Launch of the fifth *ISS* crew rotation flight on a Soyuz-FG, bringing the Expedition 11 crew of Sergei Krikalev and John Phillips, plus 8-day visitor Roberto Vittori from ESA
DART	April 15	United States	Airborne launch of NASA spacecraft "Demonstration of Autonomous Rendezvous Technology" on a Pegasus XL, but mission failed in space and was not completed
Progress M-53/18P	June 16	Russia	Crewless logistics cargo/resupply mission to the *ISS*, on a Soyuz-U rocket
Suzaku (Astro-E2)	July 10	Japan	Launch of the "Red Bird of the South" x-ray astronomy satellite on an M-V rocket from JAXA's Uchinoura Space Center (Japan's fifth x-ray satellite)
STS-114 (Discovery)	July 26	United States	The much awaited shuttle return-to-flight mission to *ISS* after loss of *Columbia* in 2003, with successful tests of new safety equipment during a 13d 21h mission with three EVAs
FSW 21	August 2	P.R. of China	Successful launch of the recoverable *Fanhui Shi Weixing* (*FSW*) 21 imaging satellite on a Long March 2C (CZ-2C) rocket
MRO	August 12	United States	NASA launch of the *Mars Reconnaissance Orbiter* to the Red Planet on an Atlas 5 rocket, to arrive in Mars orbit in March 2006, joining five other active Mars spacecraft
FSW 22	August 29	P.R. of China	Successful launch of the recoverable *Fanhui Shi Weixing* (*FSW*) 22 imaging satellite on a Long March 2C (CZ-2C) rocket
Progress M-54/19P	September 8	Russia	Crewless logistics cargo/resupply mission to the *ISS*, on a Soyuz-U rocket
Soyuz TMA-7/ISS-11S	September 30	Russia	Launch of the sixth *ISS* crew rotation flight on a Soyuz-FG, bringing the Expedition 12 crew of William McArthur and Valery Tokarev, plus 8-day space tourist Gregory Olsen (U.S.)
Shenzhou 6	October 12	P.R. of China	Second flight of a crewed spacecraft on a CZ-2F, this time carrying two taikonauts, Fei Junlong and Nie Haisheng, safely to orbit and return on a flight lasting 115h 32m
Venus Express	November 9	Europe/Russia	Soyuz/Fregat launch at Baikonur of the first European/ESA exploration probe to orbit the planet Venus, to arrive on April 11, 2006, for observation of the planet from orbit
Progress M-55/20P	December 21	Russia	Crewless logistics cargo/resupply mission to the *ISS*, on a Soyuz-U rocket
GIOVE-A	December 28	Europe/Russia	Soyuz-FG launch at Baikonur of first European-built test satellite for the European GNSS (Galileo Navigation Satellite System), to replace the U.S. Global Positioning System (GPS) for European users

TABLE 2. Successful launches in 2005 (Earth-orbit and beyond)	
Country	Number of launches (and attempts)
United States (NASA/DOD/commercial)	16 (16)
Russia	23 (26)
People's Republic of China	5 (5)
Europe (ESA/Arianespace)	5 (5)
Japan	2 (2)
India	1 (1)
TOTAL	52 (55)

in **Table 1**, and the launches and attempts are enumerated by country in **Table 2**.

International Space Station

During 2005, the *ISS* marked the fifth anniversary of continuous crewed operations (November), during which NASA and *ISS* partner scientists have gathered vital information on the station that will help with future long-duration missions, as the station has a unique microgravity environment that cannot be duplicated on Earth.

There was continuing debate about the provision of assured crew return capability after the Russian obligation to supply *Soyuz* lifeboats to the station expired in April 2006. Of much greater significance to the continuation of *ISS* assembly and operation was Russia's shouldering the burden of providing crew rotation and consumables resupply flights to the station, after the loss of *Columbia* in 2003 brought shuttle operations to a standstill that lasted until the return-to-flight mission of STS-114/*Discovery* in July 2005. The reduction in resupply missions to the station, which now could be supported by only Russian crewless automated Progress cargo ships, resulted in station crew size being reduced from three to a two-person "caretaker" crew per expedition (also known as increment), except for 10-day stays by visiting cosmonaut/researchers or commercial "tourists" arriving and departing on the third seat of *Soyuz* spacecraft.

During 2005, three two-person crews lived on the station, each one working with ground teams to do their part to keep the station safely operating while accumulating knowledge for future deep-space missions of the Vision for Space Exploration. Crews performed in-flight outfitting, preventive maintenance, replacements, and repairs on station equipment, and conducted on-going science and research activities. They also conducted four spacewalks to install external components such as an antenna, make electrical and data connections, remove and replace payloads and experiment containers, and launch a nanosatellite.

Expedition 11 was launched to the *ISS* in April with the two-man station crew of Russian Commander Sergei Krikalev and American Flight Engineer John Phillips plus Italian Visiting Crewmember 8 cosmonaut/researcher Roberto Vittori on a *Soyuz* TMA spacecraft. The two members of Expedition 10, Leroy Chiao and Salizhan Sharipov, plus Vittori, returned on the previous *Soyuz* that had served as a contingency crew return vehicle for almost the duration of its certified lifetime of 200 days. The replacement crew, Expedition 12, came 6 months later in a fresh *Soyuz* TMA, consisting of American Commander William McArthur and Russian Flight Engineer Valery Tokarev, to continue station operations into 2006.

Following the recommendations of an independent advisory panel of biological and physical research scientists called REMAP (Research Maximization and Prioritization), NASA in 2002 had established the formal position of a Science Officer for one crew member aboard the *ISS*, responsible for expanding scientific endeavors on the station. Expedition 10 Commander Chiao, Phillips (Expedition 11), and McArthur (Expedition 12) were the sixth, seventh, and eighth Science Officers.

By end-2005, 51 carriers had been launched to the *ISS*: 17 shuttles, 2 heavy Protons (FGB/*Zarya*, SM/*Zvezda*), and 32 *Soyuz* rockets (20 crewless Progress cargo ships, the DC-1 docking module, and 11 crewed *Soyuz* spaceships).

Progress M-52. Designated ISS-17P, the first of four crewless cargo ships to the *ISS* in 2005 lifted off on a Soyuz-U rocket at the Baikonur Cosmodrome in Kazakhstan on February 28. As all Progress transports, it carried about 2 tons of resupply for the station, including maneuver propellants, water, food, science payloads, equipment, and spares.

Soyuz TMA-6. *Soyuz TMA-6*, ISS Mission 10S (April 14–October 10), was launched in Kazakhstan, once again a flawless success of the Soyuz launcher. The fifth crew rotation flight by a Soyuz because of the shuttle standdown, it carried Expedition 11, the two-man station crew of Krikalev and Phillips, plus Vittori, who flew for the European Space Agency (ESA). *TMA-6* docked to the *ISS* on April 16, replacing the previous crew return vehicle, *Soyuz TMA-5/9S*, which when it undocked on April 24, 2005, had reached an in-space time of 192 days. In it, Expedition 10 crew members Chiao and Sharipov plus Vittori landed in Kazakhstan.

Progress M-53. ISS-18P was the next crewless cargo ship, launched in Baikonur on a Soyuz-U on June 16, and arriving at the station with fresh supplies on June 18.

Progress M-54. ISS-19P, the third automated logistics transport in 2005, lifted off on its Soyuz-U on September 8, docking at the *ISS* on September 10.

Soyuz TMA-7. *Soyuz TMA-7*, ISS mission 11S (September 30–April 8, 2006), was the sixth *ISS* crew rotation flight by a Soyuz. Its crew constituted Expedition 12, the sixth caretaker crew of McArthur and Tokarev, plus the third "space tourist," United States businessman Gregory Olsen (Visiting Crewmember 9), after another flawless countdown of the Soyuz-FG. On October 3, *TMA-7* docked smoothly to the *ISS*, achieving successful contact and capture. Hatch opening and crew transfer were nominal. Seven days later, the previous crew return vehicle, *Soyuz TMA-6/10S*, undocked from the FGB nadir port, where it had stayed for 177 days (179 days in space), and landed safely in Kazakhstan with Krikalev, Phillips, and Olsen.

Progress M-55. ISS-20P, the fourth automated logistics transport in 2005, lifted off on its Soyuz-U on December 21, docking at the *ISS* on December 23.

United States Space Activities

Launch activities in the United States in 2005 continued the decrease from the previous year. There were 16 NASA, Department of Defense (DOD), and commercial launches, out of 16 attempts (2004: 19 of 19 attempts, 2003: 26 of 27 [loss of *Columbia*]; 2002: 18 of 18). In February 2005, Sean O'Keefe stepped down as NASA Administrator. After his deputy, Fred Gregory, took over as acting administrator, President Bush in March selected Michael Griffin, the head of the space department at Johns Hopkins University's Applied Physics Laboratory, for the position of NASA Administrator.

Space shuttle. Because of the loss of Orbiter *Columbia* on the first (and only) shuttle mission in 2003, operations with the reusable shuttle vehicles of the U.S. Space Transportation System (STS) came to a halt for the remainder of the year and for 2004, as NASA and its contractors labored on intensive return-to-flight efforts, which led to resumption of flights in 2005: Over 2 years after the fatal last flight of *Columbia*, the space shuttle returned to the skies with the Summer liftoff of *Discovery* (STS-114). During the standdown and continuing after return-to-flight, resupply and crew rotation flights to the *ISS* were accomplished solely by Russian Soyuz and Progress vehicles.

Discovery, on its thirty-first flight, lifted off on July 26, 2005, on *ISS* Mission LF-1, carrying the crew of Commander Eileen M. Collins, Pilot James M. Kelly, and Mission Specialists Andrew S. W. Thomas, Charles J. Camarda, Wendy B. Lawrence, Stephen K. Robinson, and Japanese astronaut Soichi Noguchi, plus 13,483 kg (29,725 lb) of equipment and supplies. Docking at the station took place on July 28, flawlessly flown by Commander Collins, piloting the Orbiter from the aft flight deck control stick. Prior to final approach (**Fig. 1***a*), the *Discovery* performed a scheduled R-Bar Pitch Maneuver at approximately 180 m (600 ft) distance under the *ISS*, a 360° backflip to allow digital imagery of its thermal protection system from the *ISS* by Phillips and Krikalev. After the docking and the regular leak checks of the docking adapter (pressurized mating adapter), hatches were opened and station occupancy increased to nine persons. During the next 9 days, about 7000 kg (15,000 lb) of cargo was transferred from the shuttle's freight container, the Multipurpose Logistics Module (MPLM) *Raffaello*, to the *ISS* and approximately 3900 kg (8600 lb) from the station to the shuttle for return to Earth. Crew members also performed three successful spacewalks (Fig. 1*b*). STS-114 tested new safety equipment and repair procedures and included a first-of-its-kind spacewalking heat shield repair (removal of loose tile gap fillers). *Discov-*

(a)

(b)

(c)

Fig. 1. The space shuttle's return-to-flight mission. (*a*) Shuttle *Discovery*, photographed from the *International Space Station* (*ISS*) during rendezvous and docking operations on July 28, 2005. The Multipurpose Logistics Module (MPLM) *Raffaello* is visible in the shuttle's cargo bay. (*b*) Astonaut Stephen K. Robinson, anchored to foot restraint on the *ISS*'s Canadarm2, participates in the mission's third session of extravehicular activity (EVA) on August 3. (*c*) *ISS* photographed from *Discovery* following undocking of the two spacecraft on August 6. (*NASA*)

ery undocked on August 6 (Fig. 1*c*) and returned to Earth on August 9 with 11,395 kg (25,121 lb) of equipment in its cargo bay, touching down smoothly at Edwards Air Force Base after 219 orbits, having been redirected to California after three Kennedy Space Center wave-offs due to inclement weather conditions. Total duration of the 9.3-million-kilometer (5.8-million-mile) journey in space was 13d 21h 32m 22s. With more work required on redesign of the foam-based insulation on the space shuttle's external tank, new sensors for detailed damage inspection, and a boom to allow astronauts to inspect the vehicle externally during flight, the second return-to-flight test flight was delayed until well into 2006.

Advanced transportation systems activities. NASA announced plans for its next-generation spacecraft and launch system to support the long-range Vision for Space Exploration announced by President Bush in 2004, the Crew Launch Vehicle (CLV), which will be capable of delivering crew and supplies to the *ISS*, carrying four astronauts to the Moon, and supporting up to six crew members on future missions to Mars. The new Crew Exploration Vehicle (CEV) will be shaped like an Apollo capsule but will be significantly larger. There will also be a crewless series of heavy cargo lifters, which eventually may approach the lifting capability of the Saturn V of the 1960s. These systems are scheduled to take the place of the space shuttle by the end of 2010.

Space sciences and astronomy. In 2005, the United States launched two civil science spacecraft (two less than in the previous year): *Deep Impact* and *Mars Reconnaissance Orbiter*.

Deep Impact. The *Deep Impact* mission was launched by NASA on January 12 on a Delta 2 rocket, targeted for Comet Tempel 1, a jet-black, pickle-shaped, icy dirt ball traveling at 10.1 km/s (6.3 mi/s). The purpose of the comet probe was to show how the interior of a comet is different from its surface, by excavating as deep a hole as possible to get down to the primitive material of the comet's origin. The *Deep Impact* spacecraft was composed of two probes mated together—named Flyby and Impactor. Flyby was about the size of a small car and instru-mented to monitor the impact, with two cameras—a high-resolution camera that was tightly focused on the resulting crater and a medium-resolution one for taking wider views. Impactor was a 370-kg (820-lb) copper-fortified probe designed to produce maximum effect when it hit the comet. It also carried a medium-resolution camera that recorded the probe's final moments before it collided with the comet. Arrival at Comet Tempel 1 was on schedule, July 4. The comet slammed into Impactor, which had recorded images and gathered data while Flyby passed 500 km (310 mi) away, observing the impact, the ejected material (**Fig. 2**), and the structure and composition of the comet's interior. Most of the data were stored on Flyby and radioed back to Earth after the encounter. Early results confirmed long-standing knowledge that a major ingredient in comets is water ice. However, it had not been known whether the ice was contained mainly inside or if it could be found on the surface as well. Data from *Deep Impact* provided the first evidence that water ice can indeed exist on a comet's exterior: Tempel 1 has a surface area of roughly 110 km² (45 mi²), or 1.1×10^8 m² (1.2×10^9 ft²). The area taken up by the water ice, however, is only 2.8×10^4 m² (3×10^5 ft²). The rest of the comet surface is dust.

Gravity Probe-B. Gravity Probe-B (GP-B) is a NASA mission to test two predictions of Albert Einstein's theory of general relativity. After its launch on April 20, 2004, the spacecraft, orbiting 644 km (400 mi) above the Earth, used four ultraprecise gyroscopes to test Einstein's theory that space and time are distorted by the presence of massive objects. To accomplish this, the mission measured two factors: how

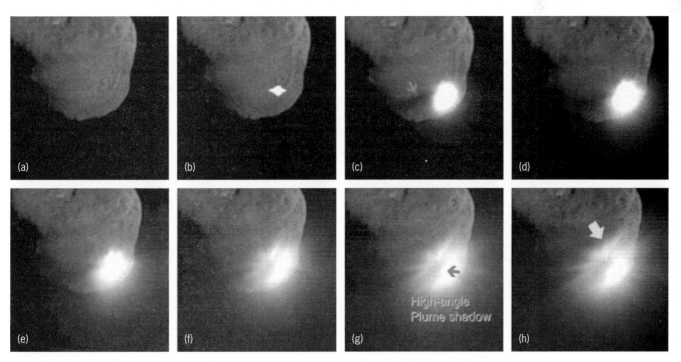

Fig. 2. Sequence of eight images, (*a*)–(*h*), depicting the development of the ejecta plume when *Deep Impact*'s Impactor colloided with Comet Tempel 1 on July 4, 2005. Images, taken by the high-resolution camera on the Flyby part of the spacecraft, were spaced 0.84 s apart. Arrows in *c* and *g* highlight shadows due to the opacity of the ejecta. Arrow in *h* indicates zone of avoidance in the up-range direction. (*From M. F. A'Hearn et al., Deep Impact: Excavating Comet Tempel 1, Science, 310:258–264, 2005, reprinted with permission from AAAS*)

space and time are warped by the presence of the Earth, and how the Earth's rotation drags space-time around with it. In 2005, after *Gravity Probe-B* had orbited the Earth for more than 17 months, scientists finished collecting data. Fifty weeks' worth of data has been downloaded from the spacecraft and relayed to computers in the Mission Operations Center at Stanford University in California. Data analysis and validation are expected to take approximately 1 year.

MESSENGER. NASA's *MESSENGER* (Mercury Surface, Space Environment, Geochemistry, and Ranging), launched on August 3, 2004, is scheduled to become the first spacecraft to orbit the planet Mercury, beginning in 2011. *MESSENGER* is only the second spacecraft sent to Mercury, after *Mariner 10* flew past it three times in 1974–1975 and gathered detailed data on less than half the surface. The spacecraft is on a 7.9-billion-kilometer (4.9-billion-mile) journey that includes 15 trips around the Sun. On August 2, 2005, *MESSENGER* returned to Earth for a gravity boost. Next, it is scheduled to fly past Venus twice, in October 2006 and June 2007, using the tug of Venus's gravity to resize and rotate its trajectory closer to Mercury's orbit. Three Mercury flybys, each followed about 2 months later by a course correction maneuver, will put the spacecraft in position to enter Mercury orbit in March 2011. During the flybys—set for January 2008, October 2008, and September 2009—*MESSENGER* will map nearly the entire planet in color, image most of the areas unseen by *Mariner 10*, and measure the composition of the surface, atmosphere, and magnetosphere. It will be the first new data from Mercury in more than 30 years—and invaluable for planning *MESSENGER's* year-long orbital mission.

Swift. NASA's *Swift* satellite, launched on November 20, 2004, was designed and built with international participation (England, Italy) to solve the 35-year-old mystery of the origin of gamma-ray bursts (GRBs). During 2005 *Swift* has already achieved every prelaunch predicted advance in GRB science. Researchers have discovered the farthest GRB ever seen, identified counterparts to short GRBs, discovered new GRBs at a rate of 100 per year, and explored a new time interval in GRB light curves (which revealed the unpredicted phenomena of GRB flares and rapid x-ray afterglow declines). The observatory and the instruments have continued to work well, and *Swift* has conducted about 20,000 successful slews to targets. Through coordination of observations from several ground-based telescopes, *Swift*, and other satellites, scientists solved the mystery of the origin of powerful split-second flashes of light called short GRBs. To track these mysterious bursts, *Swift* carries a suite of three main instruments: the Burst Alert Telescope (BAT), the X-Ray Telescope (XRT), and the UltraViolet/Optical Telescope (UVOT). The flashes are brighter than 10^9 Suns, yet last only a few milliseconds. They had been too fast for earlier instruments to catch. Scientists believe the bursts are related to the formation of black holes throughout the universe: the "birth cries" of black

holes. Updated orbital lifetime predictions for *Swift* indicate that it may remain in orbit up to 2022.

GALEX. *GALEX* (Galaxy Evolution Explorer), launched in 2003, is an orbiting space telescope for observing tens of millions of star-forming galaxies in ultraviolet light. *GALEX* is tailored to view hundreds of galaxies in each observation. Thus, it requires a large field of view rather than high resolution. During 2005, astronomers using *GALEX*'s sensitive ultraviolet detectors observed star formation in progress in a spiral galaxy nearly 7 million light-years away, a member of a group of galaxies known as the Sculptor Group. *GALEX* also surprised astronomers with an image of the galaxy NGC 4625, which was once thought to be rather plain but is actually endowed with a pronounced set of young spiral arms.

Spitzer Space Telescope (SST). Formerly known as SIRTF (Space Infrared Telescope Facility) and launched on August 24, 2003, the *Spitzer Space Telescope* is the fourth and final element in NASA's family of Great Observatories. The observatory carries an 85-cm (33-in.) cryogenic telescope and three cryogenically cooled science instruments capable of performing imaging and spectroscopy in the 3.6–160-micrometer range.

In 2005, *Spitzer* discovered some of life's most basic ingredients in the dust swirling around a young star, IRS 46, about 375 light years from Earth in the constellation Ophiuchus. The ingredients—gaseous acetylene and hydrogen cyanide, precursors to DNA and protein—were detected in the star's terrestrial planet zone, a region where rocky planets such as Earth are thought to be born. The findings represent the first time that these gases have been found in a terrestrial planet zone outside the solar system. Ophiuchus harbors a huge cloud of gas and dust in the process of a major stellar baby boom. Like most young stars, IRS 46 is circled by a flat disk of spinning gas and dust that might ultimately clump together to form planets. Organic gases such as those seen around IRS 46 are found in the solar system, in the atmospheres of the giant planets and Saturn's moon Titan, and on the icy surfaces of comets. They have also been seen around massive stars by the European Space Agency's *Infrared Space Observatory* (ISO), though these stars are thought to be less likely than Sun-like stars to form life-bearing planets. A long time ago, such organic material may have brought life to Earth.

In 2005, *Spitzer* also captured the first light ever detected from two planets orbiting stars other than the Sun, picking up the infrared glow from the Jupiter-sized planets. The findings mark the beginning of a new age of planetary science in which extrasolar planets can be directly measured and compared.

RHESSI. RHESSI (Reuven Ramaty High Energy Solar Spectroscopic Imager), launched in 2002, continued its operation in Earth orbit, providing advanced images and spectra to explore the basic physics of particle acceleration and explosive energy release in solar flares.

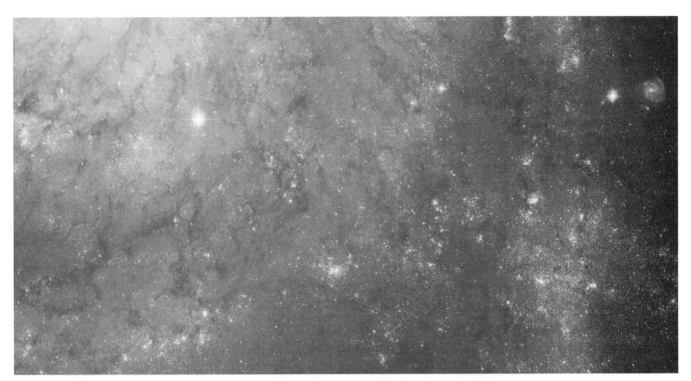

Fig. 3. Small portion of mosaic of *Hubble Space Telescope* images of the galaxy M101, showing detailed structure. Individual dust lanes of the spiral arms and bright hot regions that are areas of active star formation are visible. Several bright stars are actually members of the Milky Way Galaxy and happen to be in the line of sight. A background spiral galaxy, millions of light-years behind M101, appears at the far right. (*NASA; ESA; Hubble Heritage Team: Space Telescope Science Institute/AURA*)

Hubble Space Telescope. Fifteen years after it was placed in orbit, the *Hubble Space Telescope* (*HST*) continued to produce imagery and data useful across a range of astronomical disciplines to expand our knowledge of the universe. In 2005, *HST* images were assembled in the largest and most detailed photo of a spiral galaxy, the face-on spiral galaxy Messier 101 (M101), that has ever been released from *Hubble* (**Fig. 3**). *Hubble*'s high sensitivity and sharp view also uncovered a pair of giant rings girdling the planet Uranus (**Fig. 4**). The largest is twice the diameter of the planet's previously known ring system, discovered in the late 1970s. *HST* also spied two small satellites, named Mab and Cupid. On October 28, within a day of its closest approach to Earth (64 million kilometers or 43 million miles), *HST* took pictures of Mars showing a large regional dust storm on the Red Planet. *Hubble* also discovered that Pluto has three moons. The discovery could offer insights into the nature and evolution of the Pluto system and the Kuiper Belt. Moreover, *HST*'s resolution and sensitivity to ultraviolet light helped researchers look for important minerals on Earth's Moon that could be critical for a sustained human presence. *See* PLUTO.

In 2005, design activities continued on the *HST*'s successor, the giant *James Webb Space Telescope* (*JWST*), planned for launch in 2011. Meanwhile, plans are in development for a space shuttle mission to *Hubble* to effect some maintenance and repairs.

Chandra Observatory. Launched on shuttle mission STS-93 on July 23, 1999, the massive *Chandra X-ray Observatory* uses a high-resolution camera, high-resolution mirrors, and a charge-coupled-device (CCD) imaging spectrometer to observe x-rays of some of the most violent phenomena in the universe which cannot be seen by the *Hubble*'s visual-range telescope. In 2005 Chandra astronomers found the most powerful eruption in the universe, generated by a supermassive black hole in the hot, x-ray-emitting galaxy cluster MS 0735.6+7421 having swallowed a mass of about 300 million Suns in an eruption lasting for more than 100 million years. *Chandra* also found evidence for a swarm of black holes around the supermassive black hole Sagittarius A* (Sgr A*) at the center of the Milky Way Galaxy (**Fig. 5**). Other highlights were observations of the fiery ring surrounding the stellar explosion of Supernova 1987A, and an unusual view of the northern polar region of Earth, showing the Auroras Borealis (Northern Lights) dancing in x-ray light.

Cassini/Huygens. NASA's 5.4-metric-ton (6-ton) spacecraft *Cassini* continued its epic 6.7-year, 3.2-billion-kilometer (2-billion-mile) journey to and inside the planetary system of Saturn. During 2005, the spacecraft remained in excellent health, having successfully entered orbit around Saturn on June 30, 2004. On December 25, 2004, the *Huygens* probe detached from NASA's *Cassini* orbiter to begin a 3-week journey to Saturn's moon Titan. The *Huygens* probe, built and managed by ESA, was bolted to *Cassini* and rode along during the nearly 7-year journey to Saturn largely in a "sleep" mode. After 20 days and a 4-million-kilometer (2.5-million-mile) cruise, the probe safely landed on Titan on January

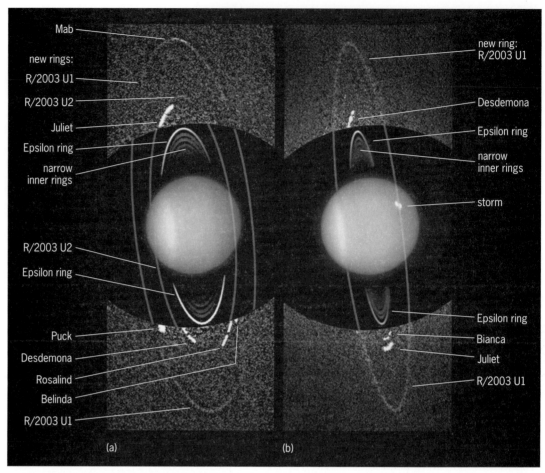

Fig. 4. Composite images of Uranus system from several observations by the *Hubble Space Telescope*, showing a pair of newly discovered rings encircling the planet. Background speckle pattern is noise in the images. (*a*) Composite image from *Hubble* images in 2003. The outermost ring, R/2003 U1, is twice the radius of the previously known ring system. It is probably replenished by dust blasted off the newly discovered satellite Mab, embedded in the ring and visible as a bright streak at the top of the outer ring. Only a faint segment is visible of the second newly discovered ring, R/2003 U2, approximately halfway between R/2003 U1 and the inner ring system. Other satellites of Uranus are also identified. (*b*) Composite image from *Hubble* images in 2005. (*NASA; ESA; M. Showalter, SETI Institute*)

14, 2005, becoming the first human-made object to explore on-site the unique environment of this moon, whose chemistry is assumed to be very similar to that of early Earth before life formed. By end-2005, its mothership *Cassini* had performed a number of close passes of some of Saturn's moons, starting with the "yin-yang" moon Iapetus, followed by, again, Titan, then Enceladus, Mimas, and Dione. Radio and plasma wave data from *Cassini*, along with ultraviolet images from *Hubble*, showed that Saturn's mysterious auroras are fundamentally different from those of Earth and Jupiter. Jets of fine, icy particles streaming from Enceladus on *Cassini* images provided unambiguous evidence that the moon is geologically active, while Dione was revealed as a pale, icy moon with a heavily cratered surface.

WMAP. NASA's *Wilkinson Microwave Anisotropy Probe* launched in 2001, measures the temperature fluctuations of the cosmic microwave background radiation (CMBR) with unprecedented accuracy. The CMBR is the afterglow light left over from the big bang, bathing the whole universe. Results to date indicate that the universe is 13.7×10^9 years old, with a margin of error of close to 1%; the first stars ignited 200 million years after the big bang; light gathered in *WMAP* pictures is from 379,000 years after the big bang; and the universe consists of 4% atoms, 23% cold dark matter, and 73% dark energy. The data place new constraints on the dark energy, which now seems more like a cosmological constant than a negative-pressure energy field called quintessence (the latter, however, is not ruled out). Scientists peering back to the afterglow have developed new evidence for what happened within the first 10^{-12} s of the universe, when it suddenly grew from submicroscopic to astronomical size; new data from *WMAP*, gathered during 3 years of continuous observations, provide further evidence to support this scenario, known as inflation. *See* DARK ENERGY; DARK MATTER.

Genesis. The solar probe *Genesis* was launched in 2001 into an orbit about the first Earth-Sun Lagrangian libration point L1 about 1.5 million kilometers (932,000 miles) from Earth and 148.5 million kilometers (92.3 million miles) from the Sun. *Genesis* collected solar wind material for 27 months, amounting to a total of $\sim 10^{20}$ ions or ~ 0.4 milligram. In April

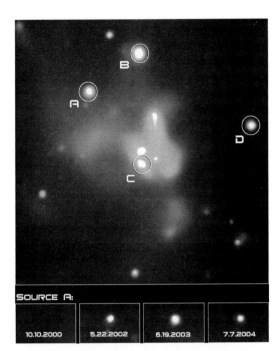

Fig. 5. *Chandra* x-ray image of the region around the supermassive black hole at the center of the Milky Way Galaxy, Sagittarius A* (Sgr A*). Four bright variable x-ray sources, marked by circles A–D, are visible within 3 light-years of Sgr A*, which is the bright source just above source C. The lower panel illustrates the variability of one of these sources. This variability, which is present in all the sources, is indicative of an x-ray binary system where a black hole or a neutron star is pulling matter from a companion star. These sources are circumstantial evidence for the presence of a swarm around SgrA* of 10,000 or more stellar-mass black holes and neutron stars, which are believed to have migrated there over several billion years. (*NASA/Chandra X-ray Center/UCLA/M. Muno et al.*)

2004, the sample collectors were deactivated and stowed, and the spacecraft returned to Earth, where the sample return capsule was to be recovered in midair by helicopter over the Utah Test and Training Range on September 8, 2004. However, *Genesis*' return did not go according to plan. The vessel entered Earth's atmosphere as scheduled, but its parachutes failed to deploy and the capsule crashed into the Utah desert at nearly 320 km/h (200 mi/h). After the crash, the 180-kg (400-lb) capsule was recovered and transported by helicopter to a nearby Army base equipped with a clean room for analysis. Scientists have since reported that a large amount of material within the *Genesis* scientific collectors remained intact and will provide useful information about the origin and development of the solar system, representing essentially a "piece of the Sun." *Genesis* samples are being analyzed with a host of new technologies and instrumentation. The cause of the failure of the parachute deployment system was determined to have been the installation of G-switch sensors in an inverted orientation, rendering them unable to sense sample return capsule deceleration during atmospheric entry and initiate parachute deployments.

ACE. In 2004, the *Advanced Composition Explorer* (*ACE*) continued to observe, determine, and compare the isotopic and elemental composition of several distinct samples of matter, including the solar corona, the interplanetary medium, the local interstellar medium, and galactic matter. By end-2005, *ACE* had been positioned in a halo orbit around Libration Point L1 for more than 8 years, and things were still working very well, with the exception of the SEPICA (Solar Energetic Particle Ionic Charge Analyzer) instrument. Due to failure of the valves that control gas flow through the instrument, active control of SEPICA's proportional counter is no longer possible. In mid-2006, researchers did not expect to deliver any SEPICA data from later than February 4, 2005, unless one of the valves opened by itself, as has happened a few times in the past.

Stardust. In January 2004, NASA's comet probe *Stardust*, launched in 1999, passed by Comet P/Wild 2, collected particles, and began its 2-year trek back to Earth. After the Wild flyby, the sample collector, deployed in late December 2003, was retracted, stowed, and sealed in the vault of the sample reentry capsule. Images of the comet nucleus were also obtained, with coverage of the entire sunlit side at a resolution of 30 m (100 ft) or better. On January 15, 2006, *Stardust* returned safely to Earth. The probe released its sample return capsule the night before at 9:57 p.m. Pacific time, and it entered the atmosphere at 1:57 a.m. The drogue and main parachutes deployed at 2:00 a.m. and 2:05 a.m., respectively, and at 2:10 a.m. the capsule carrying cometary and interstellar particles successfully touched down in the desert salt flats of the U.S. Air Force Utah Test and Training Range.

Ulysses. In 2005, the joint European/NASA solar polar mission *Ulysses* celebrated its 15th launch anniversary. The *Ulysses* spacecraft has already traveled 7×10^9 km (4.3×10^9 mi), and was still going strong in mid-2006. During this exploratory voyage, *Ulysses* has opened new windows on the heliosphere, the vast region of space carved out by the Sun's influence. *Ulysses* is the first, and probably the only, space probe to be placed in a polar orbit around the Sun. From this unique vantage point, *Ulysses* is able to directly study the previously unexplored regions of space above the Sun's poles. *Ulysses* carries a comprehensive suite of sophisticated scientific instruments, several of which are of a kind never flown in space before. In addition to enabling the mission's "core business"—providing the first survey of the solar wind in four dimensions (three spatial dimensions and time)—this combination has enabled scientists to make many ground-breaking discoveries, some in areas that were not even imagined when the mission was first planned. *Ulysses* firsts include: first direct measurements of interstellar dust and neutral helium gas, first measurements of rare cosmic-ray isotopes, first measurements of "pickup" ions of both interstellar and near-Sun origin; first direct observations of comet tails at large distances from the Sun; and first observations of particles from solar storms over the solar poles.

Voyager. The Voyager missions, begun in 1977, continue their quest to push the bounds of space exploration. The twin *Voyager* 1 and 2 spacecraft opened new vistas in space by greatly expanding our knowledge of Jupiter and Saturn, and *Voyager 2* then extended the planetary adventure when it flew by Uranus and Neptune. In 2005, *Voyager 1* reached a distance of 14×10^9 km (8.7×10^9 mi) from the Sun, entering the solar system's final frontier, the heliosheath beyond the termination shock of the solar wind, a vast turbulent expanse where the Sun's influence ends and the solar wind crashes into the thin gas between stars. Close on the heels of its sister ship, *Voyager 2* also continues the ground-breaking journey. At end-2005, *Voyager 1* was escaping the solar system at a speed of about 3.6 AU (astronomical units) per year, while *Voyager 2* was covering about 3.3 AU per year. There are currently five science investigation teams participating in the Interstellar Mission, directly supported by five active instruments aboard: Magnetic Field (MAG) investigation; Low Energy Charged Particle (LECP) investigation; Ultraviolet Spectrometer (UVS) investigation; Cosmic Ray (CRS) investigation; and Plasma Wave (PWS) investigation. Two other instruments are collecting data but do not have official science investigations associated with them: the Planetary Radio Astronomy (PRA) subsystem and, on *Voyager 1* only, the Ultraviolet Spectrometer (UVS). Both spacecraft are expected to continue to operate and send back valuable data until at least 2020.

Mars exploration. The main event in 2005 for NASA's Mars program was the launch of yet another crewless exploration probe, Mars Reconnaissance Orbiter (MRO), joining five other spacecraft currently studying Mars: *Mars Express*, *Mars Odyssey*, *Mars Global Surveyor* (*MGS*), and the two Mars Exploration Rovers, *Spirit* and *Opportunity*. This is the largest number of active spacecraft to study another planet in the history of space exploration.

MRO is a multipurpose spacecraft designed to conduct reconnaissance and exploration of Mars from orbit. It was built by Lockheed Martin under the supervision of NASA's Jet Propulsion Laboratory (JPL). It was launched on August 12, 2005, on an Atlas V launch vehicle, targeted to attain an elliptic Martian orbit in March 2006 for subsequent aerobraking maneuvers to achieve a lower circular orbit. *MRO* contains a host of scientific instruments such as the High Resolution Imaging Science Experiment (HiRISE) camera, the Compact Reconnaissance Imaging Spectrometer for Mars (CRISM), and the Shallow Subsurface Radar (SHARAD), which will be used to analyze the landforms, stratigraphy, minerals, and ice of Mars. It will pave the way for future spacecraft by monitoring daily weather and surface conditions, studying potential landing sites, and testing a new telecommunications system. *MRO*'s telecommunications system will transfer more data back to Earth than all previous interplanetary missions combined, and *MRO* will serve as a highly capable relay satellite for future missions.

The six-wheeled rover vehicle *Spirit*, launched on June 10, 2003, landed on January 3. 2004, in Gusev Crater. At end-2005, it and its twin, *Opportunity*, had explored the surface of the Red Planet for a full Martian year (687 Earth days). Both rovers' original mission was scheduled for only 3 months. Though beginning to show signs of aging, *Spirit* and *Opportunity* were still being used to their maximum remaining capabilities. On *Spirit*, the teeth of the rover's rock abrasion tool are too worn to grind the surface off any more rocks, but its wire-bristle brush can still remove loose coatings. The tool was designed to uncover three rocks, but it exposed interiors of 15 rocks. During its extensive travels, *Spirit* discovered the composition of rock outcrops altered by water.

Opportunity launched on July 7, 2003, touched down on January 25, 2004, on Meridiani Planum, halfway around the planet from the Gusev Crater site of its twin. The rover's steering motor for the front right wheel stopped working during 2005. A motor at the shoulder joint of the rover's robotic arm shows symptoms of a broken wire in the motor winding. But *Opportunity* can still maneuver with its three other steerable wheels. Its shoulder motor still works when given extra current, and the arm is still usable without that motor. Among its discoveries is evidence that water once flowed across the Martian surface.

NASA's *Mars Odyssey* probe has been orbiting Mars since 2001, and was approved for an extended mission through September 2006. *Odyssey*'s camera system obtained the most detailed complete global maps of Mars ever, with daytime and nighttime infrared images at a resolution of 100 m (328 ft). About 85% of images and other data from *Spirit* and *Opportunity* have reached Earth via communications relay by *Odyssey*, which also aided *MRO* by monitoring atmospheric conditions during the months in 2006 when the newly arrived orbiter used calculated dips into the atmosphere to alter its orbit into the desired shape. At end-2005, a new view of the largest canyon in the solar system, composed of hundreds of photos from the *Mars Odyssey* orbiter, offered scientists and the public an online resource for exploring the entire canyon in detail (**Fig. 6**). This canyon system, named Valles Marineris, stretches as far as the distance from California to New York. Steep walls nearly as high as Mount Everest give way to numerous side canyons, possibly carved by water. In places, walls have shed massive landslides spilling far out onto the canyon floor. Using *Odyssey* photography, a simulated fly-through using the newly assembled imagery was made available to the public online.

MGS has returned more data about Mars than all other missions combined. In September 2004, *MGS* started its third mission extension after 7 years of orbiting Mars, using an innovative technique to capture pictures even sharper than most of the more than 170,000 it had already produced. One dramatic example from the spacecraft's Mars Orbiter Camera (MOC) showed wheel tracks of *Spirit* and the rover itself. In 2005, *MGS* also took pictures that were the first ever of spacecraft orbiting Mars taken by another spacecraft orbiting Mars: *Mars Express* was passing about 250 km (155 mi) away when the *MGS* MOC photographed it on April 20. The next day, the

Fig. 6. View of Valles Marineris on Mars, the largest canyon in the solar system, merging hundreds of photographs from the *Mars Odyssey* orbiter. *(NASA)*

camera caught *Mars Odyssey* passing 90–135 km (56–84 mi) away. All three spacecraft are moving at almost 11,000 km/h (7000 mi/h), and at 100 km (62 mi) distance the field-of-view of the MOC is only 760 m (830 yards) across. If timing had been off by only a few seconds, the images would have been blank.

Two gullies appear in an April 2005 image of a sand-dune slope where they did not exist in July 2002. The MOC team has found many sites on Mars with fresh-looking gullies, and checked back at more than 100 gullied sites for possible changes between imaging dates, but this is the first such find. Some gullies, on slopes of large sand dunes, might have formed when frozen carbon dioxide, trapped by windblown sand during winter, vaporized rapidly in spring, releasing gas that made the sand flow as a gully-carving fluid. At another site, more than a dozen boulders left tracks when they rolled down a hill sometime between the taking of images in November 2003 and December 2004. It is possible that they were set in motion by strong wind or by seismic activity (a "marsquake"). The orbiter is healthy and may be able to continue studying Mars for 5 to 10 more years.

Earth science. In 2005, as in 2004, NASA launched one Earth science satellite, *NOAA-18*. The NASA-centered international Earth Observing System (EOS) continues to operate, with *Aqua* as the first member of a group of satellites termed the Afternoon Constellation (or sometimes the A-Train). The second member launched was *Aura*, the third member was *PARASOL* in December 2004, and the fourth and fifth members were *CloudSat* and *CALIPSO* in April 2006. Expected upcoming missions are *OCO* and *Glory*, with the placement of *Glory* not yet determined. Once completed, the A-Train will be led by

OCO, followed by *Aqua*, then *CloudSat*, *CALIPSO*, *PARASOL*, and, in the rear, *Aura*, about 15 min behind *Aqua*.

NOAA-18. On May 20, *NOAA-18*, an environmental satellite for the National Oceanic and Atmospheric Administration (NOAA), was launched on a Boeing Delta 2 expendable rocket into a circular polar orbit of 870 km (544 mi) altitude and 98.73° inclination. Twenty-one days after spacecraft launch, operational control of *NOAA-18* was transferred from NASA to NOAA. With the objective to improve weather forecasting and monitor environmental events around the world, *NOAA-18* collects data about the Earth's surface and atmosphere. The data are input to NOAA's long-range climate and seasonal outlooks, including forecasts for El Niño and La Niña. *NOAA-18* is the fourth in a series of five polar-orbiting Operational Environmental Satellites (POES), with instruments that provide improved imaging and sounding capabilities. *NOAA-18* has instruments used in the 1982-established international Search and Rescue Satellite-Aided Tracking System, called COSPAS-SARSAT. NOAA polar-orbiting satellites detect emergency beacon distress signals and relay their location to ground stations, so rescue can be dispatched. SARSAT is credited with saving approximately 5000 lives in the United States and more than 18,000 worldwide.

Aura. *Aura*, launched on July 15, 2004, is NASA's third major EOS platform, joining its sister satellites, *Terra* and *Aqua*, to provide global data on the state of the atmosphere, land, and oceans, as well as their interactions with solar radiation and each other. During 2005, observations from *Aura* showed explosive volcanic eruptions injecting gases and ash into the Earth's atmosphere, creating hazardous conditions for passing aircraft and the potential for climate effects. Two large explosive eruptions occurred at Manam (Papua New Guinea) on January 27–28 and at Anatahan (Mariana Islands) on April 5–6. Other observations by *Aura* showed that thunderstorms over Tibet provide a main pathway for water vapor and chemicals to travel from the lower atmosphere, where human activity directly affects atmospheric composition, into the stratosphere, where the protective ozone layer resides. *Aura* data also indicate that the Antarctic ozone hole's recovery is running late: the full return of the protective ozone over the South Pole will take nearly 20 years longer than scientists previously expected.

ICESat. ICESat (Ice, Cloud, and land Elevation Satellite), also an EOS spacecraft, is the benchmark mission for measuring ice sheet mass balance, and cloud and aerosol heights, as well as land topography and vegetation characteristics. Launched on January 12, 2003, into a near-polar orbit at an altitude of 600 km (373 mi) with an inclination of 94°, the spacecraft carries only one instrument, the Geoscience Laser Altimeter System (GLAS). GLAS sends short pulses of green and infrared light downward 40 times a second, all over the globe, and collects the reflected laser light with a 1-m (39-in.) telescope, yielding elevations. It also fires a fine laser beam of light that spreads out as it approaches the Earth surface to about 65 m (213 ft) in diameter. On its way to the surface, those photons or particles of light bounce off clouds, aerosols, ice, leaves, ocean, land, and more, providing detailed information on the vertical structure of the Earth system. In 2005, the GLAS instrument reached 10^9 measurements in orbit. The previous maximum measurements in space was approximately 670 million by the MOLA-2 Laser onboard the *MGS* mission.

Aqua. Launched in 2002, the 1750-kg (3858-lb) NASA satellite *Aqua*, carrying six instruments designed to collect information on water-related activities, has been circling Earth in a polar, Sun-synchronous orbit of 705 km (438 mi) altitude. During its 6-year mission, *Aqua* is observing changes in ocean circulation, and studies how clouds and surface water processes affect the climate. In 2005, NASA and NOAA scientists used experimental data from *Aqua*'s Atmospheric Infrared Sounder, a high-spectral resolution infrared instrument that takes three-dimensional pictures of atmospheric temperatures, water vapor, and trace gases, to improve the accuracy of medium-range weather forecasts in the Northern Hemisphere. In particular, they found that incorporating the instrument's data into numerical weather prediction models improves the accuracy range of experimental 6-day Northern Hemisphere weather forecasts by up to 6 h, a 4% increase. These data have meanwhile been officially incorporated into the NOAA National Weather Service operational weather forecasts.

GRACE. Launched in 2002, the twin *GRACE* (Gravity Recovery and Climate Experiment) satellites continued to map the Earth's gravity fields by taking accurate measurements of the distance between them, using the Global Positioning System (GPS) and a mi-

crowave ranging system. During 2005, the satellites exchanged positions. The maneuver was initiated on December 3, and *GRACE-2* passed *GRACE-1* on December 10. The swap was done to mitigate the risk of loss of thermal control over the K-band antenna horn of *GRACE-2* (and subsequent spurious K-band range signals) due to atomic oxygen exposure. (As the trailing satellite, *GRACE-2* had been flying "forward" with the antenna horn exposed to impacting atomic oxygen.) After the two satellites switched position, a special data collection campaign for intersatellite separation between about 70 km (43 mi) and about 170 km (106 mi) got underway. The mean inter satellite separation is usually limited to the range between 170 and 270 km (106 and 168 mi), but the closer mean separation is expected to enhance the high-frequency signal content in the K-band range measurements. The project is a joint partnership between NASA and the German DLR (Deutsches Zentrum für Luft und Raumfahrt).

Department of Defense (DOD) space activities. The increased use of satellites for communications, observations, and—through the GPS—navigation and high-precision weapons targeting is of decisive importance for the military command structure. In the Afghanistan and Iraq conflicts, orbiting assets have ably demonstrated that space-based intelligence, surveillance, communications, weather, missile warning, and navigation tools give commanders great advantages and leverage for each of the military services. Highlights of military space activity in 2005 included the launch of the final US Air Force Titan 4B rocket on October 19, carrying a 19-ton Advanced KH-11 digital imaging reconnaissance spacecraft (*USA-186*), two successful missile interceptor tests, and the successful test firing of an antimissile airborne laser.

In 2005, there were six military space launches (2004: 5; 2003: 11), carrying seven payloads: two Titan 4B/IUS vehicles from Cape Canaveral, Florida, with two observation and imaging satellites (*USA-182*, *USA-186*), one Atlas 3B/Centaur with two sigint (signal intelligence) satellites, two Minotaur (Minuteman-2 ICBM+Pegasus) rockets with technology development/demonstration payloads, and a Delta 2 launch vehicle carrying a GPS navigation satellite.

Commercial space activities. In 2005, commercial space activities in the United States took an additional downturn over prior years, after the slump in the communications space market caused by failures of satellite constellations for mobile telephony in 2001/2002 and a slight recovery in 2003. Of the 16 total launch attempts by the United States in 2005 (19 in 2004, 26 in 2003), five (31%) were commercial missions (NASA: 5; military: 6). In the launch services area, Boeing sold three Delta-2 vehicles, while competitor ILS/Lockheed Martin flew one Atlas 3B (with Russian engines) and two Atlas 5/Centaurs. The partnership of Boeing, RSC-Energia (Russia, 25% share), NPO Yushnoye (Ukraine), and Kvaerner Group (Norway) successfully launched four Russian Zenit 3SL (SeaLaunch) rockets carrying the *XM Radio 3*,

Spaceway 1, *Intelsat A-8*, and *Inmarsat 4F-2* comsats from the Odyssey sea launch platform floating at the Equator (first launch 1999).

Russian Space Activities

With financial support by a slowly improving national economy growing slightly over previous years, Russia and the Ukraine in 2005 showed increased activity in space operations from 2004, launching seven different carrier rockets. Out of 55 launch attempts worldwide in 2005, 26 space launches were attempted by Russia, placing it solidly in the lead of spacefaring countries, including the United States. Its total of 26 attempts, achieving 23 successful launches, was 3 more than its previous year's 23 attempts (1 failure): Five Soyuz-U, six Soyuz-FG (two crewed), seven Protons (-K and -M), four Zenit-3SL (sea launch, counted above under U.S. Activities), one Molniya-M (failed), three Kosmos-3M, one Volna (failed submarine launch of a test version of a Euro-Russian inflatable atmospheric reentry shield), one Dnepr, and two Rokots (one failed).

The upgraded Soyuz-FG rocket's fuel injection system provides a 5% increase in thrust over the Soyuz-U, enhancing its lift capability by 200 kg (441 lb) and enabling it to carry the Soyuz-TMA spacecraft, which is heavier than the Soyuz-TM ship used in earlier years to ferry crews to the *ISS*. Soyuz-TMA was flown for the first time on October 30, 2002, as *ISS* mission 5S. It was followed in 2003 by Soyuz *TMA-2* (6S) and *TMA-3* (7S), in 2004 by *TMA-4* (8S) and *TMA-5* (9S), and in 2005 by *TMA-6* (10S) and *TMA-7* (11S).

In 2005, Russia's federal cabinet (Duma) gave final approval to a 10-year civil space plan that prioritizes satellite telecommunications, navigation, and Earth observation, under the purview of the Federal Space Agency Roskosmos. Other priorities outlined in the plan include the development of the Angara family of heavy-lift rockets, new modifications of the Soyuz-2 launch vehicle, and development of a new space crew capsule, such as the proposed six-seat Klipr (Clipper), to replace the current three-seat Soyuz-TMA craft.

The Russian space program's major push to enter into the world's commercial arena by promoting its space products on the external market, driven by the need to survive in an era of severe reductions of public financing, increased in 2005. First launched in July 1965, the Proton heavy lifter, originally intended as a ballistic missile (UR500), by end-2005 had flown 238 times since 1980, with 14 failures (reliability: 0.941). Its launch rate in recent years has been as high as 13 per year. Of the seven Protons launched in 2005 (2004: 8), six were for commercial customers (*AMC-12 and -23, Ekspress AM-2 and AM-3, DirecTV 8, Anik F1R*), the seventh for the state/military (three *GLONASS/Uragan* navsats). From 1985 to 2005, 186 Proton and 414 Soyuz rockets were launched, with 10 failures of the Proton and 10 of the Soyuz, giving a combined reliability index of 0.967. Until a launch failure on October 15, 2002, the Soyuz rocket had flown 74 consecutive successful missions, including 12 with human crews onboard; subsequently, another 26 successful flights were added, including 6 carrying 17 humans.

European Space Activities

Europe's efforts to reinvigorate its faltering space activities after the long decline since the mid-1990s have not yet materialized, and remain at a low level compared to astronautics activities of NASA, DOD, Russia, and China. Ongoing efforts by the European Union (EU) on an emerging new European space strategy for ESA to achieve an autonomous Europe in space under Europe's new constitution that makes space and defense an EU responsibility still remained largely unresolved. However, 2005 brought the probably greatest mission success in ESA history with the landing of the *Huygens* space probe on Saturn's moon Titan.

After the December 2002 failure of the new EC (enhanced capability) version of the Ariane 5, designed to lift 10 tons to geostationary transfer orbit, enough for two big communications satellites at once, European industry accomplished a quick comeback with the successful first launch of an Ariane 5 ECA in 2005 as one of 5 flights (out of 5 attempts) of the Ariane 5-G (generic, 3 flights) and –ECA (2 flights) rockets (2004: 3), bringing its program total to 25. The five heavy-lift vehicles of 2005 carried a total of 10 satellite payloads: 7 commercial comsats, 1 weather satellite (*MSG-1/Meteosat*), and 2 technology test satellites.

In 2005, the most significant space undertaking besides the Titan landing for the 15 European countries engaged in space continued to be the development of the *Galileo* navigation and global positioning system. Starting in 2008, it will enable Europe to be independent of the U.S. GPS. In 2005, the program reached a milestone with the successful Baikonur launch, on December 28, of *GIOVE-A*, the first of two *Galileo* test-bed satellites, on a Russian Soyuz-FG/Fregat.

In the space science area, in 2005 there were the *Huygens* touchdown on Titan and two new European spacecraft launches: the *Cryosat* mission and the planetary probe *Venus Express*.

Huygens. After a 7-year interplanetary journey covering 3.5×10^9 km (2.2×10^9 mi), the European probe *Huygens* separated from its U.S. *Cassini* mothership on December 25, 2004, and started on its 4-million-kilometer (2.5-million-mile) ballistic path toward Titan, largest of Saturn's natural satellites, arriving on January 14. After a parachute-assisted descent, during which its main instruments analyzed the atmosphere for more than 2 h, the probe landed successfully on semisolid ground, likened to wet sand (possibly methane sludge). *Huygens* and its batteries survived for almost 5 h, twice as long as expected, transmitting data and 350 pictures to Earth via the receding *Cassini* mothership.

Cryosat. ESA's *Cryosat*, built to carry out a 3-year mission to monitor changes in the elevation and thickness of polar ice sheets and floating sea ice with very great precision, was lost when its Russian

Rokot launcher failed after liftoff on October 8, 2005. In early 2006, the ESA member states, in a meeting of the ESA Earth Observation Programme Board, approved the building and launching of a recovery mission, *Cryosat-2*.

Venus Express. The 1240-kg (2734-lb) *Venus Express* spacecraft was launched on November 9 aboard a Russian Soyuz-Fregat launch vehicle from the Baikonur Cosmodrome in Kazakhstan. The planetary explorer reached Earth's cloud-shrouded sister planet in a record time of 153 days on April 11, 2006. *Venus Express* is the first mission to visit Venus since NASA's 1989–1994 *Magellan* mission, but unlike the *Magellan* probe, *Venus Express* will not be capable of imaging the surface of Venus with high spatial resolution through cloud-penetrating imaging radar. Instead, it is equipped with several instruments designed to study Venus in new ways. *Venus Express* was built mostly with spare parts and designs from the European *Mars Express* and *Rosetta* missions. Its instruments are a plasma analyzer (ASPERA-4/Analyzer of Space Plasmas and Energetic Atoms), a magnetometer (MAG), three spectrometers (PFS/Planetary Fourier Spectrometer, SPICAV/Spectroscopy for Investigation of Characteristics of the Atmosphere of Venus, and VIRTIS/Visible and Infrared Thermal Imaging Spectrometer), a radio sounder (VeRa/Venus Radio), and a digital camera (VMC/Venus Monitoring Camera).

Rosetta. Launched on an Ariane 5 on March 2, 2004, the *Rosetta* probe is scheduled to rendezvous with comet 67 P/Churyumov-Gerasimenko in 2014 and release a landing craft named *Philae*. It is hoped that on its 10-year journey to the comet the spacecraft will pass by at least one asteroid. Highlights in 2005 included *Rosetta*'s observation of the NASA *Deep Impact* probe's encounter with Comet Tempel-1, using four active instruments (ALICE, MIRO, OSIRIS, and VIRTIS). Three of the remote sensing instruments were active continuously from June 29 to July 14. All instruments operated very well, and their science data were collected as planned and were analyzed. A few problems that occurred with the commanding timing of OSIRIS and with the MIRO instrument were overcome in both cases within about 24 h, with minor impact on the overall instrument operations and data return. The exercise was the first scientific planning and operations scenario over a large scale and an extended period of time for the *Rosetta* mission, providing important experience and a wealth of lessons learned. They will be very useful in designing spacecraft operations around *Rosetta*'s comet encounter.

Envisat. ESA's operational environmental satellite *Envisat*, launched in 2002, is the largest Earth Observation spacecraft ever built. The 8200-kg (18,100-lb) satellite circles Earth in a polar orbit at 800 km (500 mi) altitude. Because of its polar Sun-synchronous orbit, it flies over and examines the same region of the Earth every 35 days under identical conditions of lighting. The 25-m-long (82-ft) and 10-m-wide (33-ft) satellite is equipped with 10 advanced instruments.

SPOT 5. Launched in 2002, the fifth imaging satellite of the commercial Spot Image Company in 2005 continued operations in its polar Sun-synchronous orbit of 813 km (505 mi) altitude. Unique features of the *SPOT* system are high resolution, stereo imaging, and revisit capability.

INTEGRAL. ESA's *INTEGRAL* (International Gamma-Ray Astrophysics Laboratory), a cooperative project with Russia and United States, launched in 2002, continued successful operations in 2005. Its instruments have produced the first all-sky map of the 511-keV line emission produced when electrons and their antimatter equivalents, positrons, collide and annihilate. The nature of the sources responsible for the antimatter is a key area for further investigation by *INTEGRAL*. One intriguing possibility is the annihilation or decay of an exotic form of dark matter.

The *INTEGRAL* imager (IBIS) has detected a new persistent soft gamma-ray source, IGR J18135-1751, which is coincident with one of several sources of extreme energy in the inner part of the Milky Way Galaxy. These objects may be the sites of cosmic high-energy particle accelerators.

XMM-Newton. Europe's *XMM* (X-ray Multi Mirror)-*Newton* observatory, launched in 1999, is the largest European science research satellite ever built. Scientific results based on *XMM-Newton* data are being published at almost 300 papers per year, comparable to the *Hubble* telescope. Among its discoveries, it characterized for the first time x-ray spectra and light curves of some classes of protostars (stars being born) and provided an unprecedented insight into the x-ray variability of the coronas of stars similar to the Sun. With its capability to respond as quickly as 5 h to target-of-opportunity requests for observing elusive gamma-ray bursts, this space observatory detected for the first time an x-ray halo around a burst, consisting of concentric ringlike structures centered on the burst location. *XMM-Newton* is shedding new light on supernovae remnants, as well as on neutron stars. In particular, it discovered a bow shock aligned with the supersonic motion of a neutron star (called Geminga), and detected hot spots indicating that the configuration of a neutron star's magnetic field and surface temperatures are much more complex than previously thought.

Smart-1. *Smart-1* (Small Missions for Advanced Research in Technology 1), launched on September 27, 2003, was Europe's first lunar spacecraft. It was intended to demonstrate new technologies for future missions, in this case the use of solar-electric propulsion as the primary power source for its ion engine, fueled by xenon gas. On November 15, 2004, the spacecraft encountered its first perilune, after 332 orbits around the Earth. The ion drive was fired on that day to brake the spacecraft into lunar orbit, after which over several months its engine was fired repeatedly to lower the spacecraft into an operational orbit whose height ranged from approximately 300 to 3000 km (200 to 2000 mi). This was achieved by January 13, 2005, heralding the beginning of its science program in March 2005, using spectrometers for x-rays and near infrared as well as a camera for color imaging. On September 3, 2006, *Smart-1* ended its journey and exploration of the Moon by crashing into the lunar surface in an area called the Lake of Excellence.

Mars Express. *Mars Express* was Europe's entry into the ongoing and slowly expanding robotic exploration of the Red Planet from Earth as precursors to later missions by human explorers. The *Mars Express* orbiter entered Martian orbit on December 25, 2003, and moved into its operational near-polar orbit in January 2004. Highly successful operations and close-up imagery of the Mars surface were carried out during 2004 and 2005, supported by a sophisticated instrument package comprising the High Resolution Stereo Camera (HRSC), Energetic Neutral Atoms Analyzer (ASPERA), Planetary Fourier Spectrometer (PFS), Visible and Infrared Mineralogical Mapping Spectrometer (OMEGA), Sub-Surface Sounding Radar Altimeter (MARSIS), Mars Radio Science Experiment (MaRS), and the Ultraviolet and Infrared Atmospheric Spectrometer (SPICAM).

Asian Space Activities

China, India, and Japan have space programs capable of launch and satellite development and operations.

China. With a total of five launches in 2005 (2004: eight), China remained solidly in third place of spacefaring nations after Russia and the United States, following the successful orbital launch in 2003 of the first Chinese taikonaut, Lt. Col. Yang Liwei in the spacecraft *Shenzhou 5* (Divine Vessel 5). In 2005, the China National Space Administration (CNSA) scored another huge success with the launch of the 7700-kg (17,000-lb) two-seater *Shenzhou 6*, carrying taikonauts Fei Junlong and Nie Haisheng. The spacecraft was launched on October 12 from the Jiuquan launch site in the Gobi desert, and the crew returned safely on October 17, after 75 orbits and about 3.2 million kilometers (2 million miles), lasting 115 h 32 min. During the flight, both crew members left their seats, floated in space, and at one time also took off their 10-kg (22-lb) spacesuits. The *Shenzhou 6* spaceflight was designed to further China's human spaceflight experience as it works toward developing a crewed space station, and to serve as a symbol of national pride, demonstrating China's technological prowess.

The launch vehicle of the *Shenzhou* spaceships is the human-rated Long March 2F rocket. China's Long March (Chang Zheng, CZ) series of launch vehicles consists of 12 different versions, which by the end of 2005 had made 88 flights, sending 100 payloads (satellites and spacecraft) into space, with a 90% success rate.

Five major launches in 2005 served to demonstrate China's growing space maturity (2004: 8 successful launches; 2003: 6; 2002: 4; 2001: 1). On April 12, a CZ-3B launched the Chinese *Apstar 6* comsat, on July 5 a CZ-2D launched the *SJ-7* science research

satellite, followed on August 2 and 29 by the recoverable imaging satellites *FSW-21* [*Fanhui Shi Weixing* (Experimental Recoverable Satellite)-21] and *FSW-22*, winding up a record of 47 consecutive launch successes for the Long March. In its two-stage 2C version, Long March has a liftoff weight of 192,000 kg (423,000 lb), a total length of 41.9 m (137.5 ft), a diameter of the rocket and payload fairing of 3.35 m (11 ft), and a low-earth-orbit launching capacity of 2200 kg (4800 lb). The 3B version has a liftoff weight of 425,800 kg (938,700 lb), a total length of 54.9 m (180 ft), and a payload capability to low geosynchronous orbit of 4500 kg (9900 lb).

An important payload for China was launched on October 27 on a Russian Kosmos-3M: the *Beijing-1* spacecraft, carrying a new mapping telescope. *Beijing-1*, also known as Disaster Monitoring Constellation-4, provides 4-m (13-ft) resolution. Among other uses, Chinese planning officials intend to employ the satellite imagery for their preparations of the 2008 Summer Olympic Games.

India. While China's second human spaceflight was the highpoint in Asian space developments in 2005, India also achieved some milestones, continuing its development programs for satellites and launch vehicles through the Indian Space Research Organization (ISRO, created in 1969). Main satellite programs are the INSAT (Indian National Satellite) telecommunications system, the IRS (Indian Remote Sensing) satellites for Earth resources, the METSAT weather satellites, and the GSat series of large experimental geostationary comsats. India's main launchers are the PSLV (Polar Space Launch Vehicle) and the Delta 2-class GSLV (Geosynchronous Satellite Launch Vehicle). In 2005 (as in 2004), the country conducted only one launch, but it was a landmark: On May 5, a PSLV launched the *Cartosat-1* mapping satellite, designed to take black-and- white pictures with a resolution of 2.5 m (8 ft), India's highest-resolution imaging satellite to date. At 1560 kg (3440 lb), *Cartosat-1* was the heaviest payload launched to date on the PSLV, which also carried the *Hamsat* (*Vusat*) satellite for amateur radio operators, from India's new launch pad at the Satish Dhawan Space Center in Sriharikota.

India is working on plans to explore the Moon, with the intent to send a crewless probe there in 2007. ISRO calls the Moon flight project *Chandrayaan Pratham* (First Journey to the Moon or Moonshot One). The 525-kg (1157-lb) *Chandrayaan-1* would be launched on a PSLV rocket. ISRO in 2005 broke ground in Bangalore on a deep-space tracking station in preparation for the mission.

Japan. Japan's space program also made some notable accomplishments during 2005, leading off with the successful return to flight, on February 26, of the H2-A launch vehicle, whose failure on November 29, 2003, had destroyed a pair of high-priority surveillance satellites. The payload in the February flight was the weather and air-traffic management satellite *MTSat-1R* (which later developed technical problems). On July 10, Japan's space agency JAXA launched the x-ray astronomy satellite *Suzaku* (Red Bird of the South, formerly *Astro-E2*) from JAXA's Uchinoura Space Center on an M-V rocket, the fifth in a series of Japanese x-ray astronomy satellites.

Also notable was the performance of the spacecraft *Hayabusa* (Peregrine Falcon, formerly *Muses-C*). Launched on May 9, 2003, the probe made a successful touchdown on the asteroid Itokawa on November 11, and a second on November 25, to collect samples to bring back to Earth. However, *Hayabusa* subsequently encountered technical difficulties, including loss of contact, that cast some doubt on its ability to return to Earth. Contact was reestablished in December. If restoration of cruise conditions can be accomplished, *Hayabusa*'s return would still be delayed by 3 years to 2010, with the sample to be recovered in a reentry capsule parachuted to the surface in the Australian outback.

Other Countries' Space Activities

On September 8, Telesat Canada launched its advanced *Anik F1R* satellite from Baikonur, Kazakhstan, on a Russian Proton-M carrier with a Briz-M upper stage. *F1R* provides valuable capacity for Canadian direct-to-home satellite television, along with a range of other telecommunications and broadcasting services. The spacecraft also features a navigation payload that will make North American air navigation more reliable and accurate. Construction of the *Anik F3* satellite was underway, with launch scheduled in late 2006.

Also in 2005, Canada's *Radarsat-1* "eye-in-the-sky" celebrated its tenth anniversary. Launched on November 4, 1995, it was expected to operate only 5 years. The quality of its images exceeded the standards of the time, and they are still surpassing the standards. *Radarsat-1* catalogs the vast expanse of Canada's Arctic, providing 3800 images per year to the Canadian Ice Service, the largest of its 600 clients. Over the years, it has claimed 15% of the world's Earth observation market for Canada. *Radarsat-2*, equipped with a synthetic aperture radar (SAR) with multiple polarization modes, is under development for launch in 2007 on a Russian Soyuz launch vehicle. Its highest resolution will be 3 m (10 ft) with 100-m (330-ft) positional accuracy.

For background information *see* ASTEROID; COMET; COMMUNICATIONS SATELLITE; COSMIC BACKGROUND RADIATION; GALAXY, EXTERNAL; GAMMA-RAY ASTRONOMY; GEMINGA; HUBBLE SPACE TELESCOPE; INFRARED ASTRONOMY; MARS; MERCURY (PLANET); METEOROLOGICAL SATELLITES; MILITARY SATELLITES; MOON; REMOTE SENSING; SATELLITE ASTRONOMY; SATELLITE NAVIGATION SYSTEMS; SATURN; SCIENTIFIC SATELLITES; SOLAR WIND; SPACE FLIGHT; SPACE PROBE; SPACE STATION; SPACE TECHNOLOGY; SUN; VENUS; X-RAY ASTRONOMY in the McGraw-Hill Encyclopedia of Science & Technology.

Jesco von Puttkamer

Bibliography. *Aerospace Daily; AIAA Aerospace America*, November 2005; *Aviation Week & Space Technology*; ESA, *Press Releases*; NASA Public Affairs Office, *News Releases*; SPACE NEWS.

Speciation

Biological diversity arises through speciation—the evolution of two species from a single common ancestor. Under the standard (allopatric) speciation model, new species begin as geographically isolated populations that evolve independently, gradually accumulating genetic differences that render them reproductively isolated from one another. Should such diverging populations later come into secondary geographic contact, incompatible courtship signals could prevent them from forming hybrids (prezygotic isolation) or, if hybrids are formed, incompatible genetic interactions could cause them to be sterile or inviable (postzygotic isolation). Both pre- and postzygotic forms of reproductive isolation prevent genetic exchange, allowing the recently diverged species to remain distinct.

Despite being one of the oldest questions in evolutionary biology, the genetic details of how new species evolve are just beginning to be fully understood. This article reviews some of the recent advances in our understanding of one aspect of speciation: the molecular genetics and evolution of hybrid sterility and inviability. The review necessarily focuses on fruit fly species of the genus *Drosophila* simply because this is where most progress on hybrid sterility and inviability genes has occurred.

Identifying hybrid incompatibility genes. Hybrid sterility and inviability are caused by incompatible genetic interactions: alleles from one species can have deleterious effects when placed in the foreign genetic background of another species (**Fig. 1**). However, these so-called hybrid incompatibilities do not evolve for the purpose of causing reproductive isolation, but evolve as simple incidental by-products of divergence. This fact raises questions about the identities of hybrid incompatibility genes: What are their "normal" functions within species? Why did they diverge? Did natural selection or genetic drift (random fluctuation of gene frequencies) cause their divergence? Are certain classes of genes more likely to cause hybrid fitness problems than others? Determining the molecular identities of hybrid incompatibility genes can provide answers to these questions. Identifying hybrid incompatibility genes has, however, proven difficult. At least part of the problem is that speciation geneticists must do genetics in species hybrids—hybrids that are often sterile or inviable. Thus the very reproductive isolation that we are interested in studying actually gets in the way of doing straightforward genetics. Speciation geneticists have therefore resorted to three alternative strategies.

First, speciation geneticists have used introgression studies in which small chromosomal regions are moved from one species into the genetic background of another through repeated backcrossing. This method only works if at least one of the hybrid sexes is viable and fertile, as in *Drosophila mauritiana* and *D. simulans*, which produce sterile hybrid males but fertile hybrid females (**Fig. 2**). Second, during the last 20 years, speciation geneticists have discovered and exploited several hybrid rescue mutations in *Drosophila*. The best characterized of these is the *Hybrid male rescue* (*Hmr*) gene from *D. melanogaster*. Normally, in crosses between *D. melanogaster* and its closely related species, *D. simulans* and *D. mauritiana*, the wild-type *X*-linked *Hmr*^{mel} causes hybrid male lethality at the larval-pupal transition. However, in crosses involving the rescue mutation, *Hmr*^{rescue}, hybrid males are viable (**Fig. 3**). Rescue mutations thus appear to be rare compatible alleles at loci whose wild-type alleles cause hybrid fitness problems. By studying the differences between wild-type and rescue alleles, speciation geneticists simplify their task by doing the genetic mapping and manipulations within species and then testing the incompatibility phenotypes in species hybrids. Third, speciation geneticists have used small chromosomal deletions for complementation tests in hybrids to map recessive incompatibility alleles. For example, some deletions from *D. melanogaster* unmask otherwise recessive hybrid inviability genes from *D. simulans* that kill deletion-bearing hybrids but not hybrids lacking the deletion (**Fig. 4**).

Molecular genetics of hybrid incompatibilities. Four hybrid incompatibility genes have been identified so far. The first was found in the platyfish genus, *Xiphophorus*. Platyfish (*X. maculatus*) are polymorphic for a variety of pigmented spots, called macromelanophores. In certain backcross hybrids between spotted platyfish and unspotted swordtails (*X. helleri*), macromelanophore tissue cells proliferate uncontrollably, becoming lethal melanomas. The gene responsible for these melanomas is a recently duplicated gene that encodes a cell signaling protein, *Xmrk-2*, a receptor tyrosine kinase. Remarkably, *Xmrk-2* is present in some *Xiphophorus* species but absent in others—*X. helleri*, for example, completely lacks the *Xmrk-2* duplicate gene. In *X. maculatus*, *Xmrk-2*'s tumorigenic properties are controlled by tumor suppressor loci. Thus, when *Xmrk-2* is moved onto the genetic background of sister species lacking the appropriate tumor suppressor variants, it becomes overexpressed in macromelanophore tissue and causes cancer.

The second gene is the *Drosophila* hybrid male sterility gene, *Odysseus* (*Ods*). When introduced into a largely *D. simulans* genome by introgression (that is, by repeated backcrossing through fertile hybrid females; Fig. 2) or by germ line transformation, the *D. mauritiana* allele of *Ods* causes sterility in hybrid males. *Ods* encodes a duplication of a putative homeobox (a highly conserved region of protein found within genes important for regulation of development) transcription factor that, like *Xmrk-2*, is misexpressed in hybrids.

The third gene is the *Drosophila* hybrid inviability gene, *Hmr*, which encodes a Myb (Myeloblastosis)–related protein with predicted DNA-binding properties. Hybrid males carrying the *D. melanogaster* wild-type allele *Hmr*^{mel} die, but those carrying the rescue allele *Hmr*^{rescue} do not. *Hmr* was first mapped to the middle of the *X*-chromosome in *D.*

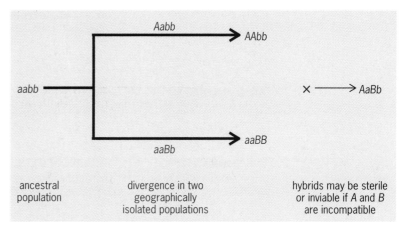

Fig. 1. Evolution of hybrid incompatibilities. The ancestral species is fixed for the two-locus genotype, *aabb*. During the independent evolution of two geographically isolated populations of this species, the *A* mutation arises and spreads to fixation in one population (*AAbb*), and the *B* mutation arises and spreads to fixation in the other population (*aaBB*). In principle, these evolutionary changes could be beneficial mutations favored by natural selection, or they could be effectively neutral changes that spread by genetic drift. After these two mutations become fixed in their respective populations, we have no guarantee that the *A* and *B* mutations will function together properly if brought together in a hybrid genetic background—that is, the *A* and *B* mutations may be incompatible with one another. The critical feature of the model is that the *A* and *B* mutations have never been in the same genetic background before and thus have never been "tested" together by natural selection. If *A* and *B* are incompatible in a way that disrupts embryogenesis or gametogenesis in hybrids, hybrids may be inviable or sterile, respectively.

melanogaster; later *Hmr*^{mel} transgenes were shown to cause lethality when introduced into hybrids carrying the *Hmr*^{rescue} allele, proving that *Hmr*^{mel} kills hybrids. Interestingly, alleles at another locus, called *Lethal hybrid rescue* (*Lhr*), on chromosome 2 of *D. simulans* can also rescue these hybrid males. This raises the intriguing possibility that the wild-type alleles *Hmr*^{mel} and *Lhr*^{sim} are the two epistatically (suppressing the effect of one gene by another) interacting partners in this hybrid incompatibility.

The fourth gene is the *Drosophila* hybrid inviability gene, *Nup96*, which encodes an essential protein that functions at nuclear pore complexes, the massive molecular channels that regulate all protein and RNA traffic between the nucleus and cytoplasm. *Nup96* was discovered by deletion mapping in *D. melanogaster–D. simulans* hybrids (Fig. 4). When the *D. simulans* allele of *Nup96* is uncovered in hybrid males carrying a *D. melanogaster X* chromosome, they die. Two findings show that the incompatible partner for *Nup96*^{sim} is a recessive factor on the *D. melanogaster X*. In one finding, otherwise genotypically identical hybrid males that carry a *D. simulans X* instead of a *D. melanogaster X* do not die (that is, *D. simulans Nup96* is compatible with the *D. simulans X* chromosome, as expected). In the other, *Nup96*^{sim} does not kill hybrid females that carry both *D. melanogaster* and *D. simulans X* chromosomes (that is, whatever it is on the *D. melanogaster X* that kills hybrid males can be masked by the *D. simulans X*).

Few patterns have emerged from our currently small sample of incompatibility genes: two are duplications and two are single-copy genes; two have DNA-binding properties, one is a cell signaling protein, and one is a nuclear pore protein. One pattern,

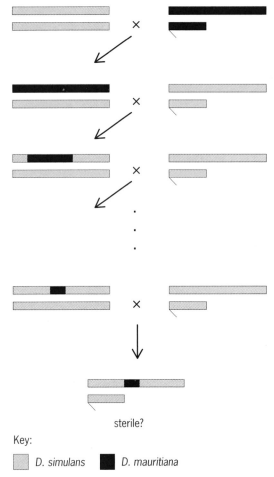

Key:

▨ *D. simulans*   ■ *D. mauritiana*

Fig. 2. Introgression studies introduce small parts of the genome from one species into the genome of a second species by repeated backcrossing through, in this example, fertile females. (Only sex chromosomes are shown: females have two X chromosomes; males have one X and one Y chromosome.) Recombination reduces the size of the foreign genomic fragment with each generation of introgression. After several generations, the introgression can be tested for its ability to cause hybrid male sterility.

sibly the *Xiphophorus* gene as well, this rapid evolution has the hallmark signatures of recurrent positive natural selection on protein coding sequences—amino acid–changing substitutions in these genes have accumulated much faster than can be explained by neutral evolution [for example, all four genes show regions where $K_a/K_s > 1$ (the ratio of nonsynonymous to synonymous substitutions, where a value greater than 1 indicates positive selection pressure)]. This may be the most important finding from the last decade of work in speciation genetics: positive natural selection at the molecular level is a pervasive cause of the functional divergence underlying the origin of species.

Evolution of hybrid incompatibility genes. Although the hybrid incompatibility genes identified so far have evolutionary histories of positive selection, we have not yet pinpointed the precise causes of selection. Speciation geneticists are now focusing on three general explanations. Differing ecologies can cause species to diverge as they adapt to their respective environments. Similarly, sexual selection may drive the rapid evolution of genes involved in gametogenesis (the formation of reproductive cells), thus causing the evolution of hybrid sterility. Finally, all eukaryotic genomes harbor selfish genes (or genomic parasites) that try to manipulate reproduction to enhance their own transmission at the expense of other genes. Meiotic drive elements, for instance, offer just one extreme, but not uncommon, class of selfish genes: Normally, heterozygous individuals transmit alternative alleles to progeny in equal proportions; meiotic drive elements, however, hijack gametogenesis by excluding or killing gametes that carry alternative alleles and thereby monopolize transmission. While these meiotic drive elements enjoy transmission advantages, they also compromise the fertility of their bearers, leading to a genetic conflict between the selfish gene and the rest of the genes in the genome. This sets the stage for an evolutionary arms race in which drive elements evolve transmission advantages and the rest of the genome evolves suppressors. One of the more exciting recent discoveries in speciation genetics is the increasing evidence that exactly these sorts of genetic conflicts,

however, is clear: all four have experienced extraordinarily rapid rates of molecular evolution compared to the other, mostly slowly evolving genes in the genome. For the three *Drosophila* genes, and pos-

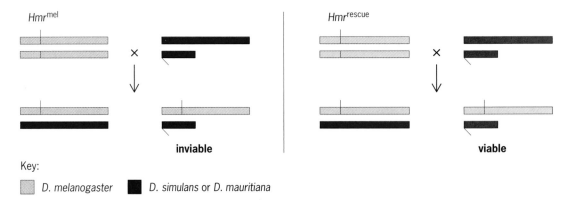

Key:

▨ *D. melanogaster*   ■ *D. simulans* or *D. mauritiana*

Fig. 3. The hybrid rescue mutation, *Hmr*, restores the viability of normally dead hybrid sons from species crosses between *Drosophila melanogaster* females and *D. simulans* (or *D. mauritiana*) males. Since rescue mutations are compatible alleles at otherwise incompatible loci, standard genetic mapping within *D. melanogaster* can localize the gene causing the incompatibility.

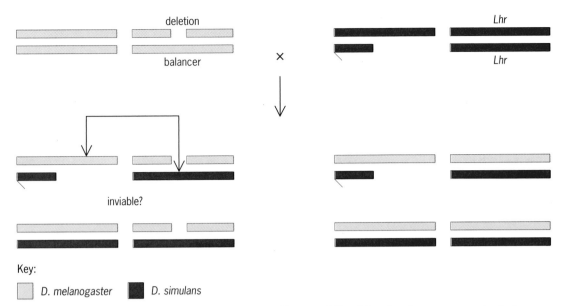

Key:

☐ *D. melanogaster* ■ *D. simulans*

Fig. 4. Deletion mapping can be used to map recessive hybrid incompatibility alleles. (Sex chromosomes and a single representative pair of autosomes are shown.) In species crosses involving heterozygous deletion-bearing *Drosophila melanogaster* females and *Lhr*-bearing *D. simulans* males, four hybrid genotypes are produced: males inheriting the deletion and those not inheriting the deletion (these inherit dominantly marked balancer chromosomes); and females inheriting the deletion and those not inheriting the deletion. If the deletion uncovers a recessive allele from *D. simulans* that is incompatible with the recessive *X*-linked allele from *D. melanogaster*, deletion-bearing hybrid males, but not their siblings, will be inviable.

between genomes and their parasites, cause perpetual molecular divergence at essential genes that, in turn, cause incompatibilities in hybrids.

For background information *see* ALLELE; GENETICS; HYBRID DYSGENESIS; MUTATION; ORGANIC EVOLUTION; SPECIATION; SPECIES CONCEPT in the McGraw-Hill Encyclopedia of Science & Technology.

Daven C. Presgraves

Bibliography. J. A. Coyne and H. A. Orr, *Speciation*, Sinauer Press, Sunderland, MA, 2004; H. A. Orr, J. P. Masly, and D. C. Presgraves, Speciation genes, *Curr. Opin. Genet. Dev.*, 14:675–679, 2004; H. A. Orr and D. C. Presgraves, Speciation by postzygotic isolation: Forces, genes and molecules, *BioEssays*, 22:1085–1094, 2000.

Sphingosine 1-phosphate

Sphingosine 1-phosphate (S1P), a bioactive lipid produced from the degradation of sphingomyelin, activates a family of G protein–coupled receptors to regulate many biological responses [G proteins, also known as GTP (guanosine 5'-triphosphate)-binding proteins, bind GTP and serve as intermediaries in intracellular signaling pathways]. S1P receptors (S1PRs) on vascular endothelial cells are needed for vascular maturation during development, angiogenesis (origin and development of blood vessels), and normal vascular homeostasis. In vascular smooth muscle cells, S1P regulates cell migration, contractility, and proliferation. Recent work has illustrated the functional role of S1P in the immune system; egress of T- and B-cells (thymus- and bone marrow–derived lymphocytes, respectively) from lymphoid organs requires the signaling of S1P1R at the nexus of the immune/vascular interface. Moreover, S1P has functional roles in the digestive, endocrine, reproductive, cardiovascular, pulmonary, and neuronal systems. In addition to the receptor-dependent function of S1P, the metabolic fate of S1P may also play an important role in cell physiology. Knowledge of S1P biology is important since pharmacologic modulation of this system may provide novel opportunities for the therapeutic control of many human diseases.

Formation and role. Sphingomyelin (SM) is an abundant phospholipid present in eukaryotic membranes. Metabolism of SM by the sphingomyelinase (SMase) enzyme results in the formation of sphingolipid metabolites such as ceramide, sphingosine, S1P, and ceramide 1-phosphate. Such metabolites are thought to play important roles in intracellular signal transduction, intracellular metabolism, and extracellular signal transduction. All eukaryotes express SM metabolic enzymes and therefore produce sphingolipid mediators. However, extracellular signaling of sphingolipid mediators appears to have evolved only in vertebrates as receptors for S1P have appeared only in vertebrate genomes.

SMase action results in the formation of ceramide, a hydrophobic sphingolipid associated with cell stress and apoptotic (cell death) responses. Ceramide formation likely occurs in cellular membranes; however, ceramide transfer proteins are thought to transport this lipid mediator between different subcellular compartments. Hydrolysis of ceramide by the ceramidase enzymes results in the formation of sphingosine, which occurs in trace amounts in tissues. This is presumably due to the high activity of sphingosine kinase (Sphk) enzymes, which phosphorylate sphingosine into S1P. S1P is also likely to be formed in cellular membranes. It may be transported from the cytosolic face of the membrane to the extracellular face, and then ultimately chaperoned into the extracellular compartment. In some cases, formation of S1P could also occur in the extracellular face of the cell membrane as the metabolic enzymes, SMase, ceramidase, and Sphk, can be found as secreted ectoenzymes (enzymes located on the external surface of a cell). Many mammalian cells can form S1P and release it into the extracellular milieu. Mammalian blood is highly enriched in S1P, and many hematopoietic (blood-forming) cells as well as vascular cells are capable of secreting this lipid mediator.

Intracellular S1P has multiple fates. It is dephosphorylated by S1P-specific phosphatases and lipid phosphate phosphatases which have a broader substrate specificity. As such, dephosphorylation of S1P would lead to formation of sphingosine, which can be further converted to ceramide. In addition, S1P is cleaved by a specific S1P lyase, resulting in hexadecenal and phosphoethanolamine which are intermediates in the phospholipids biosynthetic pathway.

In the extracellular compartment, S1P is found as a protein-bound bioactive lipid. In human plasma, high density lipoprotein (HDL) and albumin fractions contain most of the S1P. Protein-bound S1P is biologically active; S1P likely exists in equilibrium between the plasma protein compartment, the membrane-bound compartment, and the receptor-bound compartment. The mechanism by which S1P is concentrated in the plasma is not known, but high plasma concentrations of S1P are physiologically important in activating immune and vascular cells. The large concentration of S1P in plasma (0.1–1 μM) may be important for physiological tonic signaling under basal conditions. However, cell activation results in further increases in local S1P production. For example, activation of vascular and immune cells with cytokines (peptides released by some cells that affect the behavior of other cells, serving as intercellular signals) and growth factors, or of platelets with thrombotic mediators, can cause acute activation-dependent production and release of S1P. In such scenarios, S1P may participate in inflammatory and immune responses.

Receptors. The first S1P receptor, S1P1R, to be identified was originally cloned as a transcript induced by phorbol myristic acetate in vascular endothelial cells. It is a prototypical member of 5 G protein–coupled receptors that are widely expressed and are coupled differentially to the G protein subtypes. Coordinated signaling of S1PRs will therefore activate a number of cellular signal transduction pathways and thereby modulate cell behavior. For example, S1P1R is exclusively coupled to the Gi pathway and activates the phosphoinositide 3-kinase (PI3K) pathway, protein kinase Akt, and the small GTPase Rac. In contrast, S1P2R is coupled primarily to the $G_{12/13}$ pathway to activate the small GTPase Rho, even though it is capable of coupling to the G_i and G_q pathways as well. Most cells express one or more S1PRs and therefore respond to S1P. Immediate responses such as calcium transients, second messenger [cyclic adenosine monophosphate (cAMP), inositol 1,4,5-trisphosphate (IP3), and diacylglycerol (DAG)] changes, and phosphorylation changes are induced. In addition, profound changes in the cytoskeleton are induced. For example, activation of Rac and Akt by S1P1R induces microtubule polymerization and formation of cortical actin at the lamellipodia (cytoskeletal actin projections) of cells. In addition, cellular adhesive mechanisms are induced by S1PRs. Cell-cell adhesion mediated by cadherins and cell-matrix adhesion mediated by integrins are induced by these receptors by inside-out (intracellular to extracellular) signaling mechanisms. Ultimately, gene expression changes are induced by transcriptional alterations. Together, these events constitute global changes in cell behavior that is regulated by the function of S1PRs. However, receptor signaling is limited by downregulation of receptors. S1P1R is rapidly internalized by a clathrin-mediated endocytosis mechanism. Such desensitization mechanisms may be important in the physiologic regulation of S1P function.

S1PRs play important roles in the regulation of vascular development and homeostasis. S1P1R is essential for vascular development as knock-out mice that lack this receptor die at midgestation due to defective interaction between endothelial cells and mural cells (contractile cells that surround vascular endothelial cells), which leads to unstable vessels and hemorrhage. S1P1R is also important for tumor angiogenesis as downregulation of expression of this receptor with small interfering RNA (siRNA) leads to reduced angiogenesis and tumor growth rate in vivo. FTY720, an S1PR modulator which acts as a functional antagonist, also inhibits tumor angiogenesis and metastasis in mouse models, suggesting that this receptor may be a novel therapeutic target in cancer. S1P2R inhibits endothelial cell migration and angiogenesis. In addition, it couples to the tumor suppressor PTEN (phosphatase and tensin homolog deleted on chromosome 10) and thereby antagonizes the intracellular signaling of S1P1R. In resistance vessels, S1P2R and S1P3R activate the Rho GTPase pathway and the G_q/PLC (phospholipase C)/calcium pathways, respectively, and induce vasoconstriction. In contrast, activation of S1P3R and S1P1R on endothelial cells activates the endothelial nitric oxide synthase enzyme and stimulates the production of the vasodilator nitric oxide (NO). Thus, S1PRs regulate vascular tone in vivo in various vascular beds, and S1P1R in vascular smooth muscle cells induces migration of endothelial cells. This receptor is overexpressed in pathologic vascular lesions, suggesting its potential role in vascular disease. S1P1R inhibits vascular permeability by regulating the assembly of adherens junctions (anchoring junctions that connect the cytoskeleton of a cell to the extracellular matrix or to the cytoskeleton of surrounding cells) in endothelial cells. Indeed, agonism (that is, activation) of S1P1R in pulmonary vasculature may provide a novel therapeutic strategy to control pulmonary edema in respiratory pathologies.

Effects. In the immune system, S1P is a potent regulator of T- and B-lymphocyte egress from lymphoid organs into the lymph and peripheral circulation. Treatment of the S1PR modulator, FTY720, results in profound and reversible lymphopenia (an abnormally reduced number of lymphocytes in the peripheral blood). FTY720 is phosphorylated by sphingosine kinase-2 into FTY720P, which acts as a potent agonist on S1PR1 on vascular endothelial cells and on lymphocytes. Agonism of S1P1R on the vascular endothelial cells closes the "vascular gate" and thereby inhibits lymphocyte egress. This is due to the action of S1P1R to activate the small GTPase Rac and to induce the assembly of adherens junctions, which are composed of vascular endothelial (VE)–cadherin molecules. FTY720P also downregulates the S1P1R on lymphocytes and acts as a functional antagonist to desensitize the lymphocytes to the chemotactic gradient of S1P. Thus, FTY720 treatment reduces effector T-cell responses in autoimmune disease and in host response to transplanted grafts. It is currently undergoing clinical trials for multiple sclerosis and rejection of kidney transplants. In addition, S1P has multiple profound effects on various immune cells such as T-regulatory cells and dendritic cells (specialized cells of the lymphoid reticuloendothelial system that present antigens for detection by lymphocytes).

S1P also has potent physiologic effects on various organ systems. In the reproductive system, it promotes male and female germ cell survival. In the central nervous system, it promotes survival of oligodendrocytes (glial cells responsible for elaborating myelin in the central nervous system) and regulates neurons. In the intestinal system, S1P may be used as a metabolic intermediate in lipid absorption and transport. In addition, sphingosine appears to act as a tumor suppressor. Since S1P is produced by most cells and S1PRs are expressed ubiquitously, future work will reveal the physiological functions and pathological roles of S1P in various organ systems.

For background information *see* CELL MEMBRANES; CYTOKINE; GLYCOLIPID; LIPID; PHOSPHOLIPID; SPHINGOLIPID in the McGraw-Hill Encyclopedia of Science & Technology.　　　　　　　　　　Timothy Hla

Bibliography. J. Chun et al., International Union of Pharmacology, XXXIV: Lysophospholipid receptor nomenclature, *Pharmacol. Rev.*, 54:265–269, 2002; T. Hla, Genomic insights into mediator lipidomics, *Prostaglandins Other Lipid Mediat.*, 77:197–209, 2005; T. Hla, Physiological and pathological actions of sphingosine 1-phosphate, *Semin. Cell Dev. Biol.*, 15:513–520, 2004; M. Maceyka, S. Milstien, and S. Spiegel, Sphingosine kinases, sphingosine-1-phosphate and sphingolipidomics, *Prostaglandins Other Lipid Mediat.*, 77:15–22, 2005; J. D. Saba and T. Hla, Point-counterpoint of sphingosine 1-phosphate metabolism, *Circ. Res.*, 94:724–734, 2004.

Sport biomechanics

Biomechanics, broadly defined, is the application of the laws of physics in the study of biological organisms, including humans. The general goal of sport biomechanics research is to develop a detailed understanding of specific mechanical sport performance variables to enhance performance and reduce injury incidence. This translates to investigating specific sport skill techniques, designing improved sport equipment and apparel, and identifying practices that are predisposing to injury. Given the increasing sophistication of training and performance at all levels of sport competition, it is little surprise that informed athletes and coaches are turning to the research literature on the biomechanical aspects of their sports for a competitive edge.

Sport shoe design. All major sport shoe companies employ biomechanics researchers to design and test shoes. The design of sport shoes has very definite implications for both injury prevention and performance. However, proper shoe design requires a sound understanding of the requirements of each sport, and sometimes the requirements of a player's position within the sport, in terms of the range of common loading patterns over the shoe and over the foot within the shoe. Forces most likely to cause overuse-type injuries are the vertical ground reaction forces (GRFs) sustained with each footfall, and particularly the peak vertical GRFs. During running, peak vertical GRFs are 3–5 times the body weight. Also at issue is the portion of the shoe/foot sustaining the peak vertical GRFs during sport/exercise performance. In sports involving cutting maneuvers, the amount of friction generated between the shoe sole and playing surface is a concern, as too much friction increases the likelihood of knee injuries. Thus, aerobic dance shoes are constructed to cushion the metatarsal arch in the forefoot. Football shoes for use on artificial turf are designed to rotate easily to minimize the risk of knee injury and running shoes are specialized for training and racing, as well as running on snow and ice. In fact, sport shoes today are so specifically designed for their intended purposes that wearing the wrong shoe can contribute to developing an injury. A further complication is that the ground or playing surface, the shoe, and the human body are collectively an interactive system. The structure of the athlete's foot is also a consideration. Accordingly, there are running shoes designed for those with high arches, those with low arches, and those who tend to pronate (roll the foot inward) during running. Related research has identified injury-related behaviors on the part of the athlete. For example, recreational runners who engage in training errors such as a sudden increase in running distance or intensity, excess cumulative milage, or running on cambered surfaces are at heightened risk for injury.

Building a better figure skate. The extremely popular sport of figure skating dates back to the pre–Civil War period when skaters first started performing jumps and spins on the ice. The sport has been tradition-bound, with the basic design of the figure skate having undergone only minor changes since 1900. The conventional figure skate consists of a rigid boot with a screwed-on steel blade. Since the nineteenth century, the boots have become stiffer to enhance ankle stability. Today, with skaters performing more and more technically demanding jumps, there has been a concomitant increase in overuse injuries. The rigid boot is primarily to blame. With motion at the ankle severely restricted, skaters are forced to land their jumps nearly flat-footed, with a tremendous amount of landing shock transmitted upward through the musculoskeletal system. The incidence of lower-extremity stress fractures among skaters has dramatically increased, and several prominent skaters, including Tara Lipinski and Rudy Galindo, have undergone double hip replacements at young ages.

To address this growing problem, biomechanists J. Richards and D. Bruening at the University of Delaware have designed and tested a new figure skating boot (**Fig. 1**). Following the design of modern-day alpine skiing and in-line skating boots, the new boot has an articulation at the ankle that allows ankle flexion while limiting potentially injurious sideways movement.

Wearing this new boot, skaters have the ability to land toe-first, with the rest of the foot hitting the ice more slowly. This extends the landing time, spreading the impact force over a longer time and diminishing the peak force translated up through the body by about 30% (**Fig. 2**).

Although the new figure skating boot design was designed to reduce the incidence of stress injuries in skating, it may also benefit performance. Allowing skaters to move through a larger range of motion at the ankle may well enable higher jump heights along with the ability to execute more rotations while the skater is in the air. Experienced skaters who try the new boot are finding that using it effectively requires a period of adaptation. Those who have been skating in the traditional boot for many years tend to have reduced strength in the muscles surrounding the ankle

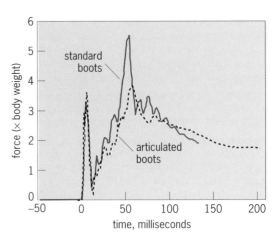

Fig. 2. Force versus landing time curves for standard and articulated figure skating boots.

since the traditional boot does not allow use of these muscles. Improving ankle strength is likely to be necessary for optimal use of a boot that now allows ankle motion. But just as the biomechanically designed speed skate (Klapskate) with a hinge near the toes revolutionized the sport of speed skating during the 1998 Winter Olympics, the articulated figure skating boot is likely to have a large effect on the sport of figure skating.

Enhancing Olympic performance. Many countries, including the United States, sponsor research on the biomechanical, physiological, and psychological aspects of elite performance of Olympic sports. In the United States, this falls under the auspices of the Sports Medicine Division of the United States Olympic Committee (USOC). The general goal of USOC biomechanics research is to investigate the ways in which mechanical factors limit the performances of elite athletes training for Olympic and other international competitions. Typically, these studies are pursued in direct cooperation with the national coaches and advisory boards of the sport to ensure the applicability of the results. USOC-sponsored biomechanics research has been fruitful in producing a wealth of new knowledge about the mechanical aspects of elite performance in various sports. For example, elite-level performances in the long jump, high jump, and pole vault have been found to be characterized by having large horizontal velocity going into takeoff and a shortened last step that facilitates continued elevation of the total-body center of mass. Many important innovations in sport equipment and apparel have also resulted from findings of experiments done in chambers, called wind tunnels, which simulate the air resistance present during particular sports. Examples include the aerodynamic helmets, clothing, and cycle designs used in competitive cycling, and the ultrasmooth body suits worn in swimming, track, skating, and skiing. Wind-tunnel research has also identified optimal body configuration during events such as ski jumping. Because of continuing technological advances in scientific analysis equipment, the role of sport biomechanists in

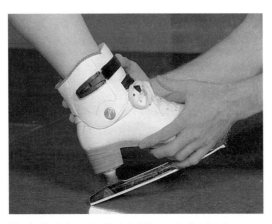

Fig. 1. Redesigned (articulated) figure skating boot.

contributing to improvements in performance at the elite level is likely to be increasingly important in the future.

Golf. The influence of biomechanics is also making its way into sports popularly enjoyed by recreational as well as professional athletes, including golf. Golf is played by 10–20% of the adult population in many countries. Computerized video analyses of golf swings based on software designed by biomechanists are commonly available at golf clubs and equipment shops. The science of biomechanics can play a role in optimizing the distance and accuracy of all golf shots, including putting, through the analysis of body segment angles, joint forces, and muscle-activity patterns. A common recommendation based on biomechanical principles is to maintain a single center of rotation to impart force to the ball. To maximize the distance of a shot, the golfer must strike the ball directly with a high club head velocity. Optimizing club head velocity requires a coordinated, sequential series of motions, including rapidly stretching hip, trunk, and upper arm muscles during the backswing, maximizing trunk rotation early in the downswing, uncocking the wrists when the lead arm is about 30° below the horizontal, and transferring weight from the back foot to the front foot during the downswing/acceleration phase. Research shows that maximizing X-factor, or the difference in the rotational orientation of the upper torso and the rotational orientation of the pelvis at the top of the backswing, contributes to upper-torso rotation velocity and ultimately to ball velocity. To accurately hit chipping and putting shots, the golfer should grip the club lower and shorten the backswing. Consistency in the spatial and temporal aspects of the trunk, shoulder, arm, and wrist movements tend to result in successful golf shots.

Anterior cruciate ligament injuries. Anterior cruciate ligament (ACL) rupture is a serious and relatively common sport injury that has received recent scrutiny from biomechanists. ACL injuries are a growing cause of concern, occurring frequently in sports involving cutting maneuvers such as soccer, football, basketball, and alpine skiing. These noncontact injuries have severe consequences for the athlete, requiring surgical replacement and rehabilitation, and predisposing the individual to early-onset osteoarthritis at the knee.

For unknown reasons, females sustain ACL injuries at a 4–6 times higher rate than do males. Possible contributors to this phenomenon include gender-based differences in hormonal effects on ligament strength and stiffness, neuromuscular control, lower limb biomechanics, ligament strength, and fatigue. Research has also shown that females tend to have less activation of the hamstring muscles and greater activation of the quadriceps muscles and tend to land from a jump with less flexion at the hip and knee as compared to males. These factors all tend to decrease kinetic energy absorption during landing and may increase the ground reaction forces and torques associated with ACL injury.

Strategies recommended for reducing the incidence of ACL injury include programs for improving neuromuscular control at the knee during standing, cutting, jumping, and landing. Such programs include proprioceptive balance-board exercises and plyometric training, as well as teaching avoidance of high-risk movements, positioning, and landing techniques. This type of training has been shown to increase active knee stabilization through altering muscle firing patterns, resulting in decreased landing forces, improved balance, and a concomitant decrease in the incidence of ACL injury among female athletes.

For background information *see* BIOMECHANICS; JOINT (ANATOMY); MUSCULAR SYSTEM; SKELETAL SYSTEM; SPORTS MEDICINE; WIND TUNNEL in the McGraw-Hill Encyclopedia of Science & Technology.
Susan J. Hall

Bibliography. R. Bahr and T. Krosshaug, Understanding injury mechanisms: A key component of preventing injuries in sport, *Brit. J. Sports Med.*, 39:324–329, 2005; S. J. Hall, *Basic Biomechanics*, 5th ed., McGraw-Hill, 2006; V. Zatsiorsky (ed.), *Biomechanics in Sport: Performance Enhancement and Injury Prevention*, Blackwell Science, 2000.

Spyware

There is a dispute in the security industry regarding the definitions of spyware and adware. Spyware collects and transmits information about users and/or organizations without informed consent or knowledge, while adware, according to the Anti-Spyware Coalition, is software with executable applications for delivering advertising content that may be unexpected and unwanted by users.

In reality, each anti-spyware company has its own definition of spyware. These different definitions make it difficult to investigate and prosecute spyware cases. The U.S. Federal Trade Commission urges the business community to agree on a definition of spyware. Experts agree that software secretly installed onto a personal computer (PC) is spyware, but applications bundled with purchased products that are disclosed in the end-users' agreement are not spyware.

There is a fine line between spyware and adware. Many companies use adware to promote products and services. Companies distribute adware via affiliates and software bundlers. Some adware companies may pursue invasive techniques to push adware onto users' computers. To this extent, the behavior of adware is similar to spyware. Legitimate adware informs PC users about the existence of adware, whereas rogue affiliates exploit a browser's vulnerability to install adware on users' PCs. Thus, the same program could be spyware and adware. Even though some adware companies have policies against rogue behavior by affiliates, consumers still need to be cautious in using adware.

When comparing adware with spyware, some define spyware tools as those that steal passwords and personal data, while others consider a program spyware if it installs itself without a user's clear knowledge and permission. Spyware may be legitimate but annoying programs for marketing purpose that users have installed on their PCs unwittingly, or the programs may be destructive such as browser hijacking. However, both spyware and adware consume computer's resources, slow Internet connections, spy on users' computing activities, and even forcibly redirect Web browsers. Usually, adware comes with an uninstaller and can be easily removed from a system, whereas spyware is more persistent and difficult to remove even with spyware removal tools.

The spyware problem is widespread. Consumers can unknowingly download or install password stealers, keystroke recorders, screen capture programs, and numerous additional software applications that surreptitiously monitor and/or transmit their activities. There are many different ways for this to occur. Channels for spyware include checking e-mail, browsing the Web, using instant messenger, and peer-to-peer networks. Activities such as downloading free software, music, or games are sources of spyware.

Spyware problems. According to the National Cyber Security Alliance, more than 80% of computers are infected with spyware, and IDC ranks it as the fourth greatest risk to business network security. Spyware causes problems such as sluggish desktop performance, removal of data/programs, loss of personal information, loss of organizational data/intellectual property, and identity theft. With spyware infections, users experience computer malfunction and redirection to pornographic Web sites, gambling Web sites, or other unwanted Web sites. Enterprises face security, privacy, and productivity risks from spyware infections. In addition, enterprises are concerned with the possible industry espionage problems introduced by spyware.

Regulations. Spyware represents a serious violation of privacy for individuals and organizations because information is collected and sold without consent. Currently, technological developments surpass laws and regulations. Spyware programs have resulted in a number of legal challenges. The Federal Trade Commission suggests that the private sector and the government resolve the spyware issue separately with a common understanding of the definitions of spyware.

State and federal government have taken antispyware issues seriously and are determined to tackle the growing problem. Currently, more than 50% of the states have introduced or passed antispyware laws. California is a pioneer in enacting extensive legislation in the privacy, security, antispam, and antispyware areas. California's legislation profoundly affects national security policy and practices for fighting identity theft, protecting personally identifiable information, and regulating the Internet. In particular, Senate Bill 1436, the Consumer Protection against Computer Spyware Act, outlaws the installation of software on a California consumer's computer without consent. In addition, several federal bills—Spy-ACT, I-Spy ACT, and Spy Block Act—regulate spyware.

Preventive methods. Preventive methods should be taken to reduce the risk of spyware infection. Personal firewalls, antivirus, antispam, and antispyware software must be installed and kept current. Online transactions should be made only on secure sites. The Web address for a secure site starts with "https" instead of "http", and a closed padlock appears in the status bar. In addition, computer users should be diligent at using hard-to-guess passwords and changing them regularly and be wary of e-mail requests to confirm a credit card, bank, or financial account.

Antispyware tools. There are many antispyware tools on the market. To effectively detect, remove, and prevent spyware infection, it is better to install multiple antispyware tools because no single tool is 100% effective. There are many popular free tools for consumers as well as good commercial antispyware software. In addition to the antispyware software available in the market, security vendors are building antispyware features into their products.

According to a survey conducted by Computerworld in 2005, information technology executives reported that enterprise antispyware products are 92% effective in detecting spyware, 84% effective in removing spyware, and 78% effective in preventing spyware. These statistics show that antispyware tools still need improvement. When more vendors develop antispyware tools, spyware writers will surely put more efforts on changing their programs to avoid detection. In order to fight spyware effectively, antispyware vendors need to develop sophisticated tools for detecting and removing spyware. Combating spyware requires technologies, regulations, and consumer education.

Consumer education. Consumers are increasingly frustrated by the growing threats that spyware has created, including serious privacy invasions. Ultimately, protecting privacy should up to the individual, although businesses and governments have their responsibility. Consumers need to protect their identities and take actions against Internet threats, as well as continuously educate themselves on spyware and Internet security.

There are Web sites available to educate consumers about spyware and other malicious software. These organizations include stopbadware.org (http://stopbadware.org/), Anti-Spyware Coalition (http://www.antispywarecoalition.org), Onguard Online (http://onguardonline.gov/index.html), GetNetWise Coalition (http://privacy.getnetwise.org/browsing/). Stopbadware.org was founded by Harvard University and Oxford University to provide reliable, objective, and useful information to help consumers understand downloadable software safety. The Anti-Spyware Coalition is composed of people from industry, academia, and consumers groups dedicated to spyware definitions and best practices. OnGuard Online is maintained by the Federal Trade Commission and works with the

Department of Homeland Security, U.S. Postal Inspection Service, Department of Commerce, Technology Administration, Securities and Exchange Commission, nonprofit organizations, and industry groups to alert the public about Internet fraud and computer security. The GetNetWise coalition is formed of industry and public interest groups to provide safe Internet use information to the public.

Although industry and government have joined forces to combat spyware, the fuzzy line between intrusive spyware and legitimate online-marketing programs presents difficulty in forming laws and regulations in consumer protection. Therefore, consumer education is essential, and users' safe surfing and PC hygiene habits play a critical role in protecting their privacy. If consumers are motivated and diligent to teach themselves about spyware and the methods to combat it, they will be able to effectively protect themselves.

For background information *see* COMPUTER PROGRAMMING; COMPUTER SECURITY; INTERNET; SOFTWARE ENGINEERING; WORLD WIDE WEB in the McGraw-Hill Encyclopedia of Science & Technology.

Xiaoni Zhang

Bibliography. B. L. Delaney et al., California privacy and security legislation affects entire nation, *Intellectual Prop. Technol. Law J.*, 17(3):21–24, 2005; L. Hatlestad, Spyware: From bad to worse, *VARBusiness*, 21(21):46, 2005; G. Morgan, In assessing risk, don't ignore spyware threat, *Amer. Banker*, 171(25):2, 2006; Spyware rising, *Computerworld*, 39(44):26–26, 2005; X. Zhang, What do consumers really know about spyware?, *Commun. ACM*, 48(3):111–114, 2005.

Stable isotope geochemistry

The origin and fate of elements and their compounds (such as in fluids, metals, nutrients, organics, gases, and pollutants) in planetary, Earth, and environmental sciences is most effectively traced using stable-isotope geochemistry. Isotopes of an element are atoms having the same number of protons but different numbers of neutrons. All but 35 elements have more than one stable isotope that can be used to trace the element through various natural systems. In contrast to unstable (radioactive) isotopes or isotopes produced from the decay of another element (radiogenic), stable-isotope geochemistry uses isotopes whose abundances do not change with time. For example, oxygen has three stable isotopes, each having eight protons, but 99.763% of oxygen atoms have eight neutrons (^{16}O, where 16 is the atomic number and refers to the sum of protons plus neutrons), while 0.0375% have 9 (^{17}O) and 0.1995% have 10 (^{18}O) [see **table**]. The three isotopes of oxygen share the same general chemical properties, but differ in mass and therefore form bonds with slightly different energies, which results in differential partitioning of the light and heavy isotopes of oxygen among various compounds.

Isotopes are measured using mass spectrometers that ionize and separate the isotopes on the basis of their differing masses. As it is easier and more precise to determine isotope ratios than to measure the absolute abundance of a single isotope, isotopic compositions are normally reported as the difference in the ratio of the less abundant heavier-mass isotope to the more abundant lighter isotope (for example, the ratio $^{18}O/^{16}O$), relative to this same isotope ratio in an international standard (for hydrogen and oxygen, this standard is ocean water). These differences are expressed in units of parts per thousand (per mil) and reported as delta (δ) values (see table).

Differential partitioning of the isotopes of an element among various phases depends primarily on four factors: (1) percent difference in mass of the isotopes, such that light elements like hydrogen, carbon, nitrogen, oxygen, and sulfur have the largest relative mass differences in their isotopes; (2) temperature, with low temperatures resulting in greater differential partitioning between phases, and isotope ratios reflecting the temperature at which the phases "equilibrated;" (3) kinetics, or the rate of reaction, with the lowest-mass isotopes of an element reacting faster than the heavier isotope; and (4) bonding environment, with the strongest bonds of an element preferentially incorporating the heaviest isotope, resulting in enrichment of the heavy isotope in more oxidized forms of elements (such as carbon in

Stable isotopes of the light elements: abundances and terrestrial variation of stable isotopes used in traditional stable-isotope geochemistry					
Element	Isotope	Number of protons	Number of neutrons	Relative abundance %	Terrestrial variation per mil
Hydrogen	^{1}H	1	0	99.984	$^{2}H/^{1}H = 700$
	^{2}H	1	1	0.0156	
Carbon	^{12}C	6	6	98.89	$^{13}C/^{12}C = 100$
	^{13}C	6	7	1.11	
Nitrogen	^{14}N	7	7	99.64	$^{15}N/^{14}N = 50$
	^{15}N	7	8	0.36	
Oxygen	^{16}O	8	8	99.763	$^{18}O/^{16}O = 100$
	^{17}O	8	9	0.0375	
	^{18}O	8	10	0.1995	
Sulfur	^{32}S	16	16	95.02	$^{34}S/^{32}S = 100$
	^{33}S	16	17	0.75	
	^{34}S	16	18	4.21	
	^{36}S	16	20	0.02	

carbon dioxide, sulfur in sulfate, and hydrogen in water) relative to coexisting reduced forms (such as organic carbon, sulfide minerals, and methane) and in solid phases relative to liquid or gaseous phases. Given that hydrogen, carbon, nitrogen, oxygen, and sulfur are common light elements in most natural materials (including those in the biosphere), and that they tend to occur in a variety of bonding environments, their isotope ratios vary significantly and are most useful as tracers (see table).

Extraterrestrial materials. The widest ranges in isotope ratios of light elements are recorded in meteorites, particularly primitive meteorites. Stable isotopes in some portions of meteorites preserve the isotopic compositions of presolar materials that did not completely mix prior to the formation of our solar system. Isotopic effects from processes that occurred during condensation and accretion also result in variable isotopic compositions.

Hydrogen, the most common element in the universe, shows extreme variations in a multitude of phases in primitive meteorites, including hydrous phases and organic material. These variations are interpreted to reflect variations in the hydrogen isotopic composition of primitive materials in the solar nebula. Probably the most significant isotopic heterogeneities in meteorites occur with oxygen, where isotopically "normal" oxygen appears to have mixed with an exotic component aberrantly rich in the lightest isotope, ^{16}O. This component could have been ^{16}O-rich dust formed from stellar nucleosynthetic processes and carried into the early solar system, or it could reflect a local chemical process. Ratios of $^{17}O/^{16}O$ and $^{18}O/^{16}O$ are so distinct that they can be used to classify many meteorites.

Isotopic heterogeneities are recorded on every scale in our solar system, occurring on the micrometer scale in some primitive meteorites and on the scale of the solar system by isotopic differences between the planets. Relative to the Earth, Mars has atmospheric water that is enriched in the heavy isotope of hydrogen because of gravitational loss of isotopically light water. These are based on direct measurements of the atmosphere on Mars, but much of what we assume about the isotopic composition comes from analyses of meteorites that are believed to originate from Mars. If indeed these are pieces of Mars, oxygen isotopes originated from mixtures of components distinct from those that condensed to form the Earth and the Moon.

Interactions between hydrosphere and geosphere. The hydrogen and oxygen isotopic composition of water in various reservoirs on the Earth are distinct, so that interactions of the hydrosphere with the geosphere and biosphere can be traced with isotopes. The ultimate origin of water in the meteoric cycle is through evaporation of ocean waters, with the water vapor that forms being depleted in the heavy isotopes of hydrogen and oxygen. As this water vapor accumulates and cools over the continents, the dominant control on the isotopic composition of precipitation is temperature, so that spatial variations over the globe are predictable. Precipitation in high latitudes is more depleted in the heavy isotopes of hydrogen and oxygen relative to lower latitudes, and the relationship between hydrogen isotopes and oxygen in meteoric water is constant.

The relationship between precipitation and temperature is a cornerstone in climate research. Our understanding of global climate change over the past million years is deduced primarily from variations in the hydrogen and oxygen isotope ratios of ice cores from Greenland and Antarctica. Similarly, the change in oxygen-isotope ratios in carbonate shells precipitated by marine organisms has been used to infer general cooling of the oceans over the past 60 million years, with distinct warming and cooling events at specific times related to changes in plate tectonics, ocean circulation, and global glaciations.

Stable isotopes of hydrogen and oxygen have played a strategic role in developing models for the formation of ore deposits, and helped shape exploration strategies. Not only do isotope ratios of ore and alteration minerals provide temperature estimates of various stages in the formation of a deposit, but they also record the isotopic composition of the fluids involved. For example, the isotopic composition of water expelled from most magmas is relatively invariant and distinct from most ground waters, although both played a role in generating many major copper deposits. In contrast, isotope ratios of fluids associated with uranium, lead-zinc, or copper deposits in ancient sedimentary basins are derived primarily from seawater modified by interactions with rock units in the basins during burial.

Interactions between biosphere and geosphere. Stable isotopes are ideal for tracing the flow of elements between the geosphere and biosphere. As elements move from inorganic reservoirs in the geosphere into organisms, there is discrimination against the heavy isotopes because the light isotopes react faster. Hence, organic matter tends to be depleted in ^{13}C relative to atmospheric carbon dioxide or marine carbonates that precipitated from waters in equilibrium with the atmosphere (see **illustration**).

Variation in the carbon isotope ratio of various reservoirs on the surface of the Earth.

Photosynthesis results in organic matter that is significantly depleted in ^{13}C relative to atmospheric carbon dioxide (see illustration). Different photosynthetic pathways produce different degrees of depletion, with most trees and shrubs (C3 plants) being more depleted in ^{13}C than grasses and maize (C4 plants). Organisms that eat plants and soil carbonate produced from the decay of this plant matter inherit this carbon isotopic composition. Abrupt increases in the carbon isotope ratios of soil carbonate and mammalian teeth 6–8 million years ago record the global spread of C4 ecosystems. The global cycle of carbon, particularly in the oceans, has been elegantly elucidated through temporal analysis of atmospheric carbon dioxide, which has been increasingly ^{13}C-depleted during the past 200 years through the burning of petroleum and coal depleted in ^{13}C relative to preindustrial atmospheric carbon dioxide (see illustration).

Other isotope systems provide additional understanding of global cycles. Nitrogen isotope ratios record food-chain information in animals and the influence of marine food sources brought into freshwater systems by salmon. Hydrogen isotope ratios in bird feathers have been used to record the environments where molting has occurred and carbon isotopes reflect habitat quality. The isotopic composition of light elements in the hair and teeth of animals can indicate where the organism consumed water and what it has eaten as part of forensic reconstructions.

Outlook. Use of isotopes in the fields of Earth and environmental science has increased significantly during the past 10 years. Greater numbers of research disciplines are embracing the recent technological advancements in isotope analysis, including of use of more nontraditional stable isotopes such as calcium and iron, among others. The benefits of integrating isotopes into pure and applied research programs are boundless, given their unrivaled ability to provide quantitative evidence of the origins, transformations, and dispersion of elements and compounds.

For background information *see* COSMOCHEMISTRY; GEOCHEMISTRY; GEOLOGIC THERMOMETRY; ISOTOPE; MASS SPECTROMETRY; MASS SPECTROSCOPE; METEORITE; PALEOCLIMATOLOGY; PHYSIOLOGICAL ECOLOGY (PLANT) in the McGraw-Hill Encyclopedia of Science & Technology.   Kurt Kyser

Bibliography. J. Hoefs, *Stable Isotope Geochemistry*, 5th ed., Springer-Verlag, Berlin, 2004; C. M. Johnson, B. L. Beard, and F. Alberede (eds.), *Geochemistry of Non-traditional Stable Isotopes*, Reviews in Mineralogy & Geochemistry, vol. 55, Mineralogical Society of America, 2004; T. K. Kyser (ed), *Stable Isotopes in Low Temperature Fluids*, Mineralogical Association of Canada Short Course Series, vol. 13, 1987; J. W. Valley and D. R. Cole (eds.), *Stable Isotope Geochemistry*, Reviews in Mineralogy & Geochemistry, vol. 43, Mineralogical Society of America, Geochemical Society, 2001; Z. Sharp, *Principles of Stable Isotope Geochemistry*, Prentice Hall, 2006.

Step-out technologies (chemical engineering)

The manufacture of commodity and specialty chemicals forms a crucial part of the foundation of the global economy. Manufactured and refined chemicals are the raw materials for many other goods, including everything from integrated circuits to textiles, to foodstuffs, to packaging. A few, such as gasoline, are used directly by consumers. The chemical industry is the second largest industrial consumer of energy in the United States, using 25% of all industrial energy and 7% of all domestic energy, according to the Energy Information Administration.

Most technologies used today in chemical and petrochemical industries are decades old. Most oil refining processes, for example, were developed between 1940 and 1960. Since then, progress has been evolutionary rather than revolutionary. Engineering science related to these processes formed the core of much research in chemical industries and academia for most of the twentieth century and shaped the chemical engineering curriculum. Today, the chemical industries are generally considered mature, and new advances are less common. Incremental improvements continue, but are not likely to lead to a dramatic competitive edge.

Yet, much scope remains for radically different, step-out technologies that offer the potential for significant savings in energy and manufacturing costs. The economic benefits of new technologies are the single greatest driving force for change in the chemical industries. Environmental benefits, in the form of reduced pollution, and improved process safety are others.

Step-out technologies in the chemical industry most often take the form of new processes, as opposed to new products. They yield a different or more efficient method for producing a compound or energy, resulting in savings in capital, energy, or raw material costs.

Despite the large potential benefits from applying new technologies, the chemical industry will continue to be very slow to adopt radical new ideas and processes. The hesitation is due primarily to the large financial risks that must be undertaken to commercialize an unproven technology. Given this situation, it is sensible to begin the effort to develop any step-out technology by first focusing on its potential attractiveness to the chemical industry. Some questions to consider are:

1. Will the new technology expand the economically available raw material base?

2. Will it mitigate an environmental risk?

3. Will it lower the capital and/or operating cost of the plant?

4. Will it lead to the same product(s) from less expensive raw materials?

Expanding the economically available raw material base. Few areas of the chemical industry have a greater impact on society than the production of fuels. The use of clean fuels leads to significant reductions in pollution by small-source emitters, such as automobiles. Gas-to-liquids (GTL) technology converts natural gas into transportation-grade fuel using the steam reforming and Fischer-Tropsch reactions. The GTL product is essentially free of sulfur and aromatic compounds, resulting in less particulate and sulfur oxide (SO_x) emissions. As a result of the complexity of the reactors resulting from extreme heat-transfer requirements and the expense of the syngas (a mixture of CO and H_2) plant, GTL is economically viable only for very large plants, producing in excess of 20,000 barrels per day (BPD) of liquid products; hence, it is suitable for only a small number of very large gas fields around the world. Step-out technologies that can lower the economically viable GTL plant size by an order of magnitude will enable conversion of "stranded" natural gas into clean fuels and products. An economically viable 2000-BPD GTL plant provides the opportunity to monetize stranded gas in more than half of the known gas fields in the world, containing more than 95% of the known gas.

A promising step-out GTL technologies is being developed by Gas Reaction Technologies. In this process, methane is first brominated to enhance its reactivity, followed by a reaction on a solid "cataloreactant," leading to a liquid hydrocarbon product.

Mitigating an environmental risk. Solid-acid-catalyzed isoparaffin alkylation is another technology that has been promising for many years but is only now starting to see some progress toward commercialization. Alkylation produces a clean, high-quality blending component for gasoline. In alkylation, isobutane and butenes are reacted to form branched octanes. Currently, alkylation is done using liquid-acid catalysts, either hydrofluoric or sulfuric acid. Much research has gone in to replacing these potentially dangerous liquid acids with benign solid acids, but the attempts have been stymied by rapid catalyst deactivation through coke formation. Three solutions to the problem have recently been proposed.

UOP is offering a technology based on a slurry reactor to reduce olefin concentration in the reactor and therefore reduce coking. The process requires the solid catalyst to be continuously moved from the slurry reactor to a regenerator, similar to catalytic reforming reactors. Albemarle, ABB Lummus, and Fortum offer a swing-bed system with multiple reactors to permit one reactor to operate while the others are regenerated. Exelus has taken a different approach and has developed a long-life catalyst that permits using two fixed-bed reactors. Although none of these processes have yet been put into full commercial operation, they promise to replace the use of dangerous liquid-acid systems in the near future.

Lowering plant capital and/or operating cost. The current need for reduced carbon dioxide (CO_2) emissions has led to the development of numerous technologies with environmental benefits. The ability to capture and sequester CO_2 during combustion of fuels or after steam reforming, in which H_2 and CO_2 are produced by the water-gas shift reaction, promises to reduce greenhouse gas emissions from power plants and other sources. The ability to remove CO_2 also has reaction engineering benefits in hydrogen production. In the steam-reforming of hydrocarbons, such as natural gas, the reaction must be run at about 900°C (1652°F) in order to shift the equilibrium in favor of hydrogen. The high reaction temperature leads to metallurgical issues (and hence expensive materials of construction) and high energy consumption. Removing one of the products (CO_2) reduces the equilibrium barrier and permits the reaction to be run at much lower temperatures (and hence less expensive materials of construction). Sixty percent of the cost of a GTL plant is associated with steam reforming of the natural gas; lowering the temperature of this reaction will have a huge impact on the economics.

L. S. Fan and coworkers at The Ohio State University have engineered a porous CaO-based sorbent for CO_2 and sulfur dioxide (SO_2). The material functions by chemically reacting with the acidic gases, creating calcium carbonate or sulfate. Once isolated, the material can be regenerated, releasing a sequestration-ready stream of pure CO_2. This material is being demonstrated for use in power plants to reduce CO_2 and SO_2 emissions, and in hydrogen production. Air Products and Chemicals has taken a similar approach using a proprietary sorbent for CO_2 removal during steam-reforming of natural gas in their sorption-enhanced reaction process. By shifting the equilibrium in favor of hydrogen, the cost of its production can be greatly reduced.

Products from less expensive raw materials. This consideration is nicely illustrated by the synthesis of styrene, which is an important monomer used in a variety of plastic products. In terms of monomer production rate, styrene ranks fourth in the United States, behind ethylene, vinyl chloride, and propylene. Global styrene demand was estimated at 25 million metric tons in 2005. Ninety-eight percent of industrial processes used for the production of styrene are based essentially on a mature, well-proven two-step process: the alkylation of benzene ($940/metric ton) with ethylene ($1100/metric ton) to produce ethylbenzene, and the subsequent dehydrogenation of ethylbenzene to styrene. This process is already operating at overall process selectivities in excess of 96 mol %, and there is very little room for major improvements.

An alternative route to styrene monomer production involves the side-chain alkylation of toluene ($650/metric ton) with methanol ($330/metric ton). Assuming a selectivity of only 80% for the alternate synthesis route, a mid-size styrene plant (250,000 metric tons/year) operator stands to save over $80 million a year in raw material costs alone through a new step-out process based on toluene and methanol.

Side-chain alkylation of toluene with either methanol or formaldehyde has been investigated for three decades by researchers in both industry and academia. However, low yields of styrene due to the high methanol decomposition rates have prevented this new route from being an economically viable alternative to styrene production. Recently, Exelus engineered a new catalyst that affords over 80% selectivity, which opens the pathway for a new breakthrough process that can displace the current technology.

Other examples. A number of step-out technologies have recently emerged that could significantly alter the way some chemicals are made. Two new unit operations of note are microreactors and ceramic membrane–based separations. Microreactors use many small channels to move, mix, and react fluid reagents. Velocys has developed several microscale technologies for reactions and separations, with potential benefits, including tight control of temperature and residence time. Industrial-scale quantities of products are made by using massively parallel reactors.

Air Products continues to develop their ITM (ion transport membrane) technology. This technology uses oxygen anion–conducting ceramic membranes to generate a pure stream of oxygen from air. If successful, this technology could reduce the economically viable scale and cost of oxygen plants.

Several new synthesis routes have made the transition to commercial reality recently. Olefin metathesis reactions, in which the groups attached to the double bonds of two molecules are exchanged to make two new molecules, have exploded due to the commercial availability of water-stable Grubbs (ruthenium-based) catalysts. These new homogeneous catalysts are being used to make higher olefins, novel polymers, and biological molecules. Hydrogen peroxide-based routes to propylene oxide are being implemented by several major companies. This new synthesis route replaces the chlorohydrin process, which uses chlorine gas. The new route offers a cleaner, less dangerous alternative to the conventional process. *See* NOBEL PRIZES.

Outlook. Step-out technologies can offer substantial economic, efficiency, and environmental gains over conventional technologies. To maximize the impact of a new technology, research should be focused on areas where the most significant gains can be made. Large gains are possible even in mature technologies if the factors that contribute the most to the cost of a technology can be altered.

Unless the new technology has the potential to differentiate itself from the current ones in an unarguably favorable way, there is little chance of it being used commercially by chemical industries. When developing step-out technologies, even if one or more of the drivers appears to be attractive, many new technologies will prove unworkable. But those that do succeed can become a standard for decades.

It is refreshing to see that in recent years the discovery of step-out technologies for the chemical industry is beginning to get the attention it deserves, both in industries and in the academia.

For background information *see* ALKYLATION (PETROLEUM); CARBON DIOXIDE; CHEMICAL ENGI-

NEERING; FISCHER-TROPSCH PROCESS; REFORMING PROCESSES; STYRENE; SYNTHETIC FUEL in the McGraw-Hill Encyclopedia of Science & Technology.

James P. Nehlsen; Mitrajit Mukherjee; Sankaran Sundaresan

Bibliography. M. A. Agee, Economic gas to liquids technologies: A new paradigm for the energy industry, *Montreaux Energy Roundtable VIII*, Montreaux, Switzerland, May 12–14, 1997; Air Products and Chemicals, Inc., ITM oxygen for gasification, *Gasification Technologies 2004*, Washington, DC, October 3–6, 2004; B. T. Carvill et al., Sorption enhanced reaction process, *AICHE J.*, 42:2765, 1996; B. Halford, Olefin metathesis, *Chem. Eng. News*, 84(10):18, 2006; M. V. Iyer et al., Multicyclic study on the simultaneous carbonation and sulfation of high reactivity CaO, *Ind. Eng. Chem. Res.*, 43:3939, 2004.

Stress and depression

In 1992, G. P. Chrousos and P. W. Gold defined stress as a state of threatened homeostasis, or instability to which a person (or animal) reacts with an adaptive response to preserve its internal equilibrium. Mild, brief, and controllable challenges could be perceived as pleasant or exciting and could be a positive input for emotional and intellectual development. However, as defined by H. Selye in 1976, more intense, persistent, and uncontrollable situations of threat or perceived threat may lead to maladaptive responses. In humans, these maladaptive responses include the development of mood disorders.

The link between stress and depression has long been observed, where chronic exposure to stressful life events has been associated with the development of depressive symptoms in certain individuals, under certain conditions. Depression is often preceded by stressors, or stressful situations. However, the onset of depression has been shown to depend not only on the characteristics of stressful life events but also the psychological and biological resources of each individual to cope with them (that is, an interaction between stress and individual vulnerability).

Hormonal involvement. A key regulator of the stress response is the hypothalamic-pituitary-adrenal (HPA) axis (see **illustration**). In response to stress, inputs from the central and peripheral nervous system signal the paraventricular nucleus (PVN) of the hypothalamus to increase the synthesis and release of corticotropin-releasing hormone (CRH). CRH in turn increases the synthesis and release of adrenocorticotropic hormone (ACTH) from the anterior pituitary. Arginine vasopressin (AVP), which is synthesized within the paraventricular nucleus, is also an important regulator of ACTH, and acts synergistically with CRH to enhance ACTH release. Peripherally released ACTH stimulates secretion of glucocorticoids (cortisol). Glucocorticoids in turn negatively feed back to reduce the synthesis and release of CRH and ACTH, and also feed back at higher brain centers (specifically the amygdala and hippocampus)

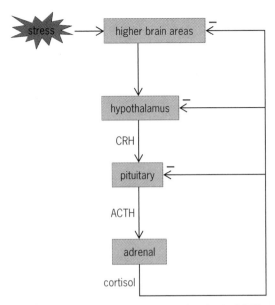

Diagram of the hypothalamic-pituitary-adrenal axis. ACTH, adrenocorticotropin; CRH, corticotropin-releasing hormone.

to modulate the neural inputs to the hypothalamus. This finely tuned system allows for a rapid response to stress and a rapid return to baseline.

Many findings have been reported indicating that the hypothalamic-pituitary-adrenal system is hyperactive in depression. Approximately 50–60% of patients with major depression show distinct baseline changes in hypothalamic-pituitary-adrenal axis activity. These changes include increased levels of circulating ACTH, increased urinary cortisol secretion, increased levels of CRH in cerebrospinal fluid, and an increased number of CRH-secreting neurons in the hypothalamus. Both increased CRH drive and decreased sensitivity to negative feedback have been hypothesized as underlying pathophysiology contributing to these abnormalities.

HPA axis hyperactivity, with consequent hypercortisolism (overproduction of cortisol), observed in patients with major depression, represents one of the most consistent findings in biological psychiatry. In general, depressive patients carry higher levels of circulating cortisol. It has been proposed that increased levels of cortisol are involved in the characteristic mood changes observed in depression. Normalization of circulating cortisol levels in depressed patients was correlated with successful clinical treatment of depression. Furthermore, hypercortisolemic depressed patients treated with antiglucocorticoid interventions experience alleviation in their depressive symptomatology.

Elevated stress hormones, particularly elevated glucocorticoids, lead to changes in the function and morphology of the brain. Based on initial studies in rodents and later in primates, R. M. Sapolsky proposed that severe and prolonged hypercortisolism may be neurotoxic, and in the case of depressive illness there is a high degree of lifetime recurrence of major depression, which suggests that an underlying pathophysiologic process is involved. Animal models have shown that repeated stress causes damage and/or leads to atrophy of the hippocampus, prefrontal cortex, and amygdala, all of which are brain areas associated in the pathophysiology of mood and anxiety disorders.

Stressful life events. Epidemiological studies have revealed that stress or emotional trauma, particularly when experienced early in life, is associated with increased risk to develop depression. The most prominent types of early-life stress in humans are sexual, physical, and emotional maltreatment, as well as parental loss. Other forms of early life stress include accidents, chronic illness, and natural disasters. The marked effects of early-life stressors on health and adaptation throughout the life span are believed to be mediated by the plasticity of the developing brain as a function of experience.

During critical developmental periods, certain brain regions are also particularly sensitive to adverse experiences, which may then lead to major, sometimes irreversible abnormalities. For instance, long-lasting hyperreactivity of the corticotropin-releasing hormone neurons, resulting in increased stress responsiveness and reflecting a glucocorticoid-resistant state, is commonly seen in depressed individuals. Aversive experiences, both in utero and in the neonatal period, result in sustained hypothalamic-pituitary-adrenal axis activation and in sensitization of emotional responses to subsequent stress. Maternal stress beginning at infancy and subsequent stress during childhood are accompanied by a sensitization of the child's hypothalamic-pituitary-adrenal axis response to subsequent stress exposure. Therefore, it is possible that stress or emotional trauma during development permanently shapes brain regions that mediate stress and emotion, leading to altered emotional processing and heightened responsiveness to stress, which in the genetically vulnerable individual may evolve into depression.

Preclinical studies have provided direct evidence that early-life stress leads to heightened stress response and alterations in neural circuits that persist into adulthood. P. M. Plotsky and M. J. Meany reported that adult rats that were separated from their mothers for about 3 hours/day within the first 2 weeks of their life exhibit up to a threefold increase in ACTH and glucocorticoid response to psychological stress when compared to control rats. These rats also revealed reduced feedback sensitivity to the hypothalamic-pituitary-adrenal axis, as well as multiple changes within the central nervous system that likely underlie physiological and behavioral sensitization to stress. Similar to the rodent model, a number of studies have evaluated the effects of prolonged deprivation or repeated separations of infant nonhuman primates from their mothers or peers. When tested as adults, nonhuman primates exposed to prolonged periods of maternal separation or social deprivation exhibit increased fearfulness, anxiety, social dysfunction, and aggression, as well as altered neurochemical and autonomic function. In rodents and nonhuman primates, early-life stress induces numerous changes in multiple neurocircuits that are involved in neuroendocrine, autonomic, and behavioral responses to stress. If similar changes also occur in humans exposed to early life stress, these changes may result in an enhanced risk to depression and anxiety disorders.

Sex differences in depression. Data from the National Comorbidity Survey, a population-based epidemiological study, show that the lifetime prevalence of a major depressive disorder is 21.3% in women and 12.7% in men. This sex gap begins in early adolescence and continues through the mid-50s, approximating the span of the childbearing years in women.

The sex-dependent gap in depression prevalence has led many to hypothesize that reproductive hormone status may, in part, be responsible for the increased vulnerability to depression/depressive symptoms experienced by women. The prevalence of major depression increases during the reproductive years, especially during times when sex hormone levels show rapid fluctuations, such as premenstrual periods, postpartum periods, and perimenopause. Gender also affects the response of the stress system. Women secrete more basal cortisol than men, and in response to the dexamethasone-CRH test women also have increased cortisol secretion in comparison to men. Ovarian steroids have been found to increase hypothalamic-pituitary-adrenal axis activity by increasing CRH production and downregulating glucocorticoid receptors in the hypothalamus, anterior pituitary, and hippocampus (brain areas responsible for inhibiting the hypothalamic-pituitary-adrenal axis response).

Several researchers have posited a second hypothesis of why women may be more susceptible than men to developing depression. Specifically, women are significantly more likely than men to report a stressful life event in the 6 months prior to the onset of a major depressive episode. Women may be especially vulnerable because they are disproportionately subject to certain kinds of severe stress, especially child sexual abuse, adult sexual assaults, and domestic violence. However twin studies performed by K. S. Kendler reported many similarities in the stress exposure of fraternal male-female twin pairs that went on to develop depression. Specifically the degree to which stress raised the risk of depression and the kinds of stress that tended to provoke depression were similar. More interestingly, Kendler found greater heritability of depression in women of male-female twins reared in the same environment, and the genes influencing the development of depression were not the same in men and women. Therefore, there appears to be an impact of gender on the genetics that influence the development of depression.

Recent findings and future directions. More recently, a functional polymorphism in the promoter region of the serotonin transporter (5-HTT) gene was found to moderate the influence of stressful life events on depression. A decrease of serotonin metabolism has been ascertained in the brain of a subgroup of depressed individuals, a phenomenon that may be associated with sustained stress. The serotonin transporter has received particular attention because it is involved in the reuptake of serotonin at brain synapses. Recent reports by A. Caspi and colleagues and Kendler and colleagues found that individuals with one or two "short" alleles at the serotonin transporter locus were more sensitive to the depressogenic effects of stressful life events than those with two "long" alleles. The short allele is associated with lower transcriptional efficacy of the promoter compared with the long allele. The serotonin system provides a logical source of candidate genes for depression, because this system is the target of selective reuptake-inhibitor drugs that are effective in treating depression.

To advance our understanding of the role of stress in mood disorders and to aid in the development of new treatment strategies, a better understanding of the mechanisms involved is necessary. Uncovering more specific genes involved in the stress response can improve our understanding of the pathophysiology of depression and lead to new treatments; or in light of recent findings regarding the 5-HTT gene, better use of current drug treatments [that is, selective serotonin reuptake inhibitors (SSRIs)]. In addition, a better understanding of how alterations of the hypothalamic-pituitary-adrenal axis lead to mood and anxiety disorders and their downstream effects (for example, cognitive impairment) could provide another treatment avenue to pursue. It is possible that pharmacological treatment, leading to the normalization of the hypothalamic-pituitary-adrenal axis, in conjunction with traditional antidepressants may be useful in treating a subpopulation of depression.

For background information *see* AFFECTIVE DISORDERS; ENDOCRINE MECHANISMS; HORMONE; NORADRENERGIC SYSTEM; PITUITARY GLAND; POST-TRAUMATIC STRESS DISORDER; SEROTONIN; STRESS (PSYCHOLOGY) in the McGraw-Hill Encyclopedia of Science & Technology.   Rose C. Mantella; Eric J. Lenze; Charles F. Reynolds III

Bibliography. G. P. Chrousos and P. W. Gold, The concepts of stress and stress system disorders: Overview of physical and behavioral homeostasis, *JAMA*, 267:1244–1252, 1992; G. Fink, *The Encyclopedia of Stress*, Academic Press, San Diego, 2000; P. M. Plotsky and M. J. Meany, Early, postnatal experience alters hypothalamic corticotrophin-releasing factor (CRF) mRNA, median eminence CRF content and stress induced release in adult rats, *Mol. Brain Res.*, 18:195–200, 1993; J. K. Rilling et al., Neural correlates of maternal separation in rhesus monkeys, *Biol. Psychiat.*, 49:146–157, 2001; R. M. Sapolsky, Stress, glucocorticoids, and damage to the nervous system: The current state of confusion, *Stress*, 1:1–19, 1996; H. Selye, *The Stress of Life*, McGraw-Hill, New York, 1976; G. E. Tafet and R. Bernardini, Psychoneuroendocrinological links between chronic stress and depression, *Prog. Neuro-Psychopharm. Biol. Psychiat.*, 27:893–903, 2003; H. M. Van Praag, Can stress cause depression?, *Prog. Neuro-Psychopharm. Biol. Psychiat.* 28:891–907, 2004.

Thalidomide

Thalidomide is the most infamous drug in the history of the world. It was formulated in the mid-1950s by Chemie Grünenthal, a West German cosmetic subsidiary, in search of new antibiotic drugs. Thalidomide had no antibiotic properties but, as no lethal dose could be found in rats, the company decided to test it in humans, where it was discovered to have a potent hypnotic effect. The drug was first marketed as a sleeping pill on October 1, 1957, under the trade name Contergan. Six months later Distillers Company Ltd. of Great Britain began to sell the drug under their trade name Distaval. The drug was marketed over the counter and was heavily advertised as the only completely safe sleeping pill. By summer 1960 thalidomide sales reached an all-time high, outselling the second-leading sleeping pill in Europe by 5:1.

Problems with use. In October 1958, Grünenthal signed a contract with Richardson Merrell to market the drug in the United States. In May 1959, Merrell began "clinical trials": 2.5 million tablets were distributed to 20,000 patients, but few records were kept as to who received the drug, let alone the outcome. On September 8, 1960, Merrell submitted an application to the Food and Drug Administration (FDA). The application was assigned to a new medical officer, Frances Kelsey. Right away Kelsey saw some problems with the drug. What had been good news to others, Kelsey saw differently. If no lethal dose could be found in rats, how could Merrell be certain that it was even being absorbed by the animals. Perhaps the drug was passing right through the rat digestive tract, and therefore the rat was not a good model for testing the drug. Merrell had no data on absorption, so the application was returned as incomplete. While Merrell was attempting to address the issue of absorption, a paper appeared in the *British Medical Journal* indicating that thalidomide was causing peripheral neuritis in some patients. Kelsey challenged Merrell on that point as well. Again Merrell had no answer. For a total of 14 months Merrell tried to get thalidomide passed by the FDA—but could not get past Kelsey.

On December 25, 1956, a baby girl was born without ears. Her father worked for Grünenthal and had brought home samples of a new drug for his pregnant wife who had been having trouble sleeping. By the latter part of 1960 reports of a rare limb defect, phocomelia ("seal limbs"), were appearing in the medical literature of West Germany. In summer 1961, Widukind Lenz, a pediatrician in Hamburg, was asked by a lawyer, whose wife and sister had both delivered malformed babies, to investigate the situation.

In the meantime, in Australia an obstetrician, William McBride, had discovered that Distaval was effective in treating morning sickness in his patients. The drug was advertised by Distillers as "safe during pregnancy" although no tests had been run to back that statement. In May and June 1961, three of McBride's patients delivered malformed babies. By mid-June McBride had concluded that Distaval was to blame and quit prescribing the drug for his patients. He has, since that time, maintained that he sent a letter to the British medical journal *The Lancet*. He did not keep a copy of the letter, and *The Lancet* has no record of having received such a letter. During the summer, several of McBride's patients who had been on Distaval delivered normal babies and McBride began to doubt his earlier conclusions. Then, on September 13, 1961, McBride delivered his fourth thalidomide baby. He arranged a meeting with the Distillers Australian representatives and sent a letter (or another letter) to *The Lancet*. This letter was received and published (December 16, 1961).

While this was happening in Australia, Lenz was investigating what appeared to be an emerging epidemic in Germany. By October 1961, he had identified eight cases of phocomelia in Hamburg. On November 16, exactly the same day that McBride was meeting with Distillers in Australia, Lenz wrote a letter to Grünenthal expressing his concern over the epidemic and its ties to Contergan. That weekend, Lenz expressed the same concern in a meeting of pediatricians. By the end of November, Contergan had been withdrawn from the market. A year letter, Distaval was off the market, and the FDA application in the United States had been withdrawn. It is estimated that approximately 10,000 infants were born deformed as a result of thalidomide, of whom 5000 survived past childhood.

Legislation. Since late 1959 the U.S. House and Senate had been considering legislature directed at controlling drug quality and pricing. By summer 1962, when news of the thalidomide disaster in Europe reached the United States press, Congress was in a deadlock. News of the thalidomide disaster changed everything. It became clear to everyone what uncontrolled drug production and distribution could lead to. In August 1962, Frances Kelsey was given the President's Award for Distinguished Federal Civilian Service (the highest honorary award that the federal government can grant to career employees) for heroically keeping thalidomide out of the United States market, and Senate Bill 1552 passed by a vote of 78:0, establishing FDA regulations still in effect today.

Treating leprosy. However, thalidomide did not go away in 1962: it had more life and many more surprises to reveal. In 1964, Jacob Sheskin, director of the Hansen Disease (leprosy) Center at the Jerusalem Hospital, used thalidomide to treat a patient with severe ENL (erythema nodosum laprosum; severe boil-like skin lesions occurring in some leprosy patients). The patient had not been able to sleep for 72 hours, and no sleep aid had been effective. The hospital pharmacy had some undiscarded thalidomide. Upon treatment the patient not only slept soundly, but the ENL lesions were significantly reduced. With additional treatment, the lesions disappeared. Sheskin published a study of the effect of thalidomide on ENL patients in Venezuela. This was followed by a large World Health Organization study involving 4552 ENL patients. The study demonstrated a 99% improvement after treatment with thalidomide. As a result of these studies and the successful treatment of ENL with thalidomide, it has been estimated that 90% of Hansen disease centers worldwide became obsolete and were closed. On July 16, 1998, the FDA released thalidomide, for the first time, for prescription sale in the United States to treat ENL. Today, thalidomide remains the drug of choice for treating ENL.

Other treatment uses. Prescription for thalidomide to treat ENL represents only a fraction of its use in the United States. It has been found useful in treating a number of inflammatory disorders, from lupus to inflammatory bowel disease. In 1991 Gilla Kaplan and others linked thalidomide to tumor necrosis factor (TNF)-α levels in the circulation. It is well known that elevated TNF-α levels are closely tied to inflammatory diseases. Thalidomide reduces TNF-α levels, thus reducing the severity of dozens of diseases. Thalidomide is also used to treat several types of cancers, especially multiple myeloma. In 1994 Robert D'Amato and others published a paper demonstrating that thalidomide inhibits angiogenesis (blood vessel formation). This effect is thought to be associated with both antitumor and teratogenic (birth defect–causing) functions of thalidomide. Many other effects of this unique drug in the treatment of disease are currently under investigation.

Cause of birth defects. However, the risk of birth defects with thalidomide is still very real. In 1996 E. E. Castilla and others reported 34 cases of thalidomide-induced birth defects among ENL patients in South America. In an attempt to prevent additional thalidomide-induced birth defects in the United States, the FDA introduced S.T.E.P.S. (System for Thalidomide Education and Prescribing Safety). Under this system, any female using thalidomide must first demonstrate a negative pregnancy test and must be on two forms of contraception. Any male taking thalidomide must use one form of barrier contraception.

A search for the mechanism by which thalidomide causes birth defects has been ongoing since the thalidomide epidemic was recognized. T. Stephens summarized 24 proposed mechanisms in 1988 and 10 additional proposed mechanisms in 1997. The approach since then has been to eliminate proposals that do not stand scrutiny and to combine the remainder into a unified model of thalidomide's action. Nineteen proposed mechanisms have been rejected.

In 1999 T. Parman and others demonstrated that thalidomide is bioactivated by embryonic prostaglandin H synthase (PHS) to a free-radical intermediate that produces reactive oxygen species (ROS), which can cause oxidative damage to DNA and other cellular macromolecules. In 2004 J. M. Hansen and C. Harris demonstrated that DNA binding of nuclear factor kappa- B (NF-kappaB), a redox-sensitive transcription factor and key regulator of limb outgrowth, was significantly attenuated in thalidomide-treated rabbit limb cells. Reduced DNA binding of NF-kappaB resulted in the failure of limb cells to express fibroblast growth factor (FGF)-10 in the limb mesenchyme, which in turn attenuated expression of FGF-8 in the apical ectodermal ridge (AER). Hansen and Harris proposed that failure to establish an FGF-10/FGF-8 feedback loop between the mesenchyme and AER results in the truncation of limb outgrowth.

In 2000 Stephens and others proposed that thalidomide affects the following pathway during development: insulin-like growth factor I (IGF-I) and FGF-2 stimulation of the transcription of alphav and beta3 integrin subunit genes. The resulting alphav-beta3 integrin dimer stimulates angiogenesis in the developing limb bud, which promotes outgrowth of the bud. The promoters of the IGF-I and FGF-2 genes, the genes for their binding proteins and receptors, as well as the alphav and beta3 genes, lack typical TATA boxes in their control regions, but instead contain multiple GC boxes (GGGCGG). Thalidomide, or a breakdown product of thalidomide, specifically binds to these GC promoter sites, decreasing transcription efficiency of the associated genes. A cumulative decrease interferes with normal angiogenesis, which results in radial aplasia (defective development resulting in the virtual absence of a tissue or organ) and truncation of the limb.

Intercalation into G-rich promoter regions of DNA may explain why the *R* enantiomer of thalidomide is not teratogenic while the *S* enantiomer is, although both apparently cause equal oxidative damage. Research is under way that attempts to unite the intercalation model and the oxidative stress model. The tissue specificity of thalidomide and its effect against only certain neoplasias may be explained by the fact that various developing tissues and neoplasias depend on different angiogenesis or vasculogenesis pathways, only some of which are thalidomide-sensitive.

For background information *see* ANGIOGENESIS; CONGENITAL ANOMALIES; LEPROSY; PHARMACEUTICALS TESTING; PHARMACY; PREGNANCY in the McGraw-Hill Encyclopedia of Science & Technology.

Trent Stephens

Bibliography. J. Ashby and H. Tinwell, Is thalidomide teratogenic?, *Nature*, 375:453, 1995; R. J. D'Amato et al., Thalidomide is an inhibitor of angiogenesis, *Proc. Nat. Acad. Sci. USA*, 91:4082–4085, 1994; D. Neubert, Never-ending tales of the mode of the teratogenic action of thalidomide, *Teratog. Carcinog. Mutagen.*, 17:i–ii, 1997; R. W. Smithells and C. G. H. Newman, Recognition of thalidomide defects, *J. Med. Genet.*, 29:716–723, 1992; T. D. Stephens and R. Brynner, *Dark Remedy: The Impact of Thalidomide and Its Revival as a Vital Medicine*, Perseus Publishing, 2001.

Type 2 diabetes

Diabetes mellitus is a disorder characterized by the presence of excess glucose in the blood and tissues of the body. The word "diabetes" comes from the Greek for "a siphon," referring to the discharge of an excess quantity of urine; the word "mellitus" comes from Latin for "honey." Thus, diabetes mellitus refers to the passage of large amounts of sweet urine. Diabetes is a disease in which the pancreatic hormone insulin is either not produced or not properly used by the body. It is characterized as a chronic metabolic disorder with hyperglycemia (high blood sugar) and abnormal energy metabolism. Diabetes is caused by a combination of genetic, autoimmune, and environmental factors.

There are two main types of diabetes: type 1 (insulin-dependent diabetes) and type 2 (non-insulin-dependent diabetes). Type 1 diabetes affects 5–10% of all people diagnosed with this disease. This type of diabetes is caused by the immune system's destruction of the β cells of the pancreas.

Of all the people who have diabetes, 90–95% have type 2 diabetes. In the past, type 2 diabetes was also termed adult onset diabetes because of its late onset (usually occurring after age 40). However, the term "adult onset" is no longer appropriate as there is a startling increase of type 2 diabetes in children as young as 10 years.

Clinically, a person may have type 2 diabetes for many years without being aware of it. Loss of vision, which usually takes years to develop, may be the first sign of the disease. The risks for developing type 2 diabetes include family history of type 2 diabetes, obesity, high blood pressure, high levels of cholesterol, a sedentary lifestyle, as well as development of diabetes during pregnancy (gestational diabetes). The causes of this disease are multifold. In type 2 diabetes, a defect in insulin action (insulin resistance) is combined with a defect in insulin secretion, which leads to hyperglycemia. Impaired insulin action has been demonstrated in muscle, fat, and liver tissues. Although there is some evidence for a genetic link in the development of type 2 diabetes, it is increasingly clear that insulin resistance is an acquired defect in most cases of type 2 diabetes.

Obesity. Obesity is the most common cause of insulin resistance and type 2 diabetes. Simply being overweight (BMI > 25, where body mass index [BMI] is a measure used to evaluate body weight relative to height) raises the risk of developing type 2 diabetes by a factor of 3.

The location of body fat can significantly affect one's risk for developing diabetes. Doctors have long recognized that people who are apple-shaped—with their fat concentrated in the abdomen—are at much higher risk for diabetes and metabolic syndrome than those whose fat is mainly subcutaneous, that is, distributed beneath the skin primarily in the buttocks and thighs. In view of this finding, more physicians are suggesting that the waist-to-hip ratio is a more accurate measure of obesity and a better predictor of insulin resistance and metabolic syndrome than the widely used body mass index.

Diagnosis. The diagnosis of type 2 diabetes is made by using any of the following criteria: a random blood glucose level greater than 11.1 mmol/L (200 mg/dL); a fasting (8-hour) blood glucose level greater than 7 mmol/L (126 mg/dL); blood glucose levels greater than 11.1 mmol/L (200 mg/dL), measured 2 h after an oral glucose tolerance test (the ingestion of 75 g of glucose following an overnight fast).

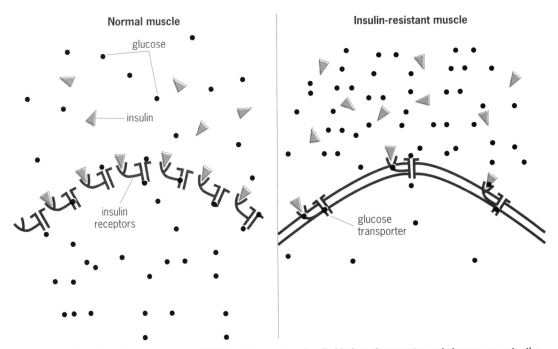

A normal muscle cell may have as many as 20,000 insulin receptors. Insulin binds to the receptor and glucose can enter the cell: an insulin-resistant muscle may have as few as 5000 insulin receptors. This keeps blood sugars high and muscle glucose low.

Treatment. There are two main forms of treatment of type 2 diabetes: alterations in diet and physical activity, and oral hypoglycemic agents (drugs that lower blood glucose). There are four types of agents, with different modes of action.

1. The sulfonylurea types (for example, glibenclamide) stimulate the pancreas to produce more insulin, which may increase the amount of insulin binding to insulin receptors, allowing for greater glucose uptake into cells.

2. Alpha glucosidase inhibitors inhibit the digestion of carbohydrates in food and hence reduce their absorption in the gut. An example is Acarbose, which slows the absorption of carbohydrate foods. This can help prevent rises in blood glucose levels and is most often used in combination with other medications.

3. Orlistat blocks absorption of about 30% of the fat that is eaten. This does not specifically treat the diabetes but may help diabetes management in people having major problems achieving weight control.

4. The biguanide types (for example, Metformin) and the thiazolidinediones (that is, Troglitazone) reduce insulin resistance and result in greater uptake of glucose by tissues. These aim to overcome the major problem of insulin resistance in type 2 diabetes. This class of medication (usually Metformin) is often the first choice for overweight teenagers with type 2 diabetes and can be very effective in combination with healthy lifestyle measures.

Insulin resistance. Insulin resistance can be said to exist when a normal concentration of insulin elicits a subnormal biological response. Impaired insulin action in diabetes has been demonstrated in muscle, fat, and liver tissues.

Metabolic syndrome. The metabolic syndrome is a group of metabolic risk factors for cardiovascular disease. There are five cardiovascular risk factors that accompany the metabolic syndrome: (1) dyslipidemia [elevated apolipoprotein B (apo B), elevated triglyceride, small low-density lipoprotein (LDL) particles, and low levels of high-density lipoprotein (HDL) cholesterol]; (2) elevated blood pressure; (3) elevated glucose; (4) a prothrombotic state; (5) a proinflammatory state.

The likelihood of an individual developing metabolic syndrome is enhanced by underlying risk factors, including obesity, insulin resistance, lack of physical activity, and advancing age. Besides being at higher risk for cardiovascular disease, people with metabolic syndrome are at increased risk for type 2 diabetes. The two major therapeutic strategies for treatment of affected persons are modification of the underlying risk factors and separate drug treatment of the particular metabolic risk factors when appropriate. First-line therapy for underlying risk factors is lifestyle changes, that is, weight loss in obese persons, increased physical activity, and diet modification. These changes will improve all of the metabolic risk factors. Whether use of drugs to reduce insulin resistance is effective, safe, and cost-effective before the onset of diabetes awaits the results of more clinical research. Individual risk components can be addressed with different treatments, such as antihypertensive therapies and statins for lowering blood lipids.

To be considered to have metabolic syndrome, a person must have insulin resistance plus two or more of the following: central obesity [waist circumference >102 cm (40 in.) (male) or >88 cm (35 in.) (female)]; dyslipidemia (triglyerides >2.0 mmol/L (36 mg/dL) or HDL cholesterol <1.0 mmol/L (<18 mg/dL)]; hypertension (blood pressure greater than 140/90 mmHg); hyperglycemia (fasting blood glucose >6.1 mmol/L, 110 mg/dL).

Exercise. Exercise is important for managing type 2 diabetes. Exercise has many benefits, including weight control, increasing glucose uptake by cells, helping insulin work better, as well as general fitness and health. It is important to do moderate exercise regularly (preferably every day or at least four times per week). Moderate exercise is an amount which makes you puff a bit, but you should still be able to carry on a conversation during the exercise. Aim for 30 to 45 minutes of moderate exercise at least four times per week.

Physical exercise helps the body utilize glucose by "sensitizing" the body to insulin. This means that for the same amount of insulin, more glucose can enter the muscle cells.

Since obesity is one of the main contributors to type 2 diabetes, as well as other diseases such as heart disease and stroke, exercising becomes an important tool to combat this. A recent national study in the United States (Diabetes Prevention Program) demonstrated that moderate exercise helped to prevent diabetes in people at high risk for type 2 diabetes.

Exercise also is beneficial as a prevention in that it lowers blood pressure (a risk factor for developing type 2 diabetes). Hypertension can lead to vascular problems in the diabetic, including blindness, kidney failure, and gangrene of the limbs.

Many people with type 1 and type 2 diabetes have high levels of cholesterol and/or triglycerides. These can lead to hardening of blood vessels. Exercise and good blood sugar control are the best ways to reduce blood triglyceride levels. Exercise may also help remove cholesterol from blood vessel walls by increasing HDL.

The only way humans can increase insulin sensitivity is by exercising. As a result of exercise, the person is more sensitive to insulin, the insulin can work more efficiently, and a lower daily dose of medication is usually required. Regular exercise (and weight loss) allows some people with type 2 diabetes to stop insulin injections and change to oral medication. It is now believed that many of the beneficial effects of exercise on the risk of heart disease, particularly in type 2 diabetes, are due to improvements with insulin sensitivity. It is thus important to exercise regularly and even vigorously.

For background information *see* CARBOHYDRATE METABOLISM; DIABETES; GLUCOSE; INSULIN; OBESITY; PANCREAS DISORDERS in the McGraw-Hill Encyclopedia of Science & Technology.   Zoe Cohen

Bibliography. K. F. Petersen and G. I. Shulman, Etiology of insulin resistance, *Amer. J. Med.*, 119(5A):105–165, 2006.

Wind tunnel experiments of cloud particles

Atmospheric clouds are dynamic systems consisting of cloud and precipitation particles falling at some terminal speed. Updrafts in the air reduce this downward descent and sometimes even reverse the direction of certain-sized cloud and precipitation particles. This dynamical nature plays a decisive role in the processes involving phase transitions and particle growth (such as melting, freezing, and evaporation), growth due to water vapor diffusion, growth arising out of collision-coalescence, as well as gas and aerosol particle scavenging by cloud and precipitation particles.

In order to study quantitatively the growth of cloud drops and ice crystals to precipitation sizes and the phase transformation process under dynamical situations as functions of the various physical parameters that affect them, it is desirable to investigate the processes experimentally under conditions that simulate as closely as possible the environment in natural clouds. This can be achieved by freely suspending falling cloud and precipitation particles in an updraft having a speed that matches the terminal fall speed of the particles in magnitude so that the process can be observed for long periods of time. (The particles appear stationary relative to an observer.) In addition, the updraft air can be conditioned so that the temperature, humidity, aerosol content, and trace-gas pollutant content are held constant and have values representative of the natural environment. In this way, one can experimentally determine collision efficiencies and growth rates of cloud drops and ice particles and compare them to theoretically computed ones. Other experiments can be done with the help of such an arrangement, such as electrifying clouds during collision-coalescence and freezing, scavenging of pollutants by precipitation particles, and determining the freezing temperature of freely floating drops containing or contacting ice nuclei. *See* PRECIPITATION SCAVENGING.

The "conditioned" updraft can be generated in a vertically positioned wind tunnel. During the last 50 years, various wind tunnels have been constructed which fulfill most experimental requirements, with some limitations. These can be broadly classified into three categories. In the open or L-type wind tunnel, a blower forces air upward through the L into the open atmosphere. The exiting airstream can be suitably formed to produce a local velocity minimum in the middle of the flow where drops could be suspended. Because of the open construction and the flow-forming structure, neither the constancy and uniformity of the environmental conditions nor a low level of turbulence could be attained.

In an attempt to improve the conditioning of the air flowing past the cloud or precipitation particles, wind tunnels were built with closed circulation (O type). In these, environmental conditions could be precisely maintained but the level of turbulence could not be reduced, thereby rendering this design unsuitable for the free suspension of cloud and precipitation particles consisting of ice or water (hydrometeors).

To overcome the shortcomings of the earlier designs and to be able to freely suspend hydrometeors in a low turbulence and stably conditioned environment, H. Pruppacher and colleagues at the University of California, Los Angeles (UCLA), decided on a Z-form tunnel layout, following the design suggestions of L. Prandtl. In this design, air is sucked in by vacuum through the observation section of the tunnel. An extremely responsive flow-speed control was achieved using a variable-area sonic nozzle installed between the wind tunnel and the vacuum pump.

Note that a normal valve placed in the flow produces a very nonlinear regulation characteristic with respect to the opening area of the valve. This is because the flow speeds up when the area is reduced without producing a proportional change in the total mass flow. In a sonic nozzle, the speed of flow is always fixed (speed of sound in air depends only on its temperature and density.) Hence, the mass flow depends only on the area of the nozzle. The only functional condition for a sonic nozzle is that the air pressure ratio across the nozzle should be less than 0.528.

With this configuration, it was possible to float waterdrops a few tens of micrometers in size up to millimeters stably in the observation section. It was also possible to freely float ice crystals, ice-crystal aggregates (snowflakes), and graupel (snow pellets) using a modified observation section. Numerous studies of the various microphysical processes in clouds have been done in this wind tunnel.

Mainz wind tunnel. With the appointment of H. Pruppacher as Chair Professor of Cloud Physics and Chemistry at the Johannes Gutenberg-Universität in Mainz, Germany, the prospect of building a second-generation, vertical wind tunnel became a reality.

The wind tunnel built was an improved version of the UCLA wind tunnel, where operational problems of the first unit were mostly corrected. The range and precision of the controlled environmental parameters were greatly improved, as were the insulation properties to allow lower temperatures to be attained at low wind speeds. The range of speeds was increased to 40 m/s. Additionally, most of the operations were made safe, secure, and user-friendly through the implementation of process automation. Recently, additional improvements were made for low-speed and low-temperature performance of the wind tunnel.

In operation, air entering the tunnel (**Fig. 1**) first passes through a condensation dehumidifier and then through a molecular-sieve dryer to produce dry air having a dew point around −40°C. This dry air is subsequently filtered to remove aerosol particles and then cooled to the desired temperature by passing it through two temperature-regulated heat

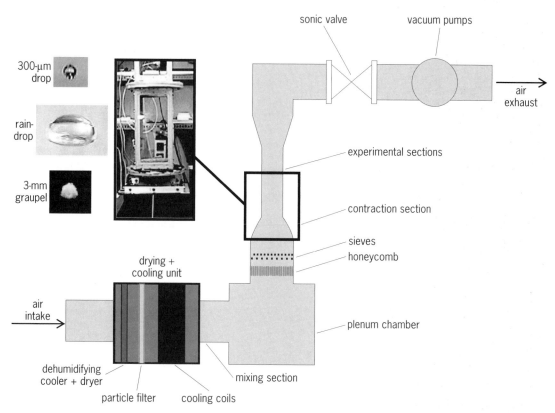

300-μm drop

rain-drop

3-mm graupel

Fig. 1. Schematic diagram of Mainz wind tunnel. Photo at the upper left show three hydrometeors in a state of free suspension and the lower observation section. Pictured as a small (almost spherical) 300-μm-radius drop, a large deformed (this is the way they really look) raindrop with an equivalent radius of 2.88 mm, and a 3-mm graupel.

exchangers. The cooled, dry air next enters a mixing chamber, where it is rehumidified and, as required by the investigation, a fine-droplet cloud of aerosols or gaseous pollutants is added.

This "conditioned" air is fed to a large quieting compartment (plenum chamber), where rapid air motions subside, and then is sucked through a honeycomb and five sieves (laminariser) to further reduce turbulence, and then through a contracting section. After passing through the contraction section, the air enters the observation section and exits through a slightly divergent diffuser into the sonic nozzle, which is located at the inlet of the vacuum pumps. To avoid corrosion when working with acidic gaseous pollutants, most of the tunnel components, from the mixing chamber on, are made of stainless steel.

For ice-phase experiments, the wind tunnel air can be cooled down to $-30°C$, and air humidification is possible up to the saturation level. The air speed can reach values up to 40 m/s, allowing the whole range of hydrometeors to be floated, starting with small drops having radii between 30 and 500 μm, large drops with equivalent radii up to 3–4 mm, snow crystals, snow flakes, graupel, and hailstones. Figure 1 also shows an actual view of the tunnel's observation section and three typical hydrometeors in their freely suspended state.

Wind tunnel experiments. Since its commissioning in 1987, various problems have been studied in the wind tunnel in the areas of cloud microphysics and

pollution chemistry. A few representative experiments will be discussed.

Collision-coalescence process. The first example is a study of the microphysics of the collision-coalescence process responsible for the growth of cloud droplets once they reach about 30-μm radii, leading to precipitation-size drops. It has long been conjectured (even though the experimental evidence has been controversial) that turbulence inside clouds enhances the collision-coalescence growth rate, thereby producing precipitation-size drops relatively quickly as compared to growth under laminar conditions.

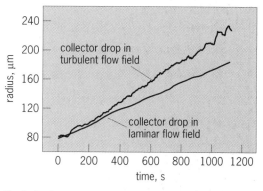

Fig. 2. Radius of a collector drop plotted against time during its continuous growth inside a small-droplet cloud under laminar and turbulent environmental conditions. The liquid water content of the cloud is ∼1 g/m³, and the droplet spectrum has a modal radius of ∼4 μm.

The pioneering experiment of the continuous growth of a collector drop in a small-droplet (about 4-μm modal radius) cloud under both laminar and turbulent conditions conclusively showed that turbulence leads to faster growth by a collision-coalescence process. **Figure 2** shows the size of a collector drops growing as a function of time under differing environmental conditions.

Ammonia uptake. The next example is an investigation of the uptake of gaseous ammonia by waterdrops. It demonstrates how the experimental results help one to learn about the complete set of reactions involved in the uptake of ammonia by waterdrops in the presence of carbon dioxide in air. The experimentally measured uptake is plotted in **Fig. 3** as a function of time. The lines represent the uptake expected from model calculations with one or more chemical reaction pathways considered.

Ice nucleation. The final example is the application of the wind tunnel to the study of the ice nucleation process. Nuclei are needed to heterogeneously

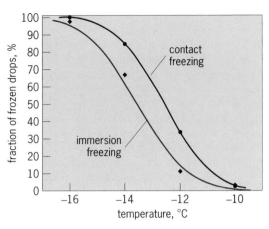

Fig. 4. Plots showing the freezing statistics of ice nucleation by alder pollen in two modes of freezing.

freeze supercooled waterdrops. As a significant fraction of atmospheric aerosols are of biological origin and some species of bacteria have already been found to function as ice nuclei at relatively warm temperatures, it was decided to investigate the ice nucleation characteristics of pollen. **Figure 4** shows the results of the ice-nucleating capabilities of alder pollen in two modes of freezing, namely, immersion and contact. In the former mode the nuclei are already present in the drops before supercooling takes place, and in the latter mode the nuclei make contact with a droplets in a supercooled state.

The statistical nature of the process required that the experiment use a large number of droplets to determine the median freezing temperature (MFT, the temperature at which 50% of the drops studied freeze) in each mode.

The plots show that pollen function as ice nuclei and do so more efficiently (around 1.5°C warmer MFT) in the contact mode as compared to the immersion mode. For quite a few other ice-nucleating materials, similar differences in the MFTs between these two modes were found.

Outlook. The usefulness has been demonstrated of a vertical wind tunnel in research in the areas of cloud physics and cloud chemistry. Although exact simulation of nature is not possible in any laboratory experiment, the information gained from experiments done under controlled laboratory conditions provides us with better understanding of the processes. In spite of operational limitations, whether useful information can be extracted from the results depends only on clever experiment design. The wind tunnel will continue to be useful for solving problems in those areas of cloud physics where experimental data are needed to validate complex computational models or to provide basic data to be used as input of such models.

[Acknowledgement: I would like to thank Professor Hans Pruppacher for his valuable suggestions in preparing the article and Ms. Nadine von Blohn for her help in preparing the figures.]

For background information *see* AEROSOL; AIR POLLUTION; ATMOSPHERE; ATMOSPHERIC CHEMISTRY;

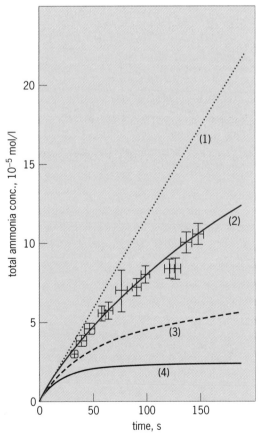

Fig. 3. Uptake of gaseous ammonia (points with error bars) by 2.88-mm-radius waterdrops at 15°C plotted as total concentration of ammonia as a function of time of exposure to the gas. Concentration of ammonia in the gas phase is 195 ppbv, and carbon dioxide concentration is 350 ppmv. The various lines represent model calculations using four different aqueous chemistry formulations: (1) assuming that carbon dioxide equilibrium is instantaneously established; (2) considering detailed carbon dioxide kinetics; (3) considering primary dissociation kinetics but ignoring high pH pathways; (4) leaving out carbon dioxide. For details, see A. Hannemann et al., 1995.

CLOUD; CLOUD PHYSICS; PRECIPITATION (METEOROL-OGY); WIND TUNNEL in the McGraw-Hill Encyclopedia of Science & Technology.   Subir Mitra

Bibliography. K. V. Beard and H. R. Pruppacher, A determination of the terminal velocities and drag of small water drop by means of a windtunnel, *J. Atm. Sci.*, 26:1066-1072, 1969; D. C. Blanchard, The behaviour of water drops at terminal velocity in air, *Trans. Amer. Geophys. Union*, 31:836-842, 1950; N. v. Blohn et al., The ice nucleating ability of pollen, Part III: New laboratory studies in immersion and contact freezing modes including more pollen types, *Atm. Res.*, 78:182-189, 2006; W. R. Cotton and N. R. Gokhale, Collision, coalescence and breaking up of large drops in a vertical wind tunnel, *J. Geophys. Res.*, 72:4041-4049, 1967; K. Diehl et al., The ice nucleating ability of pollen, Part I: Laboratory studies in deposition and condensation freezing modes, *Atm. Res.*, 58:75-87, 2001; K. Diehl et al., The ice nucleating ability of pollen, Part II: Laboratory studies in immersion and contact freezing modes, *Atm. Res.*, 61:125-133, 2002; F. H. Garner and P. Kendrich, Mass transfer to drops of liquid suspended in a gas stream, Part I: A wind tunnel for the study of individual liquid drops, *Trans. Inst. Chem. Eng.*, 37:155-161, 1959; N. R. Gokhale and W. R. Cotton, *A Vertical Wind Tunnel*, Atmospheric Sciences Center, SUNYA, Research Report, pp. 99-112, 1964; A. Hannemann, S. K. Mitra, and H. R. Pruppacher, On the scavenging of gaseous nitrogen compounds by large and small rain drops, Part I: A wind tunnel and theoretical study of the uptake and desorption of NH_3 in the presence of CO_2, *J. Atm. Chem.*, 21:293-307, 1995; M. Kombayasi, T. Gonda, and K. Isono, Life time of water drops before breaking and size distribution of fragment drops, *J. Meteorol. Soc. Jap.*, 42:330-340, 1964; R. List, A hail tunnel with pressure control, *J. Atm. Sci.*, 23:61-66, 1966; S. Matthias-Maser and R. Jaenicke, Size distribution of primary biological aerosol particles with radii >0.2 μm in an urban/rural influenced region, *Atm. Res.*, 39:279-286, 1995; H. R. Pruppacher and J. D. Klett, *Microphysics of Clouds and Precipitation*, 2d ed., Kluwer Academic, 1997; H. R. Pruppacher and M. Neiburger, Design and performance of the UCLA Cloud tunnel, *Proceedings of the International Conference on Cloud Physics, Toronto*, pp. 368-392, August 26-30, 1968; J. D. Spengler and N. R. Gokhale, A large vertical wind tunnel for hydrometeor studies, *Proceedings of 2d Conference on Weather Modification*, Santa Barbara, American Meteorological Society, Boston, 1970; O Vohl et al., A wind tunnel study of the effects of turbulence of cloud drops by collision and coalescence, *J. Amer. Sci.*, 56:4088-4099.

Yeti crab

A new crustacean decapod (order Decapoda, class Crustacea) was discovered during a submersible research cruise at a hydrothermal site of the Pacific-Antarctic Ridge (south of Easter Island, 38°S latitude,

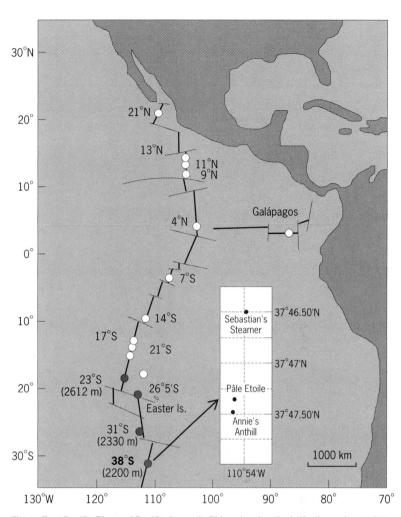

Fig. 1. East Pacific Rise and Pacific-Antarctic Ridge showing the hydrothermal vent sites, the vent sites visited during the PAR5 cruise (full circles), and the distribution of the Yeti crab at the 38°S site (detail in rectangle).

2200 m or 7546 ft depth). The PAR5 Cruise (Pacific-Antarctic Ridge, 2005) was organized during March-April 2005 by Robert Vrijenhoek of the Monterey Bay Aquarium Research Institute, California: the Research Vessel *Atlantis* was to carry the submarine *Alvin* to explore four hydrothermal vent areas of the southeastern Pacific Ridge and the Pacific-Antarctic Ridge, between 23°S and 38°S, where the spreading tectonic plates show the highest spreading rate (about 150 mm/year or 6 in./year). The principal goal of this cruise was to understand the role of the geographical barriers in the current distribution of the animal species colonizing the hydrothermal sources discovered along the ridge—in particular the role of the Easter Microplate and, farther south, Juan Fernandez Microplate (**Fig. 1**). It was also necessary to make an initial comparison between the composition of the fauna of several hydrothermal vents north and south of the ridge.

Hydrothermal vent sites are relatively small (100 m² or 076 ft²), forming discrete mounds and sulfide chimneys emitting hot fluids (360°C or 680°F or more) on the pillow basalt sea floor, along the axial valley of the mid-oceanic ridges. Dense animal communities live on or around these structures, where

Fig. 2. Unique specimen, photo taken in lab in July 2005. (*A. Fifis* © *Ifremer*)

the vent fluids range between 2–3°C (36–37°F) [seawater temperature] and 10–15°C (50–59°F). Strong bacteria production forms the base of the food chain for this unusual ecosystem.

Scientists diving in the submarine *Alvin* observed a number of large, white, "hairy" crustaceans and collected one of them by the slurp gun of the submarine, at the vent site area named Annie's Anthill (37°46.49'S, 110°54.72'W, 2228 m or 7310 ft). On board, the left fifth appendage and some setae (slender, usually rigid bristles or hairs) were dissected and preserved in ethanol for DNA analysis. The holotype (specimen of reference; **Fig. 2**) is kept at the Muséum National d'Histoire Naturelle of Paris.

Description. The morphology of the specimen is characterized by a depressed, ovoid, and symmetrical body. The carapace length (tip of the rostrum to the posterior border of the cephalothorax) is 58.6 mm (2.34 in.); the total length, including the pincers and abdomen naturally folded, is 15 cm (5.9 in.). The carapace is calcified, slightly convex, and smooth, with the rostrum well developed and triangular. The posterior half of the dorsal surface of the carapace has a few longitudinal and transverse grooves. The lateral sides of the carapace (pterygostomian region) have two longitudinal and subparallel carinae (raised ridges). The abdominal segments are smooth, not folded against the carapace, and the central article (an appendage segment) of the tail fan (telson) is folded beneath the preceding abdominal segment, with a median transverse suture and a longitudinal suture in the posterior half; the lateral articles of the tail fan (uropods) are spatulate (shaped like a spoon). The sternum is large and strongly calcified, clearly produced anteriorly. The eyes are vestigial and without pigment. The antennal peduncle is segmented, and the flagellum is of moderate length. The claws (chelipeds) are strong, subequal, and greatly elongate. The walking legs are stout, with clawlike dactyli bearing dense, yellow, corneous spinules along the flexor margin. The last pair of legs (appendages 5) is extremely reduced and chelated, inserted below the posterior portion of the sternum, with the insertion not visible ventrally.

The species has dense, long, plumose setae mainly on all surfaces of long appendages. The extraordinarily setose nature of the claws and walking legs

led to adoption of the common name Yeti crab (Yeti is the hairy abominable snowman of the Himalayas). The most numerous setae, 15 mm (0.6 in.) long, are slightly yellowish and flexible, and consistently have clusters of filamentous bacteria, mainly at their extremity. On scanning electron microscopy pictures, the bacteria show several morphotypes of probably sulfide-oxidizing bacteria, characterized by the presence of sulfide-like granulations. Other rigid chitinous setae (about 13 mm or 0.5 in. long) are barbed in their extremity, ending in a rigid spine, and they are regularly inserted in pairs along the first appendages. These setae are devoid of filamentous bacteria. These barbules appear similar to one of the sexual appendages of the crab *Austinograea yunohana*, living on the hydrothermal vents off central Japan, suggesting a copulative function. The collection of female specimens and in-situ observations could test this hypothesis.

The species is very different from other species of anomuran crabs, and the uniqueness of the differences has justified the creation of a new family of decapods. The new family is named Kiwaidae (from Kiwa, the goddess of shellfish in the Polynesian mythology), and the new species is *Kiwa hirsuta*. The closest group of species belongs to the family Chirostylidae (squat lobsters), being characterized by a smooth, triangular, and depressed body. However, the new family can be easily distinguished from the chirostylids by the presence of grooves on the dorsal surface of the carapace (absent in chirostylids), vestigial eyes (well developed in chirostylids), a sternum that is strongly pronounced anteriorly (not pronounced in chirostylids), and the insertion of the last leg, which is not visible ventrally (clearly visible in chirostylids).

Molecular analyses have confirmed the clear difference between Kiwaidae and the other families of anomuran crabs. Kiwaidae is clearly isolated from other groups, being closer to the families Chirostylidae, Galatheidae, and Porcellanidae than to Aeglidae (the only anomuran crab living in freshwaters and known from Chilean rivers), in spite of strong morphological resemblances.

The phylogenetic placement of Kiwaidae suggests that it is a basal lineage. Additional species in the new family Kiwaidae will need to be collected and analyzed to establish the evolutionary history of galatheoid crabs.

Habitat and distribution. The new species occurs at densities of one to two individuals per 10 m² (108 ft²), more or less regularly spaced on the zone of pillow basalt surrounding active hydrothermal vents. Specimens were also observed on extinct chimneys and at the base of black smokers, among vent mussels, where shimmering milky water emanates, and co-occuring with galatheid squat lobsters *Munidopsis recta*, crabs *Bythograea vrijenhoeki* and *B. laubieri*, vent mussels *Bathymodiolus*, buccinid gastropods *Eosipho auzendei*, and recently described ophidiid fishes *Ventichthys biospeedoi*. Like other vent decapod crustaceans, *Kiwa hirsuta* is omnivorous. Indeed, specimens were observed in-situ consuming tissues of mussels damaged by submersible sampling activities. Furthermore, the presence on the legs of dense bacteriophoran setae colonized by mats of probably sulfide-oxidizing bacteria makes it possible to regard this species as an obligate associated with the hydrothermal vents. These bacteria could serve as a nutritional resource.

The Yeti crab was observed on three hydrothermal sites distributed on nearly 1.5 km (0.9 mi) along the Pacific-Antarctic Ridge segment. As this animal has never been observed during the 30 years of explorations at the hydrothermal vents of the East Pacific Rise, one can summize that the 38°S area is the northern boundary for the distribution of the Yeti crab. For the moment, we propose that the Juan Fernandez Microplate (Fig. 1) constitutes such a geographical barrier to the northern distribution of *Kiwa hirsuta*.

Conclusions. Exploration of deep-sea environments continues to reveal new and remarkable animal taxa. At present, 83% of the species recorded from hydrothermal vents are unknown in other environments, and many belong to new families or genera. The last time a new family of marine anomurans was discovered (apart from a nonvent family of small hermit crabs, Pylojacquesidae) was at the end of the nineteenth century. Vent ecosystems have provided us with numerous higher-level taxa that fill gaps in our understanding of the evolution and phylogeny of marine organisms. Discovery of the Yeti crab suggests that future exploration of chemosynthetic environments will continue to reveal more biological treasures.

[Acknowledgments: The authors thank R. Vrijenhoek, chief scientist of the PAR5 Cruise, and the crew of R/V *Atlantis* and DSV *Alvin* for collecting specimens. We thank also V. Martin (Ifremer) for Fig. 1 (map). The cruise PAR5 was funded by grants from the U.S. National Science Foundation (OCE-0350554 and OCE-0241613).]

For background information *see* CRAB; CRUSTACEA; DECAPODA (CRUSTACEA); DEEP-SEA FAUNA; HYDROTHERMAL VENT in the McGraw-Hill Encyclopedia of Science & Technology.

Michel Segonzac; Enrique Macpherson; William Jones

Bibliography. D. Desbruyères, M. Segonzac, and M. Bright (eds.), *Handbook of Deep-sea Hydrothermal Vent Fauna*, 2d ed., *Denisia*, vol. 18, 2006; E. Macpherson, W. Jones, and M. Segonzac, A new squat lobster family of Galatheoidea (Crustacea, Decapoda, Anomura) from the hydrothermal vents of the Pacific-Antarctic Ridge, *Zoosystema*, 27(4):709–723, 2005; T. Wolff, Composition and endemism of the deep-sea hydrothermal vent fauna, *Cahiers de Biologie Marine*, 46:97–104, 2005.

Contributors

Contributors

The affiliation of each Yearbook contributor is given, followed by the title of his or her article. An article title with the notation "coauthored" indicates that two or more authors jointly prepared an article or section.

A

Ahn, Jong-Hyun. *Department of Materials Science and Engineering, Frederic Seitz Material Research Laboratory, University of Illinois at Urbana-Champaign.* PRINTABLE SEMICONDUCTORS FOR FLEXIBLE ELECTRONICS—coauthored.

Alaee, Mehran. *National Water Research Institute, Canada.* BROMINATED FLAME RETARDANTS IN THE ENVIRONMENT.

Albrechtsen, Justin. *University of Texas at El Paso.* INTERROGATION AND TORTURE—coauthored.

Alspaugh, Mark. *President, Overland Conveyor Company, Inc., Lakewood, Colorado.* CONVEYOR DESIGN AND ENGINEERING (MINING)—coauthored.

Armitage, Lynne. *Department of Biology, University of York, United Kingdom.* PLANT HORMONE RECEPTORS—coauthored.

Asher, Dr. Robert. *Museum of Zoology, Cambridge University, United Kingdom.* AFRICAN MAMMALS.

Atkinson, Prof. John. *Departamento de Español, Universidad de Concepcion, Chile.* INTELLIGENT SEARCH ENGINES—coauthored.

Ayre, Dr. Brian G. *Department of Biological Sciences, University of North Texas, Denton.* FLORIGEN.

B

Bader, Dr. Sameul D. *Argonne National Laboratory, Argonne, Illinois.* MAGNETIC THIN FILMS—coauthored.

Bail, Dr. Sophie. *Department of Cell Biology and Neuroscience, Rutgers University, Piscataway, New Jersey.* P-BODIES—coauthored.

Blask, Dr. David E. *Senior Research Scientist, Laboratory of Chrono-Neuroendocrine Oncology, Bassett Research Institute, Cooperstown, New York.* MELATONIN.

Burnhart, Mr. George. *Norcross, Georgia.* SONIC DRILLING.

Butel, Dr. Janet. *Distinguished Service Professor, The Joseph L. Melnick Professor of Virology, Baylor-UT-Houston Center for AIDS Research Chair, Department of Molecular Virology and Microbiology, Baylor College of Medicine, Houston, Texas.* SIMIAN VIRUS 40.

Butler, Prof. Donald P. *Electrical Engineering Department, University of Texas at Arlington.* SMART SKIN—coauthored.

C

Calais, Dr. Eric. *Department of Earth and Atmospheric Sciences, Purdue University.* OCEAN BIRTH THROUGH RIFTING AND RUPTURE—coauthored.

Caldwell, Robert. *Department of Physics and Astronomy, Dartmouth College, Wilder Laboratory, Hanover, New Hampshire.* DARK ENERGY.

Cavallero, Paul. *Naval Undersea Warfare Center, Newport, Rhode Island.* AIR-INFLATED FABRIC STRUCTURES.

Celik-Butler, Prof. Zeynep. *Electrical Engineering Department, University of Texas at Arlington.* SMART SKIN—coauthored.

Chaski, Dr. Carole E. *Institute for Linguistic Evidence, Inc., Georgetown, Delaware.* FORENSIC LINGUISTICS.

Cheadle, Dr. Michael. *Department of Geology and Geophysics, University of Wyoming, Laramie.* SLOW-SPREADING MID-OCEAN RIDGES—coauthored.

Clague, Dr. John J. *Department of Earth Sciences, Simon Fraser University, British Columbia Canada.* LANDSLIDES.

Cohen, Prof. Zoe. *College of Nursing, University of Arizona, Tucson.* TYPE 2 DIABETES.

Cole, Prof. Simon A. *Associate Professor of Criminology, Law and Society, University of California, Irvine.* FINGERPRINT IDENTIFICATION.

Collins, Steven J. *Director of Australian National CJD Registry, NH and MRC Practitioner Fellow, Associate Professor, Department of Pathology, The University of Melbourne, Australia.* CHRONIC WASTING DISEASE.

Corboy, Dr. John C. *Department of Neurology, University of Colorado Health Sciences Center, Denver.* MULTIPLE SCLEROSIS—coauthored.

Currie, Prof. Philip J. *Department of Biological Sciences, University of Alberta, Canada.* NEW CARCHARODONTOSAURID.

D

Dawson, Dr. Mary R. *Section of Vertebrate Paleontology, Carnegie Museum of Natural History, Pittsburgh, Pennsylvania.* LAONASTES RODENT AND THE LAZARUS EFFECT.

D'Costa, Vanessa. *Department of Biochemistry and Biomedical Sciences, McMaster University, Hamilton, Ontario, Canada.* ANTIBIOTIC RESISTANCE IN SOIL—coauthored.

DePaolo, Prof. Donald J. *Department of Earth and Planetary Science, University of California, Berkeley.* SCIENTIFIC DRILLING IN HOTSPOT VOLCANOES—coauthored.

Dominguez-Rodrigo, Manuel. *Departamento de Prehistoria, Universidad Complutense, Madrid.* EARLIEST TOOLS.

Doshi, Dr. Mahendra. *Progress in Paper Recycling, Appleton, Wisconsin.* PAPER RECYCLING.

E–F

Ebinger, Dr. Cynthia. *Department of Geology, Royal Holloway University of London, United Kingdom.* OCEAN BIRTH THROUGH RIFTING AND RUPTURE—coauthored.

Ferreira-Cabrera, Prof. Anita. *Departamento de Español, Universidad de Concepcion, Chile.* INTELLIGENT SEARCH ENGINES—coauthored.

Flossmann, Dr. Andrea I. *Université Blaise Pascal, Laboratoire de Météorologie Physique, France.* PRECIPITATION SCAVENGING.

G

Geyer, Dr. Pamela. *Department of Biochemistry, University of Iowa, Iowa City.* INSULATOR (GENE)—coauthored.

Gherbi, Rachid. *Bioniformatics Group, IBISC and Genopole, Evry Université d'Evry, Val d'Essonne, France.* BIOINFORMATICS TOOLS—coauthored.

Goolish, Edward. *NASA Astrobiology Institute, NASA Ames Research Center, Moffett Field, California.* ASTROBIOLOGY.

Gray, Dr. Elizabeth. *Director of Conservation Science, The Nature Conservancy, Seattle, Washington.* LOGGING AND MARINE COASTAL SYSTEMS—coauthored.

Grigorenko, Dr. Elena L. *Yale University, New Haven, Connecticut.* INTELLIGENCE, RACE, AND GENETICS—coauthored.

Gur, Dr. Raquel. *University of Pennsylvania School of Medicine, Philadelphia.* SEX DIFFERENCES IN THE BRAIN—coauthored.

Gur, Dr. Ruben. *University of Pennsylvania School of Medicine, Philadelphia.* SEX DIFFERENCES IN THE BRAIN—coauthored.

Gusse, Adam C. *Department of Entomology, University of Wisconsin-LaCrosse.* FUNGAL BIOCONVERSION—coauthored.

H

Haddock, Steven H. D. *Monterey Bay Aquarium Research Institute, Moss Landing, California.* DEEP-SEA SIPHONOPHORE.

Hall, Prof. Susan J. *University of Delaware, Newark.* SPORT BIOMECHANICS.

Hampton-Smith, Rachel J. *School of Molecular and Biochemical Science, Univesity of Adelaide, Australia.* HYPOXIA-INDUCIBLE FACTOR.

Han, Dr. Jongkoo. *Department of Wood Science and Forest Products, Virginia Tech, Blacksburg.* SOLID WOOD PACKAGING MATERIALS.

Hardy, Roger. *Cranfield Impact Centre, Cranfield University, United Kingdom.* SMART RESTRAINT SYSTEM (AUTOMOTIVE).

Harley, Dr. John P. *Department of Biological Sciences, Eastern Kentucky University, Richmond.* BAYLISASCARIASIS.

Hashmi, Prof. A. Stephen K. *Institut für Organische Chemie, Universität Stuttgart, Germany.* GOLD-CATALYZED REACTIONS.

Hauck, Scott A. *Department of Electrical Engineering, University of Washington, Seattle.* FIELD-PROGRAMMABLE GATE ARRAYS.

Head, Dr. Jason. *Department of Biological Sciences, George Washington University, District of Columbia.* SNAKE EVOLUTION.

Hla, Dr. Timothy. *Department of Cell Biology, University of Connecticut Health Center, Farmington.* SPHINGOSINE 1-PHOSPHATE.

J

Jiang, Dr. J. Samuel. *Materials Science Division, Argonne National Laboratory, Argonne, Illinois.* MAGNETIC THIN FILMS—coauthored.

John, Dr. Barbara. *Department of Geology and Geophysics, University of Wyoming, Laramie.* SLOW-SPREADING MID-OCEAN RIDGES—coauthored.

Jones, Prof. Brian. *Department of Earth & Atmospheric Sciences, University of Alberta, Edmonton, Canada.* CARBONATE SEDIMENTOLOGY.

Jones, Dr. William J. *Monterey Bay Aquarium Research Institute, Moss Landing, California.* YETI CRAB—coauthored.

K

Kanungo, Dr. Tapas. *IBM Research Division, Almaden Research Center, San Jose, California.* OPTICAL CHARACTER RECOGNITION.

Kelley, Dr. Deborah S. *School of Oceanography, University of Washington, Seattle.* LOST CITY HYDROTHERMAL FIELD—coauthored.

Keyser, Dr. Kurt. *Department of Geological Sciences, Queen's University, Kingston, Ontario, Canada.* STABLE ISOTOPE GEOCHEMISTRY.

Kidd, Dr. Kenneth K. *Professor of Genetics and Psychiatry, Yale University, New Haven, Connecticut.* INTELLIGENCE, RACE, AND GENETICS—coauthored.

Kiledjian, Dr. Megerditch. *Department of Cell Biology and Neuroscience, Rutgers University, Piscataway, New Jersey.* P-BODIES—coauthored.

Klink, Dr. Katherine. *Department of Geography, University of Minnesota, Minneapolis.* LOCATING WIND POWER.

Krzycki, Prof. Joseph A. *Department of Microbiology, Biochemistry Program, The Ohio State University, Columbus.* GENETIC CODE—coauthored.

Kubodera, Tsunemi. *Department of Zoology, National Science Museum, Tokyo, Japan.* GIANT SQUID.

L

Lane, Dr. Larry S. *Research Scientist, Geological Survey of Canada, Natural Resources Canada, Calgary, Alberta, Canada.* DIGITAL GEOLOGICAL MAPPING.

Lau, Dr. Chirstopher. *Lead Research Biologist, Reproductive Toxicology Division, National Health and Environmental Effect Laboratory, U.S. Environmental Agency, Research Triangle Park, North Carolina.* PERFLUOROOCTANOIC ACID AND ENVIRONMENTAL RISKS—coauthored.

Lenze, Dr. Eric J. *Associate Professor of Psychiatry, University of Pittsburgh School of Medicine, Pittsburgh, Pennsylvania.* STRESS AND DEPRESSION—coauthored.

Lewi, Dr. Elias. *Geophysical Observatory, Addis Ababa University, Ethiopia.* OCEAN BIRTH THROUGH RIFTING AND RUPTURE—coauthored.

Lewis, Dr. James. *Department of Pharmacology, University of Texas Health Science Center at San Antonio.* ANTIFUNGAL AGENTS.

Lexer, Christian. *Royal Botanic Garden, Jodrell Laboratory, United Kingdom.* HYBRIDIZATION AND PLANT SPECIATION.

Leyser, Prof. Ottoline. *Department of Biology, University of York, United Kingdom.* PLANT HORMONE RECEPTORS—coauthored.

Li, Dr. Xingguo. *Department of Biochemistry, University of Iowa, Iowa City.* INSULATOR (GENE)—coauthored.

Linzey, Donald W. *Wytheville Community College, Wytheville, Virginia.* COUGAR (MOUNTAIN LION).

Ludwig, Kristin. *School of Oceanography, University of Washington, Seattle.* LOST CITY HYDROTHERMAL FIELD—coauthored.

M

Macpherson, Dr. Enrique. *Centro de Estudias Avanzadas, Blanes, Spain.* YETI CRAB—coauthored.

Mahapatra, Aniriban. *Department of Microbiology, Ohio State University, Columbus.* GENETIC CODE—coauthored.

Mantella, Dr. Rose C. *Department of Psychiatry, University of Pittsburgh School of Medicine, Pennsyluania.* STRESS AND DEPRESSION—coauthored.

Mazzotta, Marisa J. *Charlestown, Rhode Island.* ECOSYSTEM VALUATION.

McPherson, Dr. Ruth. *Professor, Departments of Medicine and Biochemistry, University of Ottawa Heart Institute, Ontario, Canada.* CHOLESTERYL ESTER TRANSPORT PROTEIN (CETP).

Meissner, Dr. Christian. *Department of Psychology, University of Texas at El Paso.* INTERROGATION AND TORTURE—coauthored.

Mitra, Dr. Subir. *Wind Tunnel Facility, Institute for Atmospheric Physics, Johannes Gutenberg-University, Mainz, Germany.* WIND TUNNEL EXPERIMENTS OF CLOUD PARTICLES.

Montgomery, Dr. Beronda L. *DOE Plant Research Laboratory, Michigan State University, East Lansing.* PHYTOCHROME.

Moore, Dr. Andrew L. *Department of Geology, Kent State University, Ohio.* SEDIMENTOLOGY OF TSUNAMI DEPOSITS.

Morrow, Dr. Daniel G. *University of Illinois at Urbana-Champaign.* REDUCING HUMAN ERROR IN MEDICINE—coauthored.

Mukherjee, Mitrajit. *Exelus, Inc., Livingston, New Jersey.* STEP-OUT TECHNOLOGIES (CHEMICAL ENGINEERING)—coauthored.

N

Nagase, Dr. Masao. *Engineer, Nippon Telegraph and Telephone Corporation, NTT Basic Research Laboratories, Japan.* NANOMETROLOGY.

Nehlsen, Dr. James P. *Exelus, Inc., Livingston, New Jersey.* STEP-OUT TECHNOLOGIES (CHEMICAL ENGINEERING)—coauthored.

Nelson, Kara Shabar. *Conservation Science Assistant, The Nature Conservancy, Seattle, Washington.* LOGGING AND MARINE COASTAL SYSTEMS—coauthored.

Neumann, Gregory. *Jet Propulsion Laboratory, California Institute of Technology, Pasadena.* ARCTIC SEA-ICE MONITORING—coauthored.

Nghiem, Dr. Son V. *Jet Propulsion Laboratory, California Institute of Technology, Pasadena.* ARCTIC SEA-ICE MONITORING—coauthored.

Nilsson, Dr. Dan-E. *Department of Cell and Organism Biology, Lund University, Sweden.* BOX JELLYFISH.

O

Ormsbee, Paul. *Project Engineer, Overland Conveyor Company, Inc., Lakewood, Colorado.* CONVEYOR DESIGN AND ENGINEERING (MINING)—coauthored.

Owen, Dr. Tobias. *Institute for Astronomy, University of Hawaii, Honolulu.* PLUTO.

P

Pearson, Prof. Osbjorn M. *Department of Anthropology, University of New Mexico, Albuquerque.* MODERN HUMAN ORIGINS.

Peet, Dr. Daniel. *School of Molecular and Biochemical Science, University of Adelaide, Australia.* HYPOXIA-INDUCIBLE FACTOR.

Pemberton, Dr. S. George. *Department of Earth & Atmospheric Sciences, University of Alberta, Edmonton, Canada.* ICHNOLOGY.

Presgraves, Dr. Daven C. *Assistant Professor, Department of Biology, University of Rochester, New York.* SPECIATION.

Q–R

Qui, Dr. Yin-Long. *Department of Ecology and Evolutionary Biology, University of Michigan, Ann Arbor.* PHYLOGENY OF BRYOPHYTES—coauthored.

Ratital, Dr. Purnima. *Department of Electrical and Computer Engineering, Northeastern University, Boston, Massachusetts.* REMOTE SENSING OF FISH—coauthored.

Raymond, Dr. Jason. *Microbial Systems Division, Biosciences Directorate, Lawrence Livermore National Laboratory, Livermore, California.* OXYGEN AND EVOLUTION OF COMPLEX LIFE.

Reynolds, Dr. Charles F., III. *Professor of Psychiatry, University of Pittsburgh School of Medicine, Pittsburgh, Pennsylvania.* STRESS AND DEPRESSION—coauthored.

Rogers, Prof. John A. *Department of Materials Science and Engineering, Frederic Seitz Material Research Laboratory, University of Illinois at Urbana-Champaign.* PRINTABLE SEMICONDUCTORS FOR FLEXIBLE ELECTRONICS—coauthored.

Rollins, Dr. Karen. *Department of Neurology, University of Colorado Health Sciences Center, Denver.* MULTIPLE SCLEROSIS—coauthored.

Rubinoff, Dr. Daniel. *Department of Plant and Environmental Protection Sciences, University of Hawaii, Honolulu.* SNAIL-EATING CATERPILLAR.

S

Sampathkumar, Dr. Priya. *Division of Infectious Diseases, Mayo Clinic, Rochester, Minnesota.* AVIAN INFLUENZA (BIRD FLU).

Scawthorn, Prof. Charles. *Engineering Department of Urban Management, Kyoto University, Japan.* DESIGNING FOR AND MITIGATING EARTHQUAKES.

Schlaepfer, Dr. Thomas. *Professor of Psychiatry and Mental Health, University of Bonn, Germany, and The Johns Hopkins Hospital, Baltimore, Maryland.* DEEP BRAIN STIMULATION.

Seed, Dr. Jennifer. *Senior Research Biologist, Risk Assessment Division, Office of Pollution Prevention and Toxics, U.S. Environmental Protection Agency, Washington.* PERFLUOROOCTANOIC ACID AND ENVIRONMENTAL RISKS—coauthored.

Seffah, Ahmed. *Bioinformatics Group, Université d'Evry, Val d'Essonne, Human-Centered Software Engineering Group, Computer Science and Software Engineering Department, Concordia University, Montreal, Canada.* BIOINFORMATICS TOOLS—coauthored.

Segonzac, Dr. Michel. *Ifremer, Centre de Brest, DEEP/Laboratoire Environnement, Plouzane, France.* YETI CRAB—coauthored.

Simpson, Dr. Joyce M. *Advanced Biosensors Laboratory, Center for Biological Defense, University of Florida, Tampa.* MONITORING BIOTERRORISM AND BIOWARFARE AGENTS.

Spergel, Prof. David. *Department of Astrophysical Sciences, Princeton University, New Jersey.* DARK MATTER.

Stanton, Dr. Anthony. *Tepper School of Business, Carnegie Mellon University, Pittsburgh, Pennsylvania.* INKJET PRINTING.

Stephanopoulos, Prof. Gregory. *Bayer Professor of Chemical Engineering, Massachusetts Institute of Technology, Cambridge.* METABOLIC ENGINEERING—coauthored.

Stephens, Dr. Trent. *Department of Biological Sciences, Idaho State University, Pocatello.* THALIDOMIDE.

Sternberg, Dr. Robert J. *Department of Psychology, Yale University, New Haven, Connecticut.* INTELLIGENCE, RACE, AND GENETICS—coauthored.

Stolper, Prof. Edward M. *Division of Geological and Planetary Science, California Institute of Technology, Pasadena.* SCIENTIFIC DRILLING IN HOTSPOT VOLCANOES—coauthored.

Sullivan, Mary K. *Phoenix, Arizona.* CLINICAL FORENSIC NURSING.

Sun, Dr. Yusang. *Center for Nanoscale Materials, Argonne National Laboratory, Argonne, Illinois.* PRINTABLE SEMICONDUCTORS FOR FLEXIBLE ELECTRONICS—coauthored.

Sundaresan, Prof. Sankaran. *Chemical Engineering, Princeton University.* STEP-OUT TECHNOLOGIES (CHEMICAL ENGINEERING)—coauthored.

Symonds, Deanelle T. *Department of Electrical and Computer Engineering, Northeastern University, Boston.* REMOTE SENSING OF FISH—coauthored.

T

Tallis, Dr. Heather. *Lead Scientist, Natural Capital Project, Stanford University.* LOGGING AND MARINE COASTAL SYSTEMS—coauthored.

Tarduno, Prof. John A. *Department of Earth and Environmental Sciences, Department of Physics and Astronomy, University of Rochester, New York.* SINGLE SILICATE CRYSTAL PALEOMAGNETISM.

Taylor, Dr. Andy F. S. *Department of Forest Mycology, Swedish University of Agricultural Sciences, Uppsala, Sweden.* ECTOMYCORRHIZAL SYMBIOSIS.

Thomas, Dr. Donald M. *Hawaii Institute of Technology, University of Hawaii, Honolulu.* SCIENTIFIC DRILLING IN HOTSPOT VOLCANOES—coauthored.

Tinsley, Dr. Brian A. *Department of Physics, University of Texas at Dallas.* ATMOSPHERIC ELECTRICITY AND EFFECTS ON CLOUDS.

Tyo, Keith E. *Massachusetts Institute of Technology, Cambridge.* METABOLIC ENGINEERING—coauthored.

V

van Gent, Dr. Dik C. *Department of Cell Biology and Genetics, Erasmus University Rotterdam, The Netherlands.* DNA REPAIR.

van Hest, Prof. Jan. *Radboud University Nijmegen, Department of Organic Chemistry, The Netherlands.* BIOLOGICAL-SYNTHETIC HYBRID POLYMERS.

Volk, Dr. Thomas J. *Department of Biology, University of Wisconsin-LaCrosse.* FUNGAL BIOCONVERSION—coauthored.

Volpe, Anthony F., Jr. *Symyx Technologies, Santa Clara, California.* HIGH-THROUGHPUT MATERIALS CHEMISTRY—coauthored.

von Puttkamer, Dr. Jesco. *NASA Headquarters, Office of Space Flight, Washington, DC.* SPACE FLIGHT.

W

Wang, Benjamin. *Massachusetts Institute of Technology, Cambridge.* METABOLIC ENGINEERING—coauthored.

Wang, Bin. *Department of Ecology and Evolutionary Biology, University of Michigan, Ann Arbor.* PHYLOGENY OF BRYOPHYTES—coauthored.

Wang, Prof. Pao K. *Department of Atmospheric and Oceanic Sciences, University of Wisconsin-Madison.* CHINESE HISTORICAL DOCUMENTS AND CLIMATE CHANGE.

Wei, Dr. Chunmei. *AnaSpec, Incorporated, San Jose, California.* MULTICOMPONENT COUPLING (ORGANIC SYNTHESIS).

Weinberg, Dr. W. Henry. *Symyx Technologies, Santa Clara, California.* HIGH-THROUGHPUT MATERIALS CHEMISTRY—coauthored.

Wickens, Dr. Christopher D. *University of Illinois at Urbana-Champaign and Senior Scientist, MAAD Division of Alion Science & Technology Corporation.* REDUCING HUMAN ERROR IN MEDICINE—coauthored.

Wright, Dr. Tim. *Department of Earth Sciences, Oxford University, United Kingdom.* OCEAN BIRTH THROUGH RIFTING AND RUPTURE—coauthored.

Wright, Prof. Gerard. *Department of Biochemistry and Biomedical Sciences, McMaster University, Hamilton, Ontario, Canada.* ANTIBIOTIC RESISTANCE IN SOIL—coauthored.

Y

Yee, Dr. Alfred A. *President, Applied Technology Corporation, Honolulu, Hawaii.* PRECAST AND PRESTRESSED CONCRETE.

Yirgu, Dr. Gezahegn. *Geology and Geophysics Department, Addis Ababa University, Ethiopia.* OCEAN BIRTH THROUGH RIFTING AND RUPTURE—coauthored.

Z

Zhang, Dr. Xiaoni. *Department of Information Systems, Northern Kentucky University, Highland Heights.* SPYWARE.

Zhou, Dr. Zhonghe. *Institute of Vertebrate Paleontology and Paleoanthropology, Chinese Academy of Sciences, Beijing.* CRETACEOUS BIRD RADIATION.

Index

Index

Asterisks indicate page references to article titles.